Introduction to Forensic Science and Criminalistics

Introduction to Forensic Science and Criminalistics

Second Edition

Howard A. Harris

Henry C. Lee

CRC Press
Taylor & Francis Group
Boca Raton London New York

CRC Press is an imprint of the
Taylor & Francis Group, an **informa** business

CRC Press
Taylor & Francis Group
6000 Broken Sound Parkway NW, Suite 300
Boca Raton, FL 33487-2742

© 2019 by Taylor & Francis Group, LLC
CRC Press is an imprint of Taylor & Francis Group, an Informa business

No claim to original U.S. Government works

Printed on acid-free paper

International Standard Book Number-13: 978-1-4987-5796-6 (Hardback)

This book contains information obtained from authentic and highly regarded sources. Reasonable efforts have been made to publish reliable data and information, but the author and publisher cannot assume responsibility for the validity of all materials or the consequences of their use. The authors and publishers have attempted to trace the copyright holders of all material reproduced in this publication and apologize to copyright holders if permission to publish in this form has not been obtained. If any copyright material has not been acknowledged please write and let us know so we may rectify in any future reprint.

Except as permitted under U.S. Copyright Law, no part of this book may be reprinted, reproduced, transmitted, or utilized in any form by any electronic, mechanical, or other means, now known or hereafter invented, including photocopying, microfilming, and recording, or in any information storage or retrieval system, without written permission from the publishers.

For permission to photocopy or use material electronically from this work, please access www.copyright.com (http://www.copyright.com/) or contact the Copyright Clearance Center, Inc. (CCC), 222 Rosewood Drive, Danvers, MA 01923, 978-750-8400. CCC is a not-for-profit organization that provides licenses and registration for a variety of users. For organizations that have been granted a photocopy license by the CCC, a separate system of payment has been arranged.

Trademark Notice: Product or corporate names may be trademarks or registered trademarks, and are used only for identification and explanation without intent to infringe.

Library of Congress Cataloging-in-Publication Data

Names: Harris, Howard A., author. | Lee, Henry C., author. | Introduction to forensics & criminalistics.
Title: Introduction to forensic science and criminalistics / Howard Harris and Henry Lee.
Description: Second edition. | Boca Raton, FL : CRC Press, [2019] | Revised edition of : Introduction to forensics & criminalistics /
Howard A. Harris, Henry Lee, c2008. | Includes bibliographical references and index.
Identifiers: LCCN 2018032539| ISBN 9781498757966 (hardback : alk. paper) | ISBN 9781315119175 (ebook) | ISBN 9781498758017 (epub) | ISBN 9781351632836 (mobi/kindle)
Subjects: LCSH: Criminal investigation. | Crime scenes. | Forensic sciences. | Evidence, Criminal.
Classification: LCC HV8073 .G274 2019 | DDC 363.25--dc23
LC record available at https://lccn.loc.gov/2018032539

Visit the Taylor & Francis Web site at
http://www.taylorandfrancis.com

and the CRC Press Web site at
http://www.crcpress.com

Dedication

In Memory of Margaret Lee

This book is dedicated to Margaret, wife of Dr. Henry Lee. Her devotion, dedication, love, and kindness will be remembered forever by the authors and by her many friends around the globe.

This book is also dedicated to all the highly committed professionals who labor in forensic science and law enforcement throughout the world.

Contents

Preface to Second Edition — xvii
Acknowledgments — xix
About the Book and Pedagogy — xxi
Authors — xxiii
Contributors — xxv

1 Introduction — 1
What is forensic science and what is its role in the justice system? — 4
    The role of forensic science — 4
    Definition of forensic science — 5
    The value of forensic science to society — 6
A brief history of forensic science — 9
    Early applications of forensic science — 9
    Development of forensic science laboratories and professional organizations — 10
The many faces of forensic science — 13
    Human biological and medical sciences — 13
    Natural sciences—chemistry, biology, and physics — 16
    Technology and engineering — 16
        Digital forensic examinations; often computer or processor related — 17
Nature of forensic science and the scientific method — 17
    Applicability of scientific method to forensic science and investigation — 20
Elements of forensic evidence analysis—the types of results forensic scientists produce — 20
    Evidence recognition — 21
    Classification—identification — 21
    Individualization — 22
    Reconstruction — 24
Forensic ethical concerns and proper professional practice — 25
    Professional organizations and their functions — 25
    Professional organizations and their codes of ethics — 25
    Professional practice and ethical concerns of the organizations — 26
    Ethical responsibilities as scientists — 26
    Ethics and professional behavior for forensic scientists — 26
Key terms — 27
Review questions — 28
Fill in/multiple choice — 28
References and further readings — 29

2 Physical evidence in the legal system — 31
How physical evidence is produced — 33
    Change induced at a scene — 33
    Exchange of material upon contact — 33
    Deposits, dispersions, and residues — 34
    Damage, tears, cuts, breaks, and jigsaw matches — 35
    Imprints or indentations — 36
    Striations—dynamic marking — 36

Physical evidence—classification and uses	38
Identification of persons—victim, suspect, others	39
Establish linkages or exclusions	40
Develop modus operandi (MO), leads, and suspects	40
Provide suspects or investigative leads from forensic databases	41
Corroboration—credibility: supporting or disproving statements	42
Identification and analysis of substances or materials	42
Establishing a basis for a crime and criminal prosecution—*corpus delicti*	43
The physical evidence process	43
Recognition—critical portion of evidence process requires a trained observer	43
Documentation and marking for identification	43
Collection, packaging, and preservation	44
Laboratory analysis	45
Reporting and testimony	45
The criminal justice system and process—case flow	46
Trace the evidence pathway through the system	46
Investigative agency role	46
Adjudicative agency role	48
Importance of physical evidence, particularly scientific evidence	49
Scientific and technical evidence admissibility and the expert witness	50
Admissibility versus weight	50
Requirement of relevance and reliability	50
Daubert case	51
Key terms	52
Review questions	52
Fill in/multiple choice	53
References and further readings	53
3 Crime scene processing and analysis	**55**
Processing versus analysis	57
Types of scenes	57
Types of crime scenes	58
Initial actions and scene security	58
Steps in scene processing and analysis	60
Scene survey and evidence recognition	60
Documentation	60
Scene searches	60
Evidence collection and preservation	60
Preliminary scene analysis and reconstruction	61
Release of the scene	61
Scene survey and evidence recognition	61
Scene searches	62
Documentation	62
Notes	62
Sketches	63
Photography	65
Technical aspects	66
Forensic aspects	67
Video recording	70
Audio recording	70
New technology for crime scene recording	70
Three-dimensional (3D) imaging and scene mapping	70
Total station mapping technology	71

Computer aided drawing technology	71
Crime scene data analysis computer software	72
Drone technology	72
Duty to preserve	72
Collecting and preserving physical evidence	73
Collection methods	73
Numbering and evidence description methods	74
Packaging options	74
Proper controls	75
Laboratory submission	76
Chain of custody	76
Crime scene analysis and reconstructions	76
Laboratory analysis and comparisons of evidence	76
Medical examiner's reports in death cases	77
Reconstruction—putting it all together	77
Reconstruction versus reenactment versus profiling	77
Key terms	78
Review questions	78
Fill in/multiple choice	79
References and further readings	79

4 Examination and interpretation of patterns for reconstruction — 81

Type of pattern evidence: patterns for reconstruction and individualization	83
Most reconstruction patterns are crime scene patterns	83
Importance of documentation of reconstruction patterns	84
Bloodstain patterns	85
Basis of bloodstain pattern interpretation	85
Velocity, energy, and force	85
Various blood spatter patterns	88
Factors affecting blood patterns and their interpretation	90
Glass fracture patterns	91
Determining the side of the glass where force was applied	91
Determining the order of gunshots fired through glass	92
Track and trail patterns	93
Tire and skid mark patterns	93
Clothing and article or object patterns	94
Gunshot residue patterns	94
Projectile trajectory patterns	95
Fire burn patterns	96
Modus operandi patterns and profiling	97
Wound, injury, and damage patterns	98
Key terms	98
Review questions	98
Fill in/multiple choice	99
References and further readings	99

5 Examination of physical pattern evidence — 101

Classification—types of physical patterns for comparison	103
Physical fit or direct fit match	103
Indirect match	104
Imprint, impression, and striation comparison	104
Shape, pattern, marks, and form determination	104

General principles in physical pattern comparisons	105
The process of recognition	105
The process of identification	105
Physical fit and matches	106
Exclusions, and inconclusive and insufficient detail	107
The Daubert criteria and the National Academy of Sciences report on pattern comparisons	107
Impression and striation mark comparison	108
Impressions: imprints and indentations	108
Striations	108
Collection and preservation of impressions	109
Footwear, tire, and other impressions	112
Clarification and contrast improvement techniques	112
Weapon, tool, and object marks	113
Shape, pattern, mark, and form comparisons	113
Other patterns	113
Concluding comments	113
Key terms	115
Review questions	115
Fill in/multiple choice	116
References and further readings	116

6 Fingerprints and other personal identification patterns — 117

Fingerprints and an introduction to biometrics	119
Fingerprints—an old and traditionally valuable type of evidence	119
Composition of latent print residue	120
About fingerprints—their nature and the history and development of their use	120
History and development of the use of fingerprints	122
Fingerprint classification, management of large files, AFIS systems	123
Search for, collection, and preservation of fingerprint evidence	125
Basic physical types of fingerprints	126
Initial search and detection	126
Visualization and documentation of evidence fingerprints	126
Latent fingerprint visualization methods	126
Latent print visualization on non-absorbent surfaces	127
Physical methods	127
Cyanoacrylate (crazy glue) fuming	128
Special types of illumination and combination methods	128
Latent prints on absorbent surfaces	130
Reagents for highly specialized applications	132
Use of a systematic approach to visualization of fingerprints	132
Fingerprint comparison and identification	134
The fingerprint identification profession	136
Biometrics—use if biological diversity for identification	136
Background	136
Biometrics for authentication	138
Biometrics for individualization	139
Identification of human remains and handling of mass disasters	139
Key terms	140
Review questions	141
Fill in/multiple choice	141
References and further readings	141

7 Questioned document examination — 143

- Questioned document examination—general — 145
  - Types of document evidence — 145
  - Recognition, collection, and preservation of evidence — 147
  - Writing process — 148
- Handwriting comparison—identification of writer or detection of forgery — 151
  - Development of an individual's handwriting — 151
  - Mechanical and pictorial characteristics of handwriting — 153
  - Importance of known standards to successful examination — 154
    - Authenticated collected writing — 154
    - Requested (ordered) writings — 154
  - Examination and comparison of handprinting — 155
  - Signatures — 155
  - Legal status of underlying science of writing comparison — 156
- Non-handwriting examinations on questioned documents — 157
  - Mechanical impressions—typewriter and printer comparisons — 157
  - Copying machines — 158
- Some other useful document examinations — 158
  - Detection of alterations and erasures—usefulness of paper examination — 158
  - Examination of charred documents or indented writing — 160
  - Ways to attempt to determine when a document was created — 163
- Key terms — 165
- Review questions — 166
- Fill in/multiple choice — 166
- References and further readings — 167

8 Toolmarks and firearms — 169

- Toolmarks—introduction and toolmark definition — 170
  - Toolmarks are of three types — 171
  - Residue from softer object is often deposited on the tool — 171
  - Class and individual characteristics — 171
  - Collection of toolmarks — 172
  - Examination and comparison of toolmarks — 173
- Firearms examination—background — 174
  - Basic operation and important parts of a firearm — 174
    - Major components of a firearm — 174
    - The function of the cartridge in a firearm — 174
    - Rifling of the barrel makes modern firearms accurate — 175
    - Basic physics of firearms — 178
  - Major types of firearms — 179
    - Classification of firearms by how they load cartridges — 179
    - Classification of firearms by their functions — 180
- Collection, examination, and comparison of firearms evidence — 184
  - Careful handling of firearm, bullets, and cartridge cases — 184
  - Firearms evidence examination and comparison — 185
  - Test for functionality and to obtain control bullets and cases — 186
  - Comparison of bullets, cartridge cases, or shot shell cases—the comparison microscope — 187
  - Association of cartridges or bullets to firearm or manufacturer using databases — 190
  - Comparison of badly damaged projectiles or cases — 192
- Use of firearms evidence for reconstruction — 193
  - Recovered firearm and fired evidence — 193
  - Muzzle to target distance—powder patterns — 193
  - GSR on hands—dermal nitrate, lift, swab, tape — 193

	Serial number restoration	196
	The firearms and toolmark examiner profession	198
	Key terms	198
	Review questions	199
	Fill in/multiple choice	200
	References and further readings	200
9	**Digital evidence and computer forensics**	**201**
	Raymond J. Hsieh	
	Introduction	203
	What is digital evidence?	203
	What is digital multimedia evidence?	203
	What is computer forensics?	204
	Digital-related devices or attachments	204
	How is digital evidence processed?	206
	Collection of digital (multimedia) evidence	206
	Originals versus copies	206
	Hashing of digital files	207
	Hashing algorithms—MD5 and SHA1	207
	Digital evidence chain of custody	207
	Forensically sterile conditions for suspect digital devices	208
	Analysis of electronic data	208
	Unallocated file space and file slack	208
	RAM slack	208
	File slack	209
	Deleted/un-deleted process	209
	Forensic tools	209
	Using National Institute of Standards and Technology (NIST) computer forensics tools	210
	Internet crime	210
	Computer forensics analysis of Internet data	210
	E-mail forensics	210
	E-mail messages examination	211
	How to review e-mail headers	211
	View Google mail e-mail header	211
	E-mail tracing	213
	Mobile device/smartphone forensics	213
	Smartphone basics	214
	Acquisition procedures for mobile devices/smartphones	214
	Mobile forensics tools	214
	Other free resources for iPhone forensics	215
	Actual case examples involving digital evidence	215
	BTK	215
	Death of Caylee Anthony	215
	Appendix	216
	Key terms	218
	Review questions	219
	Fill in/multiple choice	220
	References and further readings	221
10	**Blood and physiological fluid evidence: Evaluation and initial examination**	**223**
	Elaine Pagliaro	
	How biological evidence analysis has changed because of DNA typing	225
	Nature of blood	225

Collection, preservation, and packaging of biological (including blood) evidence 226
    Blood or buccal swab from known person 226
    Biological evidence from scenes 227
Test controls, substratum controls, and contamination issues 229
    Initial examination of and for biological evidence 231
Forensic identification of blood 232
    Preliminary or presumptive tests for blood 233
    Confirmatory tests for blood 234
Species determination 234
Forensic identification of body fluids 237
    Identification of semen 237
    Identification of vaginal "secretions," saliva, and urine 240
Forensic investigation of sexual assault cases 240
    Coordination of effort—SANEs and SARTs 240
    Initial investigation 241
    The forensic scientist's role 241
    Medical examination 242
    Evidence collection and sexual assault evidence collection kits 243
    Types of sexual assault cases and their investigation 247
    Drug-facilitated sexual assault—"Date-Rape" drugs 248
Blood and body fluid individuality—traditional serological (pre-DNA) approaches 249
    The classical or conventional (pre-DNA) genetic markers 249
    How does typing genetic markers help "Individualize" a biological specimen? 249
Key terms 251
Review questions 252
Fill in/multiple choice 252
References and further readings 253

11 DNA analysis and typing 255
Richard Li

Genetics, inheritance, genetic markers 257
DNA—nature and functions 258
Where DNA is found in the body—nuclear (genomic) and mitochondrial DNA (mtDNA) 261
Collection and preservation of biological evidence for DNA typing 263
Development and methods of DNA analysis 263
    Isolation of DNA 264
    The beginning—restriction fragment length polymorphism (RFLP) 265
    The polymerase chain reaction (PCR)—the first PCR-based DNA tying methods 268
Current DNA-typing methods—short tandem repeats (STRs) 270
The power of DNA to individualize biological evidence 272
Databasing and the CODIS system 274
Applications of forensic DNA typing 277
Newer DNA technologies 282
Strengths, limitations, promise, hype 284
Key terms 285
Review questions 286
Fill in/multiple choice 286
References and further readings 286

12 Arson and explosives 289

Fire and arson 291
    The nature of combustion—flaming and glowing combustion 291
    Common fuels used to produce flaming combustion 293

Measures of fuel characteristics—measures of combustibility	294
Pyrolysis of solid fuels	296
Investigating suspicious fires—arsonists' motives	296
Investigation of fire scenes—difficult crime scenes	298
Burn patterns	298
Search for possible causes of the fire	298
Recovery of ignitable liquid residues from suspicious fire scenes	299
Reasons for finding ignitable liquid residues	299
Search for places to collect debris—patterns, sniffers, and arson dogs	300
Collection of debris and evidence samples and proper packaging	301
Collection of other physical evidence	301
Laboratory analysis of debris and other samples	302
Preparation of gas or liquid samples and processing of residue samples	302
Three primary techniques for preparation of debris samples	303
Laboratory examination of specimens prepared from fire debris	304
Gas chromatography	305
Accelerant classification by pattern recognition—importance of reference collections	306
Identification of individual components, identification of ignitable liquid, and indirect identification	307
Comparison samples	309
Examination of the other physical evidence collected	309
Explosives and explosion incidents	310
Characteristics of explosives and explosions	310
The three major classes of explosives	313
Low explosives	313
Primary high explosives	313
Secondary high explosives	314
The explosive train or device	317
The role of the scene investigator	318
Laboratory analysis of explosives and explosive residues	319
Examination of an unexploded device	319
Examination of the exploded device and associated debris	320
Microscopic examination of recovered residues	320
Use of washings from residues to search for the nature of the explosive	321
Examination of the device or debris for other physical evidence	322
Key terms	323
Review questions	324
Fill in/multiple choice	324
References and further readings	325
13 Drugs and drug analysis	**327**
Nature of drugs and drug abuse	328
Introduction to drugs and drug abuse	328
Working definition of a drug	329
Nature of drug dependence	329
Physiological dependence	329
Behavioral dependence	330
Drugs and society—controlled substances	330
Major classes of abused drugs	331
Analgesic drugs (pain killers or narcotics)	331
Stimulants	333
Hallucinogen	334

Depressants, hypnotics, and tranquilizers	337
Club drugs	338
Athletic performance enhancing drugs	339
Controlled substance laws	340
Analysis of controlled substances in the forensic laboratory	342
Screening tests	342
Isolation and separation	343
Microcrystal tests	344
Chromatography (separations)	345
Spectroscopy/spectrometry	346
Qualitative versus quantitative analysis	346
Forensic toxicology	348
General description	348
Forensic toxicology on samples from the living	350
Post-mortem toxicology	350
Classes of poisons	351
Alcohol and drugs and driving	352
Driving while impaired by alcohol	352
Other drugs and driving	354
Key terms	354
Review questions	355
Fill in/multiple choice	356
References and further readings	356

14 Materials evidence — 357

Introduction to materials evidence	360
Transfer of materials evidence	360
Materials evidence can be transferred or deposited without contact	361
Clothing and vehicles are the most common sources of materials evidence	361
Collection methods for transferred materials evidence from a crime scene	362
At the scene—collection without sampling	362
Collection at the laboratory	362
Use of forceps—picking	362
Mechanical dislocation—shaking or scraping of surface material	363
Tape lifts—sticky but not too sticky	363
Laboratory examination of materials evidence	364
Initial physical examination—stereomicroscope, hand lens	364
Optical microscopy	364
Chemical microscopy	364
Polarized light microscopy	365
Instrumental methods for identification and comparison	365
Materials evidence comparisons—individualization, inclusion, or exclusion	367
The most common types of materials evidence	367
Natural and synthetic fibers	368
Natural fibers	368
Manufactured fibers	370
Laboratory examination of fibers	373
Hair—both human and animal	376
Hair structure	376
Growth phases	378
Racial characteristics of human hair	379
Collection of hair standards	379
Laboratory examination of hair evidence	380

Automotive and architectural paint and other coatings	381
Architectural paint	382
Automotive paint	382
Collection of known control paint samples	384
Laboratory analysis of paint evidence	384
Soil and dust	387
Major components of soil	388
Examples of investigative value of soil analysis	388
Collection of soil evidence	390
Laboratory examination of soil	390
Forensic glass evidence	392
Types of glass	392
Glass as evidence	393
Collection of glass evidence	393
Laboratory examination of glass	394
Miscellaneous types of materials evidence	395
Strengthening materials evidence	397
Key terms	398
Review questions	399
Fill in/multiple choice	399
References and further readings	400
Appendix A: J. Methods of forensic science—The scientific tools of the trade	401
Index	411

Preface to Second Edition

In this second edition, we have tried to update the material to include many of the improvements in the rapidly growing field of forensic science. Forensic science has become something of a household word in the past decade or so. Forensic DNA analysis, perhaps the most important development in forensic science ever, regularly makes the news. This trend probably began with cable television's interest in covering high-profile criminal cases and the O.J. Simpson trial, in particular. In recent years, network television has featured prime-time programming that has forensic science as its focus; the *CSI, Trace Evidence,* and *Law and Order* series are no doubt the most widely recognized.

The mass exposure to forensic science through media creates a danger of incorrect or misleading impressions and information through sensationalism and "artistic license." Under these circumstances, there is a need to provide good, reliable sources of information on the subject. Many college and university students, throughout the country and the world, take introductory forensic science courses. The majority of students want to learn something about the subject, and, perhaps, how investigators, police, and attorneys make use of the information forensic science can provide, even if they themselves do not intend to become forensic scientists. There has never been a forensic science textbook directly aimed at these students, who as citizens will be our future jurors, police officers, investigators, lawyers, and judges. This is that book. It has a structure reflecting the underlying philosophy that forensic science is a science and profession. Appropriate pedagogic features have been incorporated to aid the student in learning, and the authors have vast and varied experience as forensic scientists and teachers.

The book was written for students to use in an undergraduate college or university course for forensic science majors just starting out in their studies, and as an elective for criminal justice majors and others interested in our justice system who will also find it a valuable body of knowledge. It comfortably fits into a one-semester schedule and is an introduction to what is often called "criminalistics." Think of criminalistics as comprising the activities and specialty areas found in a modern, full-service forensic science laboratory.

ORGANIZATION

The whole book is organized along the lines of the criminalistics concepts of identification, individualization, and reconstruction. After introductory material and orientation to the subject, we move from crime scene investigation, reconstruction, and pattern analysis to categories of evidence for which individualization is the goal. Finally, types of evidence having identification and comparison as the primary goal of laboratory analysis are addressed.

The book begins by establishing the subject order and organization, moving on to crime scene analysis and reconstruction patterns. Generally, the book moves on to various forms of pattern evidence, then covers biological evidence analysis and forensic DNA typing. Although biological evidence fits into the category of "evidence for individualization," the subject has become so important that it warrants separate treatment. A chapter new to this edition is the digital forensics chapter. Digital evidence has become a vital tool in numerous cases these days. The book closes with identification, comparison, and sometimes quantitative analysis of chemical and trace evidence, and the vital role that laboratories are required to play.

The chapters on particular types of evidence (such as blood, drugs, etc.) all have a consistent internal organization. The subject matter and background are introduced and explained, strategies and methods for collecting and packaging that type of evidence are enumerated and explained, the forensic methods used for examination and comparison are described, and finally the results that can be expected are explained and the strengths and limitations of the tests are discussed.

FOR THE INSTRUCTOR

An Instructor's Guide with Chapter PowerPoint™ slide presentations, additional readings, and a test question bank, are available online for professors and instructors (www.crcpress.com/9781498757966).

Acknowledgments

The authors would like to thank the following reviewers for lending their valuable experience and knowledge to the creation of this edition:

Dave Streigel
Wor-Wic Community College
Salisbury, Maryland

Raymond J. Hsieh
California University of Pennsylvania
California, Pennsylvania

We also thank the colleagues at the Connecticut Forensic Science Laboratory, Connecticut State Police, and the University of New Haven.

About the Book and Pedagogy

Even though the book was written for students who have a limited science or chemistry background, some basic concepts in scientific measurement and methods are necessary to fully understand all the material. Understanding the basic concepts will help students understand the science underpinning forensic science. To aid with this, Scientific Sidebars have been included. The authors collectively have spent over 80 years practicing and teaching forensic science, and we do hope that the book conveys some of the sense of excitement and commitment that we still feel about forensic science!

Each chapter contains the following features.

Learning Objectives: Learning objectives provide students with the chapter learning goals and key knowledge points they should understand upon reading the chapter.

A Lead Case: A Lead case is presented at the beginning of each chapter to offer a real-world forensic case, from both the author's collective experience, as well as famous cases that point out unique aspects of the evidence that helped solve a crime or convict a perpetrator. Each case contains many of the topics covered in the chapter, so that the student can experience how concepts apply to real-world forensic investigation.

Case Study Boxes: Case Study boxes provide brief case descriptions occurring throughout each chapter to illustrate specific points and identify the potential utility of the evidence.

Each chapter contains Science Sidebar boxes, which take the students further into the methods and techniques that are mentioned in the text.

Photographs and Figures: Each chapter is fully illustrated with photographs, diagrams, and figures—many from the case files of the authors. Students can actually see how to dust for fingerprints and understand what the different types of bloodstain patterns look like.

End of Chapter Elements: Each chapter closes with a list of key terms, review questions, and further references.

Authors

Dr. Howard A. Harris is currently a professor emeritus at the University of New Haven, where he teaches and conducts research. From the fall 1996 until the fall of 2003, he was director of the Forensic Science Program. He was promoted to full professor in 2006 and awarded the rank of professor emeritus in 2015. He received his bachelor's degree in chemistry from Western Reserve University, his master's degree and PhD, both in chemistry, from Yale University, and holds a J.D. from St. Louis University. He was admitted to and has maintained his membership in the Missouri Bar. Dr. Harris was a research chemist for seven years for the Shell Oil Company before entering the forensic field as director of the New York City Police Department Police Laboratory in January of 1974. He held that position for just under twelve years. During that time, he was active in the field both locally and nationally. He was one of the founding members of the Northeastern Association of Forensic Scientists. He held offices in the American Society of Crime Laboratory Directors (ASCLD) culminating in the presidency. He was active in the American Academy of Forensic Sciences (AAFS), having presented many papers and an invited Plenary Lecture, and was elected a fellow. In addition to his scientific activities, he was also active in the Business of Criminalistics section of the AAFS and held a number of positions culminating in the section chairmanship. He was awarded the Mary Cowan Award for distinguished service to the Criminalistics Section in 1997.

Dr. Henry C. Lee is an internationally revered forensic scientist and investigator. He began his law enforcement career in Taiwan. After graduating from the Taiwan Central Police College, he joined the Taipei Police Department and quickly reached the rank of police captain. After coming to the United States, he earned a second bachelor's degree from John Jay College in New York City and a PhD in biochemistry from New York University. He has been a professor at the University of New Haven for 44 years and is largely responsible for building the forensic science academic program there. In 1978, Dr. Lee was appointed the forensic laboratory director and chief criminalist for the State of Connecticut, a position he held for 20 years. He served for two terms as commissioner of Connecticut State Police and Department of Public Safety of the State of Connecticut. Prior to retiring from government, he served as chief emeritus for the Division of Scientific Services. He is now a distinguished chair professor and founder of the Henry C. Lee Institute of Forensic Sciences at the University of New Haven.

Dr. Lee is an internationally recognized authority in forensic science and has played a prominent role in many of the most challenging cases of the last 58 years. Dr. Lee has worked with law enforcement agencies from 47 countries and has helped to solve more than 8000 cases. In recent years, his lectures and consultations have taken him to England, Bosnia, China, Germany, Singapore, Croatia, Brunei, Thailand, the Middle East, South America, Africa, and other locations around the world. Dr. Lee's testimony figured prominently in the O. J. Simpson, Jason Williams, Peterson, and Kennedy Smith trials and in convictions of the "Woodchipper" murderer, as well as thousands of other murder cases. Dr. Lee has assisted local and state police in their investigations of other famous crimes, such as the murder of Jon Benet Ramsey in Boulder, Colorado, the 1993 suicide of White House Counsel Vincent Foster, the kidnapping of Elizabeth Smart, the death of Chandra Levy, and the reinvestigation of the Kennedy assassination.

Dr. Lee has testified at thousands of trials in his career and is a frequent guest on television shows around the global. His TV series of *Trace Evidence—Dr. Henry Lee File* was well received. He has written hundreds of scientific papers and author/co-authored 40 books on forensic science and criminal investigation. In addition, he has 30 honorary doctoral degrees and has lectured widely for colleges, universities, and criminal justice and law enforcement organizations. He has also received numerous awards, medals, and honors from professional organizations and law enforcement agencies, as well as from government agencies worldwide.

Contributors

Raymond J. Hsieh is a full professor in Cyber Forensics/Forensic Science at California University of Pennsylvania; Dr. Hsieh earned his PhD in communication technology from the School of Informatics, at the State University of New York at Buffalo. His undergraduate major was forensic science from the Central Police University, Taiwan, and his graduate study was information technology from the Rochester Institute of Technology. Dr. Hsieh's research focuses on Big Data Analysis and Digital Multimedia Evidence. In addition to scholarly essays and book chapters, he has written *Cognitive Mapping & Comparison* (VDM Verlag, 2009). —*An Example* of *Online Legal Polices*, and *Intelligence and Security Informatics*. Dr. Hsieh is certified as a computer forensic examiner.

Richard Li earned his MS in forensic science from the University of New Haven and his PhD in molecular biology from the University of Wisconsin-Madison. After completing his PhD, Dr. Li was awarded a postdoctoral fellowship at Weill Medical College of Cornell University and subsequently worked as a research faculty member at the School of Medicine at Yale University. Dr. Li has served as a criminalist in the Department of Forensic Biology, the Office of Chief Medical Examiner of New York City. For the past decade, he has held faculty positions in forensic science programs at several universities before joining John Jay College. Dr. Li is an associate professor of forensic biology at John Jay College of Criminal Justice. Additionally, he serves as a faculty member for the PhD Program in forensic science for the College. Dr. Li's current research interests include the identification and analysis of biological specimens that are potentially useful for forensic investigations.

Elaine Pagliaro is a member of the Henry Lee Institute of Forensic Science and an adjunct faculty member at several colleges and universities, where she teaches science, forensic science, and law. Her educational background is in biology and chemistry (BA, University of St. Joseph), forensic science (MS, University of New Haven) and law (JD, Quinnipiac University). She was admitted to the bar in Connecticut and New Hampshire, where she maintains membership. Ms. Pagliaro had a 27-year career at the Connecticut Forensic Science Laboratory where she worked in criminalistics and forensic biology and was assistant director and director. She has been active in the Northeastern Association of Forensic Scientists, where she was president and was awarded a Life Membership. She has publications in forensic journals and is co-author of two books. During her time at the Forensic Laboratory, Ms. Pagliaro was involved in major forensic investigations in Connecticut and in cases of national prominence.

CHAPTER 1

Introduction

Lead Case: State v. Richard Crafts

This case is popularly known as the "Wood Chipper Case." In November of 1986, a flight attendant for Pan American World Airways named Helle Crafts returned to New York's John F. Kennedy International Airport from a routine international flight. She and another flight attendant, both of whom lived in Newtown, Connecticut, and were friends, took a limousine to their homes. The limousine dropped Mrs. Crafts off at her home, and there she and her flight attendant friend agreed to call one another later. Helle Crafts was never seen or heard from again.

Later, after the limousine ride and into early December, the flight attendant friend continued trying to contact Helle without success. Independent of this, a private investigator named Oliver Mayo, who had been hired by Mrs. Crafts to investigate possible extramarital activities of her husband, Richard, was also trying to find Helle. Mr. Mayo had gathered unequivocal, incriminating evidence against Richard of an extramarital affair. He wanted to inform his client, Mrs. Crafts, and collect his fee.

The local police did not show much interest in the case, after Helle's colleague's initial inquiries, stating that Mrs. Crafts was an adult, she hadn't been missing that long, and that she would probably turn up. Ultimately, the state's attorney's office was contacted and an investigation by the state police was initiated.

Richard Crafts was a pilot for Eastern Airlines and flew a regular New York to Miami run. He was also a part-time officer in the local police department. The couple had three children, and, because they were in the airline industry and needed to travel so much, they had a live-in nanny.

The investigation by the state police showed that the morning after Helle returned from the international trip, Richard had risen early and told the nanny to take the children to their grandparents' home. Further investigation revealed that he had rented a large, diesel powered wood chipper from a dealership a week earlier. This large model wood chipper was one of only two in the state, and the only one in the southwestern area of Connecticut. The agent at the dealership remembered Richard because he had come to rent the machine driving a small passenger vehicle. The agent had told him the car was not powerful enough to pull the wood chipper, so Richard had then gone out and rented a U-Haul truck to use to pull the wood chipper. The agent also remembered that the wood chipper had been returned in the cleanest condition that he could ever remember. Richard did own a wooded lot in Newtown. It was not, therefore, illogical for him to go and rent the chipper, except that all this activity was taking place during a major snowstorm in that part of the state. The storm had most people off the roads and at home, and many institutions were temporarily closed. A state highway snow plow driver reported seeing a U-Haul truck towing a large wood chipper headed toward Lake Zoar, a man-made lake (reservoir)—but he could not see who was driving. This activity took place the next night after Helle Crafts had returned home and gone missing.

The state police were very suspicious that Helle might have met with foul play, and that Richard might be involved, but the evidence was very sketchy. Thinking the wood chipper might somehow be involved, an extensive search of the area along Lake Zoar was conducted. Thinking the worst—that maybe Helle had been killed and the wood chipper used to dispose of her remains—the state police, with the help of criminalists from the forensic laboratory and a forensic odontologist, searched for skeletal or other remains. It was winter, and heavy snow covered the leaves that had fallen to the ground. The investigators and forensic scientists melted away the snow inch-by-inch as they searched. The leaves and debris had to be separated from things underneath them. Large quantities of leaves, debris, and anything else on the ground were placed in oil drums filled with water to float off the leaves and light plant material. The water was then emptied through narrow mesh sieves to capture any small items that might have been present on the ground.

After some days of searching, the forensic investigation team recovered:

- A human tooth
- A dental restoration
- 56 small pieces of bone and 2660 strands of human hair
- A portion of a human finger with some friction ridge skin
- A toenail painted with red nail polish

Now convinced they were handling a probable homicide case, the state police and the forensic laboratory set about to assemble a forensic team to try and establish what had happened.

A state police dive team's search in the waters of Lake Zoar resulted in the recovery of a gasoline powered chain saw. It was not very old, and its fuel tank was still half full. However, the serial number had been filed down to prevent ready identification. Serial number restoration in the laboratory revealed "E59266." Company records showed that this chain saw had been purchased by Richard Crafts a few years previous. He had used a Visa card, and the purchase record was still available. There was no question that Richard had rented the U-Haul truck and the wood chipper. Both were extensively searched for evidence. Wood chips were recovered from the back end of the U-Haul truck. The chain saw blade was carefully examined, and yielded bits of blood, tissue, fragments of head hair, and some bluish-green fibers. There was blood on some of the fibers.

The forensic issues in the case can be summarized with the following questions:

- Were the skeletal remains recovered those of Helle Crafts?
  - Could a cause and manner of death be established? Was this a homicidal death?
  - If the remains are of Helle Crafts, and if the death is homicidal, could Richard be implicated?

The forensic aspects of the investigation of this case involved many specialties: pathology, odontology, bone identification (physical anthropology), criminalistics, trace and materials evidence comparisons (such as the nail polish), wood chip comparisons, biological evidence, and comparisons of hair, fiber, toolmarks, and handwriting.

The evidence gathered is shown in Table 1.1, along with the forensic testing used for its examination and the conclusions reached. Note that some of the findings are conclusive but others are circumstantial. The tooth and restoration identity were definite, so the remains recovered on the shoreline were confirmed to be those of Helle Crafts. The pathologist ultimately ruled the death a homicide based in part on the considerable fragmentation of the body; however, there was no way to ascertain a cause of death. The bone chips and wood chips had consistent toolmarks. The wood chipper in the case had a single cutting blade, but it had been discarded before anyone knew it might be useful as evidence. The hairs were consistent with having come from the same person and with hairs from Helle's hair brush, but hair comparison is not a means of positive identification. The defense could and did argue that the hair brush was not a true "known," because its use by someone else could not be rigorously excluded. The nanny, the children, and Richard, were all excluded as sources of the questioned hairs. The fibers were consistent with a nightgown Helle had owned and worn, but no "known" was recovered or available. The polish on the recovered toenail was consistent with fingernail polish Helle owned, but it could not be proven to be the only possible source.

This is one of the most interesting cases from a forensic-science point of view, not only because of the involvement of so many different specialty areas and experts, but also because forensic scientists were directly involved in the crime scene search and in the subsequent investigation.

In 1987, the case came to trial in New London, Connecticut (the defense had asked that the venue be changed because of extensive pre-trial publicity). The trial lasted several months, and there was extensive testimony by forensic experts for

Table 1.1 Evidence table—Wood Chipper case

Item	Examination	Findings/conclusion
Tooth—Lake Zoar shoreline	Odontological and radiological: identified as Helle Crafts by comparison with ante-mortem dental X-rays	Tooth belonged to Helle Crafts
Dental crown—Lake Zoar shoreline	Odontological: identification; Criminalistics: trace metal analysis	Identified as belonging to Helle Crafts by the dentist; Trace metals linked to the laboratory that made the crown
Bone chips—Lake Zoar shoreline	Anthropological and biological	human, from the head, hands and feet only; blood type O [A match to Helle's type?]
Sum of human remains—Lake Zoar shoreline	Pathology—Medical Examiner: cause and manner of death	Homicidal death based on the recovered bone chips; cause could not be determined
Wood chips—Lake Zoar shoreline	Toolmark: compare with U-Haul wood chips and bone chips; Wood identification: link type of wood	Chips consistent with one another (as having been made by the same cutting blade); consistent with having been made by the wood chipper; Linked type of wood to the wood lot (scene)
Wood chips—U-Haul truck bed	Toolmark: compare with Lake Zoar wood chips and bone chips; Wood identification: link type of wood	
Hairs—Lake Zoar shoreline	Hair comparison: compare with hairs from chain saw and hair brush	Hairs consistent with one another, and inconsistent with Richard, the nanny, or any of the children
Hairs—chain saw	Hair comparison: compare with hairs from chain saw and hair brush	
Hairs—hair brush from Helle Crafts' dressing table	Hair comparison: compare with hairs from chain saw and Lake Zoar shoreline	
Tissue—chain saw	Biological	human, blood type O, PGM 1-1
Blood—chain saw	Biological	human, blood type O
Blood—medium velocity spatter, from box spring in bedroom	Biological	human, blood type O, and yielded a PGM isoenzyme type 1-1, same as the tissue
Blue-green fibers—chain saw	Fiber analysis: classify and characterize	Consistent with fibers from a nightgown Helle is known to have owned and worn
Chain saw serial number	Restoration: render readable	Traced through manufacturer to a dealer to Richard Crafts
Chain saw credit card receipt	Questioned documents: compared signature with authentic Richard Crafts signature	Signature on credit card receipt was Richard's
Partial finger with friction ridge skin	Fingerprints: compare with known fingerprints of Helle Crafts; Biology/Serology and DNA	Consistent with the known prints, but inked knowns lacked detail; Blood type O and female (X chromosome)
Partial toenail with toenail polish	Trace Evidence: Compare nail polish with known nail polished seized from Helle Crafts' dressing table	Chemical composition and color consistent with one another
Yellow paint sample from the U-Haul truck bed	Trace Evidence: instrumental analysis comparison	Similar to yellow paint from the chain saw

both the state and for the defendant. Every finding and conclusion were challenged. The jury finally received the case in early 1988, but after many days of deliberations, one of them refused to deliberate further. The judge declared a mistrial.

The state retried the case in 1989, this time in Norwalk, and that trial was much shorter. Richard was convicted by the second trial jury and sentenced to a long prison term. The conviction was ultimately upheld by the Connecticut Supreme Court.

> **CHAPTER ONE LEARNING OBJECTIVES**
> - The nature and role of forensic science
> - The value of forensic science to society
> - The historical development of forensic science
> - Development of forensic science and laboratories in the United States
> - Forensic science laboratory operations
> - The importance of anthropometry and fingerprint identification to the development of forensic science
> - Nature of the scientific method and how it might operate in everyday situations
> - The key role that scientific method plays in all aspects of forensic science and investigation
> - The main specialty areas of forensic science and the scope of each of them
> - Elements of forensic analysis and the types of results forensic science can provide
> - The concepts of recognition, classification (identification), individualization, and reconstruction
> - Comparisons as a basis of forensic science analysis—inclusions and exclusions
> - Professional responsibilities and ethics

WHAT IS FORENSIC SCIENCE AND WHAT IS ITS ROLE IN THE JUSTICE SYSTEM?

The role of forensic science

The role of forensic science in the justice system has changed enormously in the last 30 years. Before then, forensic science results were used primarily in the adjudication of cases and very little in the investigation of incidents. Forensic science generally served to confirm identifications and the nature of well-defined evidence items. However, in the last 30 years this role has greatly expanded. For a variety of reasons, the role of forensic science has greatly expanded in the last several years to be used significantly in both the investigative portion (Figure 1.1) of the of the justice system as well as in the adjudicative function.

Figure 1.1 Forensic science has become an integral part of investigation. (Courtesy of Shutterstock, New York.)

Perhaps the single largest factor in this change has been the development of many computerized databases. This began with the development of useful automated fingerprint identification systems. For many years, people had been using fingerprints to identify individuals and suspects. However, one needed to know to whom a particular fingerprint or set of fingerprints might belong to make an identification. Thus, in the vast majority of latent fingerprint cases—where fingerprints are recovered from a crime scene or other relevant locations—unless one had a suspect or a very limited group of suspects, the fingerprint information might not be useful. Because the number of individuals whose fingerprints were in the hardcopy fingerprint files were so large, it was practically impossible to search for an individual's fingerprints in even a somewhat limited file. In the late 1970s, computer technology improved sufficiently so that several private companies developed search systems. These systems allowed for the searching of fingerprint information recovered from a scene, or taken at the time of arrest, against large computerized databases of people who had provided fingerprints during previous arrests. Although automated fingerprint identification systems were initially private, they fairly quickly became a government sponsored activity. These automated systems developed into extremely reliable and highly useful tools in expanding the ability to make use of both 10-print fingerprint cards and latent fingerprints from crimes. Now, of course, all 50 states have access to automated fingerprint database systems, and they are interconnected through the federal system. Thus, any fingerprint information that comes into the hands of the proper authorities can be searched against, most commonly, criminal databases in the state where they are taken. If not found there, they can then be searched in the federal jurisdiction or, subsequently, in other states if no matches are found.

The success of automated fingerprint identification systems was the first major step in making forensic science much more useful in the investigative aspects of the justice system, both criminal and civil. This use of large databases has now expanded enormously, particularly with the availability of the DNA database named CODIS, which stands for Combined DNA Indexing System. Everyone recognizes that DNA evidence has been a major breakthrough, particularly, in criminal but also in many civil situations. The ability to search a DNA profile against a large database of people who have committed crimes—or against people reported as missing persons listed in variety of other databases—has moved forensic science even further into the investigative area, rather than in just adjudicative. Forensic science-based databases are now expanding in other areas, such as firearms-related databases, as well as a variety of trace evidence-based databases that have become available to the justice system.

As a result of the ever-growing availability of such forensic science-based databases, forensic science is much more commonly looked upon as an ally by police and other investigators. The availability of such databases has allowed prosecution of many both criminal and civil cases that would never have been brought to trial before their availability. With this larger role in both investigation and adjudication, and the ensuing public awareness, understanding of the value of forensic science has grown. This has been very beneficial for forensic science, because it has attracted many young and talented individuals into the field who otherwise probably would not have become forensic scientists.

This increase in public awareness has been something of a double-edged sword. The public now virtually demands the use of forensic science by the adjudicative bodies as well as by investigators. This attention has brought to light certain high-profile failures of the justice system, rare though they may be. Forensic scientists are, of course, human and can make errors or show poor professional ethics. Because of very high expectations and greater scrutiny, the role of the forensic scientist has become more difficult though certainly not unreasonably so.

Definition of forensic science

Let us start with a very simple, working definition of Forensic Science: Forensic Science is science in service of the law. *Forensic* means, "having to do with the law." Science is a way of studying questions about the natural world in a systematic way. We will discuss the "scientific method" in some detail later in the chapter.

The term *forensics* means, "debating," and, in spite of its use in popular media, it is not the same as forensic science. Forensic science is, in fact, an incredibly broad subject that people now use to cover virtually any scientific, and some technical endeavors, which have applications to the law (Figure 1.2).

This book will concentrate, primarily, on forensic science in the service of criminal law, which is science applied to criminal cases. Besides criminal matters; however, there are numerous civil and administrative matters that can sometimes benefit from scientific and technical analyses. More and more forensic scientists are now involved in

Figure 1.2 Forensic science is a significant factor helping to ensure that the justice dispensed by the justice system is true justice. (Courtesy of Shutterstock of New York.)

civil, national security, and other administrative applications. Although most governmental crime laboratories work primarily on criminal cases, there is considerable forensic science effort applied to civil matters, such as product failure liability, disputed paternity resolution, and so on. More "expert witnesses" actually work on civil and administrative matters than on criminal ones. Although this book emphasizes applications to criminal cases, the application of science to other types of legal concerns is largely analogous. We will briefly describe and discuss some of the non-criminalistic specialty areas of forensic science. They are complex enough to require special treatment. There are entire books written on each of them.

The value of forensic science to society

One way of thinking about the "value" of forensic science is to ask: How does it serve the community? Another is to consider whether the benefits exceed the costs. Yet another way of thinking about it is to ask: What are the uses of physical evidence and physical evidence analysis in our legal system? We will discuss a number of these uses shortly.

Most public, forensic science laboratories are supported by municipalities, counties, state governments, or the federal government. Municipal and county labs with a dozen employees can have operating budgets of $1–2 million per year. Large laboratories serving major cities and laboratories serving larger states have multi-million-dollar annual budgets. To put the costs of forensic science in some perspective, it should be understood that such costs are only a miniscule fraction of the total cost of our justice system. For governments to continue to be willing to fund forensic science laboratories, there must be a belief that society significantly benefits from their work.

Looking into the type of information one can get from physical evidence helps clarify its value in investigations and prosecutions.

INFORMATION OBTAINABLE FROM PHYSICAL EVIDENCE

Corpus Delicti—Elements of a crime
Support or disprove statements by witnesses, victims, or suspects
Identify substances or materials
Identify persons
Provide investigative leads
Establish linkages or exclusions

The following are the major contributions that forensic laboratories provide to the criminal justice system, some of which are pretty obvious, while others are not.

1. Aid in the Investigation of Incidents of Possible Legal Interest

   A considerable proportion of the work submitted to and done in forensic science laboratories is *post facto*; that is, only after a crime has been committed and after someone has been arrested. Traditionally, laboratory services and findings have not been optimally utilized during the critical investigative period before an arrest. Laboratories are typically overloaded with casework and sometimes severely backlogged, which can prevent prompt analysis. However, physical evidence analysis can help provide investigative leads, or keep an investigation from going down an unproductive path.

2. Help Establish the Basic Legal Elements of a Crime–*Corpus Delicti*

   In law, *corpus delicti* refers to the body or "elements" of the crime. The elements are the things that the prosecutor is obligated to prove "beyond a reasonable doubt" to gain a conviction. Some of the analyses done in forensic laboratories serve primarily to establish elements of a crime. For example, in a possession of an illegal drug case, the laboratory must establish that the white powder seized is cocaine, or that those funny looking cigarettes contain *Cannabis sativa* (marijuana) (Figure 1.3). In a potential "drunk driving" case, the laboratory has to show that the person charged had a blood alcohol content above the legally allowed limit. Identifying that semen is present on a vaginal swab, from an alleged sexual assault victim, corroborates a crucial element of a sexual assault or rape charge; namely, penetration. Proving these elements of any such crime is required for successful prosecution, and one cannot convict someone without proving **all** the required elements of the crime.

3. Support or Disprove Statements by Witnesses, Victims, or Suspects

   The outcome of many investigations relies heavily on things people say about the case. In many instances, these things can become formal statements. Eyewitness testimony, for a host of reasons, is often known to be unreliable. Witnesses can often be influenced by their perspectives, prejudices, memory flaws, and other such factors. Suspects, and sometimes even victims, may have reasons not to be completely truthful in their statements.

   Physical evidence and its analysis can play an important role as an "objective" reporter in a case, against which statements can be evaluated. Nothing is more important in gaining a proper result in our system of justice than judging the credibility of witnesses. The entire trial process depends on the trier of fact (judge, jury, or administrative officer) being able to accurately evaluate what they hear. Forensic science can play an important role in this regard. If the physical evidence and its analysis objectively demonstrate something that contradicts

Figure 1.3 Chemical identification of the drug in seized drug evidence is a critical element in controlled substances cases. (Courtesy of Shutterstock, New York.)

a statement by someone in a case, it means that the statement is incorrect. Similarly, physical evidence can support a statement by a witness, victim, or suspect.

4. Identify Substances or Materials

   In many cases, the scientific examination of physical evidence provides an identification of a substance or material. Two obvious examples are a controlled substance possession case and a counterfeiting case. Further, identifying probable accelerant material in debris from an arson case, for example a flammable liquid (accelerant), such as gasoline, which can be used to start a fire. Perhaps finding gunshot residue on the hands of an individual suspected of firing a weapon in a shooting case, is another example.

5. Identify Persons

   Reliable identification of individuals is critical to the proper operation of our justice system. Biological evidence and fingerprints (Chapter 6) are routinely used to identify persons in criminal cases. Fingerprints have served the justice system well for over 100 years. As we will also see in Chapter 6, the identification of human remains is an important activity in cases of individual death and of mass disasters.

   As discussed previously, the development of national databases has made forensic science much more helpful during an investigation. We will discuss these databases in detail in the appropriate chapters. CODIS contains DNA profiles of convicted offenders and unidentified suspects in unsolved cases. Automated Fingerprint Identification System (AFIS) contains many known fingerprints and evidentiary fingerprints not yet identified. National Integrated Ballistic Identification Network (NIBIN) contains image data from bullets and cartridge cases from known weapons seized in cases and from evidence collected in unsolved cases.

CASE ILLUSTRATION 1.1

We have two middle-aged gentlemen who perhaps look much older because they drink much too much. They have been soused for years living off odd jobs and public assistance. Whenever the check comes, they go out and buy their bottles of wine and get drunk. In between checks, they live a classic homeless existence. So, while they are not exactly what one would call model citizens, they are not really dangerous criminals either. One night these two gentlemen are sitting in a vacant cabin out in the woods. They are old friends; sort of friends and enemies. They do not trust anybody else, and so they support each other. They are sitting in this cabin rapidly approaching oblivion when, all of a sudden, a shot rings out. The window in the cabin breaks and a bullet zips by the head of one of them, misses him, and buries itself in the wall. Both of them are startled to attention, and, about a minute later, someone smashes in the door with their foot, comes in, shoots one of the two gentlemen in the head, and then runs out the door. Now, those of you who are law enforcement officers, do you believe this story, or does it seem a little self-serving? We have a likely homicide, and one of only two witnesses is dead. The survivor doesn't want to be in the position of facing a murder charge. Did he make up this nice story after he and his buddy argued, and he shot him in a drunken rage? That's the question a jury must ultimately decide. If you are his defense lawyer, you are going to clean him up a little bit and keep him off the sauce for a while so that he makes a decent appearance. But no matter what you do, when he gets up on the stand, he is not going to be one of the world's most believable witnesses. He may be a perfectly honest individual; however, once his history comes out, he is not going to play well in front of the jury.

How can one assist the jury in making the correct decision? If we want to decide whether he is telling the truth, or whether he has made up this story to cover the fact that he murdered his friend, then how can one tell? Well, physical evidence for starters. If it's collected properly, and the scene is carefully processed, it can tell us a lot about what might have actually happened. Here are a few examples. The first thing the defendant said, in his description of the events, was that a shot rang out, the window broke, and the bullet just missed his buddy. What can one look for? First of all, one can look at the glass that's lying on the floor. In Chapter 4, on pattern evidence, we will discuss the fact that it is often possible to tell from which direction a window was broken. Did the bullet come from the outside in, or did it go from the inside out. Because, obviously, anyone who has watched crime shows on television, is aware of this common scenario. An investigator would often say; "Aha, this was staged. They really didn't break in through the French door because the glass is on the wrong side." When one looks at the physical evidence, one can probably tell that the window was broken from the outside in. Does that prove the defendant is telling the truth? It really does not, but it doesn't hurt.

(Continued)

CASE ILLUSTRATION 1.1 (Continued)

Secondly, one should look for a bullet or bullets. If our defendant is telling the truth, one should find two bullets or a bullet in the body and a second bullet or hole in the cabin. When one determines where they are, one should be able to do a little bit of trajectory reconstruction, as will be discussed in Chapter 4. An investigator should be able to tell whether one of those bullets at least came through that window. This also provides possible corroboration for the defendant's story. Thirdly, one can look at the door. Forensic labs get a lot of interesting things from doors all the time. What is on the door? One might find a dust print from somebody's foot having kicked in that door. Now, if that somebody was wearing a shoe with any kind of a patterned sole, it might leave an image of its pattern on the door. Taking a cross section of our society today, you will find a high percentage of individuals wearing pattern shoes. There are hiking boots, waffle stompers, lots of sneakers, and a variety of other patterned soles. Even some shoes, that look like dress shoes, can have patterned bottoms. The point being, one can look at the defendant and reason that he is likely to have no more than one or two pair of shoes. You can examine them to see if they are patterned shoes, and if one matches the pattern on the door. If there is no match, then he probably isn't the one who kicked in the door. None of the evidence mentioned so far can actually determine much about who committed the crime, nor much about what might have occurred.

The physical evidence **does** tell us; however, that if all these things collectively check out, it may be that the defendant is telling the truth. He probably is not knowledgeable enough to fool the experts who look at these things. The net result is that his credibility is enormously improved. This is a key role that forensic science can play in making the justice system work better. It is a fairly subtle role, because it does not scream "guilty or not guilty," but this type of evidence can play a significant role in helping the jury come to the correct decision. Those processing a crime scene should not concern themselves with who is going to turn out to be a creditable witness or who is going to turn out to be an abysmal witness. Nor, should any of these factors determine what evidence they ought to collect based on anything like this. This is why it's particularly important to examine each scene carefully, even when the case seems simple and straightforward. The investigators may not be able to see any obvious probative value in collecting all the available physical evidence. However, in a particular case, non-probative evidence may say a lot about when, in fact, someone is or is not telling the truth.

A BRIEF HISTORY OF FORENSIC SCIENCE

Early applications of forensic science

To understand forensic science as it is today, it is helpful to take a brief look at its origins and how it has evolved over time. Some very early work on forensic medicine was published in China in AD 1250. Although many of the concepts that we think of as belonging to forensic science have been around much longer, one can argue that the formal beginnings of modern forensic science in the western world began in the period between 1800 and 1850.

Medicine has become less experience-based and more scientifically based on understanding of disease process developed from experiment and careful observation. Doctors were carefully dissecting bodies, and the microscope became available to help develop a better understanding of detailed anatomy and body functions. In ancient and medieval times, there are accounts of alleged homicidal poisonings. The Medicis are thought to have poisoned people in the 1600s. Socrates was killed by being forced to drink hemlock (which has nothing to do with hemlock trees, but contained a toxic substance later identified as coniine). Many of the forensic science specialties we recognize today can be traced back to early medico-legal institutes in Europe. Although these institutes concentrated on investigating death cases, some of the early medico-legalists also did work on the identification of blood and semen stains. Proper identification of body fluid stains has long been important to the investigation and prosecution of crimes. Common questions to ask in the identification of blood stains include:

1. Is the sample blood?
2. Is the sample animal blood?
3. If animal blood, from what species?
4. If human blood, what type?
5. Can the sex, age, and race of the source of the blood be determined?

During the middle 1800s the natural sciences chemistry, biology and physics were developing and scientific method was being refined. The recognition of the potential value of the non-medical forensic sciences, particularly criminalistics, took a giant stride with the writings of Hans Gross. In 1893, he published a book entitled *Handbuch für Untersuchungsrichter, Polizeibeamte, Gendarmen* (*Handbook for Coroners, Police Officials, Military Policemen*), which was very influential on the practice of criminal investigation. Gross was not a scientist, but rather a magistrate and law professor in Austria. Him championing the utility of the developing discipline of forensic science was very important to its acceptance by many rather skeptical police agencies. Gross is responsible for the word "criminalistics," and was one of the first people to carefully consider the value of physical evidence in investigations. In European justice systems, the magistrate had a role both as judge and as the primary investigator in a case. In that primary investigator role, he could call on the services of forensic experts, and that is what prompted Gross's interest in what we now call forensic science.

Concurrent and overlapping with this period, continuing until about 1900, was the major period of development of more systematic methods for human identification. In the 1890s, Alphonse Bertillon developed a method for criminal identification for the metropolitan police agency in Paris based on a series of body measurements (Figure 1.4).

The measurements of people arrested or incarcerated were classified and kept on file. Since many people misrepresented their identities to the police, this proved a valuable method to see if an arrested individual might be a person wanted by the police for another crime, perhaps under a different name. After a time, it became clear that these files had significant limitations, such as measurement errors and not enough independent measurements to truly distinguish each individual, as the files became large.

At about the same time, Galton, Herschel, and Henry, and others in England were studying and trying to apply fingerprints to medical diagnosis and identification (Figure 1.5). This resulted in the newly developing science of fingerprints becoming the method of choice for routinely identifying people, and this is still true today. These developments are described more fully in Chapter 6.

Development of forensic science laboratories and professional organizations

Forensic science laboratories, as we know them today, began to emerge in the early twentieth century. In Europe, they tended to grow out of the medico-legal institutes, which performed what we now think of as primarily forensic pathology functions. In 1909, a professor named Riess started a forensic photography laboratory at the University of Lausanne in Switzerland, which soon broadened its areas of expertise. In 1910, Dr. Edmond Locard (Figure 1.6) started the first forensic laboratory in Lyons, France. Dr. Locard is particularly important in the history of forensic

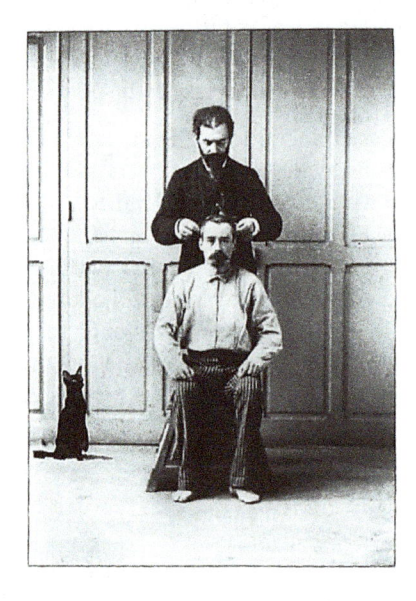

Figure 1.4 One of the first challenges faced by forensic science was establishment of a person's identity. Bertillonage was one of the first scientific methods used.

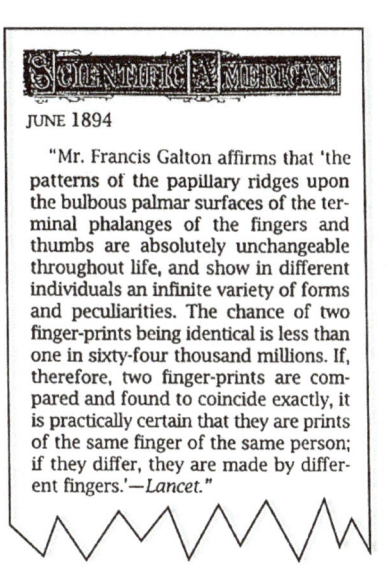

> JUNE 1894
>
> "Mr. Francis Galton affirms that 'the patterns of the papillary ridges upon the bulbous palmar surfaces of the terminal phalanges of the fingers and thumbs are absolutely unchangeable throughout life, and show in different individuals an infinite variety of forms and peculiarities. The chance of two finger-prints being identical is less than one in sixty-four thousand millions. If, therefore, two finger-prints are compared and found to coincide exactly, it is practically certain that they are prints of the same finger of the same person; if they differ, they are made by different fingers.'—*Lancet*."

Figure 1.5 Galton's stature as a scientist helped lend credibility to the early use of fingerprints for identification.

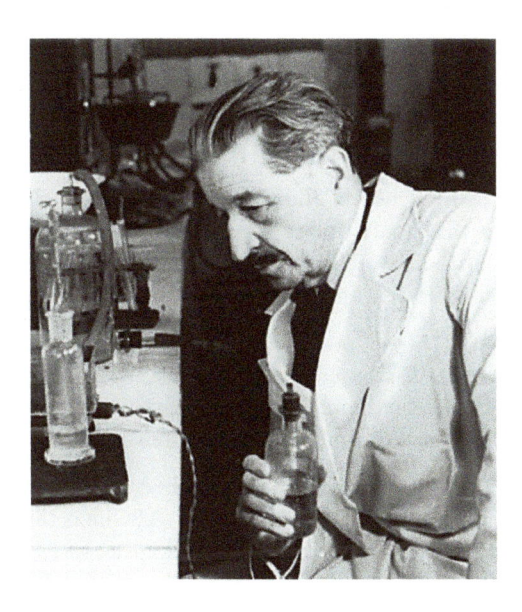

Figure 1.6 Locard was a pioneer in use of physical evidence and the founder of one of the first forensic laboratories in Europe.

science because of the Locard Exchange Principle, which will be discussed in some detail in Chapter 14. In Europe, many forensic science laboratories were, and still are, affiliated with universities. In the United States, most forensic science laboratories initially emerged in police agencies.

The development of forensic laboratories in the U.S. came little later. August Vollmer, who was the police chief in Berkley, California in 1928, became interested in the use of scientific evidence in police investigations. He was responsible for starting the forensic laboratory of the Los Angeles Police Department when he became its Chief. It is interesting that many of the pioneers in forensic science were medical doctors. An important person in the development of firearms examination, Calvin Goddard, was a military physician. He was involved in several pioneering studies that demonstrated the value of firearms identification to a skeptical law enforcement community. Following the St. Valentine's Day massacre in Chicago, Goddard was called in as a consultant and demonstrated the usefulness of examining bullet and cartridge case evidence. That led to his starting a forensic laboratory in 1929 in Chicago. It was originally a privately funded laboratory housed at Northwestern University, but, subsequently, it became the

Chicago Police Department Laboratory. The Federal Bureau of Investigation (FBI) started their laboratory in 1932, and the New York City Police Department Police Laboratory can trace its origin to about 1934. Originally, there were two detectives assigned to the New York City Police Department forensic laboratory. In the next few years, many other crime laboratories were started.

Between 1940 and 1970, the governmental responsibility to provide crime laboratory services to law enforcement became fully recognized. Between 1970 and 1980, there was a very rapid growth in the number and scope of forensic laboratories. The Law Enforcement Assistance Administration, set up by the federal government as a result of the Safe Streets Act of 1968, provided considerable funding to state and local jurisdictions to either start new laboratories or expand and improve existing ones.

The American Academy of Forensic Sciences (AAFS) was formed by a small group of interested pathologists, psychiatrists, criminalists, and attorneys led by Dr. R.B.H. Gradwohl of St. Louis in 1948. Today, American Academy of Forensic Sciences (AAFS) has sections representing 11 different forensic disciplines and specialties. AAFS has grown to a quite substantial organization with members from many countries in addition to the United States. AAFS started its peer-review journal, the *Journal of Forensic Sciences*, which many professionals feel has become the premier journal in the field. Besides the Academy, six regional associations of forensic scientists, primarily criminalists, have grown up across the country. The International Association for Identification (IAI) was formally incorporated in 1919 and fingerprint examiners, and other pattern evidence specialists belong to the IAI. Today, there are professional organizations of firearms and toolmarks examiners as well as documents examiners. Some of them are discussed in the chapters on those topics.

In the early 1970s, the American Society of Crime Laboratory Directors (ASCLD) was formed by a sizable group of crime laboratory directors with a strong assist from the FBI. One of their first projects was to develop a system of voluntary laboratory accreditation (Figure 1.7). It took 10 years to develop a workable scheme, but they created the ASCLD Laboratory Accreditation Board and began laboratory accreditation in 1982.

This has proven to be a highly successful venture, and the majority of forensic laboratories have become accredited or are actively working toward that goal. This is generally a voluntary process sought by the laboratories themselves to validate the quality of their work. In recent years, several states through legislation have made accreditation mandatory for their laboratories, and there is some pressure to mandate accreditation of all forensic laboratories.

Starting in 1994, particular concern over the complexities of DNA analysis has made accreditation of DNA sections virtually mandatory. After a study by the National Academy of Science in 2008, there has been a sizable federal bureaucracy set up to improve and regulate almost all aspects of forensic science.

Besides **accreditation** (which applies to laboratories), there has been the development of **certification** programs (which apply to individuals). The American Board of Criminalists (ABC) has the most extensive program for criminalists. There are specialty certification boards for forensic pathologists, forensic dentists (odontologists), forensic

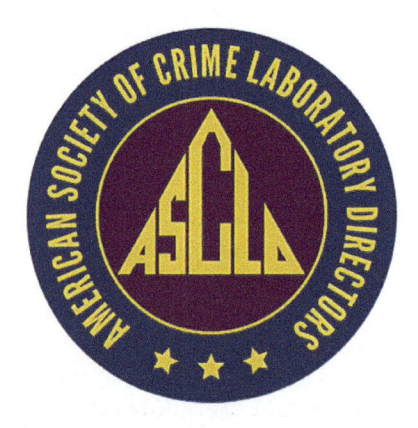

Figure 1.7 The American Society of Crime Laboratory Directors was instrumental to bringing uniformity to forensic laboratory practice and developing forensic laboratory accreditation in the United States.

anthropologists, forensic entomologists, forensic document examiners, forensic toxicologists, and other specialties. The IAI has certification programs for latent fingerprint examiners, forensic artists, crime scene photography, footwear and tire tread evidence, and crime scene investigators.

There are over 300 governmental forensic laboratories in the U.S. They are maintained by agencies of the federal government (such as the FBI, Bureau of Alcohol, Tobacco, Firearms and Explosives (ATFE), Secret Service, Drug Enforcement Administration (DEA), and so forth), or by units of local government (state, county, or city). Some states have multiple laboratories organized into a central laboratory and satellite laboratories around the state. Many labs are located within a law enforcement agency, but they can be found in prosecutors' offices, medical examiners' offices, and in departments of health. Some laboratories are very small, while others have hundreds of personnel. There are also a number of privately operated forensic laboratories, most of them specialize in DNA analysis, toxicology, engineering or questioned document examination, although virtually all forensic specialties are represented. Most large laboratories use a major portion of their analytical resources on controlled substance identification and DNA analysis.

THE MANY FACES OF FORENSIC SCIENCE

Human biological and medical sciences

Forensic science, in the broad sense of the term, encompasses many different scientific and technological specialty areas. All or most of them can have applications in both the civil and criminal justice systems. Some of the specialty areas will be described in the following sections, including some that are beyond the scope of the book and won't be discussed in detail in subsequent chapters. There are many others, such as forensic accounting, forensic meteorology, and forensic nursing (mentioned in Chapter 10) that are not discussed here. Today, "forensic" is used as an adjective to describe many disciplines in the context of applying the methods of that discipline to legal matters, and new forensic specialties are regularly being developed.

Forensic pathology is another name for forensic medicine. Forensic pathologists are Doctors of Medicine (MD) who have first specialized in pathology (the study of the nature of disease and its causes, processes, development, and consequences), then take further training in forensic pathology. Forensic pathologists are experts in determining the *cause* and *manner* of death (Figure 1.8). The *cause* of death is a medical determination—the medical explanation for why a person died. The *manner* (also called circumstances) of death is a medico-legal determination. Cause of death is a gunshot wound, asphyxiation, poisoning, and so forth. Manner can be homicide (one person kills another), suicide (a person kills himself or herself), accidental, or natural. The media routinely confuse cause and manner of death. Both the cause, or manner, of death may sometimes be *undetermined*.

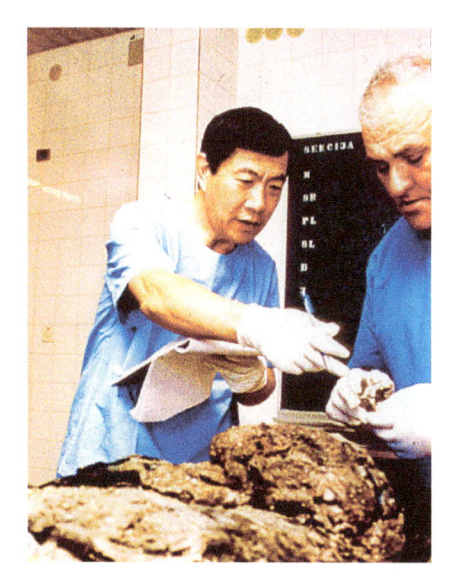

Figure 1.8 Pathology was one the first truly scientific forensic specialties. Shown here, Dr. Lee conferring with a medical examiner.

There are two "systems" of death investigation in the U.S.: the coroner system and the medical examiner system. A *coroner* is an elected official and need not have any special medical knowledge or training, since he or she can call on specialists to assist in technical determinations. A coroner has the power to convene a *coroner's inquest* and take sworn testimony at a proceeding if necessary to assist in making determinations. A *medical examiner* system specifies by law that a forensic pathologist—again, someone with extensive medical training—make appropriate determinations in cases of questioned, suspicious, or unattended deaths. Many bigger cities and some states use the medical examiner system, but many jurisdictions are still under the coroner system. There are some coroners who are forensic pathologists, and other coroners who enlist the services of a medical examiner or other pathologist.

A medical examiner's determinations are based on all available information about a death, including the scene, results of the police investigation, results from the forensic science laboratory, results from the post-mortem toxicology, in addition to the findings at autopsy.

Entomology is the branch of biology devoted to the study of insect species (Figure 1.9). When an animal or human dies, houseflies and other insects are able to detect the location of the body quickly. The adult flies will lay their eggs on or in a corpse, if they have access to it. The life cycle of many insects consists of egg, larva (or maggot), pupa (or cocoon), and adult. In some insects, there can be multiple larval stages. Entomologists know the life cycles of the insects in detail, and, thus, know how long each stage of the life cycle takes. The time is governed by temperature and by the length of daylight and darkness during each day. Forensic entomologists can examine insect eggs, larvae, or pupae from a body to determine which species of insect produced them. Eggs and larvae must be collected and reared to the adult stage to identify the species. Then, using information about the number of insect cycles, temperature, length of daylight hours, and other information from the scene, they can often "back calculate" to estimate the time of death. Since determining exact time of death is often a problem, forensic entomologists can make important contributions to cases when insect evidence is found and the time since death is an issue.

Forensic odontologists are forensic dentists. They do two major types of analyses involving human dentition. One is identifying human remains that are so altered by decomposition, fire, or explosion that they cannot readily be identified by visual means. Typically, the odontologist looks at pre-mortem and post-mortem dental X-rays. The dentition, and, in most cases, the dental work done on an individual is sufficiently unique to permit personal identification in this way. The X-rays of a decedent must be compared with pre-mortem X-rays from one or more persons suspected of being the person. The use of dental identification in mass disaster situations is a component of this and discussed further in Chapter 6.

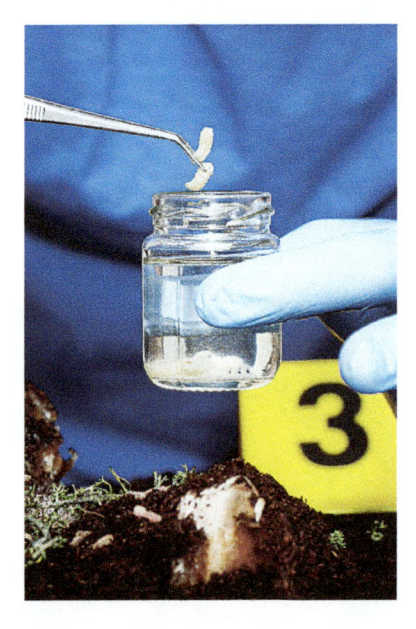

Figure 1.9 Forensic entomology is an important forensic specialty helping to discover time of death in difficult cases. (Courtesy of Shutterstock, New York.)

The second major activity in forensic odontology is bitemark comparisons. There are several different techniques for actually doing the comparisons. Bitemarks may be found on human bodies in cases of assault, sexual assault, and child abuse, and occasionally on other objects that show an impression of the teeth. If the marks are recognized, properly documented, and examined by a forensic odontologist, they can be compared with known bitemarks obtained from suspects. Often suspects who did not make the bitemark can be readily excluded. Sometimes, a suspect who did make the bitemark can be identified.

Physical anthropology is the study of the human skeleton and how it has evolved over time. Forensic anthropologists are physical anthropologists specialized in examining primarily human skeletal remains (Figure 1.10). They can quickly determine if skeletal remains are human or animal, and often can estimate approximately when they were deposited. If the remains are human, they can be "reconstructed" (laid out in proper orientation). Depending on the condition and amount of skeletal remains, forensic anthropologists can often provide estimates of the age, stature, and gender of the individual. They can also sometimes tell if the remains belong to the *Caucasoid*, *Negroid*, or *Mongoloid* race. They can spot skeletal abnormalities and skeletal trauma that may be present. Traumatic injuries can provide information about cause of death (e.g., a knife blade cut on a bone supports a stabbing death), and sometimes help in identification based on comparison with antemortem X-rays. Finding indicia of repair (plates, screws, or implants) can often provide valuable information. Thus, the anthropologist can often provide descriptive information about remains even if it is not possible to identify (individualize) the skeletal remains.

Although not a part of forensic anthropology, *per se*, we should mention that there is a sub-specialty sometimes referred to "forensic sculpture" or more technically cranio-facial reconstruction. Forensic sculptors need a skull to work with. From the skull, they attempt to reconstruct what the person's face may have looked like. The reconstruction is based on tissue thickness and other data that has been gathered from population studies. Eye color, hair color, and hairstyle are usually unknown, so it is often difficult for the sculptor to create a readily recognizable likeness of the person. Similarly, computer technology can be used to "age" a missing person's photograph. This kind of information can sometimes be helpful in determining an identity.

Forensic toxicology is the study of the effects of extraneous materials, such as poisons, toxins, and drugs, in the body. Forensic toxicologists must determine both the presence and amounts of such materials in a body and also attempt to interpret the possible effects of these materials. They must be quite knowledgeable in analytical chemistry techniques as well as biology, physiology, and pharmacology. Toxicology is discussed in Chapter 13. Forensic toxicologists who work on post-mortem specimens are often associated with medical examiners' offices. Many forensic toxicologists are also involved in testing specimens from living persons examples include: blood and breath alcohol determinations in

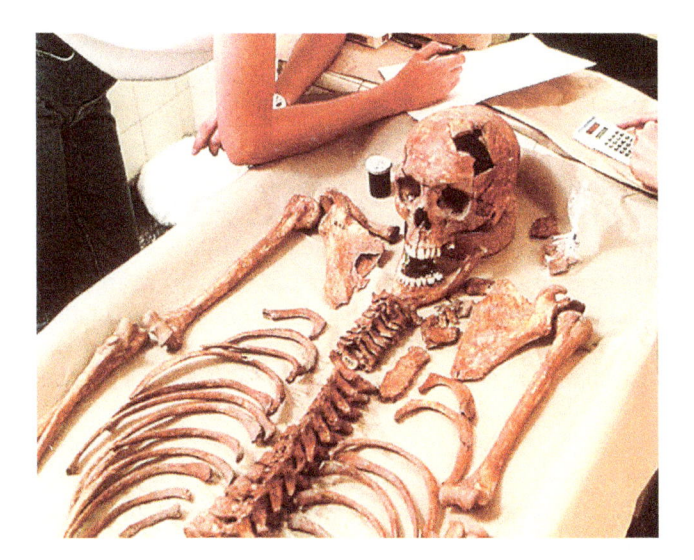

Figure 1.10 Forensic anthropology is an important forensic specialty in determining what happened to an individual in cases involving skeletonized remains.

the enforcement of "driving under the influence" laws, urine or hair testing to enforce "drug free workplace" rules, and checking for "date rape" drugs in sexual assault complainants.

Forensic psychiatrists and psychologists do similar work. The psychiatrists are medical doctors, while the psychologists are usually PhDs who have obtained a license to have a clinical practice. Forensic psychiatrists and psychologists evaluate offenders for civil and criminal competence and may be involved in offender treatment programs. They may also evaluate juveniles to assist courts in determining the best placement for them. A few of these specialists "profile" criminal cases. Profilers can sometimes provide useful information about the characteristics of an unidentified offender based on his *modus operandi* (MO), habits, and crime scene patterns. Profiling has concentrated primarily on serial murderers and serial rapists.

Natural sciences—chemistry, biology, and physics

Criminalistics is a term that sometimes is used to cover all the natural science approaches to evidence examination and is the primary subject of this book. An easy way to think about criminalistics is that it encompasses all, or most all, of the specialty areas found in full-service forensic science laboratories. It involves the examination, identification, and interpretation of items of physical evidence. In general, criminalistics can be divided into four major categories of examinations: biological evidence analysis; analysis of materials evidence; forensic chemistry, including but not limited to fire debris and controlled substance identification; and pattern evidence, including documents, firearms, toolmarks, fingerprints, footwear, and other patterns, including scene reconstruction patterns. Most of the rest of the chapters in the book are devoted to these subjects.

FOUR MAJOR CATEGORIES OF EXAMINATION IN CRIMINALISTICS

Biological	Material	Chemical	Pattern evidence
Blood	Objects	Drugs and toxic substances	Imprints
Body fluids	Pieces of objects	Paints, pigments	Fingerprints
Hair	Plastics (pieces)	Gunshot residue	Tire impressions
Tissues	Glass (pieces)	Accelerants, solvents, alcohols	Footwear impressions
Pollens		Rubber materials	Physical patterns
Wood materials		Resins	Firearms, bullets
Other plant derived material		Plastic materials	Cartridge cases
Feathers		Explosives residue	Toolmarks
		Fibers	Questioned documents
		Soil	
		Glass	
		Misc. trace evidence	

Criminalistics is as much an approach to evidence examination—a way of looking at it—as it is a collection of specialties. A criminalist is a person who thinks about a case and evidence in specific ways. A criminalist thinks about what information can be gleaned from a piece of evidence that can further the investigation or adjudication of the incident to which it is related. Criminalists are usually specialists in one of the analysis areas, but that alone is not what make them criminalists. It is possible to be a very skilled chemist or molecular biologist and not be a criminalist at all. To follow—and at intervals as we go through the book looking at different categories of evidence and analyses—we hope to give you an appreciation for those elements of forensic thinking and analysis that define criminalistics.

Technology and engineering

Forensic engineers are experts trained in one or more of the engineering specialties (often, but not exclusively, mechanical, electrical, or civil). Many are professional engineers; a professional licensure gained by qualifying for and then passing a challenging examination. Forensic engineers are involved in reconstructing automobile and some other transportation accidents, materials failure cases, and building or structure collapses. The majority of cases involving forensic engineers are civil rather than criminal.

Figure 1.11 Digital forensics is one of the newest and most rapidly growing forensic specialties and can involve myriad hardware and software. (Courtesy of Shutterstock, New York.)

Digital forensic examinations; often computer or processor related

There are two aspects to what might be called digital forensic science. You may hear this specialty called "computer forensics." This is misleading terminology as we have explained previously since "forensics" is not forensic science. One aspect is the investigative use of computer technologies and electronic records, sometimes called "digital evidence." Investigators may make use of information on computer hard drives, in pagers, cell phones, handheld devices, and other such technologies to help solve cases (Figure 1.11). Another aspect is more technical, where considerable knowledge of computer science and computer engineering may be needed to find hidden or deleted information on electronic media. Such information can be used to track down those who have committed computer crimes, such as circulating pornography or unauthorized access to confidential information residing on computer networks. This work can include Internet-based child pornography investigations, tracing the origin of computer viruses, worms, and so forth, as well as elaborate "hacking" schemes.

Forensic/Investigative Technologies—Almost any kind of technology that has or could have any application to criminal or civil investigation, can loosely be called "forensic." Often, products or technologies are called "forensic" for marketing purposes. Many technologies associated with scene investigation fall into this category. Use of various types of specialized light sources and specialized scene search techniques, like ground penetrating radar, are just two examples. Many advanced photographic and video techniques, especially digital, are coming into more frequent use in appropriate cases.

NATURE OF FORENSIC SCIENCE AND THE SCIENTIFIC METHOD

Forensic science is, first and foremost, *science*. It is important, therefore, to describe briefly how science differs from other areas of human inquiry, and how the "scientific method" works. Most scientists, who use the scientific method in their work all the time, don't consciously think about using it. The scientific method is more of a way of approaching problems than a detailed recipe. It is that particular approach, the scientific approach, which distinguishes between scientists and others.

In forensic science, the scientific method is extremely valuable in many different ways. First, as noted, forensic science is science, but the importance of the scientific method in forensic science is not limited to scientific analysis tasks. It has major applications in doing investigations, reconstructions, and many other important tasks. We will illustrate as we go along that there is a distinct parallel between the scientific method and *reconstruction* (this will be discussed in more detail in Chapters 3 and 4).

The scientific method is not esoteric, and you don't have to be a professionally trained scientist to use it. In fact, many people use it every day, without even thinking about it. There are various "formulations" of the scientific method, but it can be viewed as a four-step process.

1. Careful Observation: The importance of these two words cannot be overemphasized. The first step in the scientific method is being receptive and inquisitive. Anyone can be a careful observer. Observations of events and phenomena in the natural world and curiosity about what is behind them have been the driving force behind the development of science.
2. Make Logical Suppositions to Explain the Observations: The point of scientific inquiry is to try to understand natural phenomena and the natural world. So, the second step is to take an "educated guess" as to an explanation. The educated guess is usually called a *hypothesis*. The most important thing about the hypothesis is that it has to be made up of *experimentally testable* propositions. Predictions can be made based on the truth of the hypothesis. If the hypothesis is true, certain things that follow from it must be true. Then experiments are designed to test the predictions.
3. Hypothesis Testing—Controlled Experiments: Developing ways to test the hypothesis is the heart of experimental science and scientific method. The experiments that are devised must be *controlled*, that is, designed so that only one thing varies at a time. If the experimental design is correct, it enables the experimenter to find out the effect of that variable alone.

   As an example, a scientist might want to know how adult salmon are able to find their way from the Pacific Ocean hundreds of miles up freshwater streams and tributaries to the place they were born in order to spawn. Do the fish do this visually, by smell, or some other way? A controlled experiment might be designed in which the fishes' sense of smell was disrupted to test whether the mechanism was olfactory. Suppose the fish, whose olfactory faculties were disrupted, found their way to their spawning ground just as well as the control fish (whose sense of smell was not tampered with). The hypothesis being tested was: The fish use their olfactory sense to make the journey. The experimentally testable prediction was: If the hypothesis is true, disrupting the fishes' olfactory abilities will prevent them from migrating. The prediction was found to be false, invalidating the tested hypothesis.
4. Refine the Hypothesis—Theories and Natural Laws: Hypotheses must be continually refined; that is, re-tested over and over. The reason is that while true hypotheses generate true predictions, false hypotheses can also generate true predictions in a particular test. It may take many tests, and a long time, to discover the proper tests to show that a hypothesis is false and has to be adjusted.

   The "closed loop" of hypothesis testing continues forever. Some hypotheses that have been tested extensively by many different scientists, and found to be sound, become established as *theories*. Think of a theory as a well-tested hypothesis. Occasionally, a well-tested theory may become known as a *natural law*. But, no matter how well tested a theory is, it may still be shown to be wrong at some point. All good scientists recognized that no hypothesis, theory, or natural law is absolute. Someone may come up with another experiment that shows a flaw and requires modification of the original hypothesis or theory. That is what the "scientific method" is all about: the recognition that one is never finished, that each observation must be tested in new ways to ensure that the current theory keeps being refined. The process is diagrammed in the Figure 1.12.

   Science, because it follows the scientific method, does not deal with absolutes. Scientific "truth" is the current, best, most refined hypothesis or theory. This concept is important because forensic scientists work extensively in the legal system. References to "proof" and "truth" by lawyers do not have the same meaning as they do to scientists.

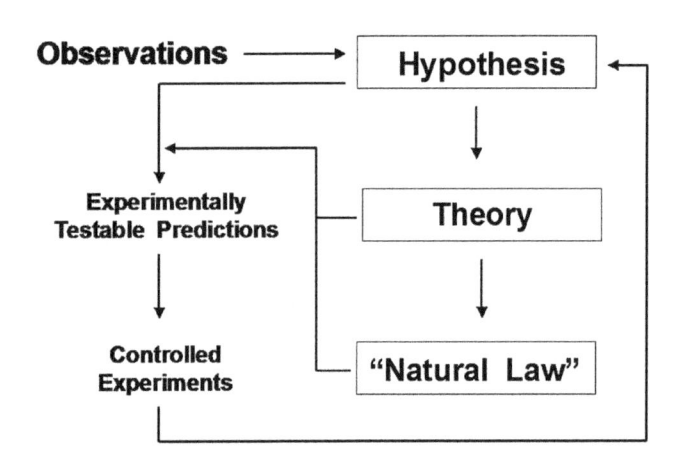

Figure 1.12 In classifying evidence comparisons, disassociation is every bit as important as association; two non-matching hairs depicted.

SIDEBAR 1.1 Everyday examples of applying the scientific method

There are many applications of the scientific method that are quite mundane. If you were to spill some water on the ground, when you returned a couple of hours later, very likely the water would have evaporated. The first person that made such an observation probably didn't understand the concept of evaporation, but knew there was a puddle there and then some time later there was no puddle there. Something had happened. We know that the water passed from the liquid phase to the gas phase and as the air circulated the water dissipated. One might also observe that on a hot day the water tends to disappear a little faster than on a cold day. Perhaps there is a correlation between temperature and the rate in which things pass from the liquid phase to the gas phase. Let us say we never had high school science so we propose a very simple set of experiments to test our hypothesis that when water evaporates heat seems to be involved. How could we test this? Hypothesis: heat has something to do with water passing from liquid phase to gas phase and dissipating. Go to the kitchen and measure out one cup of water and place it in a pot. Place the pot on a burner and turn on the burner. If you have an electric stove, you have numbered positions, so you can turn it to position two, for example. One can measure how long it takes for the water to evaporate. Let's say it takes 15 minutes for the water to disappear. One next measures another cup of water and places it in the cooled pot, puts it on the burner and sets the burner at position five (a higher heat). One measure the time until the water is gone and instead of taking 15 minutes it takes only 11 minutes. This is one simple test of the hypothesis that the more heat one puts into the water, the faster it disappears.

Another example, that virtually everyone who drives has experienced at one time or another, is the timing of traffic lights on a stretch of road with many intersections and therefore many traffic lights. Say a person is on the road late at night, when traffic is light, and they are anxious to get home. The speed limit is 30 mph, what is going to happen if they are in a big hurry and try to drive 70 mph? Besides risking being arrested, they will be stopped by virtually every traffic light. It will seem as if there is a traffic light every 15 feet and the trip will seem to take forever. It is common knowledge that there is some great force, actually the traffic department, that regulates traffic lights. In the old days, they use to put little clocks on the poles, but now computers run the world, including the traffic lights. The point is that there should be some speed at which one can travel that will allow one to go through a series of lights, maybe 10 or 15 lights, and make them all. One can guess that the speed is not going to be 70 mph, because the traffic department does not want you to go 70 mph. The "magic" speed is probably not going to be 15 mph either, because the traffic department gets judged on how well they move traffic along. How do we use scientific method to answer this question? The hypothesis is that these lights are timed for a particular speed. You could try to travel the course at a very steady speed. You could try 42 mph and see what happens. You could make 3 or 4 lights, and then gets stopped; then make a few more, and get stopped again. The next experiment could, perhaps, be at 37 mph. Carefully observe the results and try to see if you are missing fewer lights. Try a series of speeds selected based on how well you seem to be doing in avoiding red lights. You will probably find that the lights are timed fairly closely to the speed limit. It will not necessarily be the speed limit, and it also may change with the time of day. The point is that a systematic approach of selecting a hypothesis, testing it, and making adjustments based upon the tests results is a very efficient method to solve many problems. That is basically scientific method. Most people use scientific method and never think about it as being scientific method.

Here is one last, somewhat trivial, illustration, which we will call the shower caper. A family has moved into their first house, and, within the first 2 weeks, much of the kitchen ceiling almost fell on their heads while they were eating breakfast in the kitchen. The pre-teen daughter was taking a shower upstairs in the hall bathroom, which was right over the kitchen. As it turned out, it wasn't really her fault. This was about a 25-year-old house when they bought it, and it was discovered that in the hall bathroom's shower there was a big area in the grout around the soap dish that had largely disappeared. When anyone took a shower, the shower spray would hit that part of the wall and go right through the area where the grout should have been. The water ran down the wall soaking the kitchen ceiling and making the plaster soggy enough that, KABOOM, it fell into the kitchen. A few years later, they moved to a different house in a different state, and the teen-age daughter was living at college, but did occasionally come home for part of her summer vacation. The first or second time she came home, the lady of the house was in the basement in the laundry room and all of a sudden there was water pouring down through the ceiling. It happens that in the second house, the hall bathroom was over the laundry room. Remembering the experience in the other house, they asked the daughter, "What are you doing up there?" She replied, "Just taking a shower." It was time to investigate. Looking around and not being totally untrainable, they poked in the grout, surrounding the soap dish. How does one scientifically approach this problem? What is the first thing one must worry about? Is there a leak in the plumbing? If there is a leak in the plumbing, it is a potentially sizable, really expensive problem. On the other hand, if it is leaky grout, one can buy some silicone sealant, mush it around, and the problem is solved, at least for a while. Fortunately they had solved the earlier problem that easily, although it did cost quite a bit to repair the kitchen ceiling.

1. *First hypothesis*: The plumbing is leaking.
    Test: Don't turn on the shower, just turn on the water and let it run in the tub. If it is the plumbing that is leaking, when one observes the area where it was leaking in the basement, it should still leak.
    Result: Not a drop. Good, it is unlikely that one will have to tear the walls apart and put in new plumbing.
2. *Second hypothesis*: The water is leaking through the walls in the shower enclosure.
    Test: Turn on that shower blasting hard and point the showerhead to various areas in the enclosure, with somebody in the laundry room to look for water. Spraying the back wall, the sidewall and every other possible area of leakage.
    Result: No water was observed in the laundry room.

3. *Third hypothesis*: The water is coming through the floor somewhere just outside the shower enclosure.
   Test: To put the shower curtain very carefully in the shower, which not all daughters do, to make sure that water would be is confined to the shower enclosure and run the shower.
   Result: No water is observed in the laundry room.
4. *Fourth hypothesis*: The water is getting out of the enclosure onto the floor and leaking through the floor.
   Test: Open the shower curtain and splashing a little water along the edge of the tub and look for leakage from below.
   Result: There is no evidence of leakage.

That is enough scientific method for now, perhaps the problem mysteriously disappeared. A few days later their daughter is again taking a shower and again there is water pouring into the laundry room. **The fifth**, and it turns out, final hypothesis: is that since there is a toilet next to the shower, possibly water is running across floor and leaking though the area around the toilet bowl. Final test is to take several towels and put them around the base of the toilet bowl, splashed a bunch of water on the floor and look for leakage, then repeat without the towels. This test is a success in confirming the fifth hypothesis. The leak was around the toilet bowl and best of all, the problem could be solved with a little more of the silicone sealant. This is not exactly an earth shattering scientific breakthrough, but a clear demonstration of what can be accomplished through careful observation and reasoned testing of possible hypotheses. **Scientific method is something that many people use often, usually without even thinking about it.**

Applicability of scientific method to forensic science and investigation

The essence of the scientific method is the combination of observation, and the use of "feedback" from testing predictions generated by the hypothesis. Careful testing of a false hypothesis will, sooner or later, reveal the flaws in the hypothesis.

There are three important senses in which the scientific method is an important component of forensic sciences and of investigation. First, as we have noted, forensic science is *science*, and therefore follows the scientific method in building its knowledge foundation. Second, it forms the basis for event reconstruction. In the criminal case context, this may be called "crime scene reconstruction." It means using the physical evidence record, and all other available information, to determine the most likely explanation for past events, or sometimes to rule out a seemingly possible explanation. Reconstruction is discussed later in this chapter and in Chapters 3 and 4. Third, the scientific method provides a logical and productive basis for investigation. Investigators approaching a new case inevitably form a theory (hypothesis) about what might have happened. Use of a thoroughly scientific testing approach to building and correcting this theory will go a long ways toward ensuring a just outcome.

Investigators may narrow in on one suspect and look only for evidence that points to that suspect. This behavior is common in what people often call the "open and shut" case. The explanation for what happened appears so obvious in the first moments of the investigation that a more thorough investigation seems unwarranted. The problem is that, after further investigation and physical evidence analysis, things may appear quite different. If the investigators have locked onto a bad theory or wrong suspect—the opportunity to unearth evidence or the actual perpetrator may have already passed. In cases like this, it may be impossible to proceed, and the case may go unsolved. In the extreme, the result is criticism in the newspapers or on TV when it is discovered that an innocent individual was convicted or an obvious suspect was overlooked. Were the investigators to take a more scientific approach, they would look at all the evidence uncovered and take time in forming and refining their hypothesis. **The key to making this approach work is to let the data control the hypothesis or theory, not to ignore data or information that doesn't fit a pre-existing theory.**

ELEMENTS OF FORENSIC EVIDENCE ANALYSIS—THE TYPES OF RESULTS FORENSIC SCIENTISTS PRODUCE

One way to start looking at this concept is to examine the elements of a "forensic science" investigation of evidence. What types of results do forensic scientists produce? The principal elements of an overall forensic investigation involving physical evidence are recognition, classification, individualization, and reconstruction. These terms also describe the types of results forensic scientists can produce from an item or items of evidence. Indeed, one of the underlying themes of this book is the arrangement of chapters into groups based on whether the types of evidence included are, or can be used, in reconstruction, individualization, or classification. Developing a clear understanding of the significance of the various results reported by forensic scientists is one of the primary goals of this book. Without some understanding of the basic concepts underlying criminalistics, it is difficult for non-specialists to appreciate the results produced by forensic scientists.

Evidence recognition

Recognition of physical objects as evidence, or potential evidence, is the first step in a forensic investigation. As we will discuss again in later chapters, physical evidence can be important in an investigation only if is recognized first and foremost. One of the key skills an experienced crime scene investigator brings to a scene is the ability to recognize what might be useful evidence (Chapter 3).

Classification—identification

The next step in almost any examination of evidence is classification. Whatever is being examined—a glassine envelope with white powder, a hair, a fiber, a paint chip, a bloodstain—it must first be classified. Classification is the process of categorizing that object within a group of similar objects. For example, we can recognize many different objects, with quite different appearance, and still classify them all as chairs in the broad group we call chairs. It might be a very broad group or a highly specific group. Classifying things is not "individualizing" them. Relevant and careful experimentation is the cornerstone of the scientific method including the following from the Federal Rules of Investigation:

> Rule 702
> If scientific, technical, or other specialized knowledge will assist the trier of fact to understand the evidence or to determine a fact in issue, a witness qualified, as an expert by knowledge, skill, experience, training, or education, may testify thereto in the form on an opinion or otherwise, if (1) the testimony is based upon sufficient facts or data, (2) the testimony is the product of reliable principles and methods, and (3) the witness has applied the principles and methods reliably to the facts of the case.

With some types of evidence, generally the ones we will group under the heading of "chemical evidence" (illicit drugs, gunshot residue, etc.), examination in the forensic lab consists exclusively of classification. Chemical or instrumental techniques are required to establish these classifications, and the courts require that this be done to sustain a prosecution. Note that sometimes items are classified in the laboratory as a means of establishing the *corpus delicti* of a crime (see previous explanation). Demonstrating that fire debris contains an accelerant-like (flammable liquid) material helps establish that an arson may have been committed. Demonstrating that semen is present on a vaginal swab taken from a sexual assault complainant corroborates that penetration occurred. The laboratory must establish these crucial facts if cases are going to be properly proven.

An important result of the classification, or the association process, is disassociation, in other words, exclusion from the class. The "negative association" has the advantage of being an absolute in most instances. When significant differences from other items in the class are found, the object is excluded and is clearly not in the class. As one compares two objects and tries to see if they might be in the same very small class, one can only say that they are the same in all the observed properties. This examination does not preclude the possibility that some demonstrable differences could be found with further analysis. The same problem does not exist with elimination or disassociation (also see below under individualizations).

> ### CASE EXAMPLE—SIDEBAR
>
> A knife is found near the victim at the scene of a homicide. The first thing that the investigator does, sometimes largely unconsciously, is to classify that knife. It appears to be a kitchen knife with a 7" blade and a 3" handle. The blade is pointed, narrow, and not tapered. The handle is wood and fastened with two rivets where the haft of the blade is fastened into the wood of the handle. If the investigator is familiar with kitchen knives, he/she may further categorize (classify) it as a "boning knife." Collecting this knife and sending it to a crime laboratory will result in more detailed measurements of width and length and, perhaps, description of the metal of the blade and the wood of the handle.
>
> However, the initial description (classification) alone may be useful to the medical examiner in determining if the knife was the weapon used in the crime. If, however, the investigator moves into the kitchen and discovers that there is a knife block with six slits and five knives and that the empty slit is a narrow one, the classification of the knife becomes more investigatively important. If the knife found near the victim is indeed the missing knife from the block, the working hypothesis may be that this is not a pre-meditated crime, but one of opportunity. This is all done just from a careful classification of the knife. An individualization that this is the one and only knife from this set is likely to be impossible, since the manufacturer probably made tens of thousands of these sets. The block and its knives are packaged separately and sent to the crime laboratory for further examination.
>
> Careful measurement of the knife and comparison to the other knives of the set may disclose that they are not consistent (disassociation). The knife is a few millimeters wider than the slot, and the rivets in the handle are steel not brass as in the rest of the set. The hypothesis must be reevaluated. It is not necessarily false. Perhaps the original knife was lost or broken, and someone in the household purchased a replacement knife of similar type and discovered it would not fit in the slit. It was, therefore, kept in a nearby drawer. On the other hand, perhaps the killer brought a knife along to commit the murder. The important point is that a careful and complete classification of the evidence is critical to the investigation of the case and jumping to conclusions may cause serious detours on the road to a proper solution of the truth of the case.
>
> One common problem in understanding forensic results is that the word "identification" can mean "classification" in one context, but it is also used to mean "individualization" in another context. "Identifying" a small piece of material, such as a paint chip, is a classification. "Identifying" a fingerprint, or a person, is an individualization. We will use the term "identification" to mean individualization in the pattern evidence chapters and in discussing the "identification" of persons, because that is the terminology that those specialists regularly use. We will use "classification" or "classification—identification" to mean placing an item or person into a group when dealing with other items of physical evidence.
>
> It is important to realize that only pattern evidence (fingerprints, handwriting, firearms, toolmarks, tire tracks, footwear, bitemarks) and some biological evidence (blood, physiological fluids, and tissues) are commonly individualized. Individualization means we can state there was a single origin and we know what it was. This person left that blood; this shoe made that impression; this tire made this track mark; etc. We cannot individualize chemical and materials evidence, like paint, drugs, fibers, and so forth. It may be possible to "associate" these items to such a small class that, in some cases, may almost be an individualization. The discussion that follows is meant to clarify these ideas.

Individualization

Individualization can mean one of two things: (1) that in some way, by examining the various characteristics of something, it can be recognized as unique—one of a kind—among members of its class; or (2) that when a "questioned" or unknown object or item is compared with a "known" or exemplar item, they are found to have had a common origin. We discussed *class characteristics* previously. Class characteristics were those that enable us to classify something into a class of items (or people). Objects (Figure 1.13) and individuals may also possess *individual characteristics*, features that are unique to them individually and distinguish them from all other members of the class. It is the use of individual characteristics that permits evidence or persons to be individualized.

In one sense, individualization is the narrowing of classification until only one item remains in the class. Traditionally, in forensic science, this ability has generally been restricted to human identification, fingerprints, certain other patterns, and jigsaw fit matches. Recent advances in DNA technology have brought DNA profiling into the group as well (Table 1.2).

The word "individualization" implies uniqueness of an item or person among members of the class. In forensic science, one sometimes hears of "partial individualization." This sounds contradictory. What is meant is a narrowing of

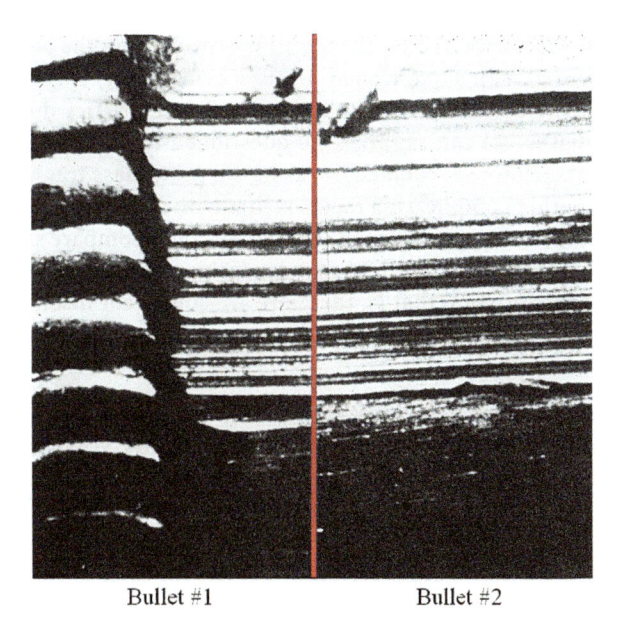

Figure 1.13 Where pattern evidence is so similar to logically rule out anything, but a common origin individualization can be inferred.

Table 1.2 DNA Alleles with a comparison of DNA profiles: a virtual individualization

LOCUS	Reference sample 1:	Crime scene sample 2:
D3S1358	14, 15	14, 15
vWA	15, 16	15, 16
FGA	22, 23	22, 23
Amelogenin	X, Y	X, Y
D8S1179	10, 13	10, 13
D21S11	29, 30	29, 30
D18S51	12, 14	12, 14
D5S818	11, 13	11, 13
D13S317	10, 11	10, 11
D7S820	10, 10	10, 10
D16S539	9, 11	9, 11
THO1	6, 8	6, 8
TPOX	8, 8	8, 8
CSF1PO	12, 13	12, 13

Note: DNA is used for individualization by showing correspondence between many markers that alone are only associations but in combination virtually insure individualization (identification).

classification, but not to the point of exclusivity. We may know, for example, that a small fragment of glass recovered from a suspect's pants cuff has the same chemical composition and optical properties as the known glass that was broken at a scene. The forensic question is: how to interpret these findings. The glass has been placed in a small class, and there are a lot of different types of glass and sources that can be excluded, but we don't know, or have any way to evaluate, how many sources there could be other than the one at the scene. Forensic scientists often call this type of finding an "inclusion." The questioned item *could be* from the known. It *cannot be excluded* as having come from the known. Sometimes we say it is *consistent with* the known. This type of issue occurs with many types of relatively common manufactured materials such as paint, fibers, paper, and many others. We know that this sort of conclusion is not a satisfying one, or one that is easy to evaluate. But oftentimes, it is the best that forensic science can do with this evidence.

> **Ethics for the Testifying Forensic Scientist**
>
> If the law has made you a witness,
> Remain a man of science.
> You have no witness to avenge,
> No guilty or innocent person to convict or save—
> You must bear testimony within the limits of science
>
> —Dr. P.C.H. Brouardel, 19th Century Medico-legalist

Figure 1.15 A forensic scientist must, above all else, maintain their objectivity and should follow upon the results of the analysis of evidence rather than speculation or conjecture.

divided into eight different regions, each of which has its own professional organization. This may sound duplicative, but it has the advantage that many examiners who belong to the regional organizations may not belong to national organization. This is because many practitioners working at the bench level feel more comfortable at regional meetings where they have similar experience to most of the other attendees. The national organization is much broader in scope having 11 different specialties, four or five of which are not usually represented at regional meetings.

In addition, to the national and regional organizations, there are specialty groups, usually at the national level—both certifying bodies and others that are narrower in their scope—designed for practitioners in a particular specialty area. These specialty organizations help members keep up with their specific field of practice. The American Academy has 11 sections that cover virtually every aspect of forensic science. The specialty organizations deal with people whose expertise is primarily in one area, such as questioned documents, forensic toxicology, criminalistics, bloodstain pattern analysis, crime scene analysis, fingerprints, and so forth.

Professional practice and ethical concerns of the organizations

One of the main reasons for each of the professional organizations to have its own code of ethics is to self-police its members. Professions are concerned with keeping their members on the straight and narrow and feel better equipped to do so than those not highly familiar with their profession. Therefore, each of the of the major organizations of forensic scientists have their own code of ethics and the ability to, in some way, sanction members who do not follow the code. In the case of serious breaches, the organization needs the right to expel a member from the organization. This is a particularly powerful ability for the professional organizations, because being expelled or dropped from membership in a professional organization is something that can be used against an individual when trying to qualify as an expert witness or testifying in court. This means that forensic experts are much more sensitive to professional regulation than practitioners in other professions who do not regularly testify in court.

Ethical responsibilities as scientists

In addition to the forensic professional responsibilities, and some of the other ethical responsibilities that are covered by the codes of ethics, forensic science practitioners also have the responsibility of being ethical scientists. That means that they must document their work and a whole variety of other ethical requirements that apply to all scientists. They must be able to document everything that they have done by taking accurate notes and providing other supporting materials to show that their conclusions are well-based. They must be very careful in their reporting and must not report conclusions that are not fully supported by what they have found. This holds true for reports in scientific publications and even in private communications. There are also important points of professional practice that forensic scientists must observe carefully that don't necessarily apply to all other professional activities.

Ethics and professional behavior for forensic scientists

Forensic scientists must be sure not to testify beyond their expertise. Forensic scientists must try to be impartial and very careful to testify only to what they actually do, not anything that involves what someone else has done. Even if they are quite aware of another's work, they may not testify concerning such work without proper attribution. It is important to avoid serious legal problems because of their role as experts in the legal system. Should they be sanctioned by a professional body or a court, it would have a serious effect on their ability to practice forensic science.

Because forensic scientists deal with the legal system, it is quite important that their reports and their other communications have a very high standard of accuracy and reliability. This is particularly true when they are testifying in court where they're held to the highest standard. In court, they can be cross-examined, and the conclusions that they draw can, and often will be, challenged. Their testimony can, in fact, affect their ability to make a living as an examiner. An examiner whose work is successfully challenged in court by cross-examination, or other credible witnesses, can be haunted by this failing for the rest or their career. Admitting that they over testified, testified to conclusions that were not fully based on their work, or that they over-stated their qualifications can also haunt them when they try to testify at a later date. Testimony from prior cases can be drawn on in some instances to challenge a witnesses' credibility, reliability, or expertise. Thus, maintaining a high level of professional practice is particularly important to forensic scientists because of their close involvement with the legal system.

Most codes of ethics require forensic scientists to come forward when they become aware of other people who are not practicing in a proper manner or are testifying to things that either are not true or that are not fully proven by the work that they did. Examiners coming forward and saying they believe that someone has committed an ethical breach, or a professional practice breach, is a major means by which the professional practice committees of professional organizations learn of infractions. It is then the ethics committee's responsibility to investigate this allegation and determine whether or not action should be taken against that particular practitioner.

Key terms

- accreditation
- antemortem
- Automated Fingerprint Identification Systems (AFIS)
- certification
- classification (identification)
- Combined DNA Indexing System (CODIS)
- controlled experiment
- coroner
- *corpus delicti*
- criminalistics
- entomology
- forensic
- forensic science
- forensics
- hypothesis
- individualization
- medical examiner
- medico-legal
- *modus operandi*
- National Integrated Ballistic Identification Network (NIBIN)
- natural law
- odontology
- pathology
- physical anthropology
- postmortem

psychiatry
recognition
reconstruction
scientific method
theory
toxicology
trier of fact
Technical Working Group (TWG) and Scientific Working Group (SWG)

Review questions

1. What is forensic science?
2. What is corpus delicti? What role does forensic science play in establishing corpus delicti?
3. List and briefly discuss three uses of physical evidence in criminal investigation?
4. What are the important ethical and professional responsibilities of forensic scientists?
5. What roles did anthropometry and fingerprints play in the development of forensic science?
6. Highlight major developments in forensic science and forensic laboratory development in the United States.
7. What is the scientific method? Why is it important in criminal investigation?
8. Briefly discuss forensic pathology, odontology, entomology, anthropology, psychiatry, and psychology, engineering, and computer science.
9. List and discuss the elements of forensic evidence analysis.
10. Describe and discuss classification (identification), individualization, and reconstruction.

Fill in/multiple choice

1. In forensic science the term "identification" refers to:
    a. the placement of an inanimate object into its proper class
    b. individualization of an inanimate object
    c. establishing the identify of a person
    d. only class determinations involving quantitative comparison
    e. a and c only
    f. b and d only
2. The two most common forensic activities performed by a forensic odontologist are _____ and _____.
3. Forensic science differs from the traditional natural sciences because the results are often used in _____.
    a. jury selection
    b. ex post facto cases
    c. jurisdictional hearings
    d. legal proceedings
4. The American Society of Crime Laboratory Directors (ASCLD) is associated with laboratory development primarily through its program of _____.
5. In a civil liability suit arising from a hit-and-run automobile incident, *physical evidence* examined at a forensic laboratory would *most likely* be used to help the jury to _____ the incident.

References and further readings

ASCLD's Comments on the Release of the NAS Report on Forensic Science, February 19, 2009, www.ascld.org. Web site of the American Society of Crime Laboratory Directors.

http://www.cstl.nist.gov/biotech/strbase/dabqas.htm. Web site of the National Institute of Standards and Technology.

http://criminalistics.com/ABC/A.php. Web site of the American Board of Criminalists.

http://www.theiai.org/. Web site of the International Association for Identification.

Inman, K. and Rudin, N. *Principles and Practice of Criminalistics: The Profession of Forensic Science.* Boca Raton, FL: CRC Press, 2001.

Kearney, J. J. Annual Report 2002, Illinois State Police, Division of Forensic Services, Forensic Sciences Command, Forensic Science Center at Chicago, Chicago, IL.

Strengthening Forensic Science in the United States: A Path Forward; Committee on Identifying the Needs of the Forensic Sciences Community, National Research Council, 352 pages, 6 x 9, paperback, http://www.nap.edu/catalog/12589.html., 2009.

Thornton, J. "Criminalistics: Past, Present and Future." *Lex et Scientia* I 1 (1) (1974): 1–44.

Thornton, J.I. and J. Peterson. The general assumptions and rationale of forensic science. In *Modern Scientific Evidence*, Faigman, D. L., Kaye, D. H., Saks, M. J., and Sanders, J. (Eds.). St. Paul, MN: West, 1997.

Thorwald, J. *Century of the Detective*. Orlando, FL: Harcourt, 1965.

Thorwald, J., Winston, R. and Winston, C. *Crime and Science: The New Frontier in Criminology*. Orlando, FL: Harcourt, 1969.

Chapter 2

Physical evidence in the legal system

Lead Case: The LeGrand Case

Reverend Devernon LeGrand had organized St. John's Pentecostal Church of our Lord in Brooklyn, New York. When he was indicted, his headquarters—a four-story townhouse—was occupied by 11 "nuns" and their 47 children, many of them fathered by LeGrand. According to police, LeGrand did most of his recruiting by seducing and impregnating young women, then threatening them or their children if they refused to beg for money on the streets. His black-clad "nuns" were often seen around Grand Central Station, and even in New Jersey. It was inside this headquarters that LeGrand had raped his victim, during August 1974, and authorities suspected that sexual assault was only the tip of the iceberg. In 1975, LeGrand was convicted, along with his 20-year-old son, Noconda, of kidnapping and rape. He was sentenced to 5–15 years in prison. This was only one of several serious encounters with the law for Referend LeGrand dating back over 10 years.

Church members Yvonne Rivera, 16, and her sister Gladys Rivera Stewart, 18, had testified for the prosecution in an earlier bribery trial, but they were missing when the district attorney (DA) sought to use their testimony in the later rape case. Informants said the girls were dead, dismembered in the Brooklyn "church." In an affidavit, it was alleged that Mrs. LeGrand had told the DA's investigators that the two Rivera sisters were told to go with the reverend, and all the other members of the household were instructed to assemble in the downstairs front room to sing hymns. Two hours later, Mr. LeGrand's daughter, Teasiene, appeared in the room and told Mrs. LeGrand, "Daddy is stomping Gladys." Reverend LeGrand's son, 26-year-old Steven LeGrand, was also charged with murdering the Rivera sisters.

The four murders, listed in Reverend LeGrand's May 1975 indictment, were all said to have occurred in the four-story townhouse Mr. LeGrand maintained as a church and residence. According to the indictments, all the victims were "beaten, stomped and dismembered." Ann Sorise, one of the two wives Mr. Legrand allegedly killed, was also shot. She was said to have been murdered in September 1963. The other wife he allegedly killed was identified as Ernestine Timmons. She was said to have been killed about May 1, 1970. Both women were in their 30s. The Rivera sisters, the indictment said, were stomped to death in October of 1975 by Devernon LeGrand and his son Steven, "acting in concert with another person." The third person was not identified. That third person was expected to testify against the two LeGrands under a grant of immunity.

In an affidavit filed with the DA, a caretaker for the LeGrands, Frank Holman, said he helped transport the dismembered bodies of the sisters from the house in Brooklyn to the Catskills for disposal near "LeGrand Acres," a large farm in Liberty, New York, maintained by Mr. LeGrand for members of his church and their children. In 1966, LeGrand's "church" had purchased a 58-acre parcel in the Catskills and converted it to a summer retreat for the faithful. LeGrand's followers returned each summer, without fail. State policed dug up the grounds of the farm in mid-December 1975, but they came away empty-handed. Three months later, on March 6, 1976, assorted bones and bits of cartilage were found in Lake Briscoe, and a Brooklyn DA raiding party turned up human bloodstains in the Crown Heights townhouse.

Mr. Holman also said he had helped burn the bodies in a metal washtub and then dumped the remains into Lake Briscoe, several miles away from the LeGrand farm. Indeed, many small pieces of burned bone were found at the edge and further out in Lake Briscoe by divers and suction dredging.

The weakness in the case was the dependence on the testimony of informants who, themselves, were deeply involved in the crimes. The prosecution needed physical evidence to lend credibility to their testimony. Corroboration, particularly in the form of physical evidence, was deemed essential to the success of the case.

The anthropologist in the Office of the Chief Medical Examiner did an impressive job of building an exhibit to substantiate that the bone fragments might be from the two missing girls. A life-size construction was made with an outline of two female bodies. The anthropologist placed the bone fragments on the outline in areas corresponding to the location on the body from which that piece had come. When he was done, the two outlines were largely covered with bone fragments. This exhibit was presented in court and placed prominently where the jury could see it.

There was concern about further corroborating Mr. Holman's testimony. He had indicated that the chopped-up bodies were doused with solvent and burned to destroy most of the flesh. In his detailed testimony, he had indicated the type of flammable liquid that had been used to burn the evidence. Some evidence samples of bones, bone burned with no added flammable liquid, and a sample of the same brand of flammable liquid were submitted to the forensic laboratory for examination. The bone samples were sealed in vials and warmed. It was discovered that traces of flammable liquid could be recovered from the headspace above the evidence bone samples, but no trace of flammable liquid was found in the samples of bone that had been burned with a torch. Further, when the known flammable liquid was examined, it was found to contain each of the flammable chemicals already identified in the evidence sample. It was certainly not possible to say that the bone had been doused with that flammable liquid, but it was possible to say that the evidence samples appeared to have been doused with a flammable liquid, which had many of the same chemical compounds as the liquid named by Mr. Holman.

Mr. Devernon LeGrand and his son Steven LeGrand were convicted of beating and stomping to death Yvonne Rivera, 16 years old, and her sister, Gladys Rivera Steward, 18, in the headquarters of the church at 222 Brooklyn Avenue in Brooklyn, and both drew prison terms of 25 years to life.

Source: www.crinezzz.net/serialkillers/L/LEGRAND_devernon_steven.php; *New York Times* articles from May 14, 1976; May 25, 1976; May 7, 1977; and September 1, 2002.

LEARNING OBJECTIVES
- How physical evidence is created during an incident
- The nature of impressions, imprints, indentations and striations
- The Locard Exchange Principle and its centrality to forensic science
- How physical evidence might be classified in ways that are useful to investigators
- The major uses for physical evidence in cases
- The steps required for the effective discovery and use of physical evidence
- Basic practices of physical evidence labeling, packaging and preservation
- Different types of laboratory analysis and their applicability to different types of evidence
- The importance of reporting and testimony to the forensic scientist's function
- How the need for social organization developed into the rule of law
- The complex pathways of the flow of evidence in the criminal justice system as an incident goes from initial report to final resolution
- Admissibility of evidence vs. its weight in a legal context
- Rules for the admissibility of scientific and technical evidence—the Frye and Daubert cases and criteria

HOW PHYSICAL EVIDENCE IS PRODUCED

The nature of and the mechanism for generating physical evidence are critical concerns for forensic scientists. Finding or identifying physical evidence is important, but not sufficient to allow one to make the best possible use of it. Knowing the process that produced it and the precise location where it was found, or associating change at an incident scene with particular evidence can all help to make the best use of the physical evidence.

Change induced at a scene

The mechanism for producing physical evidence is usually related to changes induced at a crime scene. The change could take the form of depositing something not previously there, or an alteration of something that was there. For example, finding a blood spatter pattern (Figure 2.1) in a particular area will often reveal something important about what occurred at that location.

It cannot be over emphasized that virtually any observed object or condition may prove to be useful physical evidence. Finding traces of broken glass, hairs, fibers or paint chips, just to name a few examples, can often tell a story. Even something, such as a necklace or a bracelet, which is broken during a scuffle can yield valuable information through the location of a broken piece. Although there may be no additional information from examination of the object from the necklace or bracelet, the fact that it was found lying on the floor, in a particular spot, may turn out to be an important piece of evidence. A body that was found dumped along the side of the highway may be associated with a location a distance away by a charm or link broken from that bracelet (Figure 2.2). Further, finding something disturbed from its normal location can be an important piece of physical evidence.

Exchange of material upon contact

Another way that physical evidence is generated is through the transfer of material between surfaces, which came into contact with the material. The first important statement concerning this type of evidence was formulated by Edmond Locard, who started one of the world's first forensic science laboratories in Lyons, France, in 1910. One of the guiding principles of forensic science is called the Locard Exchange Principle in his honor. A commonly expressed version of this principle is, "When two objects come into contact, there is an exchange [of] material across the contact boundary." Although his statement is probably essentially true, one must remember that one may not always be able to find

Figure 2.1 Blood spatter pattern on door and wall at a scene.

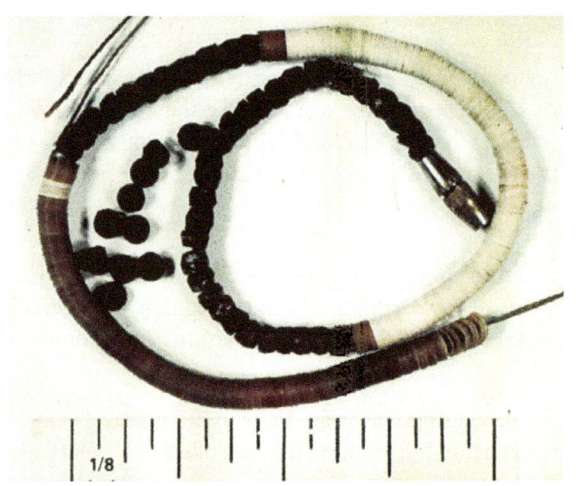

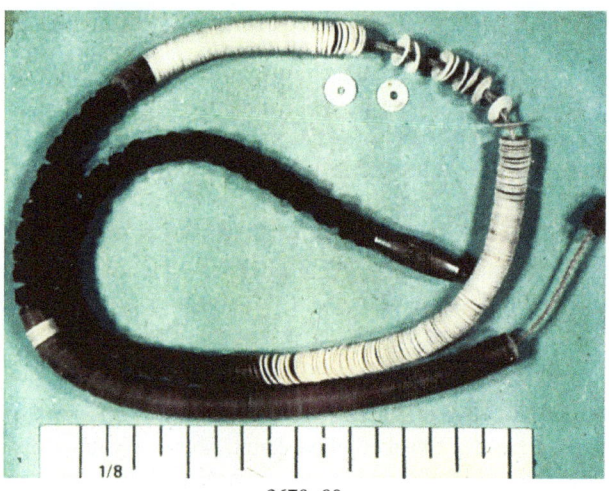

Figure 2.2 Matching beads and disks from suspect's car to victims necklace.

2678–90
PHOTOMACROGRAPH
Necklace from victim showing ten loose dark brown Beads found in vacuumings of vehicle

Scale: Approximately 1.5 X

2678–90
PHOTOMACROGRAPH
Necklace from victim showing ten loose white discs found in vacuumings of vehicle

Scale: Approximately 1.5 X

the transferred material. For example, any material exchanged may fall off, or perhaps so little material was exchanged that one cannot find it. Logically, the likelihood of a transfer increases if the contact is the result of violent contact.

An example of the Locard Principle is the finding of animal hair on the clothing of a burglar. Anyone who has a dog or cat for a pet knows that when someone walks into their home, pet hairs seem to jump up and attach themselves to the visitor's clothing. Often there is a particular piece of furniture that a pet has made their own. Even diligent efforts to clean it using a vacuum cleaner, sticky tape, or anything one can think of will usually fail to remove all the pet hair. Should anyone brush against it, sit on it, or even come near that piece of furniture, there may be a copious transfer of potential trace evidence. There are a great variety of materials that can be transferred. It is difficult to conduct properly controlled studies on how frequently different materials are transferred from an individual to a chair, for example, or between two individuals who come into contact with one another. The results of such studies have been found to be highly variable depending on the type of surfaces involved, the condition of the clothing, and many other factors. The reason it is important to try to study the frequency of chance transfers is that the information helps evaluate the significance of small traces of exchanged materials found on casework items. Certainly, someone hit with a "blunt object" is likely to transfer traces from the location of the blow to the object and from the object to the location the blow.

Deposits, dispersions, and residues

In addition to transfers resulting from direct contact (Locard Transfer Principle), one can find useful evidence traces of many types that result from a deposit rather than a direct contact transfer. Deposits can be made up of large quantities of material, such as blood or paint traveling through air and splattering onto a surface, or they can be small quantities of quite small particles, and thus much less obvious. Dust, for example, may settle on objects and then careful examination of the composition of the dust may provide useful information about where that object has been when the dust was deposited.

An interesting example is the settling of pollen on an object that is outside during the season when a particular type of plant is releasing its pollen to the winds. Pollen is a biological material that is unique to the type of plant from which it comes. Because of its small size, it is readily transported through the air and deposited over a sizable area during a specific time. The presence of a particular type of pollen on an object or vehicle can provide information on where the object or vehicle has been, and perhaps that it has been moved to several different locations. For example, by examination of a wrapped marijuana brick, an examiner might tell if the marijuana is of local origin and, if not, where it originated and, perhaps, even several other locations where it was exposed to the atmosphere.

Physical evidence in the legal system   35

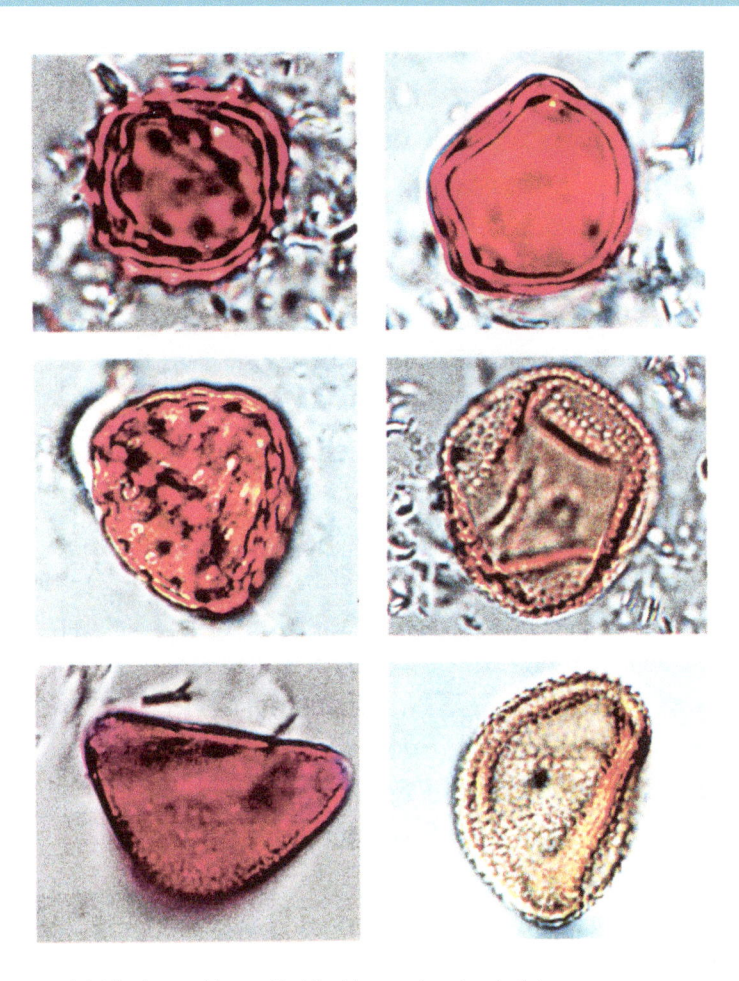

Figure 2.3 Pollen grains one might find on evidence that had been stored out of doors.

The finding of different types of pollen (Figure 2.3) on the packaging material and the actual marijuana itself may convey information about when it was harvested and where the packaged bricks were stored. There are a number of different countries that produce marijuana, and sometimes it is important to know where a particular seizure was actually grown. A pollen expert can easily determine whether marijuana was grown in India, the Middle East, the Far East, Australia, Africa, Hawaii, or the mainland of the U.S. Because the local plant life is different in each major growing region, the mix of pollens will usually indicate to an expert where it was grown. Pollen is an interesting material because it is very tiny and almost indestructible, and it can be trapped in agricultural materials and perhaps recovered intact; even long after it was deposited.

Pollen is just one possible component of dust; that material that dulls the shiny coffee table is a very complex mixture of many different types of tiny particles. Dust will differ in each location depending on what type of material is suspended in the air at that location. Particularly where man is, dust will contain man-made materials characteristic of local activity. If one is near a power plant, the dust will have residues characteristic of the power plant's fuel. If one is near (near is a broad term since fine particles can travel considerable distances before settling) a cement plant, then calcined lime will be a sizable component of the dust. In an agricultural area, the dust may indicate what crops are being grown nearby from examination of fertilizer traces.

Damage, tears, cuts, breaks, and jigsaw matches

Physical evidence is also produced as a result of damage, tearing force, breakage, cuts, and many other processes. Many of these processes produce unique two or three-dimensional surfaces, which under the right conditions, may produce individualization and allow an examiner to say that these two (or sometimes more) pieces were at one time part of the same object.

1417–78
U.S. flag from the scene (1) with torn
Portion from the suspect (2)
Scale: Approximately 1/4 X

Figure 2.4 Pieces of a torn flag that can be matched to each other in a so called "jigsaw match."

Examples might be a broken auto head light or parking lamp lenses, pieces of a knife blade, or parts of a license plate frame. Many things one might not think of can, under the right conditions, provide useful evidence. For example, a pocket torn from a piece of clothing can be compared to the color of the clothing, and the torn threads or cloth can be matched to the damaged area on the clothing. Particularly if there is a ragged edge, the two pieces can be matched against each other. If each piece has a corresponding thread count, and the shape, size, and makeup conforms exactly, the result is individualization (Figure 2.4). An expert may be able to say, "In my expert opinion, that piece of cloth, at one time, was part of that jacket or pair of pants."

Imprints or indentations

Imprints and indentations are important types of physical evidence that can easily be overlooked. A footprint in dust or blood or a clear three-dimensional footprint in mud or in snow is easily recognized. Unfortunately, some imprint and indentation evidence may be subtle or require enhancement to be visible as an evidential imprint, and, therefore, such evidence is often overlooked. A good illustration of the difficulties of recognizing pattern evidence can be found in the trial of O.J. Simpson for the murder of his ex-wife Nicole and Ron Goldman. There was a serious controversy over whether O.J. could have committed the crime alone. Dr. Henry Lee testified that when he visited the scene he observed and took pictures of bloody footwear imprints (Figure 2.5) with a different sole pattern than those that had a pattern consistent with the style of Bruno Magli shoes purchased by O.J. Simpson. Other experts denied that what Dr. Lee had observed were even footwear impressions.

Because no one recognized that these might be shoe prints, the marks were never properly documented and photographed during the crime scene processing. One can see hints of them in other pictures taken at the scene. Were these actually shoe prints and from a different shoe than the Bruno Magli pair, it could have significantly changed the theory of how the vicious murders were committed. A second set of bloody foot prints could have indicated a second participant in the crime.

An imprint is produced when an object comes into contact with a hard surface and leaves a two-dimensional representation of itself on a smooth surface in dirt, dust, blood, or some other medium. An indentation is produced by an object being impressed into a soft receiving surface, such as sand, snow, or mud, creating a three-dimensional mark. These markings are discussed further in Chapter 5. Footprints and other imprints can be problematic at crime scenes because, if the scene is not properly protected before and during processing, footprints and other marks not associated with the incident can be confused with or obscure actual evidential material.

Striations—dynamic marking

Striation markings are the result of a hard (often metal or finished wood) surface being marked by another object (tool) in motion along its surface. Sliding toolmarks (see Chapter 8) are a classic form of striation markings. They can

Figure 2.5 Markings that appear on walkway and appear to be a shoe print that was not recognized as such during the original processing.

be important in reconstruction of events, association of a tool to a particular scene, and, in some cases, can lead to individualization of a tool.

Striation is a more scientific way of referring to what are commonly thought of as scratches. Their individuality is caused by imperfections in the marking surface of a tool. In some cases, when large, they can be seen with the unaided eye but may be much too fine to be seen without magnification. Most toolmarks are a combination of both macroscopic and microscopic striations (scratches) Perhaps the most important toolmark to forensic science is that left by the inside of a gun barrel on the bullet (Figure 2.6) as it travels down the barrel. These striations result from imperfections that have been left in the barrel as a result of the rifling (gouging out of helical grooves) in the barrel (Chapter 8) during the manufacturing process, which leaves the identifiable markings. Of course, other tools like pry bars, screwdrivers, and bolt cutters all have imperfections on their surfaces that could result in striation markings.

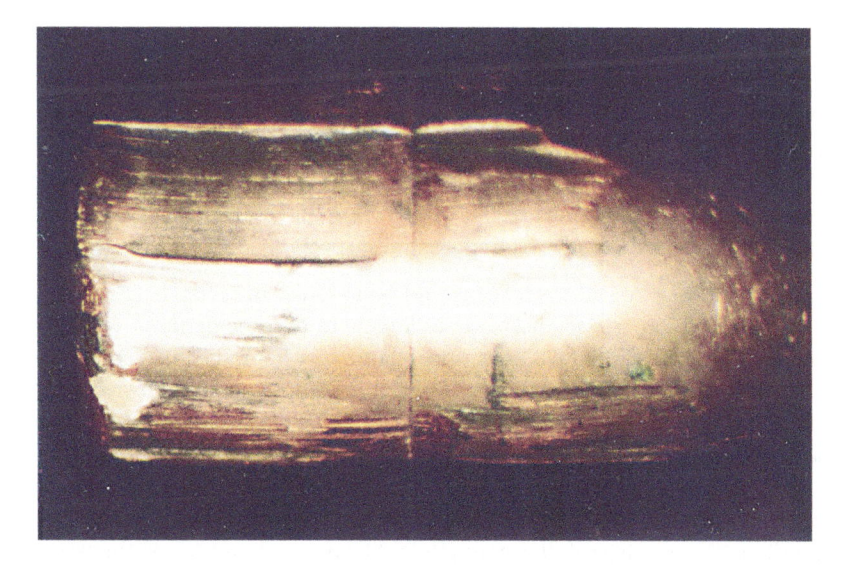

Figure 2.6 Fired bullet on which are visible lands, grooves, and stria that might be shown to match a particular weapon.

PHYSICAL EVIDENCE—CLASSIFICATION AND USES

Almost any object or condition has the potential of becoming physical evidence. There are ways of thinking about utilization of physical evidence, however, that can be of assistance to investigators and crime scene personnel. As we noted in Chapter 1 and will discuss further in Chapter 3, cases can be approached using the base tenets of the scientific method. After careful observation the next step is formulating a working hypothesis—this is an "educated guess" as to what may have happened based on the evidence available at the moment. Then, as the scene is searched and the investigation proceeds, other evidence or information is gathered that may reinforce or perhaps force a re-evaluation of the initial hypothesis. There are two potentially useful ways of thinking about physical evidence in this context. First, the evidence is going to be useful mainly for identification/classification (what it is), for individualization (providing linkages between persons and things), or for reconstruction. Second, we can go back to the various uses of physical evidence discussed in Chapter 1. Searching for evidence at a scene or in an investigation with these concepts in mind will help investigators *recognize* potentially valuable evidence.

One way to classify physical evidence is according to whether it will be used primarily for identification, individualization, or reconstruction. Often identification (classification) of some critical component of materials or transfer evidence is all that is needed to provide useful information. Finding white powdery material on a suspects clothing might not seem important. However, if the white powder is shown to be milk sugar (lactose), that could be very helpful in connecting that person to the cutting of drugs.

CASE ILLUSTRATION 2.1

A case many years ago involved a break-in at a machine shop, and some money had been stolen. Anyone who has visited a machine shop knows that the lathes used in the machining process produce little pieces of metal turnings, which will likely be all over the floor. These metallic spirals are quite characteristic and finding a couple imbedded in shoes of a suspect can prove to be excellent evidence that the individual was in the machine shop. One can show that these turnings have the same shape, appearance, and type of material, as the ones collected from the scene. Further, it can be shown by instrumental analysis of the metals involved that the metallic composition is identical to a collected control sample. These are not the type of things that a worker at a fast food restaurant, for example, would be likely to have in his shoes.

Many other types of evidence are valuable mainly because they demonstrate or disprove linkages between or among people and scenes. The purpose of collecting this kind of evidence is to submit it to the laboratory for individualization testing. Biological evidence, like blood or semen, will be compared via DNA typing to significant individuals in the incident. Fingerprints (Figure 2.7) or footprints will also be compared with known exemplars from people who might be involved.

Fired bullets or cartridge cases may be matched back to a particular firearm. Footwear or tire impressions may be matched back to a shoe or tire. Even when complete individualization is not possible, the evidence has value. Remember, often there is as much value in demonstrating exclusion as there is in demonstrating an inclusion. Hairs, fibers, glass, soil, and paint evidence all can have exclusionary value, and may, at times, have associative value in cases where they are consistent with control samples from potential suspects.

Another important application of physical evidence would be to assist in reconstructing the crime or incident of interest. Many investigators take forensic science courses largely to become better at evaluating the scene and the physical evidence it produces and to develop skill in understanding what may have happened at a scene. There are well known forensic experts who have visited thousands of crime scenes and are knowledgeable at evaluating crime laboratory reports. Often, they can look at all the evidence and give considerable insight into what actually happened. When there are no witnesses or when there are questions about the witness' truthfulness, such insight can be particularly useful. Therefore, helping to accurately reconstruct events using forensic evidence gathered at a scene is certainly an important application of physical evidence. An investigator should examine a scene for patterns that may be helpful in reconstruction. As we will discuss further in Chapters 3 and 4, certain patterns can be very useful for reconstruction. Blood spatter, glass fracture, and trail patterns are just three examples. Often, the patterns cannot

Physical evidence in the legal system 39

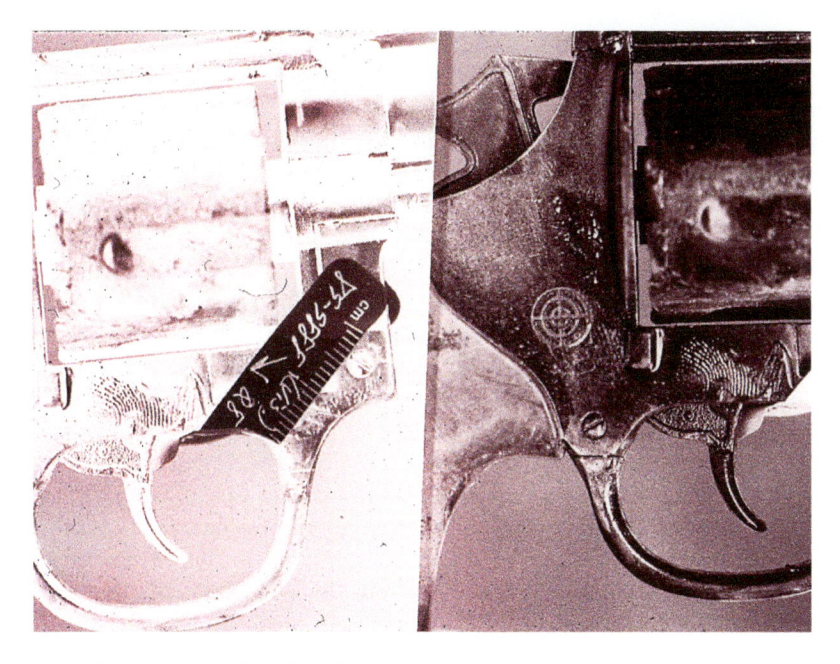

Figure 2.7 Hand gun with a fingerprint developed on the surface.

be "collected" as such, and must be "captured" or enhanced and well documented. As we noted above, reconstruction is the second useful way of looking at and thinking about physical evidence, particularly during the course of an investigation or scene search.

In addition to developing a basis hypothesis about what occurred in a specific incident, physical evidence can be used for a number of fairly specific tasks.

Identification of persons—victim, suspect, others

The role of physical evidence in helping to identify a victim or a suspect is central to forensic evidence. Fingerprints have been used for over 100 years, but nowadays investigators have come to depend on DNA, cheek swabs, blood droplets, bite marks, and a whole variety of other biological samples to help identify a victim, a suspect, or, perhaps, a witness (Figure 2.8).

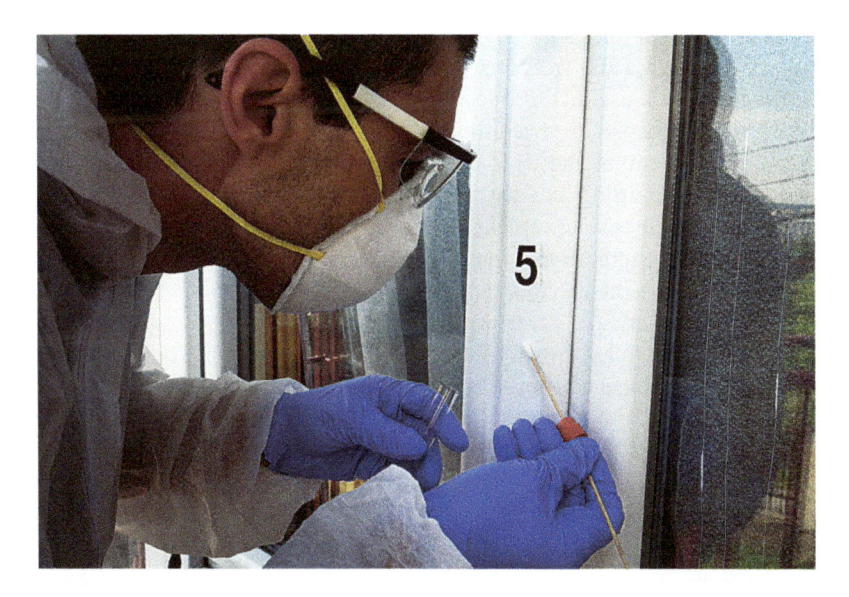

Figure 2.8 A biological evidence stain being collected from a window frame. (Courtesy of Shutterstock, New York.)

Unambiguous identification is not restricted to homicide, or even to criminal cases. In criminal cases, we tend to concentrate on identifying victims and then trying to identify suspects and, ultimately, perpetrators. Even in civil cases, people may try to conceal their identity and physical evidence can help obtain unambiguous identifications. In accidents and disaster situations, victims must be identified so the remains can be returned to families. In disputed parentage cases, the parent(s) of a child must be identified so the law can assign proper responsibility for the child's care and support.

Establish linkages or exclusions

Certainly, the classic application of physical evidence is developing linkages. That is undoubtedly one of the most important uses of physical evidence during the investigation and, particularly, at the adjudication stage. Linkage of victims, suspects, scenes, and instrumentalities can apply at many levels, from a mere possibility that two items could have a common source to individualizations. One may be able to say that an individual is the primary suspect in a sexual assault because we found seminal material in the 80-year-old victim's vaginal tract that came from the defendant. Should the victim indicate that she has not been sexually active in years, this becomes a compelling piece of evidence. These linkages can and do take a wide variety of forms. One can find paint chips on a hit-and-run victim's clothing, fiber transfers from a struggle, or soil on a suspect's shoes consistent with that at an outdoor crime scene. It is useful to find fibers from the suspect's clothing on the victim, or fibers from the victim's clothing on the suspect's clothing, but if one finds transfers in both directions the evidential value is more than twice as strong. Multiple linkages, even if weak individually, can build up to provide much stronger evidence.

The connection of a victim or suspect to a scene can be as important, in some cases, as a connection between victim and suspect. A simple illustration is the situation where someone is suspected of breaking into a bakery to commit a burglary or robbery. A possible suspect is stopped a few blocks away, and, upon being questioned, says: "No, I was never near that place," yet the white powder all over his clothing is found to be flour. Certainly, this does not prove he broke into the bakery, but it does provide a linkage, since it is observed that there are at least small amounts of flour on virtually every surface in the bakery. Another example might occur when a burglar is known to have broken in through the ceiling, is found to have some kind of insulation on his clothes, and it happens to be the same kind of insulation used in the ceiling of that particular place. It requires considerable force to break into a safe, and it is a messy job. Individuals who break into safes will usually be found to have safe insulation on their person. Most safes are primarily designed to keep important papers or money from being destroyed in a fire. Even if a building burns down, important papers inside the safe will survive the fire. To provide this protection, safes have a thick layer of insulation in the walls surrounding the safe compartment. There are a variety of things that have been used as safe insulation over the years. These are not materials that anyone is likely to have on their clothing if one has not broken into a safe. Even if the robber of the safe has tried to clean himself off, sending the clothes to a competent crime laboratory will usually disclose traces of safe insulation.

Linkages between suspects or victims and weapons, such as the connection of a hammer used in an assault to the victim or suspect, are also frequently useful. One may find a fingerprint from the suspect on the hammer or perhaps a hammer is found in a suspect's possession. Even though it has been cleaned, one is able to find a couple of miniscule spots of blood remaining that can be shown to be consistent with the victim. It seems that in every crime show or movie the investigator finds something and says: "Aha! That little statue was the one that was used to bash the victim's head." This may seem a little too convenient in the fictional context, but, in the real world, one often is able to determine the object used as a weapon in just that way. One can think of a vehicle as an instrumentality as well. Hit-and-run cases often yield considerable evidence that helps establish linkages. In a surprisingly high percentage of such cases, it is possible to tie a particular car to an accident scene or victim using a multitude of different types of physical evidence. This increases the solvability of hit-and-run cases, but it should be kept in mind that just by placing the vehicle at the scene may not establish who was driving.

Develop modus operandi (MO), leads, and suspects

Physical evidence can help investigators develop leads. For example, it can help in developing a *modus operandi* (MO) or the perpetrator's method of operation. Perpetrators of criminal acts follow behavioral patterns. One of the things that is used frequently in trying to recognize these behavioral patterns is a careful examination of the physical evidence left behind at the scene. The key is often looking at the way the crime was committed, how the house was broken into (if a burglary), how the victim was attacked (if an assault). The physical evidence left behind will provide insight into many different aspects of how an incident unfolded. This is, in many ways, similar to reconstruction, but the emphasis is more on tying together different incidents that may have been committed by the same person or persons. Such information can be extremely important to an investigation. In many situations, the ability to connect two or three seemingly unrelated incidents provides critical information to help the investigators to progress in their investigation. If one has several incidents that appear to have been committed by the same individual, searching for a common thread may allow one to significantly narrow down the suspect list. This can be illustrated by the situation where there are a number of incidents, which appear connected by an MO, physical evidence, or, now particularly, firearms or DNA results. If these incidents fall into two periods of time with a considerable period between the groups, it may be indicative of an individual who has been in jail or prison during that period. This may provide a solid investigative lead by looking at prison or jail records for persons who were incarcerated during that period and released before the recent incident.

Provide suspects or investigative leads from forensic databases

Although physical evidence had traditionally not been a major source of investigative leads this situation has been changing in recent years. Three acronyms, AFIS, CODIS, and NIBIN (Figure 2.9) are the major causes of that change.

Each of these three stands for a computerized database of information collected from physical evidence. AFIS is Automated Fingerprint Identification System, CODIS is Combined DNA Indexing System, and NIBIN is National Integrated Ballistics Identification Network. These systems have in common the ability to store data in a readily available computer database, then compare data from evidence from an incidence under investigation against the data in the database. They allow an investigator, for example, to determine if a fingerprint found at a crime scene belongs to someone who has been previously fingerprinted. CODIS allows a DNA profile to be searched against individuals who have been sampled as a result of a previous crime. NIBIN permits searches of the identification patterns on fired bullets or cartridge cases to be matched with other such evidence recovered previously. These databases also allow cases to be connected even if the source of the evidence is still not known. This can energize an investigation and give it direction. The systems now allow searching for matching data across state lines, and, since they are all rapidly growing, they will become even more useful as more and more data is added.

Before AFIS, fingerprints found at a crime scene had limited value to the investigation unless one had a pretty good idea whose fingerprint it was. The sheer volume of fingerprint cards, in even a medium size jurisdiction, precluded searching an evidence print against all those ten-print cards. Even if sufficient time and energy were devoted to such a search, there was a real probability of missing the print of interest because searching a large file manually is such a mind-numbing activity. AFIS has more than a 10-year head start on CODIS and NIBIN, but as data is rapidly

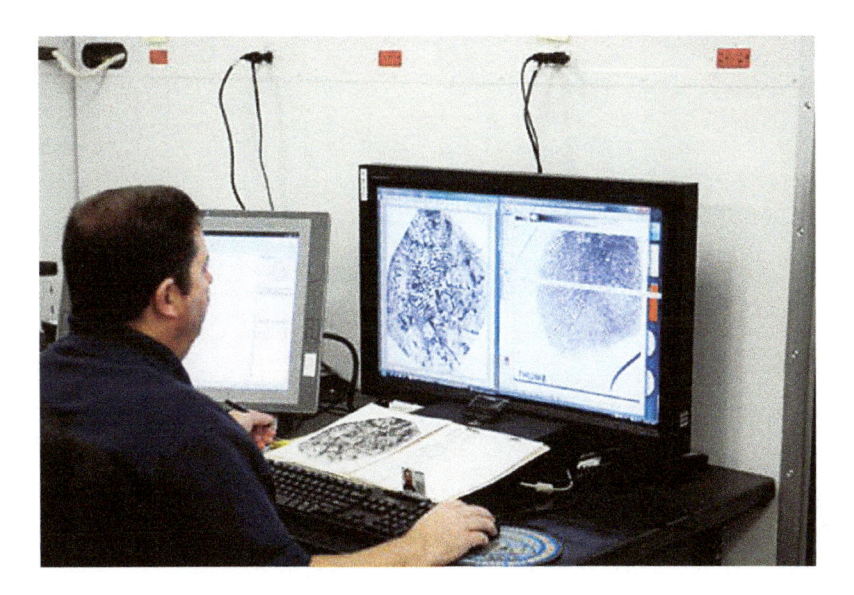

Figure 2.9 Use of Automated fingerprint identification (AFIS) to search a file of known fingerprints for a match to an evidence print.

collected and incorporated into these databases, they are rapidly becoming larger and more useful. With databases, larger is better but they require the ability to accurately search the mountain of data. Computerized systems, which bring that capability, have revolutionized the use of fingerprint, body fluid, and firearms matching and will continue to improve in their effectiveness.

Corroboration—credibility: supporting or disproving statements

One of the most useful applications of physical evidence is to corroborate statements or testimony. Corroboration can be critical to the development of a case, yet it is frequently overlooked as a role for physical evidence. The case illustration in the last chapter (Case 1.1) was used to emphasize the potential importance of physical evidence, which has no potential for proving who committed a particular crime. In such applications, its role is only to enhance credibility. Most practitioners are well aware of the common uses of physical evidence to point the finger toward or help exonerate a defendant. However, long after the evidence has been collected, a scenario may develop where physical evidence may play a critical part in breaking or corroborating an alibi or in making a witness seem, more or less, believable. This application is underutilized, at least partially, because it requires insight and creativity to identify possible applications. The initial search for possible evidence must be done thoroughly because one cannot know in advance what may turn out to be a critical piece of corroborative evidence. The ability to spot something, which is not obviously important but collect it anyway just in case, can make a significant difference in the right case.

Identification and analysis of substances or materials

Proving that an item of physical evidence falls into a particular class to support a legal action is one of the most important functions of forensic laboratories. Substances or materials, which require chemical or instrumental testing to be identified, must be subjected to the necessary testing to prove that an offense occurred. Thus, the defendant cannot be prosecuted unless someone says "in my expert opinion" this sample of white powder seized by the officer from the defendant contains a controlled substance (such as heroin or cocaine) (Figure 2.10).

Similarly, the scientific determination that a sample of blood taken from a driver contains more alcohol than the law allows when operating a motor vehicle is critical to a driving while intoxicated case. These analyses are not crucial to *solving* a crime but are critical to proving that a crime has been committed and, perhaps, who committed it. As mentioned earlier, forensic laboratory analysis has traditionally been reactive rather than proactive. This is

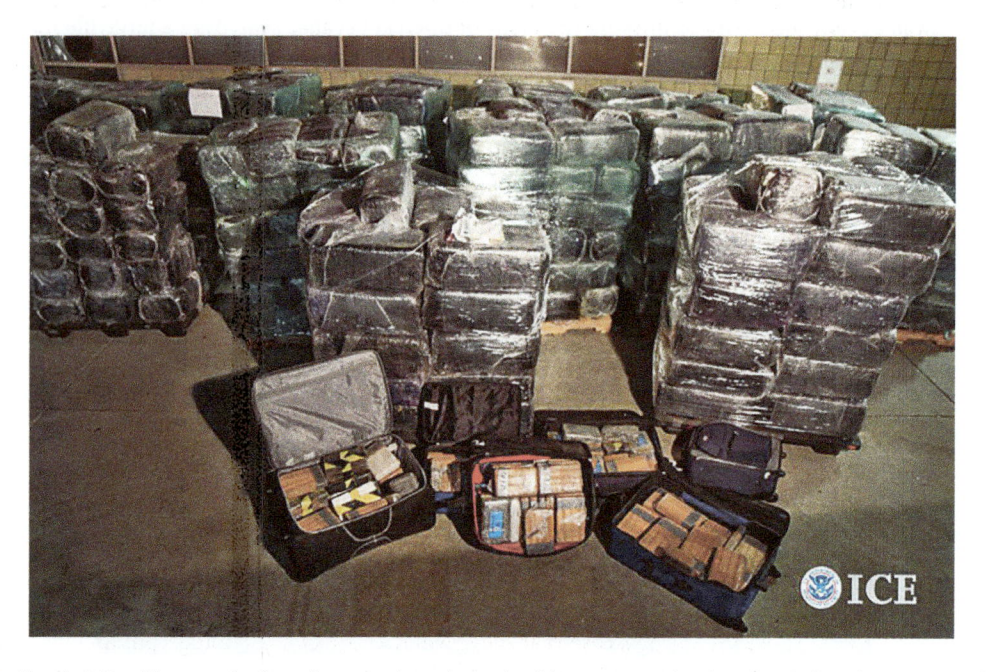

Figure 2.10 Nearly $12 million worth of marijuana and cocaine seized from a cross-border tunnel shut down south of San Diego. (Source: Department of Homeland Security.)

partly because of the nature of forensic evidence and its examination in the laboratory. Identification testing is the easiest and least time consuming. Examinations involving comparisons are more involved, in addition to requiring an exemplar specimen. Another reason forensic labs are overwhelmed with this type of examination is the disproportionately large number of controlled substance possession cases. As we have noted, the chemical identifications must be done to support the prosecutions.

Establishing a basis for a crime and criminal prosecution—*corpus delicti*

As noted in Chapter 1, some forensic testing, usually identification (classification) tests, are required to establish that a crime may have been committed, but further testing may be needed to establish the required legal elements of a particular crime. Establishing that a seized substance is controlled, that fire debris contains accelerant residue, or that semen is present on a vaginal swab taken from a sexual assault complainant, are all examples of utilizing physical evidence to establish *corpus delicti*. Often, successful civil or criminal prosecutions require that many other occurrences must be proven for the case to be made successfully.

THE PHYSICAL EVIDENCE PROCESS

Utilization of physical evidence in the justice system requires that a series of steps be taken in the proper sequence. The process of recognizing and handling physical evidence is outlined below.

Recognition—critical portion of evidence process requires a trained observer

Clearly the first step in the evidence process requires recognition that an object may be useful as physical evidence. One cannot overemphasize the fact that recognition of physical evidence is not always a routine, straightforward process. Some things that have evidentiary value are obvious, but many are not. What is or is not potential evidence in a case, depends on the scene, location, context, type of case, and other factors. A good crime scene investigator is one who has the ability to recognize possible evidence. Education and training contribute to developing this ability, but experience also plays an important role. Obviously, if an item of evidence goes unrecognized and, therefore, uncollected, that is the end of any potential value it might have had in a case.

Documentation and marking for identification

After recognition comes documentation. Proper and complete documentation is critical to establishing the legal and scientific requirements of chain of custody. The term "chain of evidence" requires that someone must be able to show that each item being offered as evidence in court is the exact same object that was collected from the victim, suspect, or crime scene. One must also be able to establish where it has been in the interim, and everyone who handled it since it was collected. In some ways it is analogous to the provenance of an art object so necessary to establishing its authenticity. Any break in that chain of evidence will likely destroy the item's value as evidence because the court may not permit it to be admitted as evidence. Documentation includes the markings used for identification. Documentation, as we will see in Chapter 3, is an important element of scene investigation. The exact location where an item was found can be very important in reconstruction of the incident. A reconstruction requires the synthesis of all the information available. Knowing where evidence was found is an important part of that synthesis.

Documentation of a scene is as important as documentation of the individual items of evidence. Many types of pattern evidence can only be recorded by proper documentation.

Recently, more attention has been paid to reinvestigating unsolved cold cases. Every sizable police agency now has a "cold case" unit. This new emphasis on reinvestigating old cases illustrates the importance of good documentation. During the initial investigative process, a good investigator may depend on his or her memory. However, when reinvestigating a case, which is 10, 15, or even 25 years old, memory is no longer a reliable tool. Those directly involved in the initial investigation may be retired or even dead. Even if they are still available and they may think they remember the facts surrounding the incident and each piece of evidence quite clearly, their memory and recollections may not be accurate. Those reviewing a case years later will need to look at photographs and read complete and accurate documentation to develop a solid understanding of what things were found in order to carry out a thorough reinvestigation.

THE CRIMINAL JUSTICE SYSTEM AND PROCESS—CASE FLOW

Trace the evidence pathway through the system

The legal flow chart in Figure 2.12 is intended as a shorthand way of looking at the role and importance of physical evidence in the legal process. Since the detailed operation of the legal system is different in each state and in the federal system, the chart has been generalized to make it applicable to as many jurisdictions as possible. The flow chart format allows one to look at how physical evidence can be used at many different stages as an incident or case moves through the criminal justice system from beginning to end.

Investigative agency role

First, there is an incident; however, at this stage it is premature to label it a crime. Perhaps someone has called a police dispatcher, a call has been made to 911, or an officer has observed suspicious activity. Further investigation must be undertaken to determine whether what occurred might be a crime. In many smaller jurisdictions, the deputies and police officers, who are the first responders, are often not as busy as big city police and have the luxury of looking into incidents that might not be carefully examined in a high crime area.

In the flow chart, the top line is quite straightforward. If one moves to the right, the box shows "undetected or unreported." Under such circumstances, there can be no action taken and that is the end. Surprisingly, a great many incidents fall into this category, even some rather serious criminal incidents. As an example, an older person is found dead at home, with no indication of foul play. No autopsy is performed, and the incident is closed as death by natural causes. If that individual had been smothered with a pillow or poisoned, a homicide could have gone undetected.

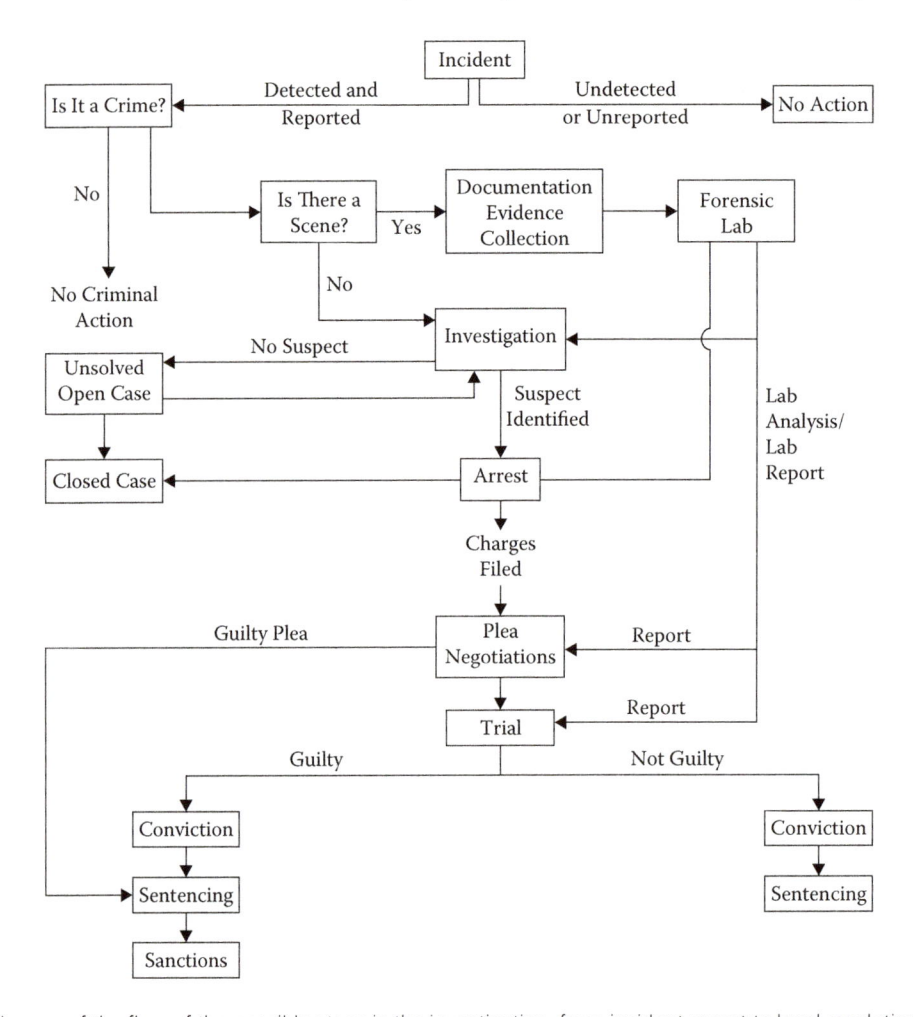

Figure 2.12 Diagram of the flow of the possible steps in the investigation, from incident report to legal resolution of the incident.

Moving to the left, "Detected and Reported" is clearly of more interest to forensic science. Could this incident be a crime? That is where the initial investigation must be done, and a decision made whether further investigation is warranted. The case illustration (Case 2.3) is a good example of such an incident. This investigation is normally done by a detective in the agency that has jurisdiction in the area.

> **CASE ILLUSTRATION 2.3**
>
> In a rural county in New York, someone noticed a considerable amount of what appeared to be bloody rags and papers in a dumpster and called the sheriff's department. The responding officers were concerned that this material might have resulted from a violent criminal incident. It was collected and brought to a nearby forensic laboratory. They were asked if the material could be examined to determine if it really was blood and, if so, was it human blood. It was quickly determined that the material tested positive for blood. However, the next test showed it was not human blood. Further testing established that it was consistent with deer blood. It became clear what had probably happened. Somebody had taken a deer out of season, gutting and cleaning it, and trying to conceal the bloody residues to avoid being fined for taking a deer out of season. Thus, it was not a homicide, but a hunting violation incident, and the prompt use of physical evidence made what could have been a time-consuming criminal investigation unnecessary.

If the initial investigation discloses a non-criminal explanation or there is no indication of criminal action, then the answer to the box is "No" and investigation stops. There may still be a civil action or someone interested for other reasons, and there may even be evidence collected and sent to a private laboratory. If, however, there is no perceived violation of criminal law, there will be no further involvement of the criminal justice system. If the police determine that a crime may have been committed, an investigation is opened. This is sometimes called "founding" the case. One of the first important questions is: Is there a crime scene? This is a particularly important concern in terms of physical evidence and not quite as simple to answer as it sounds.

Further investigation is called for if it is clear that there is a location where at least part of this incident took place and where physical evidence may be found or possible witnesses may be discovered. If one has a scene, even just a potential scene, that may be the location to initiate the investigation. With modern techniques, careful scene examination is likely to uncover useful evidence that may further help to decide if this location is, indeed, a crime scene or not. Many crimes do not have scenes, or do not produce useful crime scenes. For example, it is reported that someone had a wallet stolen from her purse. The individual reports that they had it in the morning when they left home, but, when returning home, it was no longer there. That does not mean that the incident cannot be investigated, but one is probably not going to have a scene to investigate, other than the pocketbook itself. Finding a crime scene is not only going to be quite difficult, but perhaps the wallet was lost and there was no crime at all. Even in a more serious case, such as a sexual assault, the victim may have been overcome, transported to an unknown location, assaulted, and then taken in a car to another location and dumped out. There certainly is a crime scene, but it may take a great deal of investigation before the scene is located, if it ever is.

There are two possibilities, if a scene is established; the "Yes" arrow is followed to the important scene-processing block. This block symbolizes many activities including a careful processing of the scene and full documentation of that process. This should include collecting possible physical evidence, packaging and properly identifying anything taken from that scene, and insuring that any evidence that needs scientific examination reaches the forensic laboratory or other skilled examiner for analysis. The crime scene block is usually under the control of the investigators, but those carrying out the documentation and collection of evidence, such as fingerprinting, photography, and many other activities, are usually technically trained individuals who are specialists in such matters. Much of the physical evidence collected will be processed at a forensic laboratory, and any information gathered passed on to the investigators. This information will eventually be available for use at subsequent steps in the process. The forensic laboratory block at the far right may also receive physical evidence in several other ways, depending on the type of incident and how the investigation proceeds.

If there is no identification of a crime scene at this stage, all the responsibility falls on the investigators. It is clear that the key box, the one with the most arrows in and out of it, is the one marked "Investigation." That block symbolizes control over all that will be done to investigate that incident and determine if a crime has been committed and, hopefully, come up with a strong suspect. Whoever is assigned to investigate the incident is responsible for many key decisions. Depending on the nature of the incident, that control may rest with an individual or a large team led by a high-level police officer or prosecution official who is assigned the responsibilities of supervising a task force. The investigation block symbolizes the nerve center of the investigation. The effective running of the investigation can make a tremendous difference in how likely it is to be successful and how well the criminal justices system works for this incident.

The investigation may disclose that there was no crime after all and close the investigation. More commonly, one or more suspects are identified and, if it is judged that there is enough evidence, an arrest will be made (downward arrow). If within a reasonable amount of time no suspect has been identified, the case may be transferred from an active investigation to an unsolved open case file (arrow to left). How much investigation will occur depends on many factors; to name a few, the seriousness of the crime, the resources available to the investigating agency, the cooperation of the public and the press, and many more. Depending on the nature of the incident, after a while the case may be closed (arrow down). Many incidents reside in the unsolved, open case file for a considerable amount of time before being closed. In most busy jurisdictions, after a legitimate effort has been made, and there are no further strong leads on which to work, the case will go to the cold case file.

Adjudicative agency role

Unfortunately, often new incidents to investigate are coming in every day, demanding the attention of a limited pool of investigators. Should some new evidence become available; for example, a case with suspiciously similar circumstances, information from an informant, or perhaps a new witness suddenly appears, any of these can trigger a transfer back to active status (arrow to right "new evidence"). In the last few years, many jurisdictions have formed "cold case squads" to take a fresh look at unsolved cases and apply new technology or revive the investigation with different investigators. New investigators retrace what was done and, perhaps, realize that someone, who should have been interviewed, was missed, or evidence was collected and unfortunately it went to a property room and was never sent for scientific examination. In many cases, these squads have been successful in developing suspects and moving the case back into the active stream. The arrow that says "new evidence" extends back to investigation, but if there is physical evidence it could also extend from the investigation block back to the forensic lab box. It can then produce information that can take the same three branches emanating from the "Lab Analysis Lab Report" block.

The "Investigation" block, as mentioned above, represents the control center. If the investigators are successful, they will develop a possible suspect. If the investigators have developed enough believable evidence to convince the legal authorities that there is "probable cause" to think that particular person committed the crime, then that person will be arrested. "Probable cause" is an important legal concept that says the evidence offered is sufficient to convince a reasonable person that this individual may be guilty. There are a lot of qualifiers in that description, and whole legal treatises have been written trying to define it. However, for present purposes, this definition is adequate. This decision will be made by a grand jury or a judge at an arraignment.

An important distinction to grasp is that the amount of evidence needed to supply probable cause is considerably less than the amount needed for conviction at a trial. Juries in criminal cases are instructed that they must be convinced beyond a reasonable doubt that someone is guilty before they return a guilty verdict. A suspect may be arrested on the much lesser standard of proof that is probable cause. Sometime between arrest and trial, sufficient evidence must be developed to meet the higher standard, if the suspect is to be convicted.

From the arrest block, arrows go in three different directions. The individual is interviewed, and there may be some admissions made or an alibi given. Sometimes when the alibi is checked, it is found not to be valid. Eventually, if enough information is gathered and the court system believes there is enough evidence to charge the individual, charges are filed, and the prosecution and defense processes are put into motion. If it is decided that the case is too weak to pursue, then no charges are filed, and the case is closed or returned to the open case file.

The third arrow is particularly important to our interest in physical evidence. Once a suspect is identified, often even before an arrest is made, it is possible to seek a court order to obtain evidence from that individual. This may be body fluid control samples, a warrant for a search of premises, taking of clothing for trace evidence, a search of a vehicle, and many other possible evidence sources. The evidence gathered, if suitable for analysis, must be sent to a forensic laboratory to be compared with evidentiary material gathered at the crime scene or from the victim.

Importance of physical evidence, particularly scientific evidence

It should be noted that three arrows emerge from the forensic lab block. The criminal justice system looks for a formal report when evidence is submitted to a forensic laboratory. That report can play an important role in several different aspects of a case. Whoever submitted the evidence, generally an investigator, gets a copy of the laboratory report. As indicated earlier, that report may help in the investigation and identification of a suspect. However, many laboratory reports are not issued until well after someone has been arrested since evidence obtained from the arrestee may have been critical to the laboratory examination. A copy of the laboratory report will go to the prosecutor to help in developing evidence for trial or to be used in the plea-bargaining process.

The block below "Charges Filed" is labeled "Plea Negotiations." In most busy jurisdictions, the majority of cases are settled at this stage by the charges being dropped or, more likely, the defendant pleading guilty to some lesser charge to avoid the perils, to both sides, of a trial. At this juncture, the prosecution and the defense will have the laboratory reports available as part of the information that they will use in their plea negotiations. A laboratory report that will tend to support the prosecution's case will be a strong bargaining chip for them. Conversely, a laboratory report that is weak may strengthen the hand of the defense in negotiations. Thus, the presence of physical evidence may exert a significant effect on case resolution, even in cases where there is no trial.

The third arrow from the "forensic laboratory" block points to the "Trial" block. Should the case go to trial, the forensic expert from the laboratory will usually be called to present the findings (Figure 2.13) for the jury to hear and often the laboratory report itself will be offered into evidence.

One should remember that although it is virtually always the prosecution that collects physical evidence for analysis, either side may use the results of the examination to advance their case. It is not uncommon for the defense to call for the presentation of physical evidence findings when the prosecution has decided that such presentation may not aid their case. Each side must, at this stage, weigh the evidence available and how well it will "play in court." Not every witness is highly believable and can express himself clearly and in a convincing manner. Even though a witness is absolutely truthful, his or her evidence may not be well received by a jury. Conversely, some individuals are particularly talented liars and may provide convincing sounding spurious opinions.

If plea-bargaining produces a guilty plea, the arrow goes directly to the "Sentencing" block. The time and cost of a trial are avoided. Otherwise, either the case is dropped or proceeds to trial. The "Trial" block will eventually produce a verdict of guilty or not guilty. There is a third possibility of a mistrial, but that generally just results in

Figure 2.13 Examiner being sworn in before testifying. (Courtesy of Shutterstock, New York.)

another trail and is not indicated in the diagram. If the trial results in a not guilty verdict, the defendant is acquitted and released. In the criminal portion of our justice system, if there is a trial and the person is found not guilty, there is, for all practical purposes, no appeal. Once a defendant is found not guilty, he/she is free. If convincing evidence of guilt is discovered later, it does not matter since the defendant cannot be tried more than once for the same crime.

If the trial results in a guilty verdict, the next step is sentencing followed by some kind of sanction. If one is convicted of a felony, it usually means confinement in a state or federal penitentiary. If one is convicted of a misdemeanor, it usually means confinement in a local jail for a period of less than 1 year or perhaps a fine, a sentence of public service, or any number of different sanctions. Although a defendant is charged with and tried for a felony, he/she may be convicted of a lesser crime that is a misdemeanor and therefore subject to the lesser sanctions associated with the misdemeanor.

SCIENTIFIC AND TECHNICAL EVIDENCE ADMISSIBILITY AND THE EXPERT WITNESS

Admissibility versus weight

Courts, through judges, control what evidence, both testimonial and physical, gets "admitted" in a case; that is, what evidence the trier of fact (judge or jury) can hear in determining guilt or non-guilt. Once evidence has been "admitted," it is said to "go to the weight (importance)." That means the jury or, if the defendant waives a jury, the judge in the case can consider such evidence and accord it whatever amount of weight they think it deserves in the overall case. Forensic scientists get involved in admissibility issues when a new method or test is used in a case, or when an existing method or test is challenged. Over a long period of time, a large body of law has developed around the issue of what scientific or technical evidence should be admitted. Admissibility of scientific evidence, which is new or novel, can be quite contentious.

Although the rather technical legal arguments made in admissibility issues are beyond the scope of this book, every forensic scientist should understand the basic issues. The most basic standard for admissibility is relevance, and this applies to technical and scientific evidence as well. The evidence has to be pertinent to the case and have the potential of helping the trier of fact reach a just verdict. In the 1920s, a case called Frye v. United States came before the U.S. Circuit Court of Appeals for the District of Columbia. The case involved the use of the polygraph in determining whether the defendant was being truthful. The court would not allow the polygraph results, and most courts still will not, because it is simply not considered sufficiently reliable. The importance of the Frye case was not the actual ruling, but the test the court laid out for determining whether "novel" scientific or technical evidence should be admitted. They said that the principle governing admissibility was general acceptance of the test's underlying principles by the scientific community to which the test belongs. This language sounds straightforward, but often it is not clear how to apply the principle to an individual case. What "scientific community" does polygraph testing belong to? Polygraph examiners? Neuroscientists? Electrophysiologists? All of them? What about forensic DNA testing? Is the "scientific community" forensic DNA analysts, molecular biologists, or physiologists? Each court has to decide these issues for itself. Although the "Frye" case is very old, there are still some state courts that feel it is the proper standard. As we shall see, many courts no longer use the "Frye" test.

Requirement of relevance and reliability

Some admissibility issues were clarified for the federal courts with the issuance of the *Federal Rules of Evidence*. These rules, strictly speaking, apply only to the federal courts for federal cases, and the individual states do not necessarily have to adopt the federal rules, though many have. Over the decades following Frye, many states adopted a "Frye rule," either identical or similar to the federal rule. The *Federal Rules of Evidence* were amended several times over this period and came to emphasize the dual test of **relevance** and **reliability** as the appropriate standard for admissibility. There was still controversy on what exactly constituted a proper test for reliability and relevance. One of the keys to relevance is a term mentioned before and will be mentioned many times again: Chain of evidence. This is the legal requirement that physical evidence must be unambiguously associated with the case at hand. This is done by documenting the collection and location of a piece of evidence from initial collection to the moment it is being offered in court.

Daubert case

In 1993, the U.S. Supreme Court (Figure 2.14) agreed to review a complicated epidemiology case called Merrill Dow Pharmaceuticals v. Daubert et ux. et al. The case was about whether a drug called Bendectin caused birth defects in children whose mothers took the drug to control morning sickness during pregnancy. The importance of the case is that the court took the occasion to issue guidelines for deciding the admissibility of new or novel scientific evidence.

This was significant because it was the first time the U.S. Supreme Court had ever considered this matter. The guidelines are called the Daubert criteria, and they outline several different approaches, or prongs, of the Daubert test. Strictly speaking, they only apply to the federal jurisdiction, but many states have adopted them either in detail or in spirit. The Daubert rules require that scientific tests on evidence be truly scientific. The tests must have been subjected to significant hypothesis testing of the underlying principles. Other criteria include whether the test is generally accepted (the Frye criterion), whether anything is known about its error rate, and whether there has been peer review, such as publication in a peer-review journal. Under Daubert, the judge is the "gatekeeper;" that is, he or she determines admissibility using the criteria outlined by the Supreme Court. The Daubert criteria appear to create a stiffer test for admissibility than Frye did. Adoption of Daubert standards by a jurisdiction also creates opportunities for attorneys to now challenge admissibility of evidence, long accepted under the Frye standard, using the new criteria. The Supreme Court has made it clear in a subsequent case that Daubert criteria apply equally to "technical examinations" as well as scientific evidence. Thus, the Daubert standard applies to handwriting examination, fingerprints, engineering, and many highly technical but not strictly scientific types of expert testimony.

Often, admissibility of evidence is determined by a judge following a hearing held specifically to determine admissibility. These hearings may be convened to consider whether the testing used by the laboratory conforms to Frye or Daubert standards. Many other evidentiary hearings are held to determine whether evidence chain of custody was properly intact, or a search warrant was properly obtained.

Figure 2.14 Supreme court as of January 2019: Back row (left to right): Neil Gorsuch, Sonia Sotomayor, Elena Kagan, and Brett Kavanaugh. Front row (left to right): Stephen Breyer, Clarence Thomas, Chief Justice John Roberts, Ruth Bader Ginsburg, and Samuel Alito. (Credit: Fred Schilling, Collection of the Supreme Court of the United States.)

Most forensic scientists and many other kinds of experts testify in courts as *expert* witnesses. Certainly, well qualified scientists are expert witnesses, but our legal system considers someone an expert witness if they have knowledge or special training that would be beyond the scope of the average juror. An expert might be a forensic chemist or DNA analyst, but could be a baker, a design engineer, a ship's captain, or a computer programmer, depending on the type of case. As noted above, experts must be qualified each time they testify, and if found qualified, they are allowed to give opinion testimony. Other witnesses are normally allowed to testify only as to what they personally know, saw, or heard. Becoming a good expert witness and a good communicator in the courtroom is an important part of the training of a successful forensic scientist.

Key terms

AFIS (Automated Fingerprint Identification System)
CODIS (Combined DNA Indexing System)
cold case
comparison specimens
control specimens
Daubert Evidence rule
deposit
exclusion
expert witness
Frye Evidence Rule
impression
imprint
indentation
Known (K)
linkage
Locard Exchange Principle
MO (Modus operandi)
NIBIN (National Integrated Ballistics Identification Network)
probable cause
Questioned (Q)
striation
trier of fact

Review questions

1. List and discuss the ways physical evidence is produced.
2. What is a useful way of classifying physical evidence?
3. List and discuss some of the uses of physical evidence in criminal investigations.
4. List and discuss the steps in physical evidence recognition and processing.
5. What is a "questioned" and a "known" specimen in forensic laboratory analysis?
6. What is an expert witness?
7. What role does forensic science play in the operation of the rule of law?
8. Describe the different possible steps in the "flow" of physical evidence in a criminal investigation and prosecution.

9. What is evidence admissibility?
10. What are and what have been the standards for the admissibility of forensic (scientific) evidence into court?

Fill in/multiple choice

1. When two complex, three-dimensional surfaces fit together perfectly, it is commonly referred to as a _____.
2. The results of examination of physical evidence can prove of considerable value to the operation of the criminal justice system even if a case never goes to trial by facilitating _____.
3. Physical evidence can be used in at least six different ways: (1) help reconstruct the occurrence, (2) _____, (3) provide linkages, (4) _____, (5) identify victims or suspects, and (6) _____ to help the trier of fact (jury or judge) reach a just conclusion.
4. The ability to establish the exact whereabouts of an item of evidence and under whose control it was from its collection to the courtroom and everywhere in between is known as maintaining the
   a. chain of command
   b. continuity of investigation
   c. chain of custody
   d. business records circle
5. In the American legal system, whether or not a particular expert's testimony could be used during trial was, from about 1923 to the mid-1990s, determined by
   a. the standard laid down by the *Daubert* case
   b. a "General Acceptance in the Community" standard
   c. the Locard standard
   d. a "Don't Ask Don't Tell" standard

References and further readings

Dixon, L., and B. Gill. "Changes in the Standards for Admitting Expert Evidence in Federal Civil Cases since the Daubert Decision." *Psychology, Public Policy and Law* 8 (2002): 251–308.

Horvath, E., and R. Meesig. "The Criminal Investigation Process and the Role of Forensic Evidence: A Review of Empirical Findings." *Journal of Forensic Science* 41 (1996): 963–969.

Kumho Tire Company, Ltd., et al., Petitioners v. Patrick Carmichael, etc., et al., No. 97-1709, 119 S. Ct. II 67 (1999).

Osterburg, J. W. "The Scientific Method and Criminal Investigation." *Journal of Police Science and Administration* 9 (1981): 135–144.

Peterson, J. L., J. P. Ryan, P. J. Houlden, and S. Mihajlovic. "A Review of Empirical Findings." *Journal of Forensic Sciences* 41 (1996): 963–969.

Peterson, J. L., J. P. Ryan, P. J. Houlden, and S. Mihajlovic. *Forensic Science and the Courts: The Uses and Effects of Scientific Evidence in Criminal Case Processing*. Chicago, IL: Chicago Center for Research in Law and Justice, University of Illinois at Chicago, 1986.

Petitioners v. Robert K. Joiner et al,. No.96-188, 118 S. Ct. 512, 1997.

Scientific Testimony-An online journal, www.scientific.org.

Swanson, C. R., N. C. Chamelin, L. Territo, and R. W Taylor. *Criminal Investigation*. 9th ed., Burr Ridge, IL: McGraw-Hill, 2006.

William Daubert, et ux. et al., Petitioners v. Merrell Dow Pharmaceuticals, Inc., No. 92-102, 113 S. Ct. 2786 (1993); General Electric Company, et al., Petitioners v. Robert K. Joiner et ux., No. 96-188, 118 S. Ct. 512 (1997).

Chapter 3

Crime scene processing and analysis

Lead Case: State of Connecticut vs. Duntz

In August 1985, the 234-year-old town hall in Salisbury, Connecticut, was burned down. The Fire Marshall searched the scene and determined that this fire was set by arsonists. Two local people, Earl Morey and Roy Duntz, were suspected of setting the fire. As the case developed, Morey decided to become a state witness and testify against Duntz on the arson charges. However, a day before Morey was set to testify, his body was found on the shores of Long Pond shot to death.

The crime scene was investigated by the Connecticut State Police major crime squad and Dr. Henry Lee from the laboratory. A major storm was moving into the area from the west, and the investigators worked quickly to get to the crime scene and collect any available evidence before the rain storm.

The physical evidence, laboratory and autopsy findings, and investigative information developed in the case included:

1. Two sets of tire impressions were found at the scene. One set was consistent with a tire from Morey's car. A second set was on top of the first one, which was identified as having come from a tow truck. Service request records from the tow truck company showed that it was called to the scene to tow a disabled car at 7:30 am. This information indicated that the tire impressions of Morey's car were more likely deposited at the night before dawn.

2. Two different types of shoe impressions were observed near the victim's body. One set of shoeprints had a parallel wave-type sole pattern, which was identified as coming from the victim's own shoes. A second set of shoeprints had an unusual hexagonal sole pattern. A search of the laboratory footwear files showed that this type of pattern came from a Foot Joy brand sneaker.

3. A shoeprint with a hexagonal design was found on the floor in the back of Duntz's van. This shoe imprint in the van had the same size and sole pattern as the shoe prints found at the crime scene. During a search of Richard Duntz's home, police were able to find a picture of Richard wearing the same brand of Foot Joy shoes.

4. Four spent cartridge cases were recovered from the scene. The head stamp showed that they were Smith & Wesson (S&W).

5. Bullet holes and gunpowder residue was found on the victim's clothing. Green paint deposits were found on his pants. A "nickel bag" of white powder was found in his shirt pocket. The white powder was identified as Aspirin powder by instrumental analysis.

6. A Red Devil chewing gum wrapper was found in his pants packet. Similar chewing gum wrappers were found in Richard Duntz's van. However, no positive linkage between the gum wrappers could be made.

7. No gun was found during the police search. However, they did find an elastic belt with an impression of a weapon and a gun cleaning kit. Laboratory examination of this gun impression showed that it could have come from 9 mm-type weapon.

8. Earl Morey's car was found in a parking lot with a large, fresh blood smear on the back-left side panel. Serological analysis (this case happened before DNA analysis was routine) showed these bloodstains had the same blood and isoenzyme-type as Morey. This fact indicated that the car was at the crime scene at the time Morey was shot.

9. Fifteen latent prints were discovered inside Morey's vehicle. However, none of them matched Richard Duntz's prints.
10. Police searched Duntz's home and van but did not find any of the clothes that he was reportedly seen wearing that evening.
11. The dome light of Morey's car was found to be not working. Examining the dome light, an investigator found that the light bulb was missing. During the search of Duntz's van, a light bulb was found near the driver seat. Toolmarks were found on the metal ends of the light bulb during laboratory examination. Microscopical examination of the toolmarks indicated that those toolmarks could have come from the metal clip housing of the dome light.
12. Autopsy showed that Earl Morey was shot three times. Three 9 mm bullets were recovered from his body. The bullets were fired from a weapon with a 5-right twist. The firearms examiner was able to identify that these bullets more than likely were fired from a Smith & Wesson Model 55, 9 mm pistol.
13. Information was gathered through witness interviews that Richard Duntz in fact owned a 9 mm pistol. For target practice, he fired the weapon into a tree at his friend's back yard in Upstate New York. Detectives went to New York and removed a section of the tree from the yard. The firearms examiner was able to recover several 9 mm bullets from the tree. Those bullets exhibited identical class and individual characteristics to the 9 mm bullets form Earl Morey's body.
14. Duntz's brother, Ronald, was arrested on drug charges. He later admitted that he sold a stolen 9 mm S&W Model 55 pistol to his brother, Richard.
15. A weapon of the same type was used for test fires and reconstruction experiments. The cartridge case ejection pattern helped to reconstruct the shooter's and victim's position relative to the tire marks at the time of shooting. Barrel to target distance was estimated by comparing the GSR pattern on the victim's clothing with the test firing results at measured distances.

Based on the forensic testing results and physical evidence, police arrested and charged Richard Duntz with the murder of Earl Morey. The case went to trial in 1990. The jury found Richard guilty of the murder of Earl Morey, and he was sentenced to 60 years in prison. However, in 1992, the Connecticut Supreme Court set aside the verdict, stating that the warrants used to search Richard's home and van were not justified. However, in 1994, a second trial began, and, within days, Richard accepted a plea bargain and was sentenced to prison for 15 years. He died from a heart attack a year before his scheduled release date.

> **LEARNING OBJECTIVES**
> - How crime scene processing is different from crime scene analysis
> - What are the different types of crime scenes?
> - Initial actions at a crime scene
> - Legal and scientific requirements on Crime Scene Processing
> - Establishing crime scene security and reasons for maintaining security
> - Crime Scene Safety
> - What the steps are in crime scene processing?
> - The process of evidence recognition based on hypothesis formulation
> - The schemes for searching crime scenes
> - The importance of and major methods for crime scene documentation
> - Taking notes
> - Making sketches and types of sketches
> - Technical and forensic guidelines for photography
> - Videotaping crime scenes
> - Audio Recording of scene description
> - 3-D Recording and other new Technology in Crime Scene Documentation
> - About the duty to preserve crime scene work product
>
> *(Continued)*

LEARNING OBJECTIVES (Continued)
- Different methods of collecting physical evidence and applicability to different categories of evidence
- Numbering and description of physical evidence from the scene
- Various types of packaging for different types of evidence
- Types of controls, and standards for each type of physical evidence
- Submission of physical evidence for laboratory analysis
- Crime scene analysis and crime scene reconstruction
- The difference between reconstruction and reenactment

PROCESSING VERSUS ANALYSIS

Crime scene investigation is one of the most important and most challenging area in forensic investigation. It consists of two aspects: processing and analysis. There is no way to set out a "formula" for crime scene investigation. It requires a combination of scientific knowledge, logical reasoning, systematic approach, and investigative experience. Documentation and Processing have some common guidelines and recommended procedures that we will discuss. However, there is almost always some variation in the processing depending on the individual scene, environment and the nature of the case. And some analysis is necessarily part of the evidence recognition process. There are a series of fairly standard steps that can be followed in processing most scenes.

Analysis depends on detailed observation, proper recognition and making logical connections. It also depends on the results of the laboratory analysis of evidence, study of scene patterns, and integrating all the data available from the scene and the investigation. In death cases, the data will include the findings and opinions of the forensic pathologist. Crime scene analysis and reconstruction is a distinctly scientific activity. We would say that proper crime scene analysis roughly follows the steps of the scientific method itself (Chapter 1). It requires scientific background, forensic knowledge, and experience. This is the main reason some people have argued that crime scene investigators should be trained in forensic science and criminalistics. Except in a handful of jurisdictions, crime scene investigators are not usually forensic scientists. They are police personnel or crime scene investigators. Most crime scene investigators have learned their specialty through a combination of training and experience.

Television programming about forensic science in recent years would have you believe that the same people who do crime scene processing also do the laboratory analysis, interview witnesses, investigate the case and arrest the suspects. This is almost never the case with very few exceptions. Most of the time, police investigators—sometimes specially trained crime scene technicians—conduct the crime scene processing. Generally, laboratory analysts rarely go out to scenes. There are some laboratory analysts with crime scene skills and experience as well as some crime scene investigators with considerable knowledge and experience. Sometimes, working through a scene is a collaborative effort between laboratory analysts and crime scene investigators. This tends to happen more often in major cities, with cases that have complicated scenes or special requirements.

TYPES OF SCENES

Classification of crime scenes may offer a guide to the investigator on how to proceed with crime scene processing, may suggest types of evidence more likely to be found in the type of case, and may provide a basis for an investigation logic tree. In addition, actions at scenes may be affected by where the scene is located, how much control we have over the scene, weather conditions, environmental conditions, what equipment and personnel are available, what type of case it is, and other legal or scientific issues However, classification of the crime is sometimes not a clear cut process when investigating a complex situation. Investigators should understand these classification systems and how to develop multiple alternate hypotheses. It is important to keep in mind that there is no single classification method that will satisfy all the elements of the crime scene and its investigation. It is essential that crime scene investigators develop the ability to utilize their analytical skills and logical approaches to make an initial determination regarding the number and types of crime scenes that were involved in the commission of the crime.

There are many ways to classify crime scenes, but just a couple will help us organize our thinking about this. The point here is that one should approach each different type of scene a little differently.

Types of crime scenes

Following are some common ways to classify a crime scene:

1. The original location at which the crime was committed (e.g., primary scene, secondary scene)
2. The type of crime committed (e.g., homicide, sexual assault, robbery)
3. The physical location (e.g., indoor, outdoor, vehicle)
4. The physical condition (e.g., buried, underwater)
5. The boundaries of the scene (e.g., house, train, bank, computer, car)
6. The appearance of the crime scene (e.g., organized or disorganized, passive or dynamic crime scenes)
7. The criminal activity (e.g., active, passive scenes)
8. The size of the crime scene (e.g., universal, macroscopic, microscopic scene)

Scenes could be classified broadly according to the type of crime. Remember that not every crime has a physical location of scene. Crimes like extortion, various "white collar" criminal activities, e-crime, driving-under-the-influence, prostitution, illegal activities using the Internet, dealing in controlled substances, and even some simple larcenies, may not have physical locations as "scenes" in the traditional sense. The two major categories of criminal activities that do have scenes are property crimes and crimes against persons. Property crimes are primarily larceny, burglary, robbery, and auto theft. Crime against persons are primarily assault, battery, sexual assault, robbery assault, attempted murder, and murder. Virtually all death cases occurring outside a medical facility have a scene, regardless of whether they are homicides, suicides, accidental, or natural deaths. Generally, crime against a person gets higher priority from police investigators. To some extent, the type of crime suggests looking for certain types of evidence. For instance, one would tend to expect semen or other biological fluid evidence in connection with a sexual assault investigation or scene, but not in connection with a robbery or larceny. As we will note again below; however, it is important to try not to overlook anything at any crime scene.

Scenes can also be classified broadly according to indoor or outdoor and whether they are on public or private property. Indoor scenes have built-in protection from the elements. Outdoor scenes do not and may require special setups to try and protect evidence. Further, evidence can be compromised or destroyed by rainfall, for example, before investigators even have a chance to collect it. Whether a scene is on public or private property, whether it is a private home or apartment, or a place with a lot of human traffic, all have implications for the way the scene is handled. A scene in a public area of any kind may have experienced a lot of human activity before police or investigators even arrive. In addition, these areas may be difficult to secure. Other important matters that will seriously affect any resulting legal case arising from the scene: Do the investigators have the legal right to be at the scene? Do they have the right to seize any physical evidence without a search warrant? Do they have the right to take pictures or to ask people who belong there or who own the premises to stand back? These are all questions with potentially complex answers that revolve around search and seizure laws. Moreover, these laws vary somewhat from state to state. Note that any evidence collected or seized, which is later found to have been collected or seized illegally, will be inadmissible—and that inadmissibility includes all the laboratory findings associated with it. A blood or semen stain that is linked to a suspect through DNA typing will be of no value to the prosecutor if the stain was seized illegally.

INITIAL ACTIONS AND SCENE SECURITY

Police personnel are generally trained to "render aid and assistance" as a very initial step, if this is applicable. Sometimes, crime scene safety is a high priority to protect a victim, suspect, witness, yourself, or your partner from a dangerous condition. An example of a high safety priority is response to an active shooter scene while the perpetrator may still be around. Responding to a potential bombing scene or a chemical spill are additional examples of situations where you need to maintain safety as a primary goal. Once the scene is deemed secure, and injured persons have been assisted, initial responders can turn their attention to scene security. Rendering assistance may involve an EMS response. While taking care of injured victims is clearly the first priority, both police and EMS personnel should make an effort to do as

Table 3.1 General tasks for the first responder of the crime scene

Evidence preservation/collection functions	Other/general scene functions
Scene survey	Scene safety
Scene documentation	Secure the scene
Scene search	Render aid to victim(s)
Collection of evidence	Arrest suspect(s)
Preservation of evidence	Interview witness(es)
Scene reconstruction	Communicate with superiors
Release scene	File report

little damage to the scene as possible. This could involve such actions as taking a single straight pathway in and out, not throwing bandage wrappers around at the scene, not cutting through obvious stab or gunshot wound holes in clothing, putting all bloody cloth together, and so on. These precautions do not waste any precious life-saving time and can be very helpful in preserving scene integrity. First responders should notify their supervisors as soon as possible. The supervisor must normally make the formal notification to the crime scene unit, medical examiner, forensic laboratory, prosecutor's office, and others in accordance with their training and the practices of the department. Table 3.1 provides an overview of the scene tasks.

Establishing control and security at a scene may involve arresting suspects, detaining witnesses, removing people from the premises, crowd control, vehicular traffic control, and so forth, depending on the location and situation. First responders are generally on site because they were called there, and their actions in establishing initial order and rendering aid and assistance are usually not legally questionable. As soon as the exigent emergency situation is resolved, police and investigators have to decide whether any subsequent activities are permissible without a warrant. The answer requires some knowledge of search and seizure law as applied to the particular situation. If in question, investigators should consult with the district attorney for direction.

If it is determined that they have the right to take control of the scene, a security perimeter should be established using tape, rope, or some other barrier. The purpose of scene security is to keep out as many people as possible to preserve the integrity of the crime scene for detailed investigation. The extent to which a scene can be secured depends on where it is and the type of situation. In a residential setting, it should generally be possible to establish security. The more public an area is, the more difficult security may be. In some circumstances, such as on a busy street or highway, it may not actually be possible to truly secure an area, and investigators may have to work around the traffic and activity as best they can. It is important to remember that the initial security perimeter may have to be expanded, depending on what is found or what information is developed.

The box shows a mnemonic designed by a police agency to help remind first and subsequent responders of their duties and responsibilities at a crime scene. The order of actions on the list is not necessarily the order in which they would be done at a scene.

MNEMONIC GUIDE TO PRELIMINARY STEPS AT CRIME SCENES

P roceed promptly and safely
R ender aid and assistance
E ffect preliminary notifications
L ocate witnesses
I nvestigate briefly and secure the scene
M aintain control
I nterview witnesses
N ote all conditions
A rrest suspects as appropriate
R eport fully and accurately
Y ield to continuing investigation

Over the years, various strategies have been discussed by different authors about the best ways to keep out of crime scenes people who have no appropriate role there and might introduce unwanted contamination. Anyone who must obey police directions can generally be prevented from entering scene security perimeters. But police and/or political superiors may be a more difficult problem. Collecting elimination fingerprints from anyone who enters a scene has been suggested as a potential deterrent. The prospect of having to provide a set of fingerprints to investigators might discourage someone from actually going into the scene. Today, collecting a buccal swab from anyone entering a scene as a DNA profile elimination standard could also be in order and might provide additional disincentive. Taking a few initial photos or shooting a few minutes of initial video by one investigator to bring out and show authorized people is another potential strategy.

First responders and/or follow-up investigators should gather the names and contact information of witnesses and others who may have information. They should also make notes about other potentially relevant matters, such as vehicles parked in the area.

It is worth noting here that some crimes can have more than one scene. For example, a victim might be assaulted at one location, forced into a vehicle, then taken to another location and assaulted further. Here, there are scenes at both the locations, plus the vehicle is another scene.

It is also worth distinguishing between a crime scene and what is often called a dumpsite. The primary crime scene is where the initial actions took place, and the majority of physical evidence is expected to be found there. A dumpsite usually refers to a secondary location where a body has been left, often well after and some distance from the criminal events. Less physical evidence may be expected at a dumpsite compared with that at an actual crime scene.

STEPS IN SCENE PROCESSING AND ANALYSIS

The steps in crime scene processing are briefly described below and discussed in detail in separate sections.

Scene survey and evidence recognition

Once the scene is under control (as much as it is going to be), any injured persons have been properly cared for, and the area is safe, the first action is to conduct a quick scene survey with special attention paid to potentially transient evidence, pattern evidence, and conditional evidence. The second aspect of this survey is recognition of the potential physical evidence. Recognition is the hardest thing to teach, because it depends so much on experience and is least amenable to any prescribed protocol. There are always a lot of things at crime scenes. The objective is to figure out what is relevant to the investigation and what is background.

Documentation

The next step is documentation. There are three or four different methods of documentation that must be used to insure a thorough record. The methods commonly used are: notes, sketches, photographs, and video. There should be a detailed record of the scene itself and of all relevant evidence that was recognized. A perfectly documented scene would enable someone to reconstruct (at least in their minds) every detail of the scene in proper perspective at some later time. One reason documentation is so important is that there are many patterns, which cannot be collected (in the "bag and tag" sense), but can only be documented. Documentation is one of the key duties and responsibilities of crime scene investigators.

Scene searches

There are several commonly used "formula" methods for conducting searches. They are discussed below. The method chosen depends on the type of scene and how much area it covers. A large outdoor area thought to contain a shallow grave would not be searched the same way as a house, for example. The primary consideration is to be thorough. Crime scene processing is usually a one-time thing—many times there are no second chances. It is often said, "Two searches are better than one." The point is that investigators want to ensure that the scene has been completely searched before releasing it.

Evidence collection and preservation

Once the documentation steps have been completed, the physical evidence items recognized as relevant and appropriate for collection, packaging, and preservation, can be placed into the appropriate, properly labeled containers.

Preliminary scene analysis and reconstruction

Once the scene has been properly searched and documented, investigators should conduct a preliminary scene reconstruction to determine the type of crime, the point of entrance, and point of exit. What types of evidence should be present at the scene? What evidence has been removed from the scene? Is the scene normal, or are there signs of struggle, fighting or staging? It is also important to try to find the answers to the six "W"s –what, when, where, who, how, and why.

Release of the scene

When investigators are satisfied that they have thoroughly documented the scene and associated evidence, and that they have recognized and collected where appropriate every relevant item, the scene can be released. Once a scene has been released, it is usually not possible to go back and collect anything else, because whatever it is, it may have gotten into the scene between the time of release and the time the scene is revisited. There is no longer a direct chain of custody from the original scene. Evidence found and/or collected after a scene is released could be later ruled inadmissible by a court.

SCENE SURVEY AND EVIDENCE RECOGNITION

As noted above, this step is the most dependent on the experience and training of the crime scene investigator. It may be fair to say that the ability to recognize potential evidence separates the most successful crime scene investigators from the crowd. Besides experience, the investigator should use a scientific approach and conduct a preliminary scene survey to recognize and identify the potential evidence. Table 3.2 shows this logic. Analysis of the scene and formation of an initial working hypothesis of what happened is the first step. There may be a few alternate versions of the initial hypothesis. As observations are made and data are collected, the hypothesis should be revised and refined to incorporate all the data. The data also include witness statements and other investigative information, but investigators should gauge the reliability of this information in giving it weight in the formation and revision of the hypothesis.

It is important to keep an open mind as the investigation proceeds no matter how "open and shut" the case first seems. Many cases end up developing problems later because the investigators initially locked on to one theory of the case. They then have to turn reality wrong side out, because subsequent facts or findings fail to support the theory and it is considered too late to backtrack.

At the scene and in the early stages, the working hypothesis or theory should guide investigators as they look for evidence and look for a proper direction for the investigation. Emerging findings and facts should cause investigators to continuously test and revise the working theory. It is important to realize that the final version of the theory cannot be formulated until investigators, the forensic laboratory, and the medical examiner (if it is a death case) have all completed their work.

Sometimes, it is necessary to conduct some experiments to refine the theory to its final form. Blood spatter is one example of a pattern that investigators and/or forensic scientists may have to try to replicate before deciding what may have happened at a scene. Gunshot residue (powder residue) patterns are another. Note that doing an experiment and replicating the pattern does NOT show that the experimental protocol is WHAT ACTUALLY HAPPENED at the scene. It only shows that the patterns COULD HAVE been formed in that way, and that the experimental protocol represents a reasonable, demonstrable, scientific basis for this element of the theory.

Table 3.2 Crime scene analysis and the scientific method

The scientific method	Crime scene analysis/reconstruction
Data/observations	Data/observations/witness statements
Hypothesis	Hypothesis
Experiments	Testing hypothesis against the scene pattern information/investigative information
Hypothesis testing	Further testing hypothesis against the lab and/or the medical examiner results
Theory	
More experiments	Possible experiments concerning evidence
More theory testing	Formation or origin
Best theory	Reconstruction (best theory)
Natural law	

SCENE SEARCHES

As noted earlier, there are several methods for conducting searches. Figure 3.1 shows four different methods: zone, grid, strip, and contracting and expanding spiral.

Zones and grids may be best for larger, outdoor areas. Strip searches can be used anywhere but are probably most effective in places without good visibility, such as in an area with a lot of foliage. The contracting or expanding circle and spiral patterns are applicable to a few specific situations. The zone method is ideal for indoor or vehicle searches. You might say that the "link" search is driven by the evidence. Finding a dead body that has a gunshot wound suggests looking for a gun. If the thinking is that someone from outside the location came in, did the shooting, and left, you would next logically be looking for pathways of entry and egress. You might say that "link" searches are driven by the working hypothesis of what happened in the case. Investigators should keep in mind that there is a possibility the working hypothesis is wrong; however, and that would suggest doing a systematic search of the scene before releasing it to be sure not to miss anything.

DOCUMENTATION

The classical methods for crime scene documentation include notes, sketches, and photography. They are all still necessary. Today, we can add audio recording as a potential means of taking some notes or recording information, and video recording as an additional method of scene documentation. Note that video recording does not replace or serve the same purpose as still photography.

Every piece of documentation should include the following information: date, time, location, case number, person making the record, and, in the case of evidence items, a description of the item and the item number assigned to it. In the case of sketches, there may be a sketcher and a measurer. Both names should be listed.

Notes

Notes are a way of documenting anything that cannot be photographed, sketched, or video recorded. However, it is important to take notes about everything at the scene, including things that will be sketched, photographed, or videotaped. Investigators typically take too few notes. Some of the crime scene "logs" discussed below may be considered notes. Investigators should not consider photography, sketching, or videotaping a substitute for good note taking.

The importance of first responders making notes on initial scene conditions cannot be emphasized too strongly. Often, the initial responders are first on the scene, and the only ones who observe it close to its original condition.

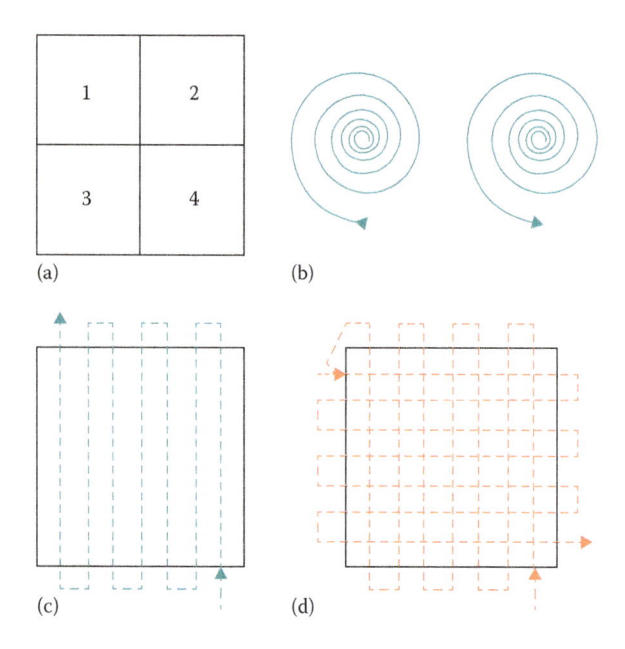

Figure 3.1 (a) Zone or quadrant, (b) contracting or expanding spiral, (c) grid or strip, and (d) double grid.

Little details can become important later. Are doors locked or unlocked, open, or closed? Are windows open or closed? Are there any odors or other ephemeral features? First responders should make notes about the scene as soon as possible after they arrive while everything is still fresh in their memories.

Notes are often used as the basis for preparing a more formal written report later. All investigators should insure that their notes support all the statements made in their final written report. Notes should be kept in bound notebooks and retained at least until the case has been fully adjudicated.

There are several logs that should be made at crime scenes, and these may be considered part of the notes since they are written documents that record information. It is good practice to have a security log, showing who was at the scene and approximately when they came and went. Some authorities go so far as to suggest that people should sign in and out every time they enter the secured area. Others have suggested that a continuously running video record of the overall scene and the activities taking place within it can serve as a security log. A photo log should be kept, and should, at a minimum, record what photos are taken and by whom. Some authorities have suggested keeping detailed records of the photographic settings used for each photo. Digital cameras used in fully automated mode render that sort of record keeping moot. The cameras can also be used manually; however, and then it would be possible to record all the camera settings. The most important thing is to have a good log of all the photos taken. Along with the photo log, an evidence log should also be kept. A scene checklist should also be used to ensure that all major tasks are completed before the scene is released.

Sketches

Sketches show the exact locations of all the buildings, streets, and permanent fixtures, as well as major items or characteristics at the scene including the physical evidence that will be seized. Sketches provide information that photographs and videos cannot, because of the distortion of distances and locations caused by the location or perspective of a photographer/videographer.

There are two general types of sketches: rough, or preliminary; and smooth, or finished. Figure 3.2a–c show examples of each. Rough sketches are made at the scene, during processing. They are usually not to scale and contain

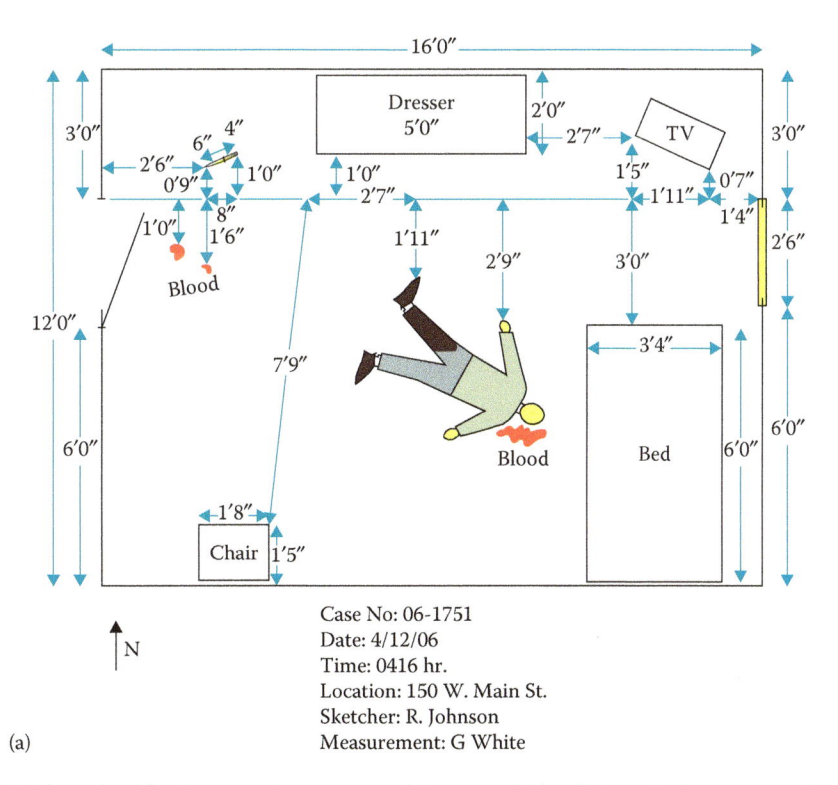

Figure 3.2 (a) Rough sketch made while at scene using a commercial app on a table with important measurements and descriptive data, *(Continued)*

(b)

1	Body
2	Dresser
3	TV Set
4	Bed
5	Chair
6	Knife
7, 8, 9	Bloodstain

Case No: 06-1751
Date: 4/12/06
Time: 0416 hr.
Location: 150 W. Main St.
Sketcher: R. Johnson
Measurement: G White

(c)

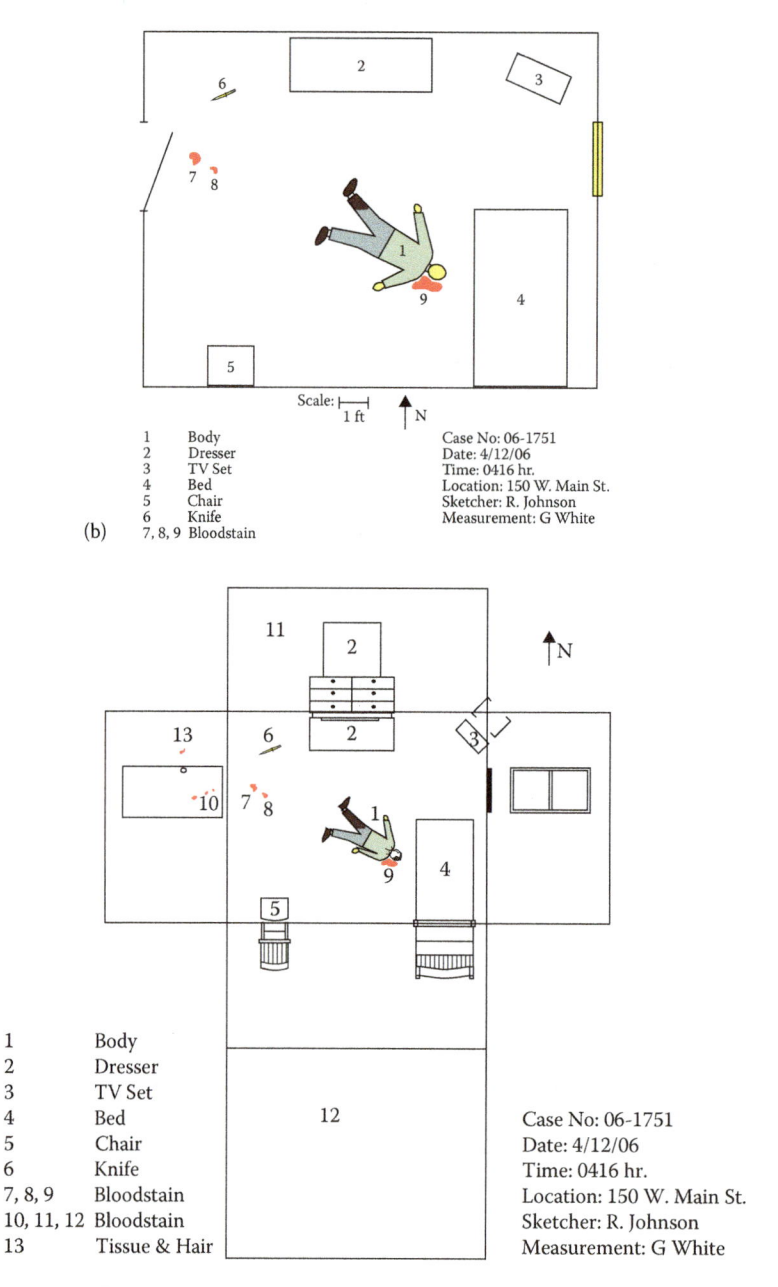

1	Body
2	Dresser
3	TV Set
4	Bed
5	Chair
6	Knife
7, 8, 9	Bloodstain
10, 11, 12	Bloodstain
13	Tissue & Hair

Case No: 06-1751
Date: 4/12/06
Time: 0416 hr.
Location: 150 W. Main St.
Sketcher: R. Johnson
Measurement: G White

Figure 3.2 (Continued) (b) more refined sketch prepared later containing the same material in a more clearly presented form and with nonessential features eliminated, and (c) larger view sketch prepared of the same material scene. This is a cross projection sketch, presenting features in a three-dimensional environment, such as a room, to be presented in a two-dimensional sketch.

measurements within the sketch. A smooth sketch is prepared later from the data contained in the rough sketch. Smooth sketches are drawn to scale. The scale should be indicated on the sketch. There are no measurements in the smooth sketch, because it is drawn to scale and it omits items or objects that are not relevant to the case.

Although we talk about crime scene sketches as if there were one sketch, in fact, there will probably be a whole set of sketches showing different scales and perspectives. One sketch might show the location of a house in relation to the street, for example. The next might show the whole house floor plan. Another could show a second story or basement floor plan. Another might show a floor plan of the garage. And there will be sketches that show the details for the rooms or areas in which most of the action occurred and/or in which most of the evidence seems to be located.

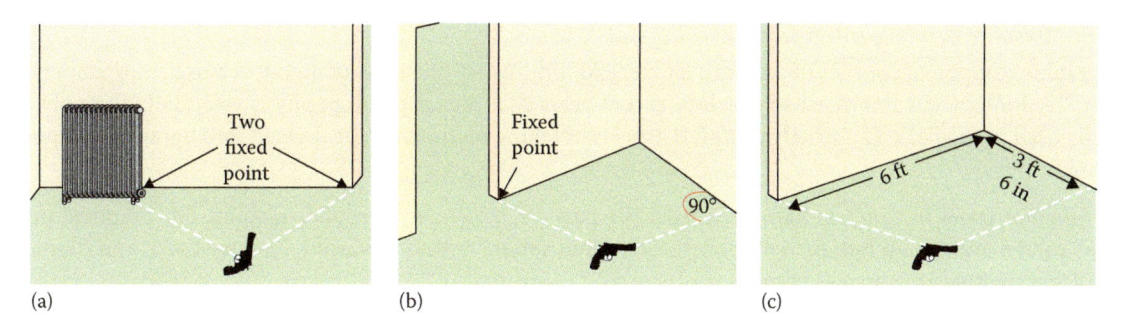

Figure 3.3 (a) Location of an item (gun) using to fixed points in the scene, (b) location of an item (gun) using a fixed point and an angle at the scene, and (c) location of an item (gun) using measurements on an XY axis drawn in the scene sketch.

There is a type of sketch that can be used to show three-dimensional character and features, if necessary. This type of sketch is called a *cross-projection* sketch and is most easily envisioned using a rectangular room as the scene. Think of the room as a box. Then think of slitting open the seams of the box at its edges and folding it open to make a pattern of flat rectangles, each representing a wall, the floor, or the ceiling. Figure 3.3 shows an example. This type of sketch could be used to show the location of bullet holes or blood patterns on walls or a ceiling, for example.

Although many sketches will be "floor plans," it is sometimes necessary to sketch and record a vertical dimension; that is, how far from the floor something is located. A blood pattern on a wall is one example.

In order to make a sketch, the dimensions of the scene have to be measured. In addition, methods have to be used to determine the exact location of objects and evidence items. There are three methods for taking measurements that then provide an exact location for objects. We can call them: triangulation on two fixed points; single fixed point with 90° walls; and the XY axis method. Figure 3.3a–c provide examples of each type.

The triangulation method may sometimes be the only useful method at an outdoor scene. The XY axis method requires that some fixed point be selected to represent the starting point (X = 0, Y = 0). Sketch preparers should be careful to select objects as fixed points or features that are not likely to move or disappear in case reference has to be made to them later.

It isn't necessary to be an artist to make a good sketch. The point is to record the scene on a scale drawing. Various vendors sell templates to help sketch preparers draw objects, like furniture, into the sketch. When doing the finished sketch, the preparer should pick a scale that is logical for the area being sketched and for the level of detail that is necessary. The same scale will probably not be used for all the sketches in the series. For example, you might be able to show the street, the house, and the driveway on one page with a scale like 1 inch = 25 feet. But in a sketch of a room, whose real-life size is 8 × 12 feet containing a body and many items of evidence, you would more likely to use a scale like 1 inch = 1 foot. Information that is not to scale may sometimes be included in a finished sketch if it is relevant.

Every sketch should show magnetic north on the drawing. A compass is an essential item in a crime scene processing kit. That way, all references to direction can be in absolute terms, rather than saying "left, right," and so on.

There are now some computer-based accident and crime scene sketch programs available to assist with producing accurate diagrams to scale. These require a bit of learning but can make sketching much easier. Hand notes should still be kept and used to verify the accuracy of the sketch produced. The cost of global positioning satellite (GPS) locators has also come down in recent years and could be useful in producing scene sketches with absolute location references.

Photography

Crime scene photography can be thought of as having two aspects: technical and forensic. The technical aspect has to do with a person's understanding of and skills in photography. The forensic aspect has to do with making good decisions about which pictures to take to accurately and appropriately capture the necessary details of the scene and physical evidence.

Technical aspects

Photographic equipment consists of cameras, lenses, and film. Today, the issue of digital versus analog cameras is also a factor. Technology has improved so much in recent years that digital photography is essentially the only method currently used in the field. The strictly technical aspects of taking pictures can be divided into lighting, sharpness, and exposure.

Many different types of cameras have been used the past for in crime scene photography. For practical purposes, cameras can be divided up into five categories: "point and shoot," instant Polaroid, 35 mm, 4 × 5, and digital. All of them, except digital cameras, used film as the recording medium.

Resolution is a major factor in digital photography. It is generally stated in megapixels. The higher the number, the higher the theoretical resolution (and probably, the more expensive the camera). Individual image resolution is generally stated in a "A × B" pixel format, where A and B are pixel numbers. The higher the numbers, the greater the resolution and the larger the image. Higher resolution images can be printed in larger and larger sizes without loss of clarity. At the higher numbers, there is less "compression." There are also various formats for digital images, such as .gif, .jpg, and .tif. Digital photographers need to understand these formats in terms of how much they "compress" the original image (with a potential for loss of detail), how large they are, and so forth. Programs like Adobe Photoshop can convert digital images from one format to another, as well as adjusting many properties and parameters of the image.

Digital images are typically stored in the camera on some sort of card that slides into a slot. These cards come in different total capacities, usually stated in megabytes (mb). The more images or the larger the images one wants to store at a time, the larger the capacity of the card should be. Images from the card can be downloaded to a computer by way of a card reader or directly from the camera by way of a cable. Generally, digital images are fairly large and occupy quite a bit of disk or storage space. This factor must be considered if many digital images need be stored for long periods. Another important consideration with crime scene images is backup provisions, in case something happens to the original images. There has been the suggestion that the integrity of digital images will tend to be suspect because they can be manipulated using computer programs such as Photoshop. Digital image and photography experts have made great strides in developing methods to ensure that the images are original and detect images that have been tampered with. Use of these strategies is very important in digital crime scene photography, because any manipulation of an image could potentially be viewed as an attempt to misrepresent evidence. There may be circumstances; however, where computer-based manipulation of an image—for example, to improve the contrast between a footwear impression and the background—would not really be any different than using traditional photographic techniques to do the same thing. Courts in the U.S. have consistently admitted photographs with such manipulations. The important factor in such a situation would be honesty about what was or was not done.

As noted above, lighting, sharpness, and exposure are the key technical elements in taking good photographs. They are equally important.

For crime scene photography, front lighting is virtually always preferred. Back and side lighting are, generally, reserved for special situations. That is one of the reasons we suggest using flash all the time. In low light situations, the flash should provide adequate illumination for a good image. Even in daylight, shadows can be a problem (such as taking a picture of something in the shade under a tree). The flash tends to "fill" in the shadow and correctly illuminate the subject of the photo (hence the term "fill flash"). Back lighting is almost always a problem. In extreme situations, like having to try and take a photo looking into a setting sun, you may have to wait until the sun has set to get the picture. An important exception to the "front lighting is best" rule is in photographing three-dimensional impression markings (indentations), such as footwear or tire impressions in soft earth (Chapter 5). Here, side lighting (usually with a flash unit) provides the best image, because a slight amount of shadow illuminates the pattern detail (Figure 3.4). A series of exposures with the flash at different angles is generally recommended. Photographing reflective surfaces also requires special lighting techniques.

Sharpness of a photographic image is a function of two factors: having the camera properly focused and holding the camera still while the lens is open. Many automatic cameras have auto-focus and, with normal objects and conditions, can focus without operator intervention. But, if there are problems, the auto-focus may have to be turned off, allowing manual focusing. The second factor is holding still. This factor becomes important when exposure times are slower. Under those circumstances, it is recommended to use a tripod.

Figure 3.4 The use of side lighting to produce shadows that can improve the visibility of a three-dimensional pattern (footprint).

Exposure is the most complicated element, because it is a function of two parameters that are separately adjustable on non-automatic cameras: the *f*-stop and the exposure time. The *f*-stop, or *f*-number, indicates the amount of light that will be allowed to reach the film or image capture device by controlling the lens aperture. The exposure time is the amount of time the lens is open and is usually expressed in fractions of a second. The larger the lens opening, the smaller the exposure time can be to admit the same quantity of light. One other factor to take into consideration is called *depth of field*. Depth of field is greater as the *f*-number is greater (i.e., as the lens aperture is closed more and more). You can think of depth of field as a range of distances from the lens where everything will be in focus. Narrow depth of field might mean that only objects between 6 and 12 feet will be in focus, where wider depth of field might have almost everything in view in focus. How much depth of field is needed for a picture depends on what is being photographed and what needs to be in focus. A picture depicting an overall view of an outdoor crime scene should have a lot of depth of field—everything should be in focus. On the other hand, depth of focus may be less important in photographing a bloody footwear imprint on a flat surface. The photographer needs to choose the *f*-number partly out of consideration for depth of field.

Forensic aspects

The forensic aspects of crime scene photography mainly cover selecting the correct subjects and objects to photograph in order to do a good job of documenting the information at the scene. This will vary for different types of evidence and from scene to scene, but there are some guidelines that can be followed.

It is generally good practice to photograph the overall scene and sub-scenes proceeding from the wider to the closer perspective—overall, midrange, close-up, as is often said (Figure 3.5a–c) Sometimes, aerial photos of an overall scene are needed, and these can be taken from an aircraft, from atop a "cherry picker" or from a drone. In some less extreme situations, the use of a stepladder can improve the perspective.

Figure 3.5 (a) Highway crime scene encompassing photo of car and body, (b) highway crime scene intermediate distance photo (c) highway crime scene close-up of involved auto license plate.

Figure 3.6 Expended cartridge case in car on driver's side with evidence number marker.

Once the overall scene and sub-scenes (such as smaller areas, rooms, etc.) are photographed, the evidence must be photographed. First, the evidence items that have been identified for collection and packaging should be photographed in their original location (Figure 3.6). A label, number plates, or markers may be placed next to an item at a scene.

The photographer should take a picture with and without this marker (the one without the marker is to document the item in place [Figure 3.7a] before any alterations took place). These photographs, along with the sketches, should permit the original location of any seized item to be reconstructed accurately.

It is very important to have good photographs of patterns that will not or cannot be collected or otherwise preserved. These may be "crime scene" patterns (Chapter 4), or they may be imprints or impressions that will not or cannot be actually collected. For patterns like tire tracks, footwear impressions, the photographer should take care that the film

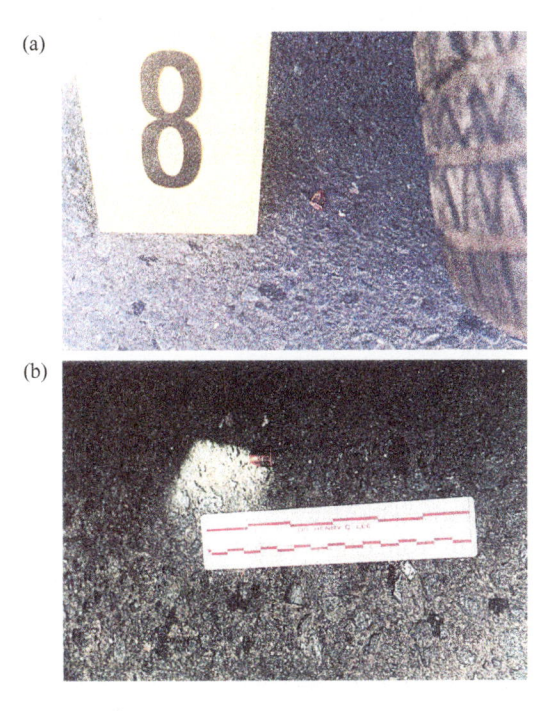

Figure 3.7 (a) Fired bullet on ground near tire with evidence number marker and (b) fired bullet on ground near tire with scale in place.

plane of the camera is parallel to the plane of the floor or ground. A tripod should be used for these kinds of shots. Multiple photos and bracketed shots are a good idea for these evidence items to ensure that there is a good picture.

There are a few other guidelines. For most evidence photos a scale (ruler) should be included in the picture (Figure 3.7b) to show the size of the item and, in some cases, to show how far it is off the floor or ground. It is again good practice to take a photo of an item without the scale and another one with the scale. Using flash, which we recommend, tends to "wash out" the numbers and markings on white scales. Scales that are gray, yellow, or any color, are available to avoid this problem.

As we noted above, the date, time, location, photographer, and so on should be noted. One strategy for "labeling" board containing the basic data and a label multiple storage cards or discs sequentially.

Video recording

Video recording of crime scenes has been feasible for a number of years since the technology became affordable. There are several different types of video cameras available, mainly differing according to what type of media they use (VHS, 8 mm, etc., or digital). Digital recording in regular or HD formats has essentially replaces other formats for video documentation.

Video recording has several potential roles in crime scene documentation. It can be used as a stationary "monitor" of all the people and activities at the scene; acting as a sort of activity and security log. More often, a scene is video recorded as another means of documentation. This type of video can be used to show others, who weren't there, the overall layout of the scene, evidence locations, patterns, and so on. "Others" could be forensic scientists at the laboratory, pattern analysts, and sometimes the jury at a trial.

Some considerations in crime scene video recording include whether to narrate, whether to have the date/time stamp turned on in the image, and the amount of video to take.

Some authorities suggest no narration and not having a live microphone during video recording. Their primary argument is that the microphones are too sensitive and pick up comments and remarks that may later seem inappropriate or insensitive. The argument for narration is that it orients the viewer and helps in understanding the video record. The narration should be slow and monotonic. References to direction should be absolute (north, south, etc.).

If the date/time stamp is on, it should be adjusted to be correct. The clock should also be synchronized with other watches at the scene so that all recorded times are consistent.

How much video is taken depends on the primary purpose. If it is used as a silent "monitor," there may be many hours of tape. As a straight-scene documentation device, one to several hours of tape may be shot. The longer the camera must be held, the more the videographer may need a tripod or monopod to help steady the camcorder.

Generally, the "overall—midrange—close-up" formula can be followed with video, just as it is with still photography. The videographer should avoid rapid camera movement and excessive zooming in and out, because these things detract from the quality and usefulness of the video record.

Rapidly advancing recording techniques may mean that "virtual reality" holographic scenes captured forever may become routine.

Audio recording

Usually, notes are written, but can also be taken using an audio recorder, especially during the initial survey stage. With an audio recorder, an investigator can document the original condition of the scene in a speedy fashion and without damage to the scene or delay the investigation. Many recording devices are commercially available. A hands-free, lapel-type recording device with a long-life battery and light weight is ideal.

New technology for crime scene recording

Three-dimensional (3D) imaging and scene mapping

Two-dimensional (2D) scene diagramming and 3D imaging software have been designed by various companies, such as 3rd Tech, Leica, and others, to obtain accurate photographic documentation and measurements using the latest laser and digital camera technology. Each area of a scene can be scanned to obtain data quickly with basic

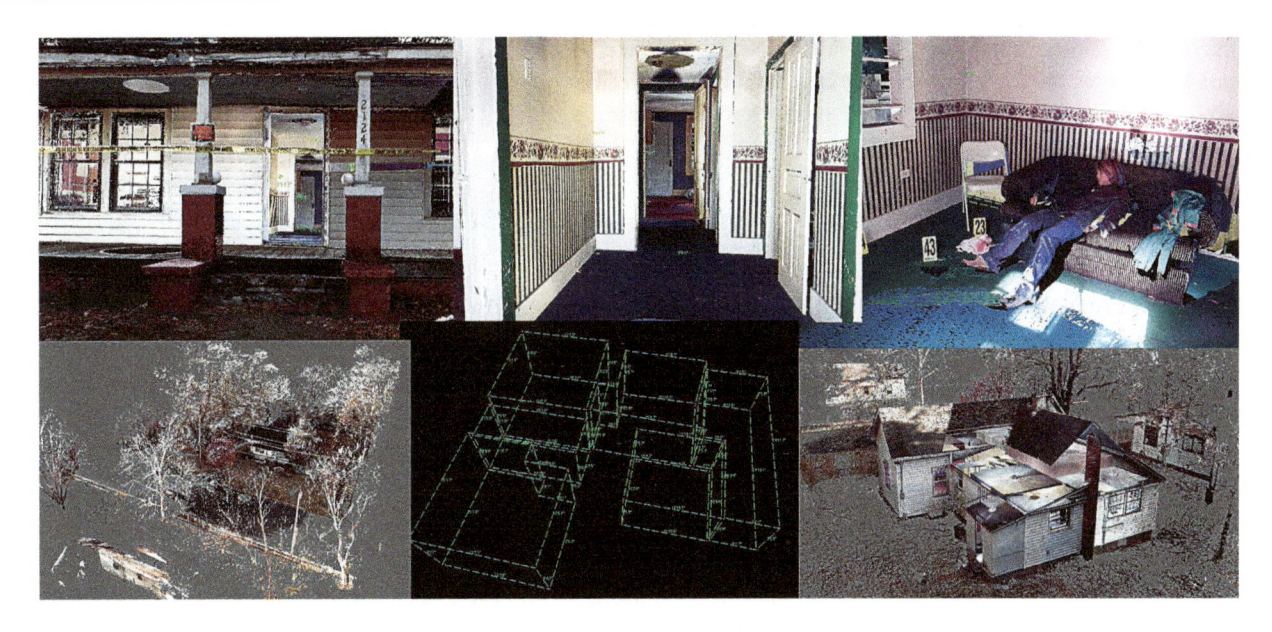

Figure 3.8 Computer generated crime scene images. (Image used with permission courtesy 3rdTech Inc. – www.3rdtech.com.)

interactions by scene investigators. This data is then used to build computer graphics models, determine measurements, and demonstrate patterns at the scene (Figure 3.8). Capturing a complete scene or object often requires scanning from more than one location to see both sides of an object or part of a surface behind another object. The software provides functions to quickly and easily align scans to produce a single scene model and fast measurement functions for finding the distance between any points in a scan, perpendicular distances, angle measurements, and intersections of lines and surfaces. The value of these systems lies in the completeness of the data obtained and the ability to obtain measurements or other data after the team has left the scene. The use of the scanned images allows any viewer to essentially revisit a crime scene at any time after the images have been taken. Many 3D laser scanning devices have been developed and can be used for scene documentation, reconstruction, visualization, and evaluation of scenes for homeland security and other related areas.

Total station mapping technology

The system uses an absolute polar coordinate system of measuring in comparison to the base line coordinate or a triangulation system. Polar coordinate refers to fixing the location of evidence by an angle and radius. The instrument transmits an infrared laser beam, which is reflected off an object or a mirrored prism. The time it takes for the beam to travel is converted into units of spatial measure by an on-board computer. Distances, angles, and other information so gathered can then be stored in a data collector and later downloaded into computers loaded with software designed to convert measurements into maps and other analytical reports.

Many 3D laser scanning devices have been developed and can be used for scene documentation, reconstruction analysis, and 3D visualization.

Computer aided drawing technology

Computer-aided drawing tools (CAD) can produce accurate and clear crime scene diagrams, which are especially useful for courtroom and reconstruction purposes. These systems are vector-based programs that use the investigators' measurements to accurately place significant items of evidence, patterns, and other scene markers. The relationship between items can be clearly shown as well.

Many of these programs contain libraries of pre-drawn objects, such as furniture, bodies, and building structures, which the investigator can place in the crime scene diagram. Data that is collected via total station or other digital 2D and 3D systems can also be imported into many of these software packages.

CAD systems are particularly useful for the diagramming of accident scenes. Additional features of the software include materials for the calculation and positioning of vehicles and related objects. CAD can also generate useful information, such as bullet trajectories and distance, which makes this system useful for visualizing various hypotheses during a crime scene reconstruction examination. These programs also give the investigator the ability to view evidence and objects from different viewpoints, which can help to support or disprove a witness's statement.

Crime scene data analysis computer software

Many comprehensive software programs have also developed. These computer software programs can be used to compile data obtained from crime scenes and can also be used as forensic data acquisition tools. These computer programs automatically sort and correlate the data to give the investigator actionable insights to the crime scene information. This technology will compile all the information and display it in various formats. It is a great time saving technology.

Drone technology

Drones have become an extremely useful and fairly inexpensive mechanism for many useful functions. There are a number of crime scene applications for drones that are coming into use and some probably will become routine. The most obvious application is taking aerial photographs of outdoor crime scenes (Figure 3.9), which has not been practical in most cases by other means. Certainly, a drone with an attached digital camera will make such applications routine. Vendors have already begun offering such packages and even video cameras for covering a significant area and capturing motion. Additionally, infrared cameras are available that could be used in low light situations and even at night. In fact, infrared imaging is well known for giving extra sensitivity to fresh biological stains and could be useful for locating such stains that might not be easily visible at ground level.

Potentially, smaller drones could be used in many indoor scenes to obtain an overhead view that is not otherwise possible. The small drones can easily operate in many reasonably high ceiling situations indoors. Of course, in scene that are too dangerous to enter, a drone could give information without endangering personnel. They would be of great use for arson or explosive scenes where there is significant damage, which might require engineering evaluation before human entry is allowed. Also, where there might be a perpetrator on scene the risk to personnel could be reduced and perhaps the drone imaging capabilities might aid in the search for possible fugitives.

Duty to preserve

In connection with the products of crime scene documentation (notes, audiotape, sketches, photographs, videotape, digital video, or images), it is important for investigators to realize that there is generally a duty to preserve these materials and recordings for a long time. How long depends on the individual jurisdiction, but, at a minimum, most jurisdictions want the materials preserved until the "case has been adjudicated." If this means the first trial that

Figure 3.9 Drone in the air over a staged scene.

yields a verdict, the time could amount to a few years. If it means until all the appeals in the case are exhausted, the time could exceed the investigator's life span. The courts have fairly consistently taken the position that the original documentation should be preserved for examination by the defendant's counsel and/or experts, and for later re-examinations of the case on appeal. Another important reason for preserving documentation and records is the increasing revisitation and reinvestigation of cases by "cold case" units.

Thus, arrangements are needed for storage and archiving of various crime scene work products.

COLLECTING AND PRESERVING PHYSICAL EVIDENCE

After scene and evidence documentation is complete, collecting and preserving physical evidence is the next step in crime scene processing.

Generally, physical evidence is collected and preserved because it will be submitted to a forensic science laboratory for analysis. Some physical items may be collected for other reasons. An example might be suspected stolen property. Some patterns may not be physically "collectible," so adequate documentation and preservation is particularly critical.

At death scenes, the medical examiner or coroner, generally, takes charge of bodies and some of the items associated with the body (clothing, etc.). Investigators should follow departmental procedures in notifying the medical examiner of a suspicious death and abide by accepted protocols in obtaining evidence associated with bodies. Often, detectives may attend the autopsy to see if the medical examiner can determine a tentative cause and manner of death and to provide the medical examiner with additional information about the scene. Final determinations may require completion of the investigation, microscopical tissue examination, completion of toxicology tests, and so forth. Clothing and other items associated with the body are generally turned over to investigators by the medical examiner for submission to a forensic science laboratory. In some cases, a sexual assault evidence kit may be taken post-mortem, as well as trace evidence on the body and sent to a forensic science laboratory for analysis.

Collection methods

There are several methods available for collecting physical evidence at scenes. The first, and the one that we recommend whenever possible, is collection of the evidentiary item intact. Collection of intact items is possible with many items and objects. Sometimes, such as with evidence on floors, walls, or other immoveable things, it may not be realistic to collect the entire item intact. In those cases, investigators have to use sampling methods—that is, a sample of the evidence must be removed from the item on which it is located (often called the "substratum"—the plural is "substrata"). These methods include:

- Use of forceps
- Tape lifting
- Shaking
- Scraping
- Vacuuming

Forceps and tape lifts, and occasionally, vacuuming may be used in the field (at a scene) under appropriate circumstances. But it is not advisable to shake or scrape items outside the laboratory, because traces of evidence could easily be lost. In connection with blood/physiological fluid stains, cutting, swabbing, or scraping can be used, as explained in detail in Chapter 10.

If investigators use forceps, they should be sure that the forceps are clean and, if necessary, new forceps used for each evidence samplings to avoid any contamination. The word "sampling" implies that there are many items or a lot of evidence, and that the investigator is going to collect a representative sample. The key word is "representative," and this activity may require some experience and judgment. Sometimes, all the evidence may be collected, as in the case of one or a few fibers or hairs. In other cases, sampling will be necessary, as in the case of a medium velocity blood spatter. The evidence pattern or deposit pattern may help guide investigators in their sampling.

Tape lifting can be a useful method for collecting evidence. It has the advantage of being thorough. But investigators should understand that the tape can sometimes cause problems for laboratory personnel trying to remove the evidence from the tape for examination.

Shaking and scraping are different versions of the same thing and, generally, apply to trace or materials evidence on clothing, or other similar items that can be processed in this way. This method should be used in the laboratory by forensic science personnel.

Vacuuming should be considered as a last resort (see Chapter 14). Here, a vacuum cleaner hose is fitted with an in-line filter device in which the filter can be readily replaced and cleaned thoroughly in between uses. The principal drawback of vacuuming is that it collects every bit of trace material deposited on the item, much of which probably has nothing to do with the case. A trace evidence examiner is then required to sort through hundreds of items on the vacuum filter to try and figure out what may be relevant.

It is good practice to thoroughly document the location of any evidence on an object or item before any collection technique is employed.

Numbering and evidence description methods

Items collected at scenes or in the laboratory are given numbers and brief descriptions on the packaging and in the evidence log. There is no universal rule for numbering evidence. Each investigator or department should develop a thorough and consistent protocol and stick with it. Obviously, the numbers and descriptions on the packaging should match those on the evidence log. It is best to use a numbering system acceptable to the forensic laboratory, which will receive the evidence, thereby avoiding the lab having to renumber the items.

Signs or markers may be placed next to evidence items at scenes to help document their original locations. The number on the sign or marker must also match the number used on the packaging and in the log.

The identity of many items being collected at scenes is obvious, such as "beer bottle," "knife," "pair of shorts," and so forth. In those cases, investigators should just use the name of the item as a description. It is wise to avoid adjectives that impugn value to an item in an evidence description—thus, "yellow metal ring" rather than "gold ring." With some items, investigators may not actually know the identity, such as with a suspected dried bloodstain. In those cases, descriptions like "blood-like substance," or "reddish-brown stains," can be used to avoid being challenged in court about how the investigator "identified" the material or substance.

Evidence packaging should include the number and description of the item, as well as the case number, date, time, and name of the collector.

Packaging options

Most evidence packaging is common sense, but there are a few principles to be followed. The majority of evidence items will be packaged in paper containers or evidence bags. The size of the container should be proportional to the size of the evidentiary item.

For most small items, particles, and objects, we recommend a "druggist fold" package as the primary container. A **druggist fold** is nothing more than a way to fold a piece of square or rectangular paper, so that it forms a leak-proof container for particle or powder type material. It can also be used as a primary container for fibers or hairs. Figure 3.10 shows the making of druggist fold step by step.

Any piece of paper can be used for a druggist fold, and the size should be determined by the quantity of evidentiary material to be packaged. The use of laboratory weighing paper or clean, white paper for evidence collection is recommended, if it is available. This paper tends to resist having particles and powders stick to it or absorb into it. Although the druggist fold is intended not to leak, small evidentiary items like powders, particles, or fibers may work their way out as the package is handled and transported. Accordingly, the druggist fold should not be used as the **only** container of the evidence. The druggist fold containing the evidence should be placed in an appropriate *secondary* container to ensure that the evidence will not be lost. For blood or physiological fluid evidence, the secondary container must be paper and must not be airtight (see Chapter 10). For non-biological evidence (see Chapter 14), the secondary container could be a plastic "zip lock" bag.

Plastic "zip lock" bags are suitable for many types of solid evidence items. The only major exception is biological evidence, which should be packaged in paper containers after thorough drying to avoid putrefaction and degradation. Plastic containers have the advantage that the contents can be seen without opening the container. Packaging is usually sealed at the scene with tamper proof evidence tape, and not opened again until a laboratory examiner does so.

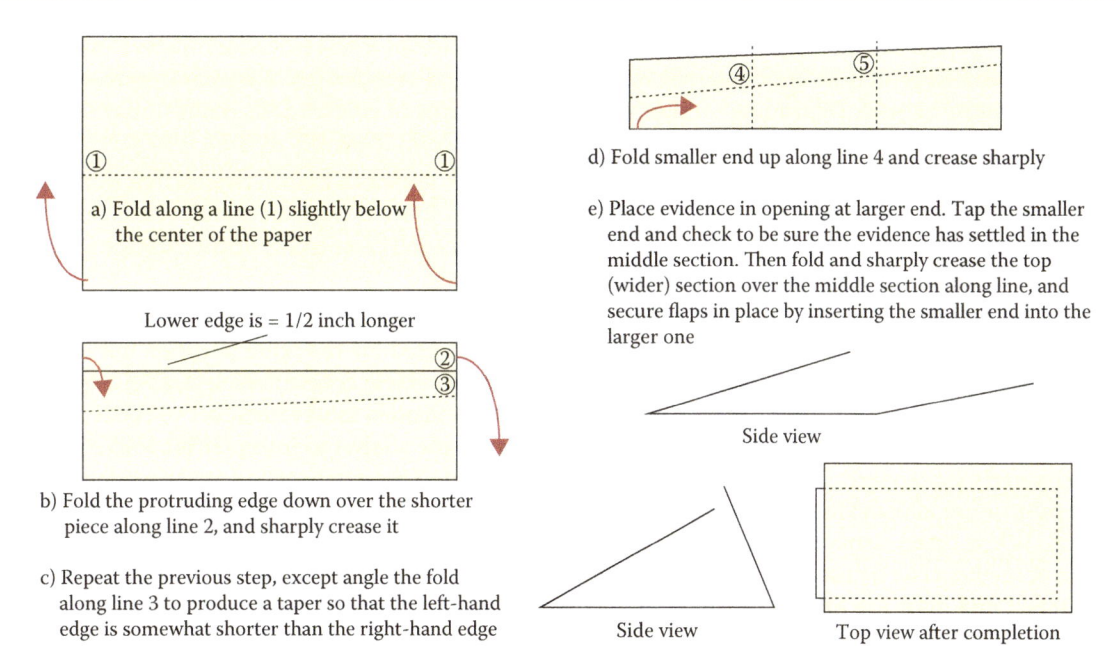

a) Fold along a line (1) slightly below the center of the paper

b) Fold the protruding edge down over the shorter piece along line 2, and sharply crease it

c) Repeat the previous step, except angle the fold along line 3 to produce a taper so that the left-hand edge is somewhat shorter than the right-hand edge

d) Fold smaller end up along line 4 and crease sharply

e) Place evidence in opening at larger end. Tap the smaller end and check to be sure the evidence has settled in the middle section. Then fold and sharply crease the top (wider) section over the middle section along line, and secure flaps in place by inserting the smaller end into the larger one

Figure 3.10 The steps in the preparation of a druggist fold to contain loose trace evidence.

We suggest that lab examiners open the package at a different location, preserving the original seal, if possible. The lab examiner can then re-seal the package with tamper proof tape. This way, both seals are intact when the packaged item is presented in court.

Paper containers do have the disadvantage that you can't see into the container. There are paper containers on the market that have cellophane "windows." Paper containers must always be used with biological evidence.

A few other types of containers are used in special situations. With fire debris thought to contain ignitable liquid residues (see Chapter 12), clean paint cans with tight-sealing lids are used as containers. Liquids should be packaged in glass vials of appropriate size. Sometimes, boxes are used to contain weapons, multiple fragments of broken glass, and so on. Investigators should be certain that firearms are rendered safe before packaging. If a firearm or other weapon is found in water, it is recommended that it be left in the same water; that is, placed in a suitable container and covered with the water for transport and submission to the lab. If items are fragile, steps should be taken to make sure that additional breakage does not occur during handling and transport.

Proper controls

Controls are very important in forensic testing. Without the appropriate control, it may not be possible to do an examination or to draw any useful conclusions. Investigators are responsible for seeing to it that controls are collected and submitted to the laboratory. The required controls differ to some extent depending on the type of evidence.

There are, generally, four different types of controls that might be important in forensic testing of evidence: knowns, alibi knowns, substratum controls, and blank controls. Some control specimens must be collected at the scene or in connection with evidence collection and processing. Investigators need to know which controls are required for which types of evidence. Remember, as noted earlier, one may not be able to return to a scene after its release, so the collection of all required control specimens the first time through is critical.

Known (Exemplar, Reference) Controls. Many types of forensic tests discussed in this book involve comparisons between a questioned (evidentiary) specimen and a known specimen (whose origin is known with certainty). The known (which may also be called the exemplar, or the reference) is essential for the comparison. Examples are known blood (or a buccal swabbing) from a person (for comparison with DNA types from evidentiary blood of criminal behavior. Both have their value in criminal investigation aspect. But both are different than crime scene reconstruction.

Scientific criminal investigators should be appropriately cynical about reenactments of past events supposedly based on witness statements and should not be drawn into participating in this kind of activity without a full understanding of the limitations and potential to mislead. It must be emphasized that it is critical to consider all the available evidence including the results from the forensic laboratory in developing any reconstruction.

Key terms

- collection techniques: forceps, tape lift, shaking, scraping, vacuuming
- contracting/expanding spiral search
- controls: blank, substratum, known, alibi known
- crime scene photography
- crime scene security
- cross-projection sketch
- depth of field
- druggist fold
- duty to preserve
- elimination fingerprints
- evidence collection and preservation
- evidence recognition
- first responder
- f-stop/lens opening
- grid search
- lens
- line or strip search
- link search
- notes
- reconstruction
- reenactment
- scene documentation
- scene search
- scene survey
- sketches
- videography
- working hypothesis
- zone or quadrant search

Review questions

1. What is crime scene processing versus crime scene analysis?
2. What are some types of crime scenes? What are the implications of the different types of scenes for crime scene investigators?
3. Describe the steps in crime scene processing and analysis.

4. What are the purposes of the initial scene survey?
5. What are some types of scene searches and to what kinds of scenes are they applicable?
6. What are the main kinds of scene documentation, and why is each necessary?
7. What are some ways of locating evidence items on a sketch?
8. What are the important principles of crime scene photography?
9. List and briefly discuss some methods for collecting physical evidence from scenes.
10. What are the main types of controls that must be available and/or used for the lab to be able to properly test physical evidence?

Fill in/multiple choice

1. Photographs of crime scenes must include overall views, midrange shots, and _____, to properly record the details of the scene and object.
2. The rough sketch that is usually prepared at a crime scene is used to _____.
    a. help identify the victim
    b. reduce the number of photos
    c. precisely locate evidence
    d. help visualize the scene
3. All unauthorized individuals should be _____ a crime scene during its processing.
4. The three most common and classical methods for document a crime scene are (1) _____, (2) _____, and (3) _____.
5. The best method for collecting evidence items from a scene is
    a. tape lifting
    b. submit intact the item that contains the evidence
    c. vacuuming
    d. using sterile forceps

References and further readings

Fisher, B. A. J. *Techniques of Crime Scene Investigation.* 7th ed. Boca Raton, FL: CRC Press, 2003.

Lee, H. C., T. Palmbach, and M. Miller. *Lee's Crime Scene Handbook.* New York: Academic Press, 2001.

Touch DNA: From the Crime Scene to the Crime Laboratory; *Forensic Magazine*; www.foremsicmag.com/print/7885.

Chapter 4

Examination and interpretation of patterns for reconstruction

Lead Case: State of Hawaii vs. Mathison

At midnight on November 30, 1992, the Hawaii Police in Hilo (Hawaii is the name of the largest Hawaiian island as well as the name of the state) received a call from a motorist reporting a traffic accident. Yvonne Mathison was dead on a roadside near Hilo. Her husband, Ken, a police sergeant, claimed Yvonne's death was a tragic accident. He said she jumped out of their moving van while they were having an argument. When he backed up the van to find her, he accidentally ran over her. He said he picked up her bloody body and put her inside the van while waiting for help.

A number of witnesses drove by the van that night and reported seeing suspicious behavior. Some police investigators believed that Ken Mathison had intentionally killed his own wife, but others in the department believed his version of events. The Chief and Deputy Chief of Police went to the hospital to console Mathison on the night of the incident. And, at the suggestion of Mathison himself, the police classified the incident as a negligent homicide. But several things in the van caught the investigators' eyes, including physical evidence in unexpected locations, a rope with hair on it, broken glasses, and a shoe with drops of blood.

The local pathologist had performed a thorough autopsy. The manner of the death was ruled homicide, and the cause of death was multiple head injuries. The file was reviewed by a forensic pathologist, who concluded that the injuries that killed Yvonne were not caused by the van. Because of the questions surrounding the evidence, the prosecutor asked for another autopsy by the chief medical examiner in Honolulu and re-examination of the van. He was told that the van had been returned to the owner, and the body had been cremated on the orders of Ken Mathison.

With no body and possibly no more evidence from the van, the prosecutor feared he could go no further in the case. At this point, he sought the assistance of Dr. Henry Lee. Fortunately, Mathison hadn't picked up his van from the towing company because he didn't want to pay the $335 towing charge. The van was seized as evidence and secured by investigators.

The reconstruction based on pattern evidence found inside the van enabled forensic scientists to determine the following facts and help establish the sequence of events.

1. Approximately 200 hundred bloodstains were found on the instrument panel. These bloodstains were consistent with medium-velocity impact spatters. Reconstruction showed those blood droplets traveled from left to right downward and were deposited on the instrument panel at an approximate 45° angle. These spatters resulted from the impact force of a beating. It is more likely the bloodstains were caused by blows to the victim's head when she was in the front seat.
2. Approximately 50 small bloodstains (2–4 mm in size) were found on the driver's side window and door. These bloodstains were likely produced by medium-velocity impact force and could not have resulted from Mathison putting his wife back in the back of the van.
3. Medium-velocity blood spatters found in the cargo bay suggested that the victim received additional blows in the back of the van.

4. A blood smear pattern was noticed on the van's cargo door. Enhancement of the pattern with tetramethylbenzidine (see Chapter 9) revealed a set of imprints consistent with a fist and hand moving from the top downward. This pattern indicated that the victim's bloody hands were against the cargo wall and subsequently slid down the wall onto the floor.
5. A mixture of soil and blood was found on a piece of sheetrock in the cargo bay floor area. Through detailed examination, it was determined that the blood was deposited before the soil, further undermining Mathison's story that he put his wife in the van after she was run over.
6. Bloodstains found on the floor of the van were whole blood with no indication of any dilution by water, as would have been expected if Yvonne had been outside the van in the middle of a rainstorm.
7. Three blood trails were found on the roof of the van over the cargo bay. These blood spatters were consistent with an overhead cast-off pattern. This fact suggests that the victim was hit repeatedly while she was on the floor.
8. A bloodstain was found on the driver side sun visor. This bloodstain was 4×6 inches and consistent with a contact transfer pattern.
9. A large amount of bloodstains and head hair were observed on the roof over the drive seat. These bloodstains were similar to a combination of multiple swipes and wipe patterns. Head hairs were found imbedded in blood. These patterns were a result of multiple contacts and hair swipes from a bleeding head.
10. Hairs were found on a piece of yellow rope and imbedded in blood. These hairs were microscopically similar to the victim's head hair.

These and some observations and interpretations were used to formulate a comprehensive reconstruction of events. Investigation of the exterior and interior of the van for damage patterns also showed no indication of a typical vehicle-pedestrian impact. Overall, the pattern evidence in the van did not match up with Mathison's story about what happened that night. Mathison was arrested and charged with murder. In 1995, a trial jury found him guilty of murdering his wife. He is now serving a long prison sentence.

> **LEARNING OBJECTIVES**
> - The difference between reconstruction and individualization patterns
> - Evidence patterns that can be collected, and the main purpose of which is individualization, are called individualization patterns (Chapter 5)
> - There are 10 types of major patterns for reconstruction: bloodstain, glass fractures, track and trail, tire and skid marks, clothing and article or object, gunshot residue, projectile trajectory, fire burn, MO and profiling, and injury and damage
> - The analysis of crime scene patterns involves four distinct processes: pattern recognition, pattern identification, pattern interpretation, and pattern reconstruction.
> - Some reconstruction patterns are best compared with experimentally produced patterns to aid in interpretation
> - Reconstruction patterns are generally intrinsic to scenes
> - Reconstruction patterns have to be documented; they usually cannot be collected as such
> - Blood droplets moving through the air behave predictably according to physical laws
> - A good reconstruction from blood patterns is based, in part, on knowing how many blood sources there were at a scene, which source made which pattern, and the sequence of events
> - The side of glass from which force was applied to cause the breakage can be determined
> - The order of gunshots or other impact points can sometimes be determined in glass
> - The direction and angle of each shot sometimes can also be determined
> - Foot, footwear, tire, or blood trail patterns can help reconstruct the number of persons at a scene and their movements
> - Tire and skid mark patterns are used by traffic accident reconstruction experts to estimate position and speed of vehicles
> - Clothing, article, or object patterns are based on looking for unusual or unexpected arrangements or disorder in a scene
>
> *(Continued)*

LEARNING OBJECTIVES (Continued)
- Tears, cuts or damage to clothing or other objects can also provide information for reconstruction
- Gunshot residue patterns on target surfaces can be used to estimate muzzle to target distances
- Trajectory analysis (ballistics) can help establish the positions and orientations of shooters and victims in shooting cases
- Ballistics should not be confused or equated with firearms identification
- Burn patterns at suspicious fire scenes can help establish origin and cause of fires
- Burn patterns are used by fire investigators along with analysis of the overall scene and investigation of mechanical and electrical equipment to help determine origin and cause
- MO refers to a repeat offender's habits and can be used to help connect related cases
- Criminal profiling involves statistical and psychological analysis to give insight in unsolved case and on previous offenders

TYPE OF PATTERN EVIDENCE: PATTERNS FOR RECONSTRUCTION AND INDIVIDUALIZATION

As we discussed in the previous chapter, many types of pattern evidence that appear at crime scenes are useful for reconstruction. In some of the cases, these patterns can be "collected" only by observation and documentation; that is, they cannot be "picked up" and "packaged." Ten major types of patterns, which we call "reconstruction patterns," are discussed in this chapter—as an extension of the discussion of crime scene investigation in the previous chapter.

There are several categories of pattern evidence that have the possibility of "individualization" as the primary goal. The major categories are fingerprints, palm prints, bare footprints, handwriting and other document patterns, and toolmarks and striae associated with fired bullets and casings. They are discussed in Chapters 5 through 8. This type of pattern evidence generally can be collected and brought to the laboratory for examination.

MOST RECONSTRUCTION PATTERNS ARE CRIME SCENE PATTERNS

The patterns discussed in this chapter are mainly found at scenes. If they are documented and interpreted properly, they can provide valuable information about the events—often more reliable information than witnesses provide.

Ten common reconstruction patterns are discussed in some detail. These are the most common types of patterns seen in forensic investigations, but there are other patterns that could fall into this class.

Currently, there is not a rigorous protocol for crime scene pattern analysis. The analysis of any pattern evidence consists of four distinct processes: pattern recognition, pattern identification, pattern interpretation, and pattern reconstruction.

1. Pattern Recognition: The first stage of crime scene pattern analysis is to observe and to recognize any potential physical pattern. Non-destructive techniques and forensic light sources are used to visualize the pattern followed by detailed documentation of the pattern with overall and close-up photographs.
2. Pattern Identification: The second stage of crime scene pattern analysis is to study the pattern and make preliminary identification of the pattern types by class characteristics. Chemical methods may be used to further enhance the pattern. Measurement and documentation are extremely important. Many computer software programs are commercially available to assist in the identification and documentation of the pattern.
3. Pattern Interpretation: This stage can be conducted either at the crime scene or in the laboratory by experienced personnel. A detailed study of the pattern should be conducted. If necessary, control experiments should be carried out. By comparing past experience with similar patterns, the published data and literature, and controlled experiment results, an interpretation of a pattern can be made. At this stage, the examiner should avoid learning about or considering any other type of information about the case to avoid bias in the pattern interpretation.
4. Pattern Reconstruction: Once the interpretation of the crime scene pattern is completed, the examiner can reconstruct the crime through the crime scene pattern analysis.

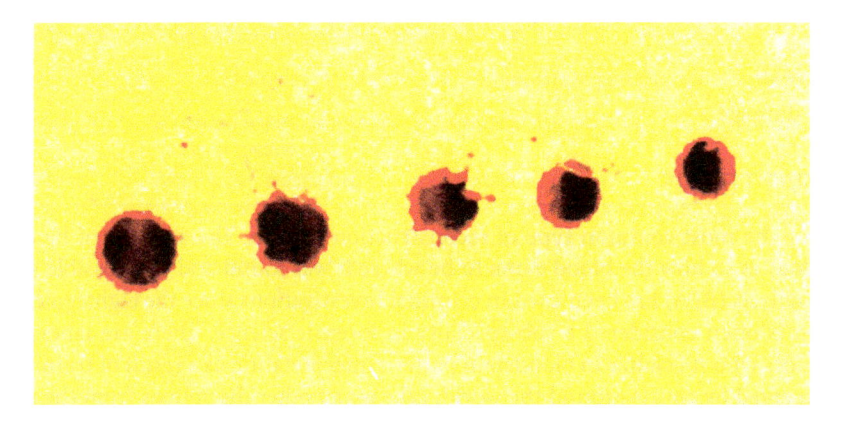

Figure 4.1 Low-velocity blood spatter pattern.

Figure 4.2 Medium-velocity blood spatter pattern.

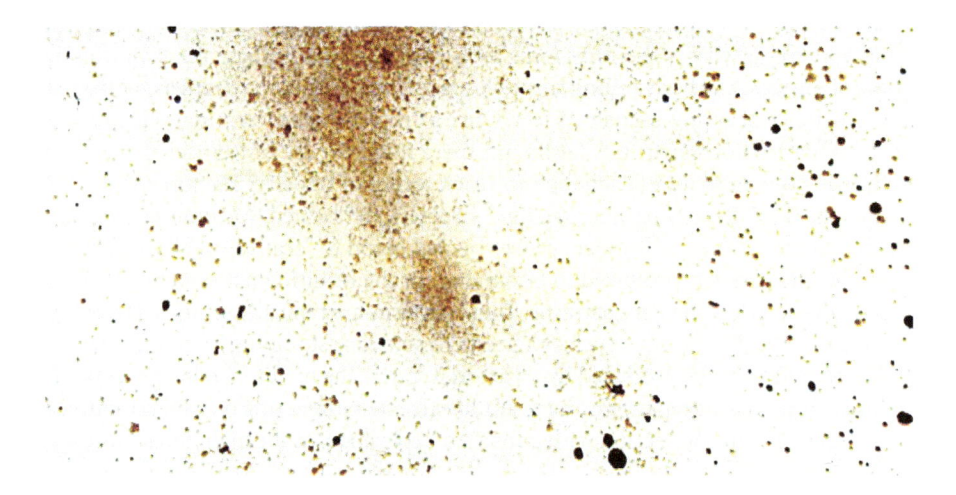

Figure 4.3 High-velocity blood spatter pattern.

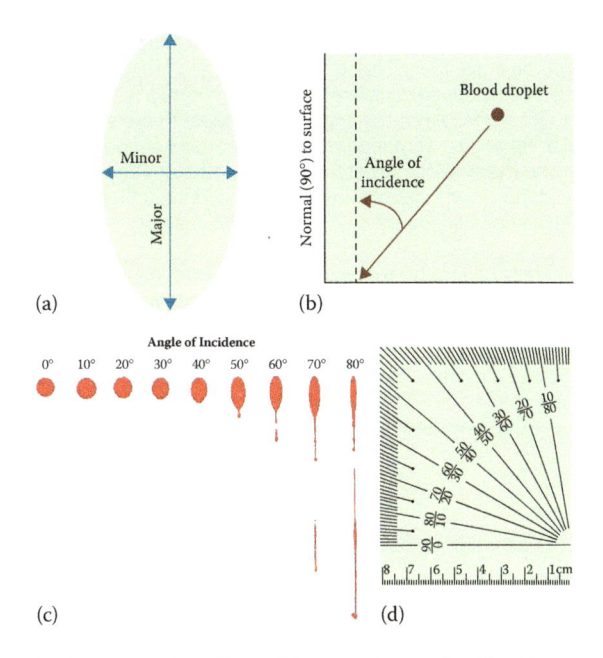

Figure 4.4 (a) Diagram of major and minor axes of an ellipse, (b) measuring angle of incidence on blood drop, (c) effect of angle on blood spot shape, and (d) special angle measuring device.

force has been applied including blood flow and pool patterns. Transfer stains include transfer patterns, smears, swipes, wipes, and other stains caused by contact. Impact stains result from blood traveling through air and then contacting a surface. These are usually seen as blood spatter, but also include cast-off, spurts, splash, and expired blood patterns. The characteristics of blood spatter depend on the speed at which the blood leaves the source and the amount of energy or force applied to the blood. The shape of a blood spatter droplet pattern indicates the angle with which it impacted the surface. A blood droplet hitting a non-absorbent surface at a 90? angle (perpendicular to the surface) results in a circular stain. As the angle changes, the resulting stain becomes more elliptical (Figure 4.4a). In fact, using simple measurements and some elementary trigonometry, the angle of incidence of a bloodstain can be estimated from the shape of the drop (Figure 4.4b). See the sidebar titled "Calculation of the Blood Spatter Angle" for a description of how this is done.

SIDEBAR 4.1 Calculation of the blood spatter angle

A blood droplet impacting a non-absorbent surface at exactly a 90° angle (exactly perpendicular to the surface) forms a circle pattern. Depending on the impact velocity, there can be satellite droplets around the outside of the circle, but the main pattern is a circle.

Note that an ellipse is an "egg-shaped," two-dimensional figure that has a long diameter and a short diameter (Figure 4.4a). As the angle of the surface relative to the direction of the falling droplet changes from 90°, the pattern becomes increasingly elliptical (Figure 4.4b).

There is a simple mathematical relationship between the properties of the ellipse and the angle of impact. The relationship is based on a trigonometric function known as the cosine. There are cosine tables in various scientific and mathematical books and handbooks that can be used to relate cosine values to angles. Many calculators and apps will also determine the angle from the cosine value. The cosine value is always a number between 0 and 1.

To figure the angle of impact for a blood droplet from its pattern, the major (long) and minor (short) diameters of the ellipse are measured as accurately as possible. Then the ratio of minor to major (width to length) is calculated. Because the measurements are used to compute a ratio, the units of measurement don't matter, meaning the measurements can be in inches, millimeters, and so on. Since the ratio is always that of a smaller to a larger number, the value will lie between 0 and 1. This ratio value is the cosine of the angle of incidence.

Expressed mathematically,

$$\cos\theta = \frac{\text{minor diameter (width)}}{\text{major diameter (length)}}$$

where we are designating the angle of incidence by the Greek letter θ (theta).

If you calculate the ratio of the two diameters in an elliptical blood pattern and look up the angle corresponding to that ratio (cosine) in a cosine table, then you get the angle of incidence.

It is important to note that the angle of incidence calculated using this formula is the angle of incidence of the blood droplet with respect to a line perpendicular to the surface, not the angle with respect to the surface itself (Figure 4.4c).

There are scales available to make the angle estimation measurement easier (Figure 4.4d). In a more complex pattern, such as a *medium velocity*, or cast-off pattern, with numerous blood spatters, a selection of the droplet pattern can be measured and the angle calculations done. Then, straight lines (string, wires, etc.) can be drawn out from the droplet patterns at the estimated angles. An example is shown in Figure 4.5. The straight lines will tend to converge at a area that was the source of the blood pattern.

Note that the measurements generally cannot be that exact, so the resulting angles obtained are estimates. The straight lines will not converge to an exact point for that reason. They will usually converge to an area, though. These pattern reconstructions can tell an investigator approximately where the blood source (say someone's head) was located at the time the pattern was formed. You may be able to tell, for example, whether a person was standing or lying down when a pattern was formed on a nearby wall, or whether a pattern in a vehicle is consistent with having originated from a person seated on the driver or passenger side.

When analyzing blood spatter, it is possible to use the angle calculations along with a straight-line projection backwards (using strings, probes, lasers, or wires) to the area of convergence and to estimate the approximate area of origin of the blood forming the pattern (Figure 4.5).

Various blood spatter patterns

There are a number of blood patterns commonly observed at scenes. We will briefly describe them in this section.

Falling droplets from a bleeding source that is stationary will generally result in blood pooling below the source. It may be possible to discern separate "droplet" patterns in the pool. If a dripping blood source is moving, blood "trails" can result. If the source is moving fast enough, the droplets may show the direction in which the source was moving (Figure 4.6a).

Contact deposit patterns result from an object coming into direct contact with a blood pool or blood source. These patterns can appear on clothing or on objects at scenes. Sometimes, it is important to try and determine what object caused the pattern.

Figure 4.5 Straight line reconstruction from medium/high blood spatter pattern.

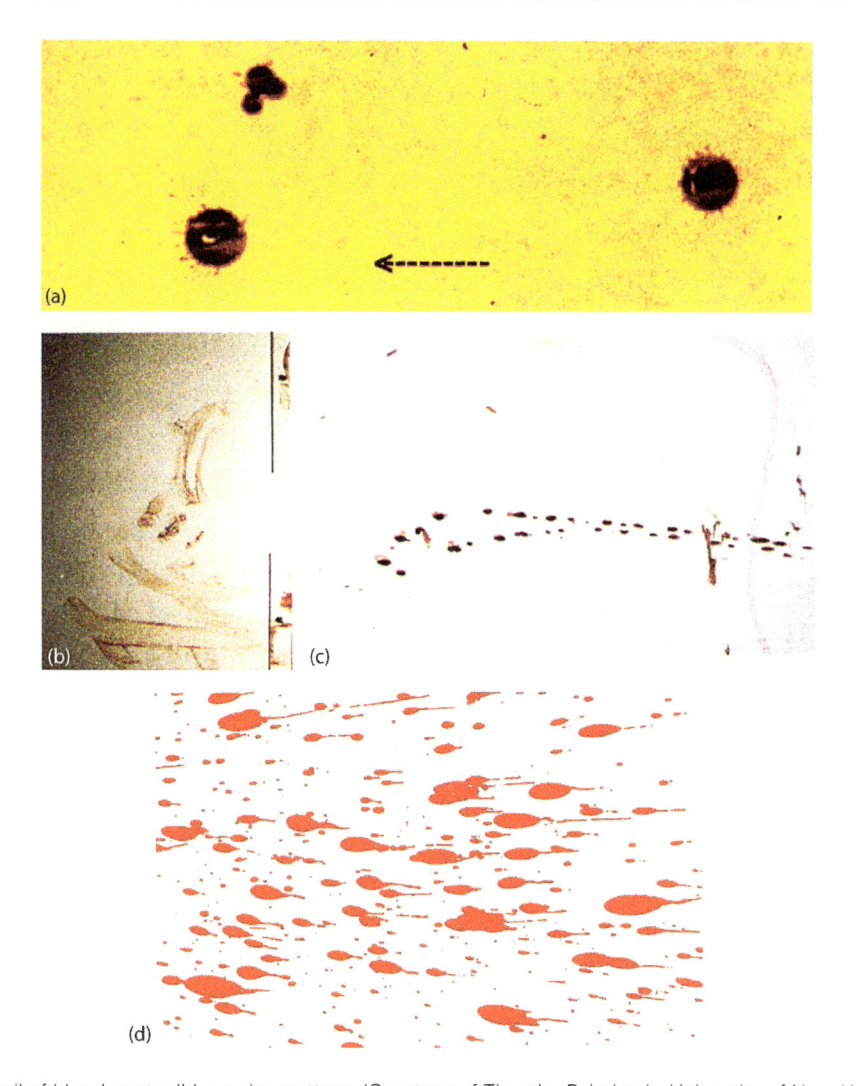

Figure 4.6 (a) A trail of blood spots, (b) a swipe pattern. (Courtesy of Timothy Palmbach, University of New Haven. With permission.), (c) cast off pattern. (Courtesy of Timothy Palmbach, University of New Haven. With permission.), and (d) this pattern was produced by several droplets of blood that were traveling from left to right before striking a flat level surface.

Wipe and swipe patterns result when an object contacts or transfers wet blood and smears it on a surface. A wipe pattern is created by an object contacting an existing bloody surface with motion. A swipe pattern is created by a bloody object contacting another surface with motion (Figure 4.6b).

An *arterial spurt* pattern results when an artery is cut or severed, and blood is pumped out of the body by the beating heart and onto a nearby surface. The repeated spurts cause a rather characteristic pattern. These patterns generally contain quite a bit of blood. In addition, an individual with a seriously severed artery is losing blood at such a rate that he or she will not be able to move too far, and without immediate and extreme medical intervention, will die fairly quickly.

Cast-off (also called *arc swing*) patterns result when a bloody object is swung through space and throws off droplets onto a nearby surface (Figure 4.6c). These patterns may be seen on ceilings or walls, even occasionally on floors. The most common action causing such a pattern is the repeated use of blunt force on a person who is bleeding.

Running patterns are just what the name says. Blood hits a vertical surface, but the volume is sufficiently high that gravity causes the droplet to run (Figure 4.7). Note that blood can only run down. As obvious as that statement is, it is sometimes quite helpful in reconstructing events from blood patterns.

Figure 4.7 Illustration of a running pattern that runs downward due to gravity.

Secondary spatter results when blood drops are falling into a pre-existing pool of blood. As each drop hits the liquid surface, it can cause small droplets to splash upward, and some of these may hit a nearby vertical surface. The most common example of this pattern may be on the cuffs of pants, which were located for a time near a blood pool into which drops were falling. Since the secondary droplets splash upward, the direction of the blood spatter pattern on the pants cuff will indicate that the blood came from the floor. This sounds unlikely until you realize that it could be secondary spatter.

Imprint and impression patterns that we will talk about in Chapters 5 and 6 (fingerprints, footprints, footwear impressions) are sometimes made with blood. In the sense that they are imprints made in blood, they are also considered blood patterns. But they would be handled like other fingerprints, footwear impressions, and others, from the standpoint of analysis and comparison. One important difference is that there are some special techniques for "enhancing" bloody imprints (discussed in Chapter 6) that may help make such patterns clearer or more visible and thereby more suitable for comparison.

Factors affecting blood patterns and their interpretation

Blood pattern interpretation can be a very helpful tool in crime scene reconstruction involving bloodshed. Several things about blood pattern interpretation should be kept in mind. Considerable training and experience is required to become skilled in this type of work. Many of the principles concerning blood patterns, which we have discussed and illustrated above, are clearest on non-absorbent surfaces. On absorbent surfaces, the patterns may not be as clear. Their shapes and angles may not even be obvious. Target surface absorbency, texture of the surface, and volume of blood are important variables in blood pattern formation. Other factors that can affect interpretation include ambient environment, which can affect the time it takes blood to clot and to dry, and wind or air currents, which could alter the resulting blood patterns. Activities of the victim, suspect, witnesses, medical personnel, and police officers can also change the appearance of a pattern and complicate its interpretation.

In some cases, experts may conduct experiments to try and replicate patterns observed at a scene. This is recommended procedure when interpreting stains on substrates or in environments that have not been carefully explored previously by the investigator. The experiments might involve varying the amount of force, distance, or motion of a source with respect to a target surface. As noted in the discussion of reconstruction in Chapter 3, successful experimental replication of a scene pattern does *not* prove that the experimental conditions are those that prevailed at the scene when the original pattern was formed. Rather, it shows that the *theory* concerning the formation of the pattern is scientifically sound.

It should also be noted that scene reconstructions involving blood patterns cannot be done accurately without knowing whose blood was shed. Generally, dried blood samples from the patterns will be DNA profiled in the laboratory (Chapter 10) to associate the bloodstain with a person. There may be times when it appears safe to assume that only one person (usually a victim) was bleeding at a scene, a blood pattern reconstruction based on such an assumption should be considered conditional until the DNA typing results confirm it. Should the blood be from more than one source, the interpretation of the pattern could be significantly affected.

A recent study on BPA funded by the U.S. Department of Justice has found significant bias in bloodstain interpretation when examiners are provided with investigatory information. Misclassification was much lower (8% for rigid surfaces, 14% for fabric surfaces) when the vignette presented with the bloodstain evidence supported the "right" answer than when it was neutral (11% for rigid surfaces, 26% for fabric surfaces). When the background information pointed toward the wrong classification, the error rate was higher still (20% for rigid surfaces, 30% for fabric surfaces).

This means that at an early stage of the bloodstain pattern analysis, additional case-specific information, such as medical findings, injury and wounds, case circumstances, investigator opinions, crime scene reports, and even witness statements considered during the pattern recognition, identification, and interpretation process has the potential to introduce bias into the pattern interpretation process. Therefore, only after the pattern is recognized and analyzed is its relevance to the case investigation considered. It is extremely important for a bloodstain examiner to analyze evidence first and make an independent interpretation of the pattern before considering other investigative information to avoid the potential introduction of bias into the pattern reconstruction process.

Glass fracture patterns

There are different formulations of glass (Chapter 13), and the type of glass can affect the amount of information obtainable from a pattern of broken glass.

Determining the side of the glass where force was applied

It can be important in an investigation to determine which side of the glass a force was applied causing the glass to fracture. Sometimes, glass can be fractured but remain essentially intact (that is, not fall apart into pieces). Automobile windshield glass (called safety plate glass) and certain types of plate glass tend to behave in that way. It is much easier to derive information from a piece of fractured glass that remained intact. At a point of impact in an intact but fractured piece of glass, it is often possible to discern a cone-shaped pattern (Figure 4.8). The smaller end of the "cone" is

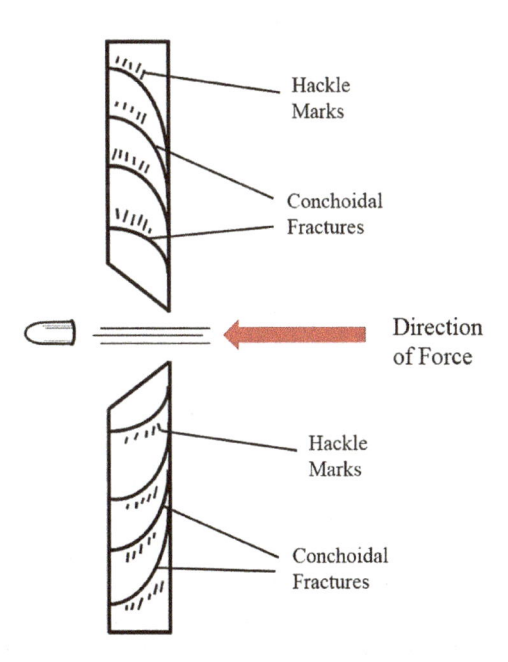

Figure 4.8 Drawing showing the cone-shaped hole on the far side of a piece of glass caused by a projectile penetrating the glass.

on the side where the force was applied. The larger side of the "cone" is the side from which fragments were broken due to the force. If the force was a projectile, like a BB or a bullet, there will be a hole in the glass where the projectile passed through, at the center of the "cone." In the case where a projectile broke the glass, we would refer to the surface corresponding to the smaller side of the cone as the "entry" and the other side as the "exit."

This same type of pattern can be formed in other objects, such as walls or even in people's skulls, as the result of extreme force or projectiles.

If the glass did not remain intact and small pieces must be examined, the task is more difficult. With numerous small pieces, the examiner must first try to determine which side of the glass is which. In a case of an exterior window, one side of the glass was exposed to the outside and the other side to the inside. Results showing weathering, dirt, and other characteristics, may help in making this determination.

The broken edges of glass fragments have a pattern of marks, called hackle marks, which result from breakage by force. These patterns form in certain predictable ways depending on the side of the glass from which the force originated. There are "rules" to help determine the "force" side of glass fragments. The 3R rule says that the force was applied from the REVERSE side to the side that shows RIGHT ANGLE hackle marks when viewing the edge of a RADIAL crack (Figure 4.9). Often, these hackle marks on the edge of a piece of glass can be matched up from two pieces to confirm the same origin. In other cases, the shape of glass pieces themselves can often be pieced together, like pieces of a puzzle, to confirm glass found at one location actually originated from another.

Determining the order of gunshots fired through glass

Among glass types, the "safety plate" glass used for automobile windshields is unusual in that it is manufactured with two sheets of glass sandwiching a sheet of plastic material. This structure is intended to keep the windshield from fragmenting in a collision or accident. Side window glass, by contrast, is tempered glass, which pelletizes if fractured. As a result, windshields generally hold together even when fractured by intentional force or by penetration of bullets, shot shell pellets, or slugs. And, because they hold together, the resulting fracture patterns are easier to analyze. Side window glass is heat tempered. It is very difficult to break but, when it does break, it forms square chunks of glass without sharp edges to reduce the possibility of injury. Because it is manufactured to disintegrate upon being forcefully impacted, its pieces are of little use in determining direction.

Glass breaks under the stress of force in a particular way, causing both *radial* and *tangential* fracture lines to form (Figure 4.10). Viewed at a 90° angle to the glass surface, the tangential fracture lines tend to encircle the projectile hole, while the radial fracture lines tend to appear to be radiating out from the center. Usually, a nascent tangential

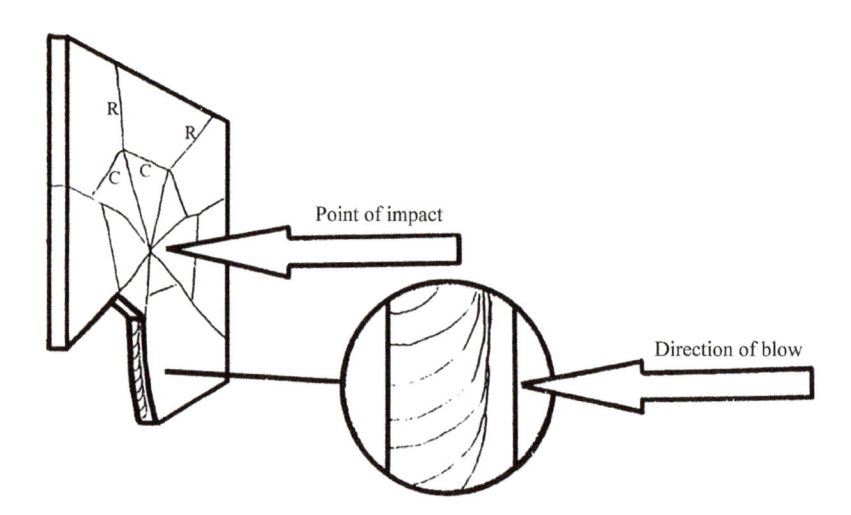

Figure 4.9 An illustration of hackle marks and shell-like, or conchoidal, fractures in response to an applied force. Hackle marks run parallel to the direction of the fracture. Conchoidal marks occur on the glass piece's edge. Both types of marks can be utilized to match glass fragments.

Figure 4.10 Three holes bullet holes in a windshield. Order of shots and holes determined through the analysis of radial and tangential fracture lines.

fracture line will not cross a pre-existing fracture line. Both radial and tangential fracture lines, which appear to stop at pre-existing lines, must belong to a breaking force that occurs *after* the pre-existing lines were formed—and thus, to a hole formed later. If there are two holes, and they are reasonably nearby so that the interactions of the fracture lines can be observed, it is often possible to reconstruct the order in which they were made (Figure 4.10).

Thus, the amount of information available from glass fracture patterns depends on the type of glass, whether it holds together as a unit even though fractured, and what type of event caused the fractures. At times, this kind of information can be valuable in reconstructing events at a scene.

We will mention it again in Chapter 13, but it may be noted here that larger pieces of broken glass that can be physically fit matched back together in the manner of a jigsaw puzzle can thereby be shown to have had a common origin. With many small pieces, this activity is much more difficult or impossible.

Track and trail patterns

As the name suggests, these patterns could be of footwear, socks or stockings, bare feet while walking, or other actions such as an object or body being dragged. These patterns often can exhibit a shape that could yield information on how many individuals were at a scene and the possible sequence of events. It may also indicate direction of travel.

These patterns can help show the relevant activity of victim(s) and suspect(s) at a scene, before, during, and after the crime has been committed. The estimation of height, weight of the individuals, and any special features of walking might also be deduced by analysis of footwear evidence.

Track or trail patterns can be fairly obvious, such as impressions in snow, soft earth, blood, or dirt. Also, they can sometimes be subtle, such as a drag mark across grass, a very diluted or washed bloody footprint, or a set of dust prints. As will be discussed later (Chapter 9), such a bloody trail can often be made visible using special chemical techniques even after a fairly thorough clean-up.

Tire and skid mark patterns

Tire and skid patterns are used in reconstructing vehicle accidents. Tire impressions that may help an expert identify the make and model tire that made the mark will be discussed in Chapter 5. The tire patterns we are discussing here refer to those that show the number, location, or direction of movement of a vehicle or vehicles at a scene.

Vehicle accident reconstruction is a specialized area of forensic science, often done by a traffic accident reconstruction officer/expert or forensic engineers (Chapter 1). Tire and skid mark patterns are only part of the data used in an overall vehicle accident reconstruction. The length of skid marks can help an expert determine the speed of the vehicle at the time the operator applied the brakes.

These patterns can also be helpful in reconstructing the presence, direction, and actions of a vehicle, or vehicles, at a burglary, robbery, or car-jacking scene. Hit-and-run scenes and scenes where a body has been removed from a vehicle and dumped are also obvious examples.

Clothing and article or object patterns

We include in this category the general state of objects, things, and clothing at a scene. These patterns must be looked at as a whole and in the context of the scene location and scene conditions.

One "rule" often cited in recognizing pattern evidence at a crime scene is "look for the unusual." In other words, look for things that seem out of place in the context of that location. You might expect a certain amount of disorder in a college dorm room; for example, that you would not expect in a private home managed by a scrupulous homemaker. Therefore, an area of considerable disarray in an otherwise very neat household will call for further investigation. After the potential pattern is recognized, the pattern should be documented by notes, sketching, and photographs without and with a scale.

Clothing or objects scattered around at a crime scene, overturned furniture, and so on, can provide indications of struggles, of force being used, or of ransacking, among other things.

Clothing out of place or damaged on a victim may give indications of assault or sexual assault. It may also indicate that a body was hastily redressed after death. Cuts, tears, or damage to clothing articles may help to determine the type of force, weapon, or MO involved in the criminal activity.

Gunshot residue patterns

Gunshot residue (GSR) is produced when a firearm is discharged. Details are discussed further in Chapter 8. The residue may be deposited on the hands of the shooter. Testing, to detect any gunshot residue, may sometimes be done by hand swabbing or sticky tape lifts taken from people thought to be involved in shooting cases. Rigorously identifying gunshot residue is not a simple problem nor is interpreting the results of GSR testing on peoples' hands. Those matters are further discussed in Chapter 8. Gunshot residue comes up here, in the reconstruction patterns, because it can be deposited on a target surface provided the target surface is not too far away from the muzzle of the weapon when that weapon is fired. The pattern formed on a target surface, usually around a bullet hole, can help in estimating the muzzle to target distance at the time of firing.

The gunshot residue that is propelled out of the muzzle of a fired weapon along with the projectile consists of powder particles (both burned and unburned), primer residue, traces of lead from the bullet (and copper or other metals from the bullet jacketing if present), dirt, lubricants, and other debris from the barrel. On a white or light-colored surface, the GSR pattern may be visible without enhancement. The residue is dark gray to black. On colored surfaces, the pattern may require chemical treatment to enhance it or photographic enhancement using infrared photography. Objects or surfaces can be sprayed with sodium rhodizonate, a chemical that reacts with lead particles or with Griess reagent that reacts with gun powder residues. These chemical reactions produce strongly colored products, which greatly enhance the visibility of the often small traces of material.

Figure 4.11 shows a series of GSR patterns at various muzzle to target distances. Usually, the "tight contact" or "loose contact" patterns, where the muzzle is essentially resting against the target surface, are quite distinctive. The patterns can be seen on clothing, on other objects, or surfaces that were in the path of a fired weapon. They can also sometimes be seen on human bodies, and pathologists use the patterns to estimate muzzle-to-body distances.

Gunshot residue particles are quite small and do not travel great distances. With all but the most powerful handguns, it would be unusual to see much GSR pattern beyond about two feet. With rifles, the distances are somewhat greater. With shotguns, the pellet spread can be used similarly to measure muzzle to target distance. Since shotgun pellets travel much farther, pellet spread can also be used to estimate firing distance to a considerable distance.

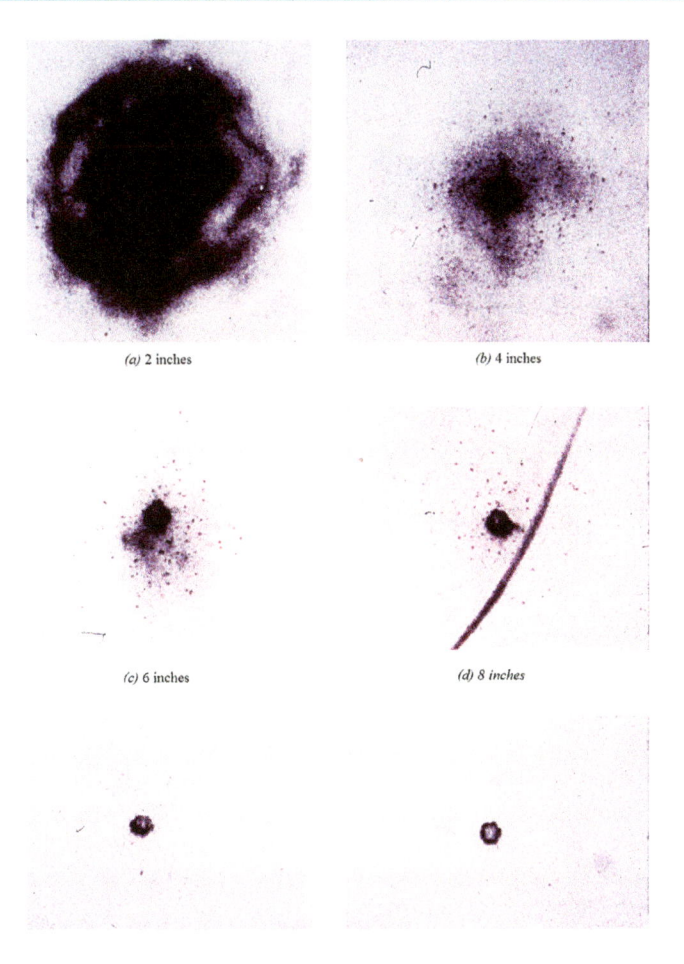

Figure 4.11 Gun powder patterns on a target from gun shots at various muzzle to target distances.

Distance estimations (that is, muzzle to target distances) using gunshot residue pattern cannot be done exactly just by looking at the pattern. There is too much variation between different weapons and different ammunition. To get a good estimate, an examiner has to conduct a series of control test fires using the same weapon and ammunition thought to have been used in the case itself with similar conditions. The test patterns are then compared to the scene/evidence pattern to make an estimate. As we have noted elsewhere in connection with reconstruction patterns, replicating the scene pattern on the firing range does not prove that the experimental muzzle to target distance is in fact the distance that prevailed at the scene. It does show; however, that the experimental distance is a reasonable estimate of the actual distance. It is extremely important that control tests be conducted independently of other aspects of the investigation and before an opinion or theory is formed.

Distance determinations based on GSR patterns are commonly done in shooting cases in which the weapon to target distance may be in dispute. They may also be done to verify a victim, suspect, or witness statement concerning the actual distance. An allegation by a shooter that "we were struggling over the weapon and it went off" is rendered suspect when the muzzle to target (bullet hole in the victim or victim's clothing) distance is found to be over three feet.

PROJECTILE TRAJECTORY PATTERNS

The physics of firearms projectile flight through the air is called *ballistics*. One often hears the laboratory activities done by firearms examiners (firearms identification work—Chapter 8) called "ballistics" in the popular media, but that is incorrect terminology. Ballistics has to do with projectile flight and path of travel. There is a sub-specialty of

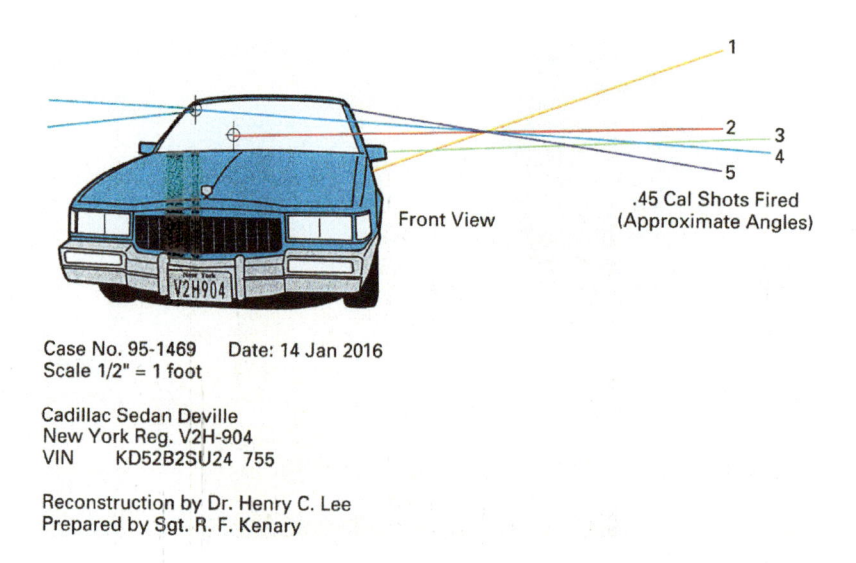

Figure 4.12 Reconstruction of the paths of five shots into an automobile.

forensic pathology called *wound ballistics* that has to do with the wounding and destructive power of firearms and ammunition to the body.

There are cases in which an effort must be made to reconstruct trajectory, or path of flight, from the muzzle of the weapon to the target. Much of the time, the distances involved are relatively short, and straight-line paths can be assumed (as long as there are no intermediate targets.). Trajectory reconstruction is often used to estimate the position of the shooter from holes or impacts made by a bullet, particularly in sniper-type situations. In addition, movements of victim or shooter can sometimes be reconstructed.

Reconstructing trajectories, when a body was the target, require consultation with the pathologist, who makes the determinations of entry and exit wounds on the body and the direction of travel of the bullet in the body. These analyses can often help establish accurate enough theories of shootings to test the veracity of statements by witnesses, victims, or suspects.

Figure 4.12 shows a trajectory reconstruction in a shooting case.

FIRE BURN PATTERNS

Fires and explosions are discussed in Chapter 11. The objective of a fire investigation is the determination of the *cause* and the *origin* of the fire. Burn patterns can be helpful in reconstructing the point or points of origin of a fire, the pathway it followed, and provide estimates of the temperatures that may have been reached or the fuel that may have been available.

Figure 4.13 shows several burn patterns that may be seen at a fire scene. The classical "inverted cone" or "V pattern" often indicate the point of origin, since fire follows the pathway of available fuel (upward and outward). Multiple "V patterns" may indicate the possibility of arson. Smoke color can provide information about the nature of the fuel that is burning. For this reason, first responders are advised to take a picture of an active fire and the smoke, if possible. Smoke stains can show the path traveled by the fire. Material melting patterns provide information about the temperature the fire reached.

Burn patterns are an important source of information for arson investigators. Along with the results of examining electrical and mechanical equipment and the scene, in general, they can help reconstruct where the fire started, and, in that way, help establish the cause and origin.

Fire patterns are discussed further in Chapter 11. Over the years, a number of myths about the meaning of certain burn patterns have developed and been perpetuated. It is important that the patterns be interpreted correctly.

Figure 4.13 Burned building showing multiple burn patterns.

MODUS OPERANDI PATTERNS AND PROFILING

Long before anyone commonly used the term "criminal profiling," police investigators were aware that repeat offenders develop habitual patterns. These habits, that can often be reconstructed from physical evidence, study of scenes or from interviewing victims, are referred to as a criminal's MO, or *modus operandi*. The MO can include language (words or phrases) used, methods of gaining entry to a scene, type of crime committed, use of or absence of weapons, use of or absence of force or threat of force, disguises used, and so forth. Such habit patterns are usually pieced together by investigators from a series of cases thought to have been committed by the same offender.

The development of criminal profiling took this basic reasoning a step further. Criminal profiling is based on the detailed analysis of numerous repeat offender cases. All the scene patterns resulting from the offender's activities are taken into account, along with the pathology and forensic science laboratory results from the scene and the evidence and a knowledge of human behavior and psychology.

Profiling has been best developed for serial murderers and rapists, because these crimes are so serious, and the offenders often difficult to apprehend. The idea is to use scenes, lab results, autopsy or medical findings, and information obtained from extensive interviews with known offenders, to be able to construct a "profile" of an as yet unidentified serial offender. There has also been some work done on serial arsonists.

It is generally fair to say that profiling can occasionally be an aid for investigators. Other avenues of investigation are usually exhausted before those specializing in profiling are brought in. Part of the reason is that there aren't many trained criminal profilers available. Another factor is that a profile does not identify a specific person. Rather, it describes a category of person. The more information available, the better the profile and the narrower the category. Investigators can use the profile as an aid in focusing the direction of the investigation and eliminating possible suspects.

Some of the profiles that are developed turn out to be quite accurate, but a profile alone is not sufficient to allow an offender to be identified and apprehended. One of the major values of profiling on a national level, for example by experts at the FBI, is the ability to discern connections between cases from different jurisdictions. Many serial offenders tend to operate in different jurisdictions; even across different states. There would typically not be a way for investigators in any one jurisdiction to be aware of the similar cases in other jurisdictions without the connection via profiling.

Detailed interviews of serial offenders by profilers have also resulted in a more useful body of information concerning the psychology of those individuals.

WOUND, INJURY, AND DAMAGE PATTERNS

Wound and injury patterns generally refer to wounds or injuries on a human body. Forensic pathologists sometimes call these "patterns of injury." They tend to be characteristic of certain events, such as a walking pedestrian being hit by a motor vehicle. Similarly, certain kinds of industrial or farm accidents involving heavy equipment produce certain injury patterns. These can help the pathologist determine that injuries observed are consistent with having been caused by specific events. Wound patterns are similar. There are wound patterns consistent with gunshots, cutting (sharp force), blunt force, and so forth.

Damage patterns often refer to patterns seen on clothing. Clothing may be penetrated by a bullet, it may be cut, or it may be scraped on the hard surface of a road or the external surface of a brick or stone building. These patterns may also help in reconstructing events.

Key terms

- angle of incidence
- ballistics
- blood spatter pattern
- contact, wipe, swipe, arterial spurt, cast off, running, and secondary blood patterns
- criminal profiling
- glass fracture 3R rule
- gunshot residue (GSR)
- high-velocity blood pattern
- individualization pattern
- low-velocity blood pattern
- medium-velocity blood pattern
- radial fracture lines
- reconstruction pattern
- tangential fracture lines
- trajectory
- wound ballistics

Review questions

1. What is the difference between individualization patterns and reconstruction patterns?
2. How are reconstruction patterns collected?
3. What are low-, medium-, and high-velocity blood patterns? What types of events are likely to produce them?
4. What can the shape of a dried blood droplet pattern tell you about the angle of impact?
5. List and briefly describe five blood patterns.
6. What can be learned from glass fracture patterns?
7. What can be learned from track and trail patterns?
8. Why are clothing and/or article patterns important and crime scenes?
9. What are gunshot residue patterns used for?
10. How can a wound or injury pattern help in reconstructing case events?

Fill in/multiple choice

1. Radial and tangential fracture pattern lines, in a broken but still largely intact piece of glass, may enable the examiner to determine _____.
   a. the direction from which the breaking force was applied
   b. the refractive index of the glass
   c. the order in which two or more distinct breaks occurred
   d. only a and c
   e. a, b, and c
2. Examination of the blood spatter patter flung from a weapon can help to determine _____.
   a. the attacker
   b. the shape of the weapon
   c. where the attack took place
   d. the time of the attack
3. If suitable controls are available, examination of the distribution of gunpowder particles, and other discharge residues around a bullet hole, permits an approximate determination of the _____.
   a. ammunition used
   b. shot shell shot size
   c. gun muzzle to target distance
   d. condition of the weapon's barrel
4. Ballistics is
   a. the path followed by a projectile from a firearm
   b. firearms identification
   c. NIBIN
   d. type of bullet
5. A blood spatter pattern consisting of well-formed individual separated droplets indicates that the stain that arose was from _____.
   a. a transfer
   b. artery spurt
   c. rapidly moving weapon
   d. dripping from above the surface

References and further readings

James, S. H., ed. *Scientific and Legal Applications of Bloodstain Pattern Interpretation*. Boca Raton, FL: CRC Press, 1999.

James, S. H., P. E. Kish, and T. P. Sutton. *Principles of Bloodstain Pattern Analysis: Theory and Practice*. Boca Raton, FL: CRC Press, 2005.

Laber, T., P. Kish, M. Taylor, G. Owens, N. Osborne, and J. Curran. "Reliability Assessment of Current Methods in Bloodstain Pattern Analysis." June 2014, NIJ Report Document No.:247180

Lee, H. C., T. Palmbach, and M. Miller. *Henry Lee's Crime Scene Handbook*. New York: Academic Press, 2001.

Monahan, D. L., and H. W. J. Harding. "Damage to Clothing: Cuts and Tears." *Journal of Forensic Sciences* 35, no. 4 (1990): 901–912.

Taupin, J. M. "Clothing damage analysis and the phenomenon of the false sexual assault." *Journal of Forensic Sciences* 45, no. 3 (2000): 568–572.

LEARNING OBJECTIVES

- There are four types of forensic analysis of physical patterns: physical fit-match; indirect pattern match; imprint, impression and striation mark comparisons; and shape, pattern, marks, and form determination
- Physical fit and match are jigsaw fit match
- An indirect match may be possible if pieces or parts are not solid enough to be fit together or if intervening pieces are missing
- Impression marks can be imprints (effectively two-dimensional) or indentations (three-dimensional)
- Striations are caused by an object moving across a surface making a dynamic impression. They are characteristic of firearms and toolmark evidence as well as many other types of toolmarks.
- There are several steps in the process of discovering, classification, or individual characteristics of patterns (determining common origin between questioned and known patterns)
- Comparisons between questioned and known patterns can lead to identification, inclusion, exclusion, or they may be inconclusive
- Requirements for knowns (exemplars) are critical and differ according to the type of pattern. The comparison process is rather similar with different pattern types.
- The process moves from pattern recognition, comparison, and identification of class characteristics toward positive individualization
- Footwear and tire impressions are the most common individualization patterns, leaving aside fingerprints, bare footprints, and document evidence (such as handwriting)
- Shape, pattern and form are compared in a manner similar to the way the human mind recognizes people and objects; shape and form patterns include handwriting and the morphology of drug crystals
- The methods of pattern evidence comparison and the conclusion of the pattern analysis have come into question by a recent NAS and PCAST Report and a few courts as not sufficiently based in science or that they do not meet the Daubert criteria for admissibility of scientific evidence into court
- Several other individualization pattern types, including bitemarks, bloodstain pattern, fingerprint, hair, firearm, and toolmark comparisons were also questioned for the strength of their scientific basis by the National Academy of Science

(a)

Figure 5.1 (a) Tractor-trailer thought to have been involved in hit-and-run on a state trooper; a close up of impression left on the dirt of the tractor trailer. *(Continued)*

(b)

Figure 5.1 (Continued) (b) Close up of the patch at the shoulder of state troopers.

CLASSIFICATION—TYPES OF PHYSICAL PATTERNS FOR COMPARISON

There are four categories of pattern evidence commonly found at crime scenes, as noted in the Chapter 4 introduction. Each of them has subcategories and classes for different types of evidence.

Physical fit or Direct fit match

Physical fit, also called direct fit, matching is the simplest, yet, often one of the most persuasive types of evidence that can be obtained. This type of matching is often referred to as primary or first-order matching (Figure 5.2). If the fractured or torn pieces are solid and from an item or object expected to fracture randomly, and if the questioned pieces can be directly fit in the rest of known pieces back together, a direct physical match is obtained.

A physical direct and fit match is a match between or among pieces of a randomly fractured, torn, or cut object. Such a match might show that the pieces were originally part of the same item. A common example of direct fit-match is a piece of broken glass or plastic lens from a hit-and-run scene that fit perfectly into the broken lens still on a suspected vehicle.

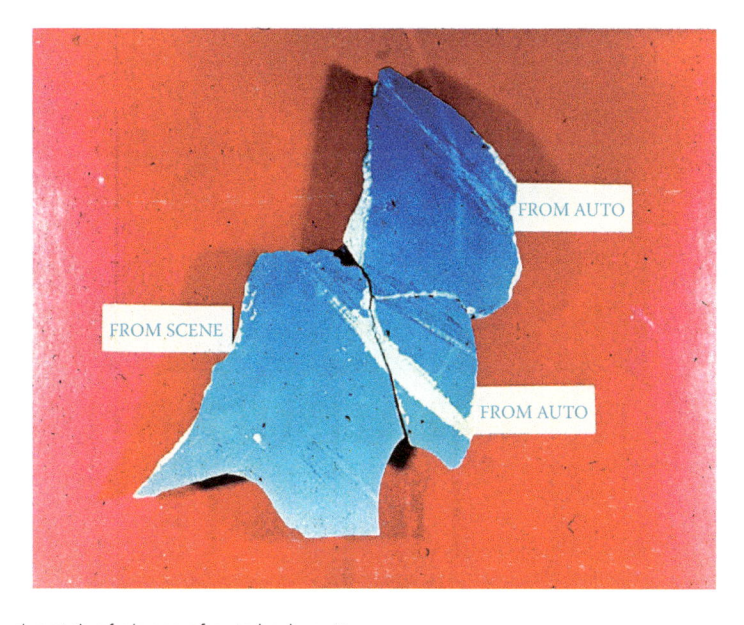

Figure 5.2 Direct physical match of pieces of auto body putty.

Indirect match

An indirect match is often referred to as a secondary or second-order matching. If some portions are missing, or if the item is not solid (like fabric) or has distorted or poorly defined edges, a secondary match may only be possible. In such cases, the match is often the result of a complex pattern that carries across the boundaries of the two objects. A common example is the grain of a wood board that has been sawn (Figure 5.3). Although the saw has removed sufficient material to make a direct physical match impossible, in most cases, the common pattern of the wood grain or growth patterns across the boundary often allows a for an indirect pattern match. Another example is where striae correspond across a break in an object.

Imprint, impression, and striation comparison

Impression marks result when a patterned object contacts a receiving surface and leaves a negative impression of itself. Generally, an impression mark is produced from direct static contact of the object with a receiving surface. Among the most familiar examples are plastic fingerprints (see Chapter 8), tire and footwear impressions. Striation marks result from a tool, objects or body parts moving relative to a receiving surface. The most forensically important striations are the markings imposed by firearms barrels onto bullets or a screw driver sliding over a metal surface. This type of mark is generally produced by a dynamic motion. In mark comparisons, a "questioned" mark from a scene is always compared with a known mark produced on a test object by the object suspected of being the source of the evidence mark and with a similar motion.

Shape, pattern, marks, and form determination

Shape, pattern, marks, and form are terms we use to denote the extraordinary ability of the human mind to discern and discriminate shapes, colors, patterns, designs, formations, figures, forms, and the subtle differences between them. A familiar example is recognizing a friend or acquaintance in a crowd. You have no trouble doing this once you know the person. Yet if you tried to explain how you did it—in enough detail to allow someone else who did not know the person to recognize him or her—you probably would find it difficult to do so. Although we cannot enumerate all the features and characteristics of the pattern in enough detail to make it uniquely recognizable to someone else, humans are innately very good at pattern recognition and matching. Recognizing a friend's face involves comparing the visual data against an existing mental image and "making a match." Various patterns and impressions often

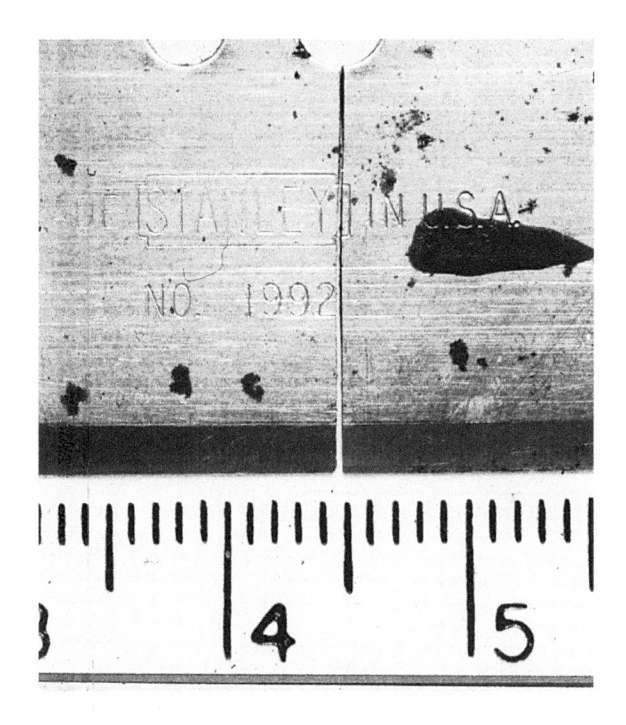

Figure 5.3 Matching stria across a break in a utility knife blade.

found at a crime scene or left on clothing, vehicles, or other objects can be recognized because of features, shapes, or forms. The experienced criminalist can recognize certain objects, weapons, numbers, figures, or shapes and know to preserve those marks.

This process is the same one forensic scientists use in comparing, hair, fiber, rope constructions, shape formations, fabric designs, or crystals under a microscope and in comparing handwriting. Later in the chapter, we will discuss why it is so difficult to explain the comparison process in detail (e.g., the process is subjective) or offer a statistical value on such a comparison. Thus, the result of such determinations has created some problems when the Daubert admissibility standards are applied; however, it should be noted that the subjectivity of a process does not make it intrinsically inaccurate. We recognize vehicles, people, and figures we know very accurately.

GENERAL PRINCIPLES IN PHYSICAL PATTERN COMPARISONS

The process of recognition

This is the first step of physical pattern comparison. It is the most basic but the most important step. The main reason for forensic scientists to be at the crime scene searching for evidence or examining physical evidence in the laboratory is to recognize potential evidence, which will be useful for linkage or to exclude an individual in a specific case. The examiner has to process the evidence using his ability to recognize features, forms, design, and the class characteristics of the pattern from the rest of materials or environment.

The process of identification

There are some general principles that apply to all types of pattern evidence analysis. As we noted earlier, pattern evidence experts use the term "positive identification" to mean they have matched questioned and known specimens well enough to say that they had the same origin.

The comparison process begins by insuring that the proper specimens are being compared side by side. Positives must be compared with positives, and negatives with negatives. If a sneaker is a positive; for example, its indentation in mud is a negative. The cast of that indentation is again a positive. The cast (the questioned) would be compared with the shoe itself (the known). On the other hand, a grease-dirt impression of the sneaker on a linoleum floor is a negative. A tape lift or photograph of the impression is still a negative. These would be compared not to the sneaker but to an inked impression made in the lab using the sneaker. In this way, a negative is compared with a negative.

Please refer to Table 4.1 in Chapter 4 for a summary of the features of the pattern comparison process.

An important step is general orientation of the impressions to be compared. With footwear, left and right orientations can get confusing when you are working with lifts. Once the orientations of questioned and known impressions are the same, the class characteristics are compared.

Class characteristics are those that identify the item as a member of a class. Brand, size, shape, and possibly the general pattern are class features. These must match if the questioned pattern (Q) and the known object (K) are really identical. Many times, forensic scientists will only be able to match class characteristics of an imprint or impression from the scene with a known object. The conclusion will be that the Q "is consistent" with the K. It is important to remember that the "class" characteristics can range from quite large to very small. Thus, the evidentiary value and its weight accorded the inclusion in the class will also vary.

Next, *individual* characteristics are compared. These are features that differ among individual members of the same class and, thus, have the ability to confer individuality. There are specific individual features used in fingerprint and firearms identification comparisons, which will be discussed in subsequent chapters. With tire or footwear impressions, the individual characteristics are usually the result of wear. Besides just modification of the pattern due to wear on the surfaces, there can be cuts, tears, abrasions, wear, and other damage patterns that are random and accidental and would not be expected to be exactly duplicated in another object from the same class. If a sufficient number of individual features are found to be the same in both the Q and K specimens, and there are **no unexplained differences**, the examiner makes a positive identification. That means the examiner is ready to give the opinion that Q and K had the same origin.

Physical fit and matches

Direct physical fit and matching is *jigsaw fit* matching and is applicable to solid materials (typically glass, plastic, or metal) that break into pieces in a random way. Careful matching can show, unequivocally, that the pieces were originally part of the same object or item. The value in a case is usually that some pieces have been recovered from a victim, a suspect, or a location (such as a suspect vehicle), while the others have been recovered at a scene. Match results, thus, place a suspect or a vehicle at the scene and link a victim, suspect, or scene together. Hit-and-run scenes may have broken pieces of head lamp, parking lamp lenses, grill parts, license plate frames (Figure 5.4), or frame pieces. If the responsible vehicle can be found before any repairs are done, it can be placed at the hit-and-run scene by physical matching the objects still remaining on the vehicle to evidence recovered from the scene. Other solid physical items that fracture or separate in a random way may also sometimes be useful as objects for physical fit-matches. Buttons, sometimes broken or cut plastic or cardboard, and broken metal objects like a knife, screwdriver, or bolts may also be subject to physical matches.

There are a few requirements for physical matches. First, the object or item should have fractured or been severed in a way that produces a complex three-dimensional surface. In other words, the object or item would not be expected to separate in exactly the same way twice. The pieces must be capable of being realigned three dimensionally. Physical matching is intuitive, and correspondingly persuasive. Another requirement is that the object is not too fragmented; otherwise reconstruction would be prohibitively difficult for a complete *direct* physical fit and match, all the major pieces must be available. If there are missing pieces in between the pieces trying to be matched, the match becomes indirect.

Indirect physical matching is fitting, or attempting to fit, severed, broken, or torn pieces of an item or object together in a way that convincingly demonstrates that the pieces were once part of the same structure. If an object or item is solid and randomly fractured, would normally be amenable to direct physical matching but there are significant missing pieces, or the cutting edge is too even and regular, any match becomes secondary. Indirect matching also applies to items like fabric, paper or items that are not solid, and therefore cannot be jigsaw fit matched, and whose edges may also be distorted during the cutting or tearing process making matching more difficult. Other examples are broken or sawn pieces of wood (Figure 5.5), torn edges along perforated papers, and matching a torn-out match from a matchbook to the remaining stubs. Secondary physical matches are often less intrinsically convincing than primary matches, and examiners must convince themselves the match really exists before reporting it.

Figure 5.6 provides a full list of the various kinds of pattern evidence.

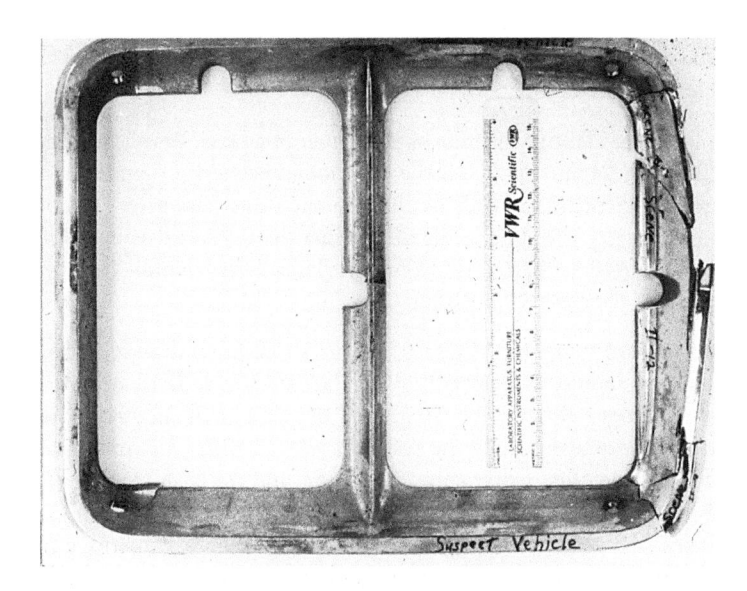

Figure 5.4 Jigsaw direct physical match between pieces of a license plate frame.

Figure 5.5 Secondary physical match using wood grain where a direct match is not possible due to wood removed in sawing process.

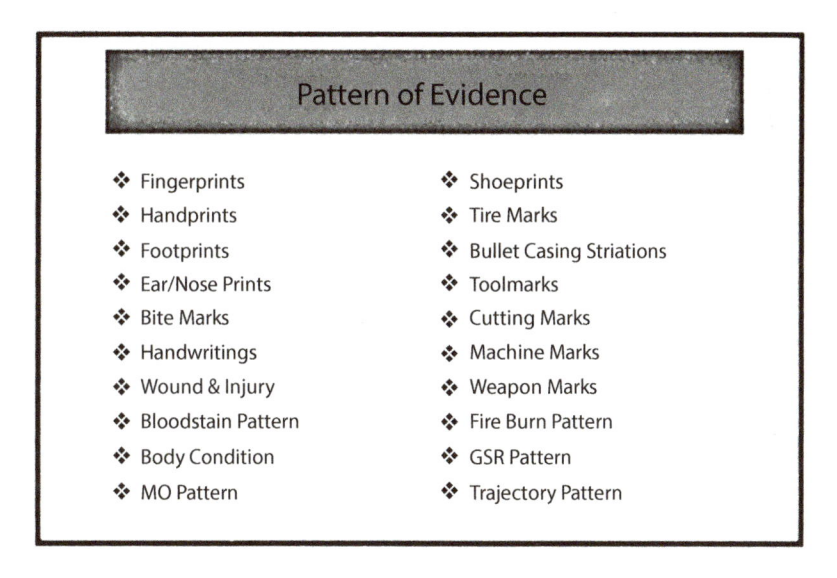

Figure 5.6 The various kinds of pattern evidence.

Exclusions, and inconclusive and insufficient detail

If class characteristics do not match, or if there are unexplained differences between Q and K, the Q may be excluded as having come from the same source as K. Exclusions are absolute. It should be remembered that sometimes an exclusion is just as crucial to an investigation or prosecution as an individualized identification.

Sometimes, a forensic examiner may find that there is not enough individual marking details to make a valid comparison. In that case, "insufficient detail for comparison" might be reported. Similarly, even if some details are present, they may be insufficient to make an identification. At the same time, there may be a certain number of class characteristics present, so the examiner cannot exclude, so these comparisons may be reported "inconclusive."

The Daubert criteria and the National Academy of Sciences report on pattern comparisons

The U.S. Supreme Court's decision in Daubert v. Merrell Dow (Chapter 2) set down new criteria for the admissibility of scientific evidence into court proceedings. In effect, the Supreme Court said that results intended to be introduced into court proceedings as scientific evidence, through expert witness testimony, must be based on a hypothesis-testing scientific method model. For evidence analysis that uses established methods of chemistry or biochemistry, like drug identification, toxicology, DNA typing, fiber identification, and so forth, it is almost self-evident that the conclusions are scientifically based, and thus admissible. With pattern evidence analysis, this is not so obvious.

Pattern evidence identification criteria rely on many decades of experience along with limited experimentation designed to show that trained, experienced examiners can, in fact, distinguish different individual patterns, even if they are expected to be quite similar, such as with bullets from consecutively manufactured gun barrels, or fingerprints from identical twins. There are any number of "experiments" of this kind in the open literature. You might call this "outcome research." That is, given a group of items of pattern evidence for comparison, can experienced examiners match the correct ones and exclude the ones that don't match—do they obtain the correct outcome? You can also look at this sort of activity as proficiency testing. It can be voluntary, or it may be required by an accrediting or certifying body. Generally, trained, experienced examiners do obtain the correct answer. It can also be shown that while trained, experienced examiners nearly always get such exercises right, inexperienced and/or untrained individuals do not. Pattern evidence examiners are typically trained one on one by senior, experienced examiners in the same field. The training is long, and considerable proficiency testing and supervised casework are typically required before a new examiner is allowed to do casework alone.

Still, the collective body of experience and training over many decades does not constitute a systematic, hypothesis-driven, experimentally based test of pattern evidence individuality. Many times, erroneous identifications were reported in cases and those erroneous identifications were testified to in court. Some courts, applying the *Daubert* standards, have found this situation troubling. At least one court has characterized handwriting comparison as a technical expertise (and admissible as such), but not reaching the threshold to be called "scientific."

There are approaches available to carry out experiments that will firmly establish the scientific grounds, once and for all, for this is what most people already believe—namely, that individualization patterns are individualizable. The issues raised by Daubert and the National Academy of Sciences findings are not new. Recently National institute of Standards and Technology (NIST) has established study groups and committees to review guidelines and establish national standards for the examination of pattern evidence. However, this process will take time to work through, and whether this is the solution will have to be seen.

IMPRESSION AND STRIATION MARK COMPARISON

We will use the term "impressions" as the broader one, which includes imprints and indentations.

Impressions: imprints and indentations

The distinction between *imprints* and *indentations* is made because the two are, generally, found on different surfaces. An *imprint* is a mark that is effectively two-dimensional. By that we mean the mark has essentially no depth and no three-dimensional characters. They are generally left on hard, flat surfaces. An *indentation* is a mark that has a distinct three-dimensional character. An object makes an imprint on a hard, receiving surface, while it makes an indentation in a softer receiving surface.

The terminology describing marks with depth and marks without depth can vary with different kinds of evidence. For instance, fingerprints without depth are often called "residue" prints, while those in soft media (like silly putty or butter) are called "impression" or "plastic" prints. Similarly, the mark a firing pin makes on a cartridge primer is usually called a "firing pin impression," although it is an indentation.

Striations

Striation marks are usually toolmarks made by a tool sliding along a receiving surface or two hard surfaces in direct, dynamic contact. A special case of "toolmarking" occurs with rifled barrels of firearms, which impart striation markings onto bullet surfaces as the bullet passes down the rifled barrel. Striation marks can also be present on a tool or object and be impressed onto a receiving surface. Grooves are cut into firearms barrels and this process leaves toolmarks inside the barrel. These markings form the basis for firearms identification. This subject is discussed in Chapter 8.

Collection and preservation of impressions

Generally, both imprints and indentations should be documented by photography before collection. The photographs should be taken to show the location of the impression at the scene, and to show the impression with an evidence number sign and ruler next to it. Depending on the impression and how it is collected, a very good documentation photograph may be of nearly equal value as the collected mark for comparison purposes. Smaller impressions can be photographed 1:1 (image at film plane is the same size as the actual object being imaged) with a 35 mm or digital camera. If possible, such a photograph should be taken. With larger impressions, some reduction will inevitably occur to fit the image onto the film. However, the larger the negative is, the less the loss of detail. That is the reason some crime scene photographers use 4 × 5 cameras for larger impressions, like footwear or tire tracks. Care should be taken to ensure that the plane of the film is parallel to the plane of the impression (to avoid distortion), and a tripod should be used. Multiple exposures are always a good idea too. With indentations, use of a side lighting technique is recommended. If the indent is shallow, grazing illumination (lighting almost parallel to the surface) will help to make it more visible and, therefore, easier to record (Figure 5.7).

Following thorough photographic documentation, imprints are generally tape-lifted, gel lifted, or lifted with an electrostatic lifting device, and indentations are cast. We note again here, as was stated in Chapter 3, that it is always best to collect the entire, intact object bearing the impression, if possible. If that is not possible, investigators must think about lifts or casts.

Tape lifting is a classical method for collecting and preserving fingerprints, usually after they have been enhanced by powder dusting or other treatment (Chapter 6). But tape lifting can also be used for footwear or other imprints. There are larger lifting tapes and different colored backings available. The technique involves overlaying the imprint with the sticky tape smoothly and avoiding air bubbles. Then the tape is lifted up in a single, smooth motion and transferred to a suitable backing surface that contrasts the color of the dusting powder. In addition, there are commercial "gel lifters" available for this purpose. They are made of material that picks up the dirt, grease, oil, and so on efficiently (Figure 5.8). For dust prints, which are often subtle and would be destroyed by tape lifting, an electrostatic dust lifter device is available commercially. It imposes a charge on a special film material (metalized mylar sheet) laid over the imprint. The dust particles that form the impression are attracted and transferred to the film and, in that way, the pattern can be made more visible and be preserved (Figure 5.9).

Figure 5.7 The use of various photography techniques can greatly enhance the illustrative quality, and evidentiary utility, of crime scene photographs. (Courtesy of Shutterstock, New York.)

Figure 5.8 Lift of shoe print in dust using gel lifter.

Figure 5.9 A shoe impression in dust, barely visible on a carpet, reveals incredible detail electrostatically lifted onto lifting film.

Indentation marks are cast using dental stone, a special form of Plaster of Paris. It comes in many grades and colors. This material is used by dentists to make accurate casts of teeth so that inlays, crowns, or other restorations can be made. The steps are shown in Figure 5.10a and b. Generally, the impression is carefully sprayed with lacquer; the spray is deflected onto the mark itself to avoid damaging any detail to preserve detail during addition of the plaster (Figure 5.10a). The dental stone is then mixed to around the consistency of pancake batter and poured into the impression using a spatula or stick as a pouring aid (Figure 5.10b and c). The mark is covered completely. The plaster is then allowed to set for a time, and a strengthening material (sticks, chicken wire) may then be placed onto the cast before more dental stone is added and allowed to set. Once the cast is set up, it can be initialed or marked and then carefully collected. The cast should be packaged so as to avoid breakage during transport. If the casting was done in soil, any soil clinging to the cast should be left in place, as a soil reference material (Chapter 14).

Examination of physical pattern evidence 111

Figure 5.10 (a) Preparing an impression for casting with dental stone; the marking is generally sprayed with lacquer. (b) The marking is "framed" to contain the dental stone. (c) The casting of an impression with dental stone. Half the dental stone may be poured, then a solid support material, such as chicken wire, may be added to give the cast more strength before pouring the last half. Identifying information such as case number, initials of the person making the case, and other such information may be scratched into the dental stone surface before it completely hardens.

There is a special casting material for indentation marks in snow, called Snow Print wax. When plaster sets up, it releases heat. Heat would obviously destroy pattern detail in snow. The Snow Print Wax® is used to stabilize the mark and insulate it from the heat of the setting dental stone (Figure 5.11). There are also some special casting materials like Mikrosil® that may be used to capture fine detail in sliding toolmarks. Many of these are silicone rubber-type casting materials similar to those used in dental work.

Lifts and casts need to be properly labeled with the case number, date, investigator's initials, evidence number, and location.

Figure 5.11 Spraying a shoe impression in snow with snow print wax to stabilize it before casting.

FOOTWEAR, TIRE, AND OTHER IMPRESSIONS

We have mentioned footwear and tire impressions in the discussions above. Those are the most commonly encountered impression evidence other than fingerprints, toolmarks, or firearms markings.

Footwear could be any shoe, slipper, sandal, sock, or stocking. Any of these can leave an impression. It is also possible to have a foot print, the impression of a bare foot. Since foot surfaces have friction ridge skin, like those on fingertips that form fingerprints, bare footprints are compared by fingerprint examiners in the same manner as fingerprints.

Most footwear and tires are mass produced, so there are typically millions of pieces with many of the same class characteristics. Sufficient wear may introduce enough individual characteristics to make an identification possible. Keep in mind that an evidentiary impression may not reflect the whole shoe or tire, just a part of it. Further, the shoe or tire suspected of making the impression must be recovered before a known impression can be produced and a comparison done.

Known footwear impressions should be produced with someone wearing the footwear such as the suspect or someone who is approximately the same size and same weight as the suspected wearer. Similarly, known tire impressions should be produced with the suspected tire on the vehicle. And the mark must be made that covers a complete rotation of the tire, since there is no *a priori* way of knowing which part of the tire left the questioned impression at the scene.

Other than fingerprints or footprints, footwear and tire impressions are the most common impression marks encountered as evidence. But impressions could in theory be made by any object or material. In hit-and-run cases, there can sometimes be fabric impressions from the victim's clothing on the bumper. This type of impression may have been more common when automobile bumpers were bare metal. In Chapter 4, we mentioned "contact transfer" blood patterns. These are similar to the patterns we are discussing here, except they may not have much microstructure so that a point-by-point comparison could be done. But the contact transfer pattern might match an aspect of an object and enable an expert to say that the pattern is consistent with having been made by blood on that object.

The Lead Case illustrations (Figures 5.1a-b) show how important impression evidence can be in associating a person with a crime scene.

CLARIFICATION AND CONTRAST IMPROVEMENT TECHNIQUES

In many cases, patterns are difficult to analyze or compare because of poor visibility. There can be lack of contrast between pattern and background, very weak pattern or interference between pattern, and complex backgrounds. Three different types of techniques can be used to address such problems and make pattern information more useful.

Several photographic techniques including side lighting and use of special illumination, such as UV, laser, or alternate light sources have been mentioned. In addition, filters can be used to increase contrast where background and pattern are of different colors. Special films and high-contrast printing papers can also help in some situations.

A number of physical and chemical techniques are also used to clarify or enhance contrast. It has been mentioned that chemical blood test reagents can be used to make very subtle blood patterns visible through color change or added image intensity of even very weak patterns. There are also chemicals that react with proteins and biological materials commonly found in dust or soil that can be used to enhance shoeprints or other such imprints. Fingerprint powders and a variety of lifters can isolate the imprint from background or enhance contrast.

Finally, many of the above can be done more quickly and more effectively using digital imaging techniques. There are a number of programs such as Photoshop and Image Pro that have tremendous power to improve the clarity and contrast of images. They work well on digital images captured either photographically or scanned on a flatbed scanner. Once the image information is digitalized, the variety of techniques available is quite great. They must be done in a well-documented way if such improved images are to be used in court. The Scientific Working Group on Imaging Technology (SWGIT) established proposed guidelines for image acquisition, handling, documentation, and enhancement in 2005. Additional guidelines will soon be published by a NIST study group.

With all the mentioned enhancement techniques, it is critical that the original image be well documented before any of the above techniques are used since some may degrade or destroy the image rather than clarify it.

WEAPON, TOOL, AND OBJECT MARKS

Many types of patterns might appear at a crime scene or be left on a body or other object. Those patterns can be produced by direct contact of two surfaces, such as toolmarks, ear prints, button prints or belt buckle prints, or produced by force, such as impact, contact, cuts, wounds, or pattern injuries.

SHAPE, PATTERN, MARK, AND FORM COMPARISONS

As noted above, patterns called "shape, pattern, mark, and form" can be compared, but the comparison between Q and K is conducted looking at the overall color, design, structure, markings, and construction of the pattern evidence. Sometimes there are individualizing features/or specific markings in the general patterns, but the individuality of the pattern consists of more than a list of features that match. Above, we likened this comparison to recognizing the face of a friend or acquaintance.

The most prominent examples of these patterns in forensic science are the microscopic structure of hair and handwriting. These are discussed in Chapters 14 and 7, respectively.

OTHER PATTERNS

A few other special types of patterns can be important in forensic science. Bitemarks, certain special features in the human skeleton, and voice patterns, are all used to help identify individuals. They will be described in Chapter 6, along with fingerprints. Patterned injuries (Chapter 4) may also be of considerable forensic use, but are primarily the responsibility of the forensic pathologist. In jurisdictions where there is no forensic pathologist available, such patterns may be compared by forensic laboratory personnel.

Tear, cut, or damage patterns on clothing, furniture, or other objects can often provide evidence through pattern analysis. For example, was a window screen cut from the inside or outside? Was damage to a victim's clothing made by tearing or cutting? Was a knife wound pattern on someone's arms self-made or made by an assailant? Was a pattern found on carpet, a car seat, or floor consistent with a particular object?

CONCLUDING COMMENTS

Pattern evidence, as we have noted, can be seen as one of the four major categories of physical evidence in terms of the scope of criminalistics—the others are chemical, trace and materials, and biological evidence.

Pattern evidence is usually thought of as physical matching exercises, imprints, striations, shapes and forms, and reconstruction patterns. Firearms, toolmarks, fingerprints, questioned documents (especially handwriting), footwear, and tire tracks are the most common examples of pattern evidence.

It is useful to remember; however, that even in chemical and biological analysis, some of the analytical methods produce patterns that must be interpreted. Infrared spectra (Chapter 13), gas chromatographic patterns (Chapter 13) and DNA profiles (Chapter 11) are some examples.

The forensic science profession in general has been subject to increasing scrutiny in recent years by the press and courts. In 2009, the National Academy of Sciences published a landmark study calling into question the scientific basis of many of the longstanding forensic identification techniques, like fingerprint comparison, bite mark comparison, hair analysis, bullet and firearm identification, and arson investigations. The NAS report stated, "The simple reality is that the interpretation of forensic evidence has not always been based on scientific studies to determine its validity. This is a serious problem." Forensic odontology (particularly bitemark patterns) was the subject of that particular criticism.

The level of certainty of general pattern evidence comparison (with the exception of fit and match) cannot be compared to DNA typing. The issue of quantifying the uncertainty in pattern analyses is unclear and the reporting of the value of a comparison and its weight currently does not meet scientific standards. The statistical models developed for pattern comparison are often confusing to examiner and jury. Interpretation of a comparison often may be subjective. Nonetheless pattern evidence still has tremendous value in forensic investigation. It is critical to establish national standards and guidelines on pattern evidence examination and interpretation. It is also equally important to provide education and training for examiners involved in examining and interpreting pattern evidence.

CASE ILLUSTRATION 5.1

Several years ago, Warwick, Rhode Island, was shocked by the finding of all three members of one family—a mother and her two daughters—brutally stabbed to death in their home. The mother was raising her daughters by herself, had no known enemies, and was not involved in any kind of criminal activity. There was no obvious motive for the homicide.

Forensic experts from the FBI and the Connecticut State Police were asked to assist in the investigation. The youngest daughter's body was found in the kitchen. The oldest daughter's body was found in the hallway, and the mother's body in the bedroom. The clothing on all three was intact, and there were no signs of sexual attack. It did not appear that rape was the motive.

A kitchen window at the back of the house had been pried open and appeared to be the point of entry. The window was high off the ground, which suggested that the perpetrator more than likely climbing in the window would have to be of above-average height and strength. Inside the house, directly below the window, a small kitchen table appeared to have just been broken, as if someone heavy had stepped on the table while coming through the window. A reddish, footwear imprint pattern was observed on the table top. In addition, a partial palm print was also found.

There was a large amount of blood in the kitchen, hallway, and bedroom. Bloody footprints were observed leading from the kitchen to the living room window, then toward the hallway. Upon closer examination, it was found that the prints were made by socks, not shoes. The sock prints were found throughout the house, as if someone, one person only, were searching every room, looking for something.

It was also discovered that blood stains were in different stages of coagulation. The youngest daughter's blood was clearly dry. The mother's blood was in an advanced stage of coagulation. But there were other drops of blood that were fresher and in a semi-liquid stage. These blood drops were consistent with low-velocity, passive dripping from a height of approximately 3–4 feet. The fresh blood drops were consistent with someone bleeding from a serious cut of the hand or finger.

(Continued)

CASE ILLUSTRATION 5.1 (Continued)

A crime scene profile indicated that the house was the primary scene, and sex was not the motive. The suspect was more than likely a strong young man, perhaps a teenager, of above average height, overweight, and with a serious cut on one hand or arm.

Police put out the word, and, in one of those examples of keen observation that would bring a pat on the back from every law enforcement agency, a town of Warwick detective noticed a teenager working on a car. What caught the officer's attention was the bandage the boy wore on his hand. The officer stopped to chat and asked the boy about the bandage. Craig Price said he had cut his hand. The police officer asked Price to come to the station.

A court order was obtained for blood, hair, fingerprints, footprint, and shoes exemplars. Laboratory testing showed the following results:

1. The partial handprint on the kitchen top matched the right hand of the suspect
2. The shoe wear imprint on the kitchen table top showed a similar size, shape, sole pattern, and other class characteristics to the suspect's shoe
3. Several latent fingerprints were found inside the house that matched the known fingerprints of the suspect
4. DNA profiling showed that several blood drops matched the known DNA from the suspect

The evidence was sufficient to allow prosecution to proceed and for a trial jury to convict this suspect of three counts of murder.

Key terms

class characteristics
direct physical match
exclusion
identification
impression
imprint
inconclusive
indentation
indirect physical match
individual characteristics
insufficient detail
physical match
shape and form
striation

Review questions

1. What are the three major types of patterns for individualization?
2. Discuss the general principles involved in pattern analysis and comparison. What is an "identification?"
3. What is the difference between imprints and indentations? What are examples of each?
4. What are the most evidentiary striation markings?
5. What are class characteristics?
6. What are individual characteristics?

7. What are shape and form comparisons?
8. Why is pattern evidence identification an issue under the Daubert standard?
9. What are the main criteria for making a direct physical match?
10. What would cause an examiner to conclude that there was "insufficient detail for comparison" in a pattern evidence comparison case?

Fill in/multiple choice

1. Whether a mark is three-dimensional or essentially two-dimensional determines if it is considered an imprint or a(n) _____ marking.
2. A marking on the surface made by the movement of an object across the surface is referred to by the technical term _____.
3. Point-by-point matching of the jagged ends of two pieces of a wooden axe handle that had broken in an irregular way is known as _____.
   a. comparison of markings
   b. direct physical matching
   c. general form recognition
   d. indirect physical matching
4. When shoe and tire marks are impressed in soft earth at a crime scene, their documentation and preservation are best accomplished by _____ and then _____.

References and further readings

Bodzniak, W. J. *Footwear Impression Evidence: Detection, Recovery, and Examination*, 2nd ed. Boca Raton, FL: CRC Press, 1999.

Bodziak, W. J. *Forensic Footwear Evidence*. Boca Raton, FL: CRC Press/Taylor & Francis, 2016.

David Sheets, H., S. Gross, G. Langenberg, P. J. Bush, and M. A. Bush. "Shape Measurement Tools in Footwear Analysis: A Statistical Investigation of Accidental Characteristics Over Time." *Forensic Science International* 232 (2013): 84–91.

Farrugia, K. J. et al. "A Comparison of Enhancement Techniques for Footwear Impressions on Dark and Patterned Fabrics." *Journal of Forensic Sciences* 58 (2013). doi:10.1111/1556-4029.12209

Scientific Working Group on Imaging Technology (SWGIT). "Best Practices for Forensic Image Analysis, Version 1.5, 3/14/05." *Forensic Science Communications*, 7, no. 4 (October 2005).

CHAPTER 6

Fingerprints and other personal identification patterns

Lead Case: California v. Gerald Mason—Fingerprints Provide Solution to a Police Killing Cold Case

This case occurred in the 1950s but was unsolved, and became a "cold" case. It was eventually solved because a latent fingerprint taken from the scene decades ago could be subjected to a nation-wide search, using Automated Fingerprint Identification System (AFIS) technology that was unavailable when the crime occurred. The case was featured on a CBS News magazine program, *48 Hours Mystery*. In July 1957, two young police officers on a routine traffic stop were gunned down in the Los Angeles suburb of El Segundo, CA.

On the night the crime occurred, four teenagers were coming home from a summer party when they decided to make a stop at Lover's Lane. "I rolled the window down. … And that's when the gun came through the window," recalled one of them. "This is a robbery." I thought, "Gotta be somebody pulling a prank." But the gun was real. The gunman came prepared with surgical tape and a flashlight. He covered the teenagers' eyes with tape and ordered them to take their clothes off. They had little choice but to do what they were told. The perpetrator came around from the driver's side to the passenger side, opened the car door, and raped one of the girls. He told the four to get out of the car, and said, "I think I'm gonna kill you. I want you to march out into the field." Then, the car door closed, and he drove away.

While making his getaway in the stolen 1949 Ford, he stopped for a red light, then, for no apparent reason, drove on through it. A patrol car parked off the side of the road saw the stop light violation and proceeded to pull the violator over. El Segundo police officer Richard Phillips and rookie officer Milton Curtis were in the police vehicle. A second patrol vehicle with officers James Gilbert and Charlie Porter drove by shortly, thereafter, but having no reason to be suspicious and still unaware of the events at the Lover's Lane, drove on. Moments later, there was a radio call in which Philips said that they'd been shot and needed an ambulance. Porter and his partner raced back to the scene. Phillips and Curtis were both dead, and the killer had simply disappeared. Hundreds of police from El Segundo and neighboring areas searched all night. They found the stolen Ford, but there was no sign of the suspect.

The crime became one of the oldest unsolved murder cases in Los Angeles County. At the time of the murders, the 1949 Ford hadn't yet been reported stolen, and the four teenagers had not yet told the police what happened to them.

Investigators arriving that morning to look at the stolen car noted bullet holes in the trunk and in the rear window. It appeared the car had been hit three times. Officer Phillips had fired six shots at the vehicle, before he died, and hit it three times. Two rounds were recovered from the vehicle's interior, but one was not. Police theorized that the killer might have been hit. Several latent prints were taken from the interior of the vehicle. At the time, however, automated fingerprint search systems did not yet exist.

In 2002, a woman called the El Segundo Police and said she had heard an uncle bragging about being responsible for the murders. A comparison of the latent prints from the vehicle with those of the uncle did not yield a match. The fingerprint examiners decided to take the search further, however. They cleaned up the latent a bit using modern digital imaging techniques and were eventually able to search for it in the nationwide Integrated Automated Fingerprint Identification System (IAFIS).

The print matched Gerald F. Mason, who had been arrested for burglary in 1956 in South Carolina. It was the only time he had ever been arrested, and it was the only record of his prints on file. Mason was easily located, still living in his hometown of Columbia, SC. He wasn't a career criminal but a retiree living with his family.

Looking through boxes of evidence in the case, collected over the years, turned up the actual murder weapon. It had been recovered in 1960 in a back yard not far from the original murder scene by a man cutting weeds. The serial number allowed the gun to be traced to Shreveport, LA. The gun was sold there in 1957 by Billy Gene Clark to someone who called himself G.D. Wilson. A "George D. Wilson" could be tracked to a nearby YMCA, but no one with the name could ever be found to match the latent print. The name was an alias but after the fingerprint led the investigators to Mason, a questioned document examiner was able to match the registration signature with the handwriting of Gerald F. Mason.

Mason was nearly 70 when El Segundo police arrested him at his home. There is no record of his ever having committed another crime. When he was examined, it was discovered that Mason had a bullet-shaped scar on his back, consistent with being hit by a bullet from Phillips' gun. After a hearing in South Carolina, Mason agreed to return to Los Angeles. He pleaded guilty to murdering officers Phillips and Curtis, and he tried to make amends before being sentenced to life in prison: "It's impossible to express to so many people how sorry I am. I do not understand why I did this. It does not fit in my life. It is not the person I know. I detest these crimes."

Why did he do it? "I didn't have a family life. I didn't have any place to go, and things were not going well for me, so I took off to California," said Mason. "I bought a gun at Shreveport with the intention of using it simply as a deterrent in so far as I was hitchhiking." When asked why he attacked the teenagers and raped a 15-year-old girl, Mason said he really didn't remember. But as to why he killed two cops in cold blood, Mason's answer was: "I thought, 'If I don't get them, they're going to get me.' So when the officer turned away from me, I shot both officers, got back in the car and drove away."

LEARNING OBJECTIVES
- Fingerprints are an old and very valuable type of physical evidence
- What friction ridge skin is and how it makes up fingerprints
- Use of fingerprints as a means of personal identification goes back to mediaeval times, but in the western tradition dates back to the nineteenth century in British India and the UK
- Fingerprints can be classified, and the most useful system is the ten-print classification system developed by Henry
- Large files of ten-print cards cannot be readily searched for individual prints
- AFIS systems contain individual print images and can be searched for individual prints efficiently and quickly
- There are established procedures for the collection and preservation of latent fingerprints and items suspected of having latents at scenes
- There are three types of evidentiary fingerprints: visible, indentation, and latent
- Methods for visualizing latent fingerprints include physical, chemical, and special illumination techniques. There are also combination methods
- Processing latent prints with maximum efficiency and results requires a systematic approach
- The approach commonly used in fingerprint comparisons and identification may be summarized as "ACE-V" for analysis, comparison, evaluation, and verification
- Fingerprint identification specialists belong to a professional specialty group that has its own professional journal and in which certification is available
- Other patterns for personal identification include palm and sole prints, bite marks, certain skeletal features, lip and ear prints, and voice identification
- There are several definitive and several less discriminating methods for the identification of human remains
- The method used to identify human remains depends on the circumstances, the condition of the remains, and the number of possible identities

FINGERPRINTS AND AN INTRODUCTION TO BIOMETRICS

Although fingerprints are much more familiar to almost everyone, biometrics is a broader subject that includes fingerprints as one of its most important subdivisions. The term "biometrics" refers to the use of measurements of some part or feature of the human body for purposes of personal identification or verification. The availability of rapid and accurate measurement and image acquisition techniques has made practical applications of biometrics possible that would never have been imagined in the early days of fingerprint research and development. Fingerprints and their application have earned the right to our initial discussion, but we will come back to biometrics later.

> **THREE BASIC FUNCTIONS**
> - Enrollment
>   - Adding biometric information to a data file
>   - Can include screening for duplicates in database
> - Verification (one-to-one)
>   - Matching against a single record
>   - Answers "Is this person who they claim to be?"
> - Identification (one-to-many)
>   - Matching against all records in the database
>   - Answers "Do we have a record of this person?"

FINGERPRINTS—AN OLD AND TRADITIONALLY VALUABLE TYPE OF EVIDENCE

Fingerprints are among the oldest and most important kinds of information (evidence) used for human identification. The individuality of fingerprints is so impressed into the public consciousness that fingerprint analogies are regularly used in advertising interchangeably with "unique" or highly individual. Although the terms "DNA fingerprints and DNA fingerprinting" are often invoked by the popular press and even some DNA scientists, to imply individuality, most forensic scientists feel it is not proper terminology. The use of these friction ridge skin patterns on fingertips as a means of personal identification dates back many centuries. A convincing fingerprint match is almost universally accepted as certain evidence that identifies a particular person.

As noted in the previous chapter, fingerprints belong to the individualization pattern category. Fingerprints are often essentially two-dimensional (residue prints) (Figure 6.1), but in soft receiving surfaces can be three-dimensional (plastic or impression prints) (Figure 6.2). One can think of fingerprints as being used primarily to help locate, identify, and eliminate suspects in criminal cases, but that is too narrow a concept of fingerprints. The initial driving force behind the development of fingerprints was not solving crimes but unambiguous identification of individuals. Fingerprints, along with dentition patterns (see later in the chapter), are also important in making unequivocal identifications of human remains when more conventional methods of post-mortem identification cannot be used. Fingerprints are one member of a class of biometric identifiers that includes retina or iris patterns of the eye, facial recognition, hand geometry, and others.

Figure 6.1 A normal two-dimensional fingerprint.

These items would be those that might have been touched or handled by people involved in the case. The purpose might be to try and visualize fingerprints, so a cold search can be done in AFIS to try and identify or eliminate possible suspects. Another purpose might be to find out who among several people involved in a case touched or handled a particular item.

At times, fingerprints may be on items that would be difficult or impractical to collect for processing at a laboratory. In addition, some crime scene investigators have knowledge and training in the visualization of latent fingerprints at the scene. Whether to attempt development of latent prints at a scene is a decision that has to be made in each case. It will be based on the practicality of submitting the intact item, the latent print knowledge and skills of the investigators and, to some extent, on what visualization techniques would be most likely to be successful, under the circumstances.

Basic physical types of fingerprints

There are essentially three physical types of prints that may be encountered at crime scenes and/or on items of evidence: visible (patent), plastic (impression), and latent. A *visible* or *patent* print is one that needs no further "visualization" or "development" to be clearly recognizable as a fingerprint. Such a print is often composed of grease, dark oil, dirt, ink, blood, or other materials visible on the surface it sits on. It may be suitable for comparison with no additional processing. A *plastic* or *impression* print is a recognizable fingerprint indentation in a soft receiving surface, such as putty, butter, silly putty, tar, and drying paint. These prints have distinct three-dimensional character, but are immediately recognizable and often require no further processing other than side-lighting for improved visibility. A *latent* print is one that requires additional processing to be rendered clearly visible and, thus, potentially suitable for comparison. The processing of latent prints to render them visible, and hopefully suitable for comparison, is called "development" or "visualization." Great progress has been made in visualization of latent fingerprints by the clever applications of chemical and physical principles, coupled with a better understanding of the composition of latent fingerprint residues.

Initial search and detection

At the crime scene, a systematic search for possible fingerprint evidence should be carried out. The scene processors should look at areas central to the suspected criminal activity. Examination of things that appear disturbed or are likely to have been handled are obvious targets. Further, points of entry to or exit from the scene are also prime targets. The processors should try to imagine any paths through the scene that the parties involved might have taken as possible locations of latent fingerprint evidence.

Detection of fingerprints may be technically assisted by use of strong lighting at a low or grazing angle to the suspect surfaces to help make latents at least slightly visible. An instrument for detection of possible latent prints is available that uses strong ultra-violet (UV) illumination combined with a filter and an image enhancement tube, like those used in night vision devises, to help to see very weak fluorescents caused by the UV illumination. A chemical method that has been used for many years works by directing iodine vapors at areas where latent prints are suspected. The basis of latent print development is discussed below. The advantage is that iodine vapor, which is deeply colored, makes the latent material temporarily visible, but quickly evaporates and leaves the print still processable by other more permanent methods.

VISUALIZATION AND DOCUMENTATION OF EVIDENCE FINGERPRINTS

Latent fingerprint visualization methods

Latent print visualization techniques are based on using physical or chemical methods to target one or more of the components of the latent fingerprint residue itself. Methods for developing latent prints were devised based on knowledge of the latent print residue composition. A method known to be capable of visualizing one of the compounds or elements present in latent residue is applied to try and target that compound or element. For the application to be successful, it has to be possible to apply the method to evidentiary fingerprints on the variety of surfaces where they are found, and without destroying the integrity of the fingerprint pattern. The methods commonly employed fall broadly into two basic groups depending on the surface on which they are found. These are non-absorbent surfaces such as metal, finished wood, painted plaster, plastic, and many other such surfaces and absorbent surfaces, where the moisture and oils composing the print are drawn from the surface into the interior, primarily paper but also cardboard, leather, and others.

Should visualization methods be applied at the scene or at a forensic laboratory? When potential latent prints are recognized at a scene, investigators and scene technicians have a choice: apply latent print development methods at the scene; or collect the relevant item or object intact and submit it to the latent print section of a local forensic laboratory.

Any item, object, or surface at a scene thought to contain latent prints should be documented by photographs and sketches. Smaller objects or items that are easy to collect and submit should be taken to the laboratory. If it is impractical or impossible to remove a surface or object, investigators may have to apply latent development techniques at the scene. Depending on their training and experience, they may want to call on latent print examiners from a forensic lab or identification unit for consultation or assistance. Any fingerprints developed at the scene must be carefully photographed and lifted if possible. If neither the developed print nor the object can be moved or collected, the photographs will be the only record of the fingerprint.

Latent print visualization on non-absorbent surfaces
Physical methods

For many years, latent fingerprints on non-absorbent surfaces were visualized primarily using fingerprint dusting powders. The powder-dusted fingerprints were then collected by tape lifting. This is still a good technique. But powder dusting may no longer be the best choice as the technique for developing such prints. The technique selected should depend on the nature of the surface and the conditions at the collection site. An evaluation of whether any of the techniques listed below might be more suitable should be done first.

Latent fingerprints on non-absorbent surfaces are most often visualized by physical methods, ones that do not involve any chemicals reactions. Powder dusting is the most well-known and frequently used physical method. Dusting works by applying some type of fine particles to the fingerprint residue, to which the powdery material tend to adhere better than to the background material. This results in a visually contrasting ridge pattern on the background. Thus, the best known physical method is powder dusting—a mainstay of latent fingerprint detection for a long time. The most common powders are finely divided particles of inorganic materials that are available in many colors. There are also a variety of brushes available that very gently applies the powder to the surface. Careful use of the proper brush and powder often results in the development of excellent prints. Black or chemist's gray powders are superior to the other colors. These powders are produced in a way that yields more uniform particle size and generally produces better contrast. Figure 6.9a illustrates the technique. Also shown is the tape lifting (Figure 6.9b) of the dusted latent print.

Tape lifting is a fine way to preserve dusted latent fingerprints, but should only be done after photographic documentation. Fingerprint tape is basically clear cellophane tape. The tape is carefully placed just above or below the powdered print and carefully pressed down over the dusted print. When the tape is lifted, the image of the print

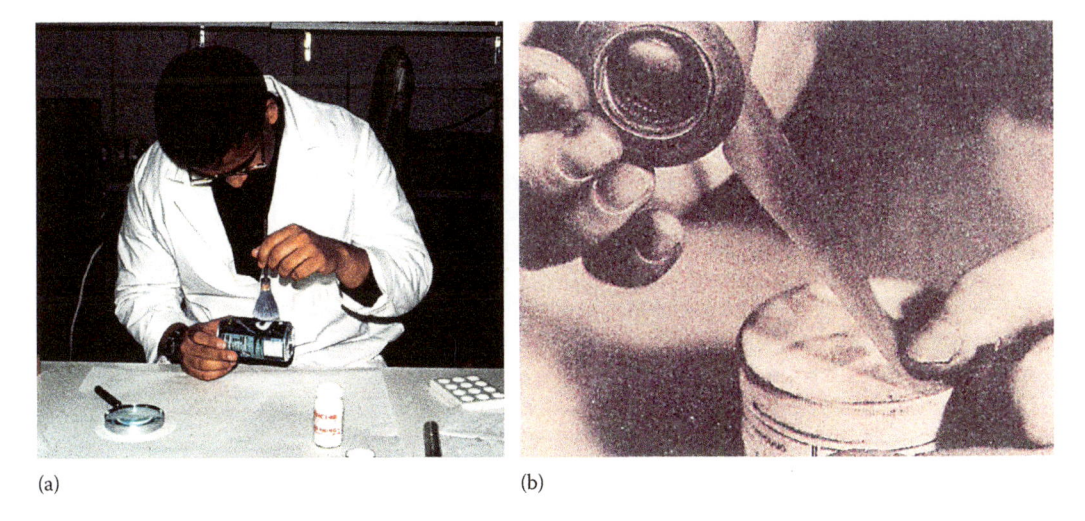

(a) (b)

Figure 6.9 (a) Applying fingerprint powder to a soda can with a typical fingerprint powder brush. (b) Using fingerprint lifting tape to remove a dusted print from a surface so that it can be placed on a card of contrasting color for easy viewing and preservation.

in powder is removed from the surface. The tape with the powder image is then mounted on a backing card that provides maximum contrast to the color of the powder (e.g., white backing for black powder). One-piece lifters (so-called hinge lifters) are commercially available for this purpose. One must ascertain if the tape might partially lift the surface material, which might partially obscure the fingerprint ridges, before trying to tape lift the print. This is done on an area of the surface close to where the print is located. The magnetic brush is a variant of the simple brush and powder combination. The original trademarked version is called the Magna Brush. Actually it is a hollow tube that contains a small retractable magnet and is not a real brush at all. The magnet in a plastic tube is placed close to the magnetic powder, fine iron filings, which is attracted to the magnet and the particles to each other to form a loose aggregate that looks and acts like the brush. The magnetic brush uses special magnetic powders that can be obtained in several colors. The principle of magnetic powder brushing visualization is the same as that for conventional powder, namely adherence of the fine particles to moisture or fatty components of the latent print residue. The magnetic brush technique is applicable to a larger variety of surfaces (especially vertical surfaces) than conventional power dusting. The magnetic brush technique also works well on some surfaces that tend to develop static charging such as some plastics. It is also a gentler technique, in the sense that there is no brush and, thus, no bristles, so it is less likely to damage the latent pattern in the brushing process.

Another physical latent print developing procedure involves small particle reagent (SPR), which is typically applied by spraying or painting on a surface or dipping a latent containing object in a solution of the reagent. The most common formulation of SPR contains molybdenum disulfide (a heavy, black very finely divided powder) suspended in a detergent solution. The particles adhere to the lipid (oily) components of the residue. SPR is most commonly used on evidence that has been wet or has been recovered from water.

Cyanoacrylate (crazy glue) fuming

A very important technique for visualizing latent prints on non-absorbent surfaces is treatment with Super Glue (cyanoacrylate esters). Super glue "enhancement" of latent print residue was first observed by scientists at the National Police Agency in Japan. The method quickly caught on and is now used by latent print examiners all over the world. In the early 1980s, fingerprint examiners at the U.S. Army Criminal Investigation Laboratory at Japan and, a little later, in the U.S. Bureau of Alcohol, Tobacco and Firearms (ATF) Laboratory introduced Super Glue fuming for latent print development in the U.S. Super Glue will vaporize (produce fumes) slowly at room temperature and more rapidly with gentle heating. These fumes interact with latent print residue by being polymerized and forming a stable friction ridge impression pattern off-white in color, usually firmly attached to the surface. Items to be processed with super glue are usually placed into fuming cabinets where the glue is induced to fill the cabinet with vapors. Glue vaporization is very slow at room temperature, so it is usually accelerated either using strong alkali or by heating. Usually some water vapor is introduced in the fuming chamber as well since it improves the fuming results in many cases. Super Glue fuming is an excellent method for developing latent fingerprints on many surfaces. Super Glue–developed prints can be further treated before examination. The simplest enhancement of a glue-developed print is powder dusting. Other post-treatments of Super Glue prints include "dye stains." Because the Super Glue ridges will selectively absorb certain dyes from their solution, they can be "dye stained" to make them luminescent or fluorescent. Depictions of the results utilizing the various methods outlined in this section are illustrated in Figure 6.10a–h.

Lastly, illumination with a laser or alternate light sources at the appropriate wavelengths for the particular dye used produces greatly enhanced visibility of the print (Figure 6.11). Gentian violet, coumarin 540 laser dye, ardrox, Rhodamine-6G, and other commercial dye mixtures have been used to produce luminescence under alternate light or laser illumination after treatment of Super Glue-developed prints.

Special types of illumination and combination methods

Sometimes, a "latent" fingerprint can be visualized simply by illuminating it from an oblique angle (see Figure 6.10h). This technique may work with white light, or with what is often called an "alternate" light source. These are special, high-intensity light sources that often have filters to control wavelength (see the sidebar). Sometimes, latent fingerprints show up better under illumination by certain narrow wavelength bands of light.

Alternative light sources are also regularly used in connection with some of the chemical treatment methods discussed below. Often, the chemical treatments result in a compound that has a strong fluorescence when illuminated with light of a particular wavelength. This principle is the basis for using alternative light sources and lasers to

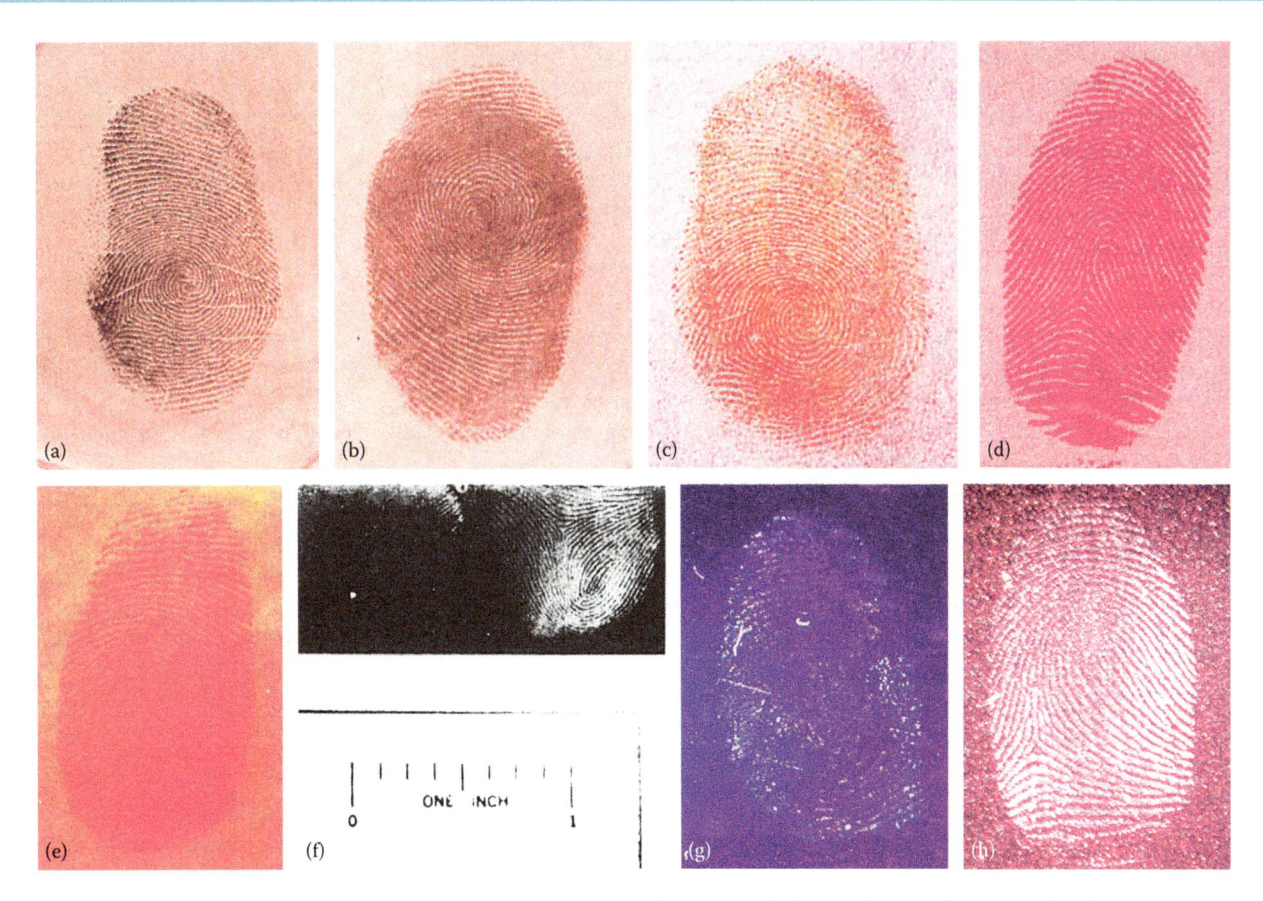

Fingerprints and other personal identification patterns 129

Figure 6.10 Eight fingerprints each visualized by a different development technique: (a) Latent fingerprint visualized by powder dusting, (b) latent fingerprint visualized by dusting with magnetic powder, (c) latent fingerprint visualized by iodine fuming, (d) latent fingerprint visualized by treatment with ninhydrin reagent, (e) latent fingerprint visualized by treatment with ninhydrin, followed by zinc chloride, then viewed under a laser, (f) latent fingerprint visualized by fuming in a closed chamber with Super Glue (cyanoacrylate), (g) latent fingerprint visualized by Super Glue fuming and washing with laser dye Ardrox to produce a fluorescent print, (h) plastic fingerprint (impression) visualized by side (oblique) lighting.

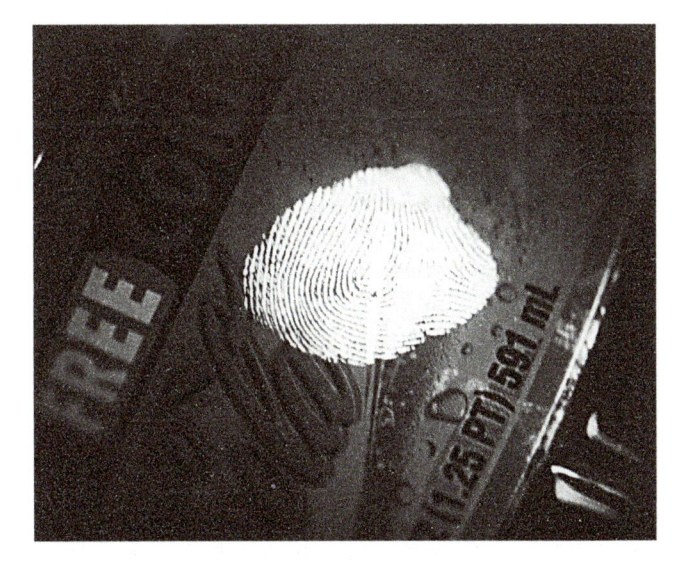

Figure 6.11 Fingerprints dusted with fluorescent powder and made to fluoresce using an alternate light source.

visualize latent fingerprints (see Figure 6.11). Lasers emit high-intensity light beams of a single wavelength. A number of methods have been developed to take advantage of the excitation wavelengths available from laser illumination. The undisputed pioneer in this field is Dr. E. Roland Menzel of Texas Tech University. He wrote one of the first books on the subject and has published many research papers. The alternate light sources were developed to provide very high-intensity illumination at fairly narrow wavelength bands and are, generally, much less expensive. The most recent development in this context has been the development of luminescent nanoparticles for latent fingerprint enhancement.

> ### CASE STUDY 6.2   BLOODY FINGERPRINT HELPS PLACE SUSPECT AT A HOMICIDE SCENE AT THE TIME OF THE HOMICIDE
>
> An elderly man was found dead in a pool of blood in his bed. He had few, if any, valuable possessions and lived on social security. Nevertheless, he was admired in the neighborhood for helping others in time of need. He had become close to a troubled youth who ran errands for him and came to see him and talk with him frequently. When his modest house was processed after the murder, the young man's prints were found everywhere. This had limited significance since he was known to be a frequent visitor. During the careful processing of the scene, a small patent print in blood was found on the back of the victim's head board in a position consistent with where one would place one's hand to hold onto the headboard while attacking the man in the bed. It was found that this print also belonged to the youth and the blood was consistent with the victim. This established the youth's presence during the attack on the victim and proved to be a major piece of physical evidence in bringing him to justice.

Bloody fingerprints have a special value because they often allow one to put a time on when a fingerprint print was made. One of the limitations on latent fingerprint evidence is that when an individual has normal access to the crime scene, it is not possible to say that a print was made at the time of a crime rather than some time before or even shortly after the incident. Blood prints must have been made at the time of bleeding and before the blood dries. Bloody fingerprints can be patent fingerprints; clearly visible without enhanced visualization. However, bloody patent fingerprints are often not easily visible and chemical treatment may be necessary to make them sufficiently clear for comparison. Blood color reagents (Chapter 10) can significantly improve visibility and even visualize fingerprints otherwise not visible. Under these conditions, use of a blood-color enhancement reagent should be considered. Also, investigators need to think about whether it will be necessary to attempt DNA profiling of the blood, which is forming the apparent ridge patterns. There is substantial published evidence that most latent fingerprint visualization procedures, including those designed for bloody prints, do not interfere with subsequent DNA profiling. But this situation is one where discussion among latent print examiners, DNA analysts, and investigators is important. Bloody fingerprint visualization reagents are usually applied as a very fine spray to keep from washing the print away. Some of the recipes produce a reagent that is not very stable in solution and, thus, has a short shelf life. These need to be prepared shortly before use. Further, there are potentially serious chemical hazards associated with some of the ingredients, so significant training and experience are required in the preparation and use of these reagents. Many bloody fingerprint enhancement reagents are based on the peroxidase reaction chemicals (phenolphthalein, leucomalachite green, tetramethylbenzidine, etc.), which will be discussed in connection with presumptive blood testing (Chapter 10). These interact with the hemoglobin portion of the blood. There are also some recipes and techniques based on general protein staining dyes like Amido Black and Coomassie Blue. They are generally less hazardous and easier to use than the peroxidase reaction chemicals.

Latent prints on absorbent surfaces

Absorbent surfaces present both problems and advantages for latent print examination. Since the fingerprint residue is absorbed into the material simple, stickiness is of no use in visualizing the fingerprint. On the other hand, it can offer stability to the residue protecting it from vaporization and air oxidation. As a result, prints can sometimes be recovered from absorbent surfaces months or even years after deposition.

Iodine fuming is one physical method that can be used to visualize fingerprint residue from an absorbent surface. Elemental iodine, which is a deep purple solid, is one of the few compounds in nature that *sublimes*; that is, it can pass from the solid to the vapor state without becoming a liquid, readily at or near room temperature. Solid iodine crystals sublime with moderate warming and produce a purple vapor that dissolves in oily materials to produce a deep brown coloration. The iodine vapor can be directed toward a latent fingerprint with an iodine fuming gun, or an object to be "fumed" can be placed in a closed cabinet, which is then filled with iodine fumes by warming a small dish of iodine crystals in the bottom of the cabinet. Iodine treatment of latent prints is usually called "iodine fuming," because the latent print residue is actually exposed to the iodine vapors (fumes). It was once thought to involve a chemical reaction, but now it is known to be just a simple dissolving of the iodine vapor in the fingerprint residue. The ridge features take on a dirty-brown colored appearance (see Figure 6.10c). The iodine-developed color is not stable, and, if removed from the presence of iodine vapor, the iodine will soon re-vaporize to return the latent print to its original colorless condition. Therefore, iodine prints have to be quickly photographed. There are also chemical methods for converting iodine prints to a permanent color that will not fade. The traditional method involved using a starch solution for that purpose, but the 7,8-benzoflavone (α-naphthoflavone) treatment is the preferred method today. Iodine fuming is used primarily on inherently valuable items precisely because of its impermanence, or where one wants to quickly visualize where a print is located before applying another visualization technique.

One of the oldest and most well established chemical procedures for visualizing latent prints makes use of ninhydrin. Ninhydrin is a chemical that reacts with compounds called amino acids—amino acids are the building blocks of proteins that are found in fingerprint residue—to form a chemical compound called Ruhemann's purple. Ninhydrin can be applied by spraying, painting, or dipping. It reacts slowly unless the process is accelerated by heat and humidity. Ninhydrin used to be made up in Freon 113 (a compound similar to those used in air conditioners), but concern over the effect of Freon on the earth's ozone layer culminated in the signing of the Montreal Protocol in 1987, resulting in a ban on most uses of Freons. Now, ninhydrin can be made up in several different solvent systems. One selects the solvent system depending on the type of absorbent material to be processed and what else is present on the evidence. Because paper and other similar absorbent materials are such important components in "white collar" and document-involved crimes (forgery etc.), ninhydrin and other such reagents are widely used.

Ninhydrin develops bluish-purple colored fingerprints (see Figure 6.10d), and is extremely useful on absorbent surfaces (such as paper). Currently, ninhydrin is often used as a primary treatment in processing. It may be followed by further treatment of the ninhydrin-developed prints with other chemicals and subsequent viewing under laser or alternative light source illumination.

Figure 6.10e shows a ninhydrin-developed print, which has been treated with $ZnCl_2$, viewed under blue-green light. These secondary treatments can improve the utility of fingerprints on absorbent surfaces.

SCIENCE SIDEBAR 6.1 Post ninhydrin and post superglue treatments

In *Scientific Tools of the Trade: Methods of Forensic Science, Part I*, the interactions of light with materials were discussed. Light impinging on a material may be reflected, transmitted, or absorbed. When light is absorbed, it brings about changes in the electronic structure of the material absorbing it. One potential change in the material is a state of temporary excitation. This excited state is not stable, and the material tends to relax back to the original stable state (ground state) fairly quickly. In the process of relaxing back, energy may be given off in the form of light that is called fluorescence or of phosphorescence. Fluoresced or phosphoresced **light** is always of lower energy (longer wavelength) than the light causing excitation of the material, because some energy is lost during the process of excitation and relaxation.

Different wavelengths of light are absorbed differently by different classes of chemicals. This behavior is the basis of spectrophotometry. A spectrophotometer is essentially an instrument that exposes a chemical or material to a range of wavelengths or light, and can detect whether the material absorbs a particular wavelength or not. The result is plotted out pictorially on paper, or on a computer monitor, and is called a spectrum. There are also instruments that can monitor fluoresced light as a function of excitation light wavelength. They are called spectrofluorometers.

Latent fingerprints visualized using ninhydrin have formed Ruhemann's purple. Ruhemann's purple can, in turn, be treated with chemicals that form chemical complexes with it. These complexes have particular light absorption-emission characteristics. In the same way, Super Glue-treated fingerprints can be treated with dyes that have particular light absorption-emission characteristics. These characteristics can be exploited in helping better visualize the treated latent prints. The light emitted is always much less intense (bright) then the illuminating light.

In the simplest case, suppose one of the complexes or dyes absorbs red light of the visible part of the spectrum and reflects green light. We could then use an instrument or device to impinge light onto the material, and look for green reflectance. The easiest way to do this might involve using a viewing filter that transmits the green light. In this kind of situation, a so-called alternate light source and viewing filter could be used. An alternate light source is really just a high intensity white light source. Filters can be used to select a particular output color of light. Similarly, filtered viewers can be used to enable the observer to see particular wavelengths of light (that are reflected by the material).

Intense white light sources can also cause fluorescence or phosphorescence in some chemicals. From what has been said above, it should be clear that a particular wavelength of the "white" light is being absorbed and, thus, gives rise to the fluorescence or phosphorescence. Alternate light source outputs can be filtered so as to select a narrow band of wavelengths containing the wavelength that is absorbed by the material of interest. Lasers are extremely intense light sources whose beams have particular vibrational characteristics while traveling through space and are of a single wavelength. For a laser to be useful as an excitation source, the material of interest (like the complex or the dye in the latent print cases) must absorb that wavelength of light. Thus, chemicals capable of being excited by certain wavelengths of light can be formed by appropriate latent print treatments. Then, lasers or alternate light sources can be used to produce reflected light, or fluoresced emissions, that can be visualized and make the ridge characteristics of the fingerprint that much clearer. Certain lasers can be used to help visualize untreated latent prints. That fact means that there is something in the latent print residue that can absorb the laser light wavelength, and then give off fluorescence or phosphorescence. The Argon laser, with an output wavelength primarily at 488 nm, has been used quite a bit in latent fingerprint enhancement techniques. The high-intensity illumination is necessary to make the much weaker fluorescence visible.

Another chemical method is the use of "physical developer." The process is obviously mis-named, because it involves chemical reactions. Physical developer visualization is similar to a photographic process, based on the production of metallic silver when material in the latent reacts with the physical developer to form a fingerprint image. The reaction involving silver ions and an oxidation/reduction reaction involving iron salts. It is thought that the physical developer reacts with lipid (fatty) material in the fingerprint residue. The procedure can be used for latent prints on paper, some nonabsorbent surfaces, and pressure-sensitive tapes. It has been reported that the physical developer sometimes works on latent prints that did not develop with ninhydrin. This result makes sense considering that this procedure is based on reactions with lipid components rather than the amino acids in the fingerprint residue.

Reagents for highly specialized applications

Fingerprints deposited on tape, especially on the sticky side surface, present a special situation. Techniques for visualization include staining with a solution containing crystal violet or with a material called "sticky side powder." Crystal violet stains skin cells "trapped by the adhesive" a deep violet. Sticky side powder is actually composed of lycopodium (a plant) pollen mixed with a detergent and water. The sticky side powder slurry is painted onto the sticky side of the tape with a brush, and the tape is then rinsed off with water. The tiny, brightly colored pollen grains preferentially adhere to the skin cells making them visible. The process can be repeated until the desired contrast has been achieved.

Another special situation is the development of latent prints on human skin; almost always on a decedent's body. The idea is to bring up the fingerprints of those who may have harmed an individual with their hands. In murder cases, especially involving manual strangulations, or, if the body has been handled in the process of moving the body, the murderer's fingerprints might be on the victim's skin. A variety of different techniques have been tried for this purpose over many years, but the success stories are few and far between. It is fair to say that a generally useful, robust procedure has not yet been devised. Tenting the area of suspected fingerprint and applying super glue vapors will occasionally work under ideal conditions.

Use of a systematic approach to visualization of fingerprints

Most latent fingerprint examiners probably use a "systematic approach" often even without expressly thinking about it. The idea is to apply latent development techniques in a way that maximizes the number of identifiable prints. The least destructive technique is applied first, and techniques are, generally, applied in an order that allows the maximum number to be used until the best set of visualized prints possible is obtained. Systematic approaches vary according to the surface (or substratum) on which the latents are located. As indicated above, porous surfaces, such as paper, for example, call for a different set of techniques applied in a different order than nonporous surfaces. Flow charts illustrating these approaches are shown in (Figure 6.12a–c).

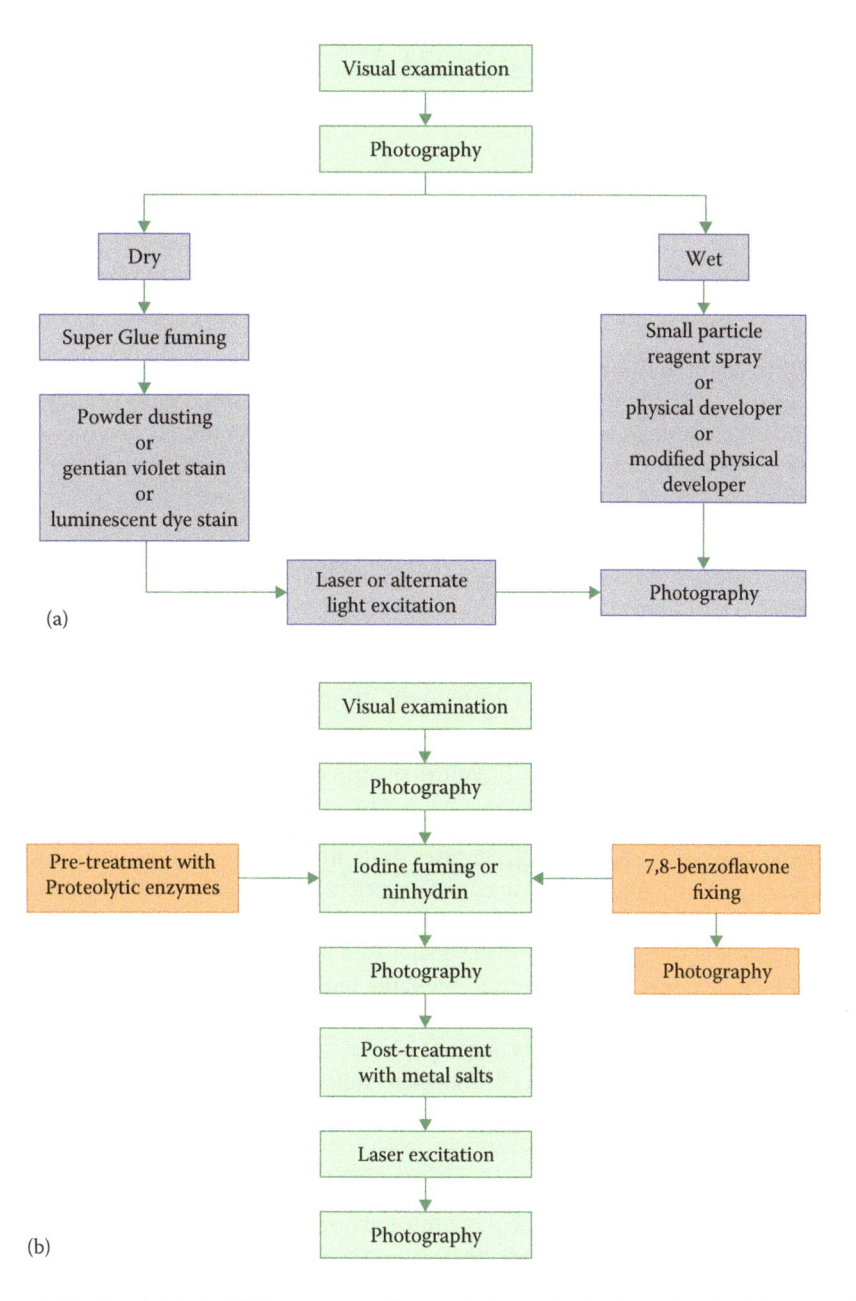

Figure 6.12 (a) Flow charts showing suggested sequence of reagents to use to develop a latent print on non-absorbent surfaces. (b) Flow charts showing suggested sequence of reagents to use to develop a latent print on porous surfaces. (*Continued*)

```
                    ┌──────────────────┐
                    │ Visual examination│
                    └────────┬─────────┘
                             ▼
                    ┌──────────────────┐
          ┌─────────│   Photography    │
          │         └────────┬─────────┘
          │                  ▼
          │         ┌──────────────────┐
          │         │       TMB        │◄─────┐
          │         └────────┬─────────┘      │
          ▼                  ▼                │
┌──────────────┐   ┌──────────────────┐   ┌──────────┐
│ Amido black  │   │    Ninhydrin     │──▶│ Heating  │
└──────────────┘   └──────────────────┘   └────┬─────┘
                                                ▼
                                       ┌──────────────┐
                                       │ Photography  │
                                       └──────────────┘
```

(c)

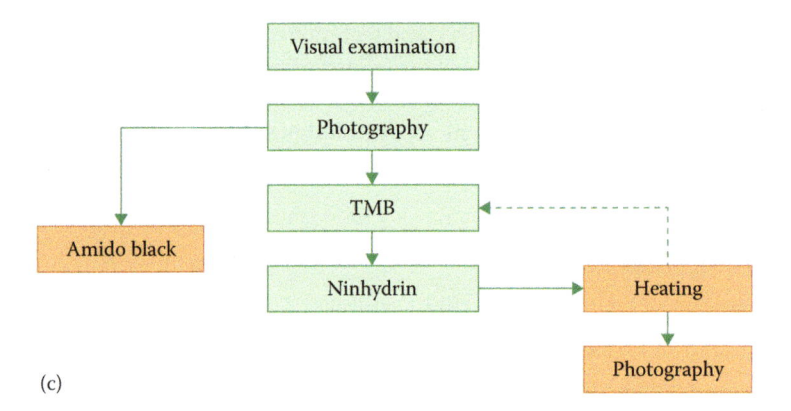

Figure 6.12 (Continued) (c) Flow charts showing suggested sequence of reagents to use to develop a latent print in blood surfaces.

Fingerprint comparison and identification

Everything we have talked about in this chapter—indeed, the reason there is a chapter on fingerprints at all—comes down to the use of fingerprints to identify persons. As noted earlier, the uniqueness of fingerprints is a matter of common knowledge. Ads and commercials commonly use phrases like "... as unique as a fingerprint." Even the term "DNA fingerprints," as undesirable and sometimes misleading as it is, was coined to reflect the notion that a DNA profile might be as "individual" as a fingerprint. And infrared spectra of pure compounds are sometimes called "chemical fingerprints."

David Ashbaugh, among others, has noted that fingerprint individuality, and therefore fingerprint identification, is based on four premises:

1. Friction ridges develop in fetuses before birth in their definitive form
2. Friction ridges remain unchanged throughout life with the exception of permanent scars
3. The friction ridge patterns and their details (minutiae) are unique to an individual
4. The ridge patterns generally fall into common groups that allow the patterns to be classified

In general, fingerprint examiners are extensively trained and required to accumulate significant experience before they are given the responsibility of making identifications in cases. Thus, in addition to the general principles and approaches used to make identifications, the knowledge, training, and experience of the individual examiner is an important factor. In law enforcement situations, identifications are always made by trained and, usually certified, examiners. Where unambiguous identification of someone arrested is the goal, inked prints from a person may be compared with a set of inked prints on file to determine if they came from the same individual. The more complex problem is where the examiner will be comparing a visualized latent print, from a crime scene or a piece of evidence, with inked prints from a known person or persons. Evidence fingerprints may not be clear and/or may represent only a portion of the finger (Figure 6.13). Therefore, the quality and amount of fingerprint ridge detail will seldom be ideal in evidence prints. If someone's fingerprints are in an AFIS system, AFIS searches can quickly narrow the number of possible matching persons in the database to a manageable size, but an examiner, not a computer, actually must make the final identification.

In examinations of latents, the first issue is determining what is called "suitability" of the latent fingerprint for identification. Here, the examiner must decide if sufficient quality and quantity of the ridge detail is present in the latent to make it possible to do a comparison with a known. This determination also requires training and experience. Once a latent is judged suitable for comparison, known prints that might match the latent must be sought. Such knowns might be obtained through an AFIS search or from certain persons who are suspected; their prints might be taken or may already be on file.

The overall process an examiner uses has been described by Ashbaugh as "ACE-V," for *analysis*, *comparison*, *evaluation*, which comprise the formal process, then *verification*. (Figure 6.14). The examiner must first analyze the latent, figure out its proper orientation, decide if there are any color reversals or other unusual circumstances, decide

Figure 6.13 Partial fingerprint most representative of the type of fingerprint found at a scene or evidence objects.

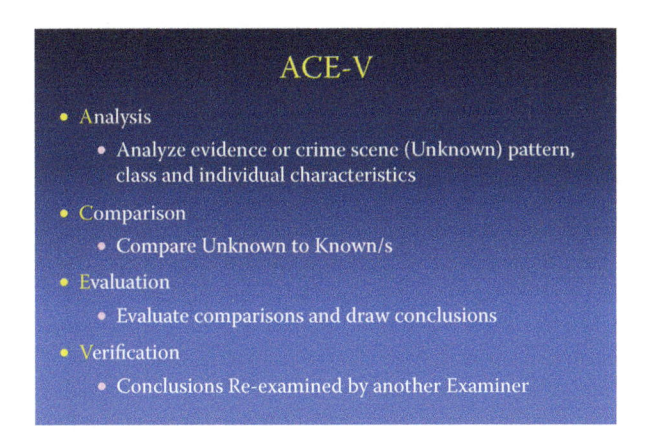

ACE-V

- Analysis
  - Analyze evidence or crime scene (Unknown) pattern, class and individual characteristics
- Comparison
  - Compare Unknown to Known/s
- Evaluation
  - Evaluate comparisons and draw conclusions
- Verification
  - Conclusions Re-examined by another Examiner

Figure 6.14 Chart of the ACE-V method; steps that should be followed in examination and comparison of fingerprints for reliable offering in court.

suitability, then proceed to the comparison. Comparison with the known fingerprints takes place at several levels. The overall pattern and ridge flow (called Level I) must be examined. Next, the minutiae (called Level II) are compared, point by point, as to type and relative location. Finally another level of detail (called Level III), consisting of pore shape, locations, numbers and relationships, and the shape and size of ridge features, is compared. Any **unexplained difference** between known and latent during this process would result in the conclusion that the known is **excluded** as a source of the latent. This is one possible outcome of the evaluation decision. If every compared feature is consistent with the known, and there are enough features that are sufficiently unique, when considered as a whole, the examiner may make an **identification**. Since peer review is a feature of most scientific endeavors, Ashbaugh notes that verification of conclusions by an independent examiner is a necessary practice. Recent legal challenges have generally supported only identifications that can be shown to have followed the ACE-V method.

Thus, submitting a latent or an item bearing latents to a fingerprint examiner—assuming there is a suspect—could result in several possible outcomes: identification (the latent was made by the suspect); exclusion (the latent was not made by the suspect); the latent is unsuitable for comparison; or inconclusive (there are not enough features to form a definite conclusion).

There has been considerable discussion in the identification literature about defining the criteria for making a fingerprint identification. For many years in some jurisdictions, a "minimum number of points" rule was followed. During that time, if the rule was "12 points to make identification," the examiner had to find that many or more points of comparison or could not make the identification. Over time, it became clear that such an absolute rule was not a proper basis for making decisions in every situation. The International Association for Identification (IAI), the organization that, by and large, sets peer standards for the fingerprint community, adopted a resolution in 1973 that no minimum number of features is required for making fingerprint identification. This position was reiterated in a slightly modified form in 1995. Most fingerprint examiners subscribe to this principle. Discussions of the criteria for making fingerprint identification lead naturally into discussions about the basis for fingerprint uniqueness. Although most courts have supported the validity of fingerprint identifications, there is still some question of this having been sufficiently scientifically proven.

AFIS System Operation—AFIS is now available in all 50 states and has taken over much of the drudgery of fingerprint searching, and ten-print cards of inked prints are scanned into the system with an optical scanner. In many locations, the scanners have either supplemented or been replaced by what is called live scan. The finger is placed directly in contact with a plate and scanned into the electronic system directly. This has the advantage of skipping the inking step and is much less subject to poor inking and smearing of the print. AFIS systems also have the advantage of maintaining sub-databases of fingerprints that have been previously searched but where no record was found that closely matched the latent. This has the huge advantage of connecting incidents involving the same individual who has not been identified as yet. Thus, if an individual is arrested at some later time for a different incident, their prints can be searched against the sub-file and the cases associated. Since many different jurisdictions use the same AFIS database, this can alert law enforcement to criminals who move from jurisdiction to jurisdiction to avoid detection. AFIS systems are also programmed to improve the quality of the prints in their databases. When an individual, who is in the database, is subsequently arrested and fingerprints taken, the AFIS software compares the database set to the new fingerprints and selects the best quality images for each finger.

The fingerprint identification profession

In the past several decades, a number of trends have worked to further professionalize fingerprint examiners. People entering the profession today have more formal education than was once true. The extraordinary progress in latent development methods, noted above, has demanded that fingerprint examiners understand much more chemistry and physics than ever before. The IAI has been a positive force, through facilitation of discussions, encouragement of research and scientific approaches, and through its professional journal, the *Journal of Forensic Identification*. Journals are commonly and routinely the primary, peer-reviewed source of original research in scientific fields, and the IAI Journal rose to fulfill that role for the fingerprint and identification sciences. The IAI has also been the primary peer standard-setting group in the identification sciences. The latent fingerprint certification program is operated by IAI.

Fingerprint examiners today have a wide range of educational backgrounds, but as we have noted, the strong trend is in the direction of more education. Individuals may be hired in some agencies primarily as AFIS technicians, to maintain systems, scan images into the system, and so forth. They do not need the level of training required for someone who wants to be a certified latent print examiner. More and more employers are requiring the IAI latent print certification as a condition of continued employment. The certification requires passing a challenging written test, demonstrating competency in comparisons, and then maintaining a record of activity and continuing education during the certification period.

Biometrics—use if biological diversity for identification
Background

Although fingerprints are one of the oldest and certainly the most widely used biometric, there is enormous activity in developing other useful biometrics. In this section, we will describe some of the many biologically diverse characteristics, in addition to fingerprints, that can be used to either unambiguously identify an individual or to select a strong candidate from a limited pool of individuals. One such technique is an image of the veins in an individual's retina; the pattern of which are unique.

The first scientific method of criminal identification—called *anthropometry* (Figure 6.15)—was devised by Alphonse Bertillon, and is sometimes called *Bertillonage*. One could look at anthropometry as the grandfather

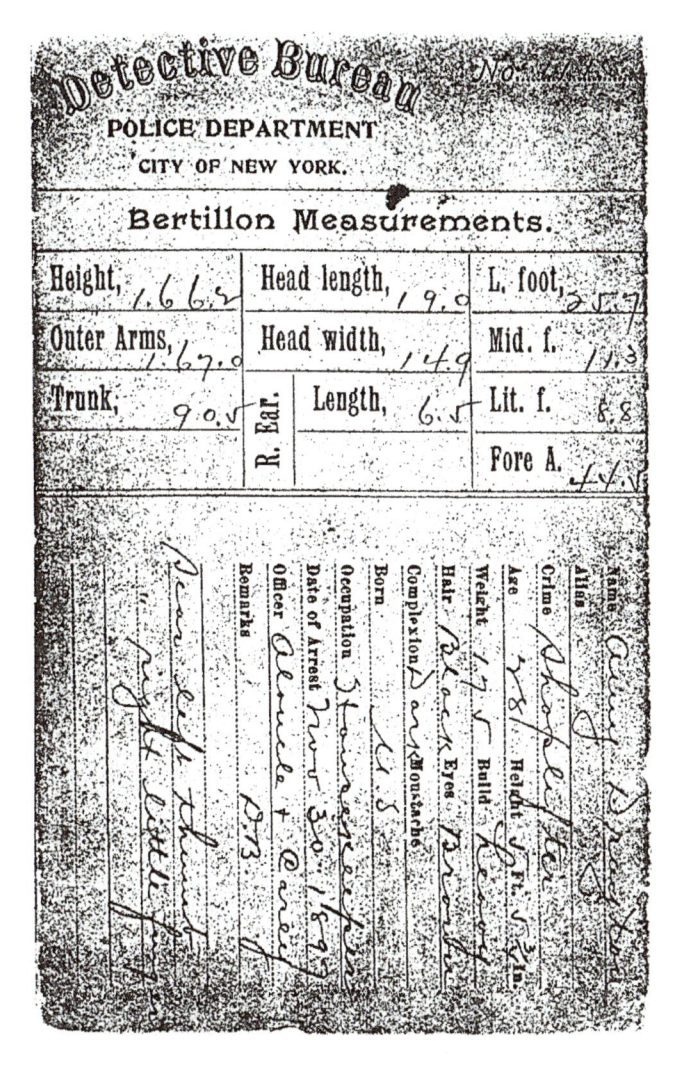

Figure 6.15 A card obtained from an individual listing Bertillon measurements that were used for biometric identification before fingerprints became common.

of biometrics. Bertillon is honored because he laid a scientific foundation for personal identification. Bertillon worked in the Police Prefecture in Paris and developed a system of identification based on body measurements, such as head size, finger length, and so forth, ultimately choosing 11 characteristic body measurements. Eventually, the system proved its worth in identifying and helping to convict persons with prior arrest or conviction records. Police agencies in many places began using his system. Bertillonage was ultimately undone when it was found that different individuals could have the same or very similar anthropometric measurements. These incidents came to convince police identification agencies that fingerprints represented a more reliable basis for personal identification.

Anthropometry (Bertillonage): A system of bodily measurements devised by Alphonse Bertillon for personal identification of persons with criminal histories. The profile of measurements could distinguish people until the number of persons in the database got larger. Bertillonage was supplanted by fingerprints.

Biometrics are suitable for two distinct functions, Authentication and Identification. Authentication works with simpler biometrics since it is used to determine if a candidate belongs to a specified group of individuals. There are dozens of examples of non-biometric verification one encounters in daily life: computer passwords, magnetic strips on your credit cards containing "identifying" information, photo identification cards. Biometrics have advantages over non-biometric verifications, since you cannot lose your biometrics, forget your biometrics, nor can someone easily steal or forge one's biometrics. To avoid these problems, biometric verification methods are an attractive alternative. For some purposes, it is not necessary for a biometric method to yield individuality among everyone in the country or the world. It may be enough to distinguish several hundred, or several thousand, or several hundred thousand people, from one another. Applications like this are called Authentication. Thus, Authentication would be used to check if someone works at a particular location. The chosen biometric for each of the employees who work at a given location is collected in a database, and, when an individual wants to gain access to a secured area, their biometric is read and checked against that database. This is the most rapidly growing area of biometrics because the amount of data needed is much less, and it is less costly and invasive to operate than a full identification system. Biometrics, such as voice recognition or hand geometry, that can be easily captured and compared quickly to a fairy large database are useful for Authentication. Individualization requires a biometric that provides enough differentiation power to individualize, determine that the candidate is a particular person, and differentiate the candidate from essentially all other individuals. The availability of imaging acquisition, storage, and rapid retrieval has made many biometrics methods quite cost-effective.

Biometrics for authentication

Examples of biometrics used for Authentication such as skeletal features from X-ray photography has been very useful for "identification" for many years. Actually, in the vast majority of cases this is more properly an Authentication. Forensic (physical) anthropologists and forensic radiologists are often able to make identifications of persons from the comparison of pre-mortem and post-mortem X-ray images. Usually, this is done to identify skeletal remains. If these features are adequately visible in both X-rays, an expert may be able to identify the remains as those of the person who was X-rayed while alive. Fractured and healed fractures and other injuries can also have strong identification potential.

Dental X-rays make up a special sub-category of skeletal feature, which are extremely valuable in identifying bodies in mass disasters (Figure 6.16). The forensic odonatologist looks not only at restoration and other dental work done on the individual, but also features of the sinus cavities in skull X-rays, the shape of root channels, and other bony characteristics of the teeth and jaw in making identifications. We note that this type of analysis is different from the more usual examination of actual skeletal remains by forensic anthropologists. With skeletal remains alone, usually only class characteristics can be discerned, not identity.

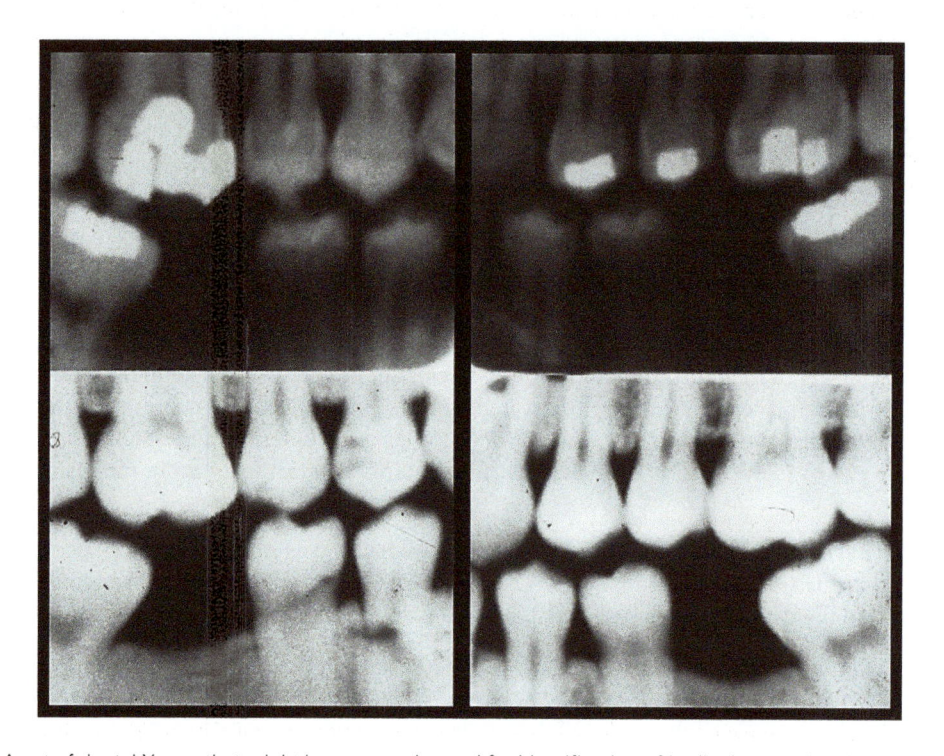

Figure 6.16 A set of dental X-rays that might be commonly used for identification of badly damaged or decomposed bodies.

Voice prints represent a record of a person's speech patterns. Voice analysts use a combination of aural (listening) and sound spectral analysis to make comparisons between questioned and known voice patterns. The sound spectrogram has been called "voice prints." This area is highly specialized, and comparisons are done by examiners having specific training and experience. A high degree of reliability is reported when the comparisons are done by appropriately trained and experienced examiners on high-quality evidence recordings. With more limited quality evidence, such as that from most telephone lines, the results are much less reliable. As computer technology for measuring voice characteristics has developed better ways of comparing voices, this has become an area of considerable research, particularly in the intelligence community.

A simple biometric, widely used for access to secure areas, is hand geometry. A candidate for entry, who has earlier submitted a hand geometry sample, places his hand on a reader, and it is quickly checked against the database of those approved to enter the secure area. Automated image recognition has received considerable attention and has already been quite successful in appropriate settings. Currently, biometrics like voice recognition, hand geometry, automated facial recognition, limited fingerprint capture, and others are being further developed for Authentication. There is much research activity in this area, so new systems are still emerging.

Biometrics for individualization

Fingerprints are still by far the most widely used biometric for individualization. However, there are several others being currently used, and many others that have seen considerable usage will undoubtedly see use in the near future. Palm and sole prints are actually quite analogous to fingerprints as they are based on friction ridges. The palms of the hands and soles of the feet have friction ridge skin just like the fingertips and, thus, can be used for identification just like fingerprints. Palm and sole prints are handled like fingerprints in terms of recognition, enhancement, collection, documentation, and comparison. However, palm and bare footprints are not generally organized and stored in databases, such AFIS, and therefore an examiner can't look for a palm or sole pattern in a file. There must be a suspect or a limited group of suspects to provide known impressions for comparison.

Patterns of the veins in the back of an individual's retina as well as the pattern of the cornea can be captured with a video camera. There is strong evidence that these patterns are unique to each individual and, therefore, can be used in the same way as fingerprints. There are commercial products that can capture such patterns quite effectively to use, certainly for Authentication and almost certainly for Identification as well. There are databases of such patterns that have been developed for specific applications, but unclassified, large databases have not been publicized.

With advances in DNA typing of bone, identification of skeletal remains as those of a particular individual is becoming more practical and, as a result, could raise the utility of bone from a primarily Authentication to an Individualization level. The existence of CODIS as mentioned in Chapter 1 and discussed in Chapter 11 greatly increases the utility of bone DNA as an Identification technique.

Identification of human remains and handling of mass disasters

A special and, particularly, a forensically important sub-topic of biometrics is the identification of remains. Earlier we briefly considered the methods used for identifying human remains. Most of the time, identifying human remains is straightforward. It becomes a problem when there is destruction of the body or when the body is fragmented. Both destruction and fragmentation are particularly common in mass disaster situations. As a society, we have traditionally placed considerable value on positively identifying the remains of the dead and providing those remains to the decedents' families for interment or cremation in accordance with their traditions and beliefs. As a result, governmental agencies often expend considerable effort to achieve reliable identifications of remains. This responsibility is often the responsibility of local medical examiners or coroners. They must rely on various other specialists to assist with this task.

Where remains consist of an intact body, identification can often be achieved by having a relative, friend, or acquaintance of the deceased identify the person by direct viewing. If there is destruction or fragmentation of a body, because of fire or traumatic events, viewing may not be useful. Where direct identification is not possible, dental and fingerprint identifications are used next. As mentioned above, forensic odonatologists can make positive identifications of persons from the X-rays of their dentition, provided pre-mortem X-rays are available. That means there must be suspected identities, because even where files of data on missing persons now exist, blind searching of such records is difficult. Fingerprint specialists can make identifications using fingerprints obtained post-mortem, again provided there is a pre-mortem record of the person's fingerprints on a card in a file or in an AFIS system. Dental and fingerprint identifications are considered definitive. The remains must have the jaws (with the teeth) or friction ridge skin intact for such records to be used. If neither of these techniques can be used, other less direct methods can be tried.

The less direct methods consist of using clothing, personal items (such as jewelry), marks, scars or tattoos, or artificial body parts or limbs. Because there is the possibility of chance duplication of scars, marks or tattoos, and of mix-ups involving belongings, many authorities will not rely on these methods alone for definitive identification.

Skeletal remains are normally examined by a forensic anthropologist. Depending on what bones are recovered, an anthropologist can provide estimates of race, sex, age, stature, and other features that may ultimately aid in making identification. Anthropological features alone will not yield identification, in the absence of highly individual injuries or medical interventions. In this day of transplants and other repair procedures, a whole new type of material for identification is available in such cases. These "repair" materials are often serialized and, therefore, readily identifiable. Thus, a pacemaker, for example, has a serial number that should be traceable to the person into whom it was installed. Occasionally, when skeletal remains cannot be identified but a skull is recovered with the remains, a forensic sculptor may be called to "reconstruct" the person's face. The reconstruction is based on data provided by the anthropologist as to gender, race, and approximate age, and upon a database of tissue thickness measurements. Unless hair was recovered, hair color is unknown to the sculptor, as are hairstyle and eye color. Different eye colors and wigs may be used to provide different possible images of the face. Typically, these images are circulated in the hope that someone might recognize the person. The Michigan State Police and Louisiana State University; for example, maintain web sites showing facial reconstruction images of unidentified persons.

In Chapter 11, we discuss the use of both nuclear and mitochondrial DNA in human identification from tissue or bone specimens. DNA profiling has been used extensively in several recent mass disaster situations, including the World Trade Center terrorist attacks on 9/11/01. Nuclear DNA identifications are definitive. Mitochondrial DNA identifications are probably not definitive alone, but with other circumstantial evidence and a limited universe of possible identities, identifications can be made. There have been great strides in improving the ability to obtain useful DNA data even on extremely small or otherwise compromised sample in the years since 9/11/01. Its success relies on an accurate database of possible identities, which can be a serious problem to obtain.

Key terms

ACE-V method
alternate light source
anthropometry (Bertillonage)
authentication
Automated Fingerprint Identification Systems (AFIS)
basic fingerprint patterns: loop, whorl, and arch
biometrics
dye stains
fingerprint classification
fingerprint development (enhancement, visualization)
friction ridge skin
laser
latent print
Level I, Level II, and Level III detail
magnetic brush technique
minutiae
ninhydrin
physical developer

plastic (impression) print
powder dusting
small particle reagent (SPR)
Super Glue (Cyanoacrylate)
tape lifting
visible (patent) print

Review questions

1. What are fingerprints? Why are they useful in criminal investigation?
2. What is an AFIS? Why is it valuable?
3. What are the main types of evidentiary prints that might be found at scenes?
4. What are some physical methods for enhancing latent fingerprints?
5. What are some chemical methods for enhancing latent fingerprints?
6. What are some special illumination methods for enhancing latent prints?
7. How might bloody fingerprints be enhanced?
8. What is the basis for fingerprint identification? What are the main principles?
9. What are some other patterns useful in person identification?
10. What are the major methods of identifying human remains in mass disasters?

Fill in/multiple choice

1. One important characteristic of friction ridge patterns is that they _____ between birth and death.
2. The three basic fingerprint patterns are (1) arch, (2) _____, and (3) _____.
3. Fingerprints found at crime scenes or on evidence can be divided into three broad categories (not classifications) based on their appearance and physical makeup: (1) _____, (2) patent (visible) and (3) _____.
4. When a latent fingerprint is compared with a known inked print, and according to the numerical rule, the number of points of comparison required to correspond before a positive identification can be declared is
   a. enough to satisfy the expert latent examiner
   b. at least eight
   c. at least twelve
   d. the pattern and the 10 minutiae
5. Computerized fingerprint search systems match fingerprints by comparing the relative positions of
   a. core and delta
   b. pattern center and all bifurcations
   c. all individualizable minutiae
   d. bifurcations and ridge endings

References and further readings

Ashbaugh, D. R. *Quantitative-Qualitative Friction Ridge Analysis: An Introduction to Basic and Advanced Ridgeology*. Boca Raton, FL: CRC Press, 1999.

Champod, C., C. J. Lennarc, L. R. Margot, and M. Stoilovic. *Fingerprints and Other Ridge Skin Impressions*. Boca Raton, FL: CRC Press, 2004.

Federal Bureau of Investigation. *The Science of Fingerprints: Classification and Uses.* Washington, DC. Department of Justice, 1993.

Gaensslen, R. E., and K. Young. "Fingerprints." In *Forensic Science: An Introduction to Scientific Investigative Techniques,* 2nd ed., eds. S. James and J. J. Nordby. Boca Raton, FL: CRC Press, 2015.

Lee, H. C., and R. E. Gaensslen. *Advances in Fingerprint Technology*, 2nd ed. Boca Raton, FL: CRC Press, 2001.

Scientific Principles of Friction Ridge Analysis & -Applying Daubert to Latent Fingerprint Identification Written & Compiled by: Thomas J. Ferriola. Identification Technician, Sebastian Police Department, Florida; www.clpex.com/Articles/ScientificPrincoplesbyTomFerriola.hjtm

Chapter 7

Questioned document examination

Lead Case: Lindbergh Kidnapping

It was called "The Crime of the Century." In 1927, Charles Lindbergh had become a national hero by being the first to complete a solo flight across the Atlantic Ocean. He had married the beautiful and aristocratic Anna Morrow, and they were perceived as a dream couple. Tragically, only 5 years later, in 1932, the kidnapping and death of their first child received even more public attention than the solo flight.

On March 1, 1932, the 20-month-old son of the Lindbergh's, was taken from his second-story nursery at the Lindbergh Estate near Hopewell, NJ. Beneath the child's window, a three-section, hand-crafted ladder was found, and a handwritten note had been left. That initial ransom note, and those that followed, clearly revealed the writer to have difficulty with English, and likely to be of German extraction. A month later, a $50,000 ransom was paid through an intermediary in St. Raymond's Cemetery in New York City. A list of the serial numbers of the ransom bills had been printed and circulated to banks. Approximately two-thirds of the ransom money was in the form of gold certificates, and, about a year after the ransom was paid, President Roosevelt ordered the turning in of all gold certificates.

A little over a month after the ransom was paid, a truck driver came upon the partially decomposed body of a child a short distance off the road about two miles from the Lindbergh estate. In September of 1934, 2-1/2 years after the kidnapping, the driver of a 1930 Dodge sedan bearing New York plate 4U13-41 gave a New York City gas station attendant named Walter Lyle a $10.00 gold note for 98 cents worth of gasoline. Lyle recorded the plate number on the bill's margin in the event the bank refused to cash it.

The finding of this ransom bill was reported to the authorities by the bank, and they determined that New York plate 4UI3-41 was registered to Bruno Richard Hauptmann of 1279 E. 222nd St., Bronx, NY. Hauptmann was arrested. A ransom bill was in his wallet, and $14,560 of ransom money was found in his garage.

In January of 1935, Hauptmann was brought to trial in Flemington, NJ. After a month and a half trial on February 13, 1935, Hauptmann was found guilty and sentenced to die for the attempted kidnapping and death of the Lindbergh baby. The trial and conviction of Bruno Richard Hauptmann was largely on the basis of scientific evidence, and it created landmarks in the utilization and presentation of scientific evidence. Unfortunately, the enormous public interest in this case created something of a circus atmosphere at the trial.

Questioned document evidence played a central role in this case. The examination and association of the 14 ransom notes was critical to the prosecution. In fact, Hauptmann was heard to say to his counsel "Dot handwriting is the worstest thing against me." Although there was considerable, important physical evidence including association of the handmade wooden ladder left at the kidnap scene to Hauptmann's house and tools, examination of cloth and thread from the baby's handmade nightshirt to help identify the decomposed remains, and, of course, the finding of the ransom money in his home, the association of the ransom notes to Hauptmann was critical. The primary thrust of the defense was that Mr. Hauptmann had an accomplice or had been "framed" by another individual. Thus, the prosecution had the job of proving two things about the 14 ransom notes: First, that they were **all** written by the same individual; and second that they were all written by the defendant, Mr. Hauptmann. This was not as simple as it might seem, although the defendant had quite characteristic writing. Examination of the notes indicated to the experienced questioned document experts that there were attempts to disguise that writing.

The highly unusual approach taken by the prosecution was to have all the material examined independently by eight document examiners from across the country. This goes against the common wisdom that having more than one examiner will give the opposing party the opportunity to exploit the inevitable subtle differences of approach and reporting to create uncertainty in the mind of the jury. This was indeed the tack that was taken in cross-examination of the experts. It is a tribute to the highly experienced group of examiners that this defense strategy was not successful. Another reason for the multiple examiners was the fact that the defense said it had 14 experts who would say that notes were not written by the same person. To counter this testimony, the prosecution chose to have their questioned document examiners distance themselves from the potential defense witnesses by explaining the difference between a questioned document examiner and a graphologist. The defense witnesses were graphologists. The testimony of Albert Osborne, the most well-known of the prosecution witnesses, concerning this difference is quoted below:

> Graphology, in this country and in England, is understood as determining from handwriting the character of the writer, as distinguished from the students of handwriting, document examiners, who examine writing for the purpose of determining whether it is genuine or not, that is; the question of forgery and also examining writing for the purpose of determining whether it can be identified as the writing of a certain individual.

> There are two classes of handwriting examiners. One class examines writing for the purpose of determining whether it is genuine or not and whether it can be identified. The other class examines handwriting for the purpose of determining whether it indicates the character of the individual who did the writing. And the questions are entirely different. In one case, it is a question of genuineness and the question of identity. The other case is a question of whether the writer is honest, whether the writer would be a good husband or wife; whether the writer likes children and dogs... Any kind of question. And the graphologists go further than that, some of them... To determine disease from handwriting... Diagnosing disease. The two classes of examiners are entirely different.

This strategy was successful, in the end, since the defense chose not to offer its parade of experts to counteract the prosecution examiners. Further, the unanimity and strength of the opinions and the demonstrative evidence presented clearly convinced the jury. The skill of these experienced expert witnesses can be seen in his conclusions as stated by Mr. Clark Sellers:

> In examining these documents, I have also kept in mind an important thing... And that is the dangers of error: what might lead to error, such as mistaking a natural characteristic for an individual characteristic; such as coming to a conclusion on too few standards or too small an amount of disputed writing.

> The character of the writing, the manner in which it was written, have been taken into consideration, having in mind many other things besides those which I have mentioned here. And, I believe this combination of characteristics, some of which have been mentioned, others not mentioned... But in order to save time I will make this general statement... That a combination of characteristics in Mr. Hauptmann's writing, may just as truly identify him as a combination of scars, moles and birthmarks, or whorls and loops in combination may identify a man by his fingerprints. So convincing to my mind that Mr. Hauptmann wrote each and every one of these ransom notes - *it is, I say, so convincing to my mind, that he might just as well have signed each and every one of them.*

Questioned document examination

LEARNING OBJECTIVES
- The wide variety of evidence that can be examined by a questioned document examiner
- The evolution of an individual's handwriting from childhood to adulthood
- The major steps in preparing a document
- Special problems involved in properly collecting and preserving document evidence
- The science and technology that underlies handwriting and handwriting comparison
- Class and individual characteristics as applicable to handwriting
- The importance, and proper methods, of collecting known writing samples
- Basic approaches to the comparison of known and evidentiary writings
- The important non-handwriting examinations performed by document examiners
- The examination of documents produced on typewriters, computer printers, and copy machines
- The examinations used to reconstruct altered documents
- Some techniques for deciphering of charred documents
- Techniques used to look for and read indented writing on a document
- The problem of trying to determine when a document was written

QUESTIONED DOCUMENT EXAMINATION – GENERAL

Questioned document examination is an important forensic activity for making connections between written communications of all types and their authors. Questioned document examination has been forced to slowly shift its emphasis from primarily handwriting to a much broader look at many other forms of communications, due to the rapid development of technology. Questioned document examination often is not fully utilized because investigators and other potential customers are not aware of the full range of services available from questioned document examiners. As the electronic age progresses, fewer and fewer forensically important communication have involved handwriting. This has not reduced the importance of questioned document examination, but shifted it towards the other areas of examination. Questioned document examination have for many years offered a great variety of examinations that can provide useful, practical information. It is important to become aware of the wide range of examinations that are available and the many skills that questioned document examiners bring to this type of evidence. Questioned document examination can make significant contributions to many investigations and prosecutions, particularly when a knowledgeable examiner has the appropriate evidence to examine.

Types of document evidence

The most well-known activities of questioned document examiners are examinations and comparisons of handwriting, typewriting, and copier output. Was the document written by this individual or not? For older documents, was the document typed on this particular machine (Figure 7.1) or not? Was a particular handout copied on a particular coping machine? These are all questions asked frequently of questioned document examiners. The output from many other mechanical printing devices, such as rubber stamps, various commercial printing processes, and computer printers, can also lend themselves to careful examination and comparison. These may also provide useful information.

In addition to the types of examinations mentioned above, there are many other technical examinations, such as looking for erasures, alterations, source and type of inks, the sequence of writing or printing, and even sometimes dating of questioned documents.

Questioned Document examiners must deal with considerations such as: Was the document written and then signed, or was it created over an existing signature? Such timing considerations can have important consequences in an investigation or legal controversy. It is not uncommon to hear of situations where someone alleges that: "I signed this piece of paper with one thing on it, and, all of a sudden, my signature now appears on the bottom of a changed document." The questioned document could be a will, a contract, or some other potentially legally binding agreement.

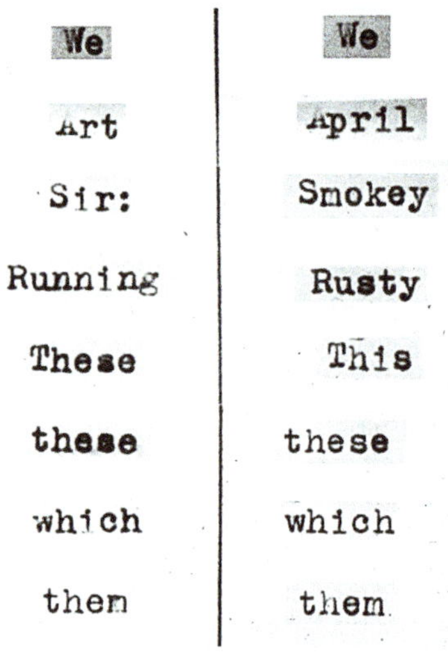

Figure 7.1 Comparison of words typed on suspect typewriter and those on the evidence document.

Document examiners are often asked to authenticate the author, the signer, or the contents of a document. The question of whether a document was properly executed can also be an important issue. For example, an individual says that they signed a particular contract, and, when the terms are satisfied, the other party indicates a different figure as being due than the one specified in the original contract. The contract presented is different than the complaining party's copy. Which, if either, is the authentic agreement? The question of whether one of the parties has altered the original contract is an important question and often not easy to answer. The authenticity of a signature is frequently an issue in court, though more commonly in civil than in criminal cases. Although forgery is a criminal offense, most cases are about money or power and are usually settled without resorting to the criminal law. On the other hand, check forgery or alteration is a very common criminal complaint (Figure 7.2). The complainant states that their checks came back from the bank, and one was not written by them or perhaps not signed by them. In general, banks do not carefully look at the signatures until there is a complaint. In one common scenario, a benefits check is not received on time, so a missing check form is filed with the employer or agency providing the benefits. The payer receives a number of such claims, and, when the checks or microfilm copies are received, the payees are asked to come in and check the accuracy of the signature.

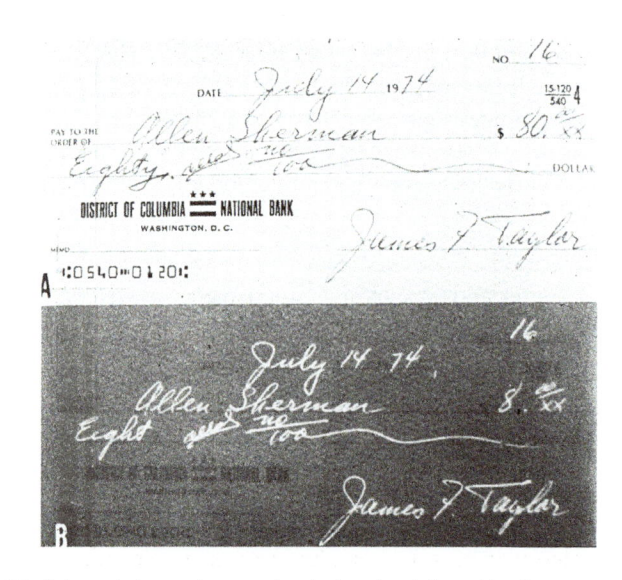

Figure 7.2 Check photo in visible light and photo of same check showing Infrared luminescence.

> **BOXED CASE ILLUSTRATION**
>
> In one case, it was discovered that there was a whole group of checks that went astray. Whoever had stolen this batch of checks, had signed the beneficiaries' check, for example, Herman Harris. The check was actually made out to H. Harris (Howard), and, whoever stole it, didn't know what the H stood for. So they just picked an "H" name. Thus, in this case, it is not a question of a forged signature, since the check does not even have the right name. There is no real question of it being forged, but someone is going to be out of some money that ended up in the thief's pocket.

A common variation of the above scenario occurs when the individual, who claims not to have received the check, is suspected of having cashed it and disguised the signature in order to claim forgery and obtain a replacement (second) check.

Recognition, collection, and preservation of evidence

A significant reason for the underutilization of handwriting examination is that handwriting examiners must be quite demanding, particularly in the quality and quantity of the known control samples they need. This is because it is very important that the control samples and evidence samples provided be sufficient in the amount of writing and need to be in pristine condition. As a rule, they must be original writings and not copies. A handwriting examiner is much more concerned with the way the handwriting was written than how it looks. For example, they look at where people stop and start their letters, how they connect letters, whether or not they connect a letter in some situations but not in others, and many subtle variations in how one writes. All of these things can be ascertained from the original writing, but often subtle differences are not evident in a copy. Information easily discerned from an original writing, often just does not show up on copies. The importance of how a document is written is illustrated by the fact that part of the initial examination is done by a handwriting examiner looking at the reverse side of the document. An examiner often can see more clearly on the back how letters were formed and where the strongest pressure was applied. Although handwriting examination can provide very important information, there is often other valuable information available from the document, as well. For example, it is crucial to preserve and protect any fingerprints that might be present. The ability to develop an identifiable fingerprint unambiguously indicates that the individual leaving the fingerprint handled the document. This may identify the writer, or it may bring a third party into the picture. Careful handling and protection of a document may also allow and examiner to find other useful evidence, particularly trace evidence. It is important with this type of evidence to think about which laboratory section should process it first. Some latent fingerprint development techniques (Chapter 6) have the potential to cause ink to run, for example. Examinations must be done in an order that permits all the potential information to be extracted.

Careful examination of a document can be very helpful, and may allow an examiner to say that a suspect may have forged another person's signature or might have changed the amount of a check. Finding a fingerprint of an individual where it should not be is often more useful evidence in a case than a handwriting opinion, which may have to be qualified because of a lack of sufficient sample or identifiable characteristics. It is important to use all the potential evidence in a document case. One must look at everything that can be of assistance; be it fingerprints, trace evidence, hairs, fibers, erasures, indented writing, or alterations. Unfortunately, there is a tendency to think about a handwritten document as a handwriting case and fail to make use of all the potential evidence available from the document evidence.

When one obtains a document that may be important to a case, one should immediately put it inside of a plastic sheet protector or protect it in some other way. If one receives a threatening letter, an extortion note, or a communication claiming credit for a bombing incident, it should not be passed around to everybody in the office. Such behavior, although tempting, can compromise a whole variety of possibly useful evidence. Although almost everyone knows that one can dust for fingerprints on a window or water glass, many people do not know that one can develop fingerprints on absorbent materials, such as paper, with a high degree of success. Therefore, individuals may fail to be as careful not to leave fingerprints on documents as they would on other surfaces. Forging a document with gloves on is often not practical.

As indicated above preserving physical evidence, fingerprints and trace evidence in particular, is very important. If one were to visit the document section in most labs and hand an examiner a piece of paper, the document examiner might, out of habit, grasp it between the knuckles of the first and second finger, where there are no friction ridges.

In addition to proper handling, gaining the most information from questioned document evidence requires submission of sufficient control or known standard writing. This is critical because, as discussed below, there is variation within an individual's writing, due to the time, due to the speed it was written, due to the implement with which it was written, and many other factors. Having sufficient known writing allows an examiner to determine how much variation is natural and, thereby, if the observed variation might be due to a different writer. It is also critical that the control writing be reliably attributed to the writer. An examiner, no matter how experienced, needs a good standard exemplar before putting his/her reputation on the line. The expert must say, "In my expert opinion this letter was written by this particular individual." To do so with confidence, a sufficient quantity and quality of known writing must have been studied and the examiner must know unequivocally that the sample came from the party in question.

CASE ILLUSTRATION 7.1

Even the best-questioned document examiners can be led astray. A number of years ago, the so-called "Hitler Diaries" were published in Germany in *Der Stern* magazine. They were reported to be his diaries up until his death in the bunker where he committed suicide. They came out of East Germany from a particular well-known collector of Hitler memorabilia. He claimed that the source was someone who was actually in the bunker with Hitler. This collector had sold many documents purported to be written by Hitler over a period of years. The "Hitler Diaries" were sold, as an exclusive to *Der Stern*, a major magazine in Germany. It is a very influential magazine with a large circulation. Before the magazine published these diaries, they took them to several document examiners for authentication. They had the paper checked, they had the handwriting examined, and the style and authenticity of the historical references were also checked. One of the document examiners that they used was Ordway Hilton, who was one of the most respected document examiners in the United States and a particular expert on handwriting.

It later came out that the diaries were forgeries. The East German dealer was found to have made a living, for many years, forging Hitler papers and selling them anywhere he could. He had obtained old paper that was identical to that used during the Third Reich and the proper inks so that one could not detect the forgeries from the materials used. He was a very talented forger, and there was little authentic Hitler writing for comparison. Hilton was very embarrassed by his failure to detect the forgeries, but it turned out that it was not actually his error. When Hilton was asked to authenticate the diaries, he was given some supposedly authentic control samples of Hitler's writing. It was these writings that he used to judge the authenticity of the diaries. He carried out his examination very carefully and reported that the same person who had written the provided control writings had written the diaries. Unfortunately, for his reputation and *Der Stern*, this same forger had forged the supposedly authentic (control writing) Hitler writings. This individual had been so successful at forging Hitler memorabilia that they were widely accepted as authentic Hitler writings. The importance of the story is that it illustrates the crucial need for carefully authenticated known writing. It is rather sad that when Hilton died a few years ago, one of the things that was mentioned in his obituary was that he incorrectly authenticated the Hitler diaries. Ordway Hilton, one of the world's best and most respected document examiners, was remembered for one of the few instances where he had reached what appeared to be the wrong conclusion.

In addition, the examiner must know the source and the circumstances of the collection of both the evidence and known control samples. One must know if the material being compared was written 3 weeks ago or 25 years ago. The examiner will have to make some allowances for differences in writing mechanics due to age. Obtaining known standards that were written about the same time as the questioned materials greatly increases the chances of a successful outcome. Document examiners must often tell the investigator to please find additional standards or more contemporary standards. There are a great variety of potential sources of known writing, but it takes considerable investigative time and knowledge to ferret them out.

Writing process

All aspects of the writing process can yield useful information. One can break a document into three basic parts: a surface for the writing, a writing instrument (Figure 7.3), and a transfer medium between the writing instrument and the surface.

Figure 7.3 Documents and forms can be made from a variety of forms with a variety of different types of writing instruments. (Courtesy of Shutterstock.)

Figure 7.4 Clay tablets once played the part of writing surface and the writing instrument a stylus to impress symbols.

The most common surface is a piece of paper. The most common writing instrument is a pen or pencil. The most common transfer media are ink or pencil lead. However, the possible variations are nearly endless. The surface may be a bathroom wall or a subway car. The instrument may be a paintbrush, a crayon, or a spray can. The transfer medium may be paint, chalk, or even blood. A rare exception would be exemplified by ancient writing on clay tablets (Figure 7.4) where a stylus was used to make marks in the soft clay. In such a case, there is no transfer medium. Even with mechanical printing, such as a typewriter, computer printer, or many others, there is a surface, a writing instrument, and a transfer medium to make the writing visible. For example, in thermal printing devices, the instrument is a hot wire that causes a chemical or physical change (transfer medium) in a coating on the surface of the paper.

Each of the three components can produce important information for a document examiner under the right condition. A document examiner can look at the paper of a questioned document and quickly tell whether it is expensive paper, cheap paper, has a high content of cotton fibers in it, and something about the surface treatment. In addition, with a little more careful examination or testing, they can tell if it was produced from mechanically pulped or a chemically pulped wood and perhaps a whole variety of other things as well. There are many other characteristics of

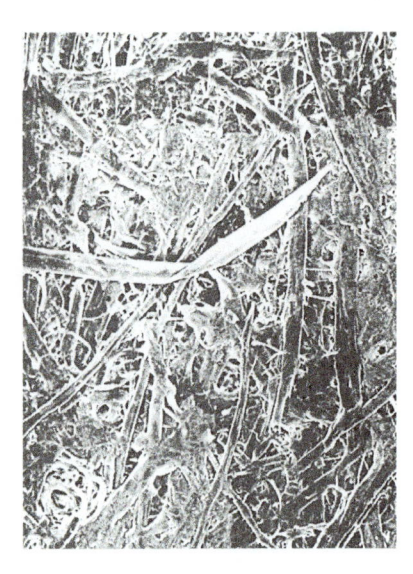

Figure 7.5 Magnified paper surface showing matted paper fibers.

paper that can provide useful information in certain cases, such as the type of surface treatment. When one is writing on normal paper, one is actually not writing directly on the paper fibers. Paper, even fairly inexpensive paper such as copier paper, has a thin surface coating or sizing on it. That coating is usually a starch or a clay type material. It is necessary because paper is a mat of interwoven fibers (Figure 7.5), and the surface of it is a bit rough. The surface treatment smooths that rough surface so that the paper will provide a nice surface for writing or printing. Every company that produces paper has its own technology for this process. They may use different things in their clay, different techniques for applying and bonding this surface finish, and many other possible variations. These variations may be useful to determine whether or not two papers were manufactured by the same company or at the same plant. That information may turn out to be important to an investigation.

As an example, let us say one is examining a five-page document. The document examiner is unable to find much evidence that it has been altered, as has been alleged. Perhaps at execution, the parties did not bother to initial each page, as is customary on legal documents, to prevent someone from slipping in an extra page. It is alleged that one of the parties did slip or substitute in an extra page. A document examiner will look at many different things, including the paper. Even if the alteration was done sometime later, it could still easily have been typed on the same typewriter, or generated from the same computer printer. But. if an additional page was added or substituted, although it may look the same, a chemical analysis might show that there are four pages of one kind of paper, and one page of a similar, yet distinguishable kind of paper. This could be a convincing piece of evidence in court to back up the claims of an added page.

The type of and exact nature of the writing instrument used can also give one useful information in many questioned document situations. There are, of course, literally hundreds of different brands of ballpoint pens. These can be divided into a much smaller number of different mechanical types. There are many manufacturers of ball pen ink, but far fewer than the number of different brands of pens. Different pen makers may be buying their ink stocks from the same company. Visual examination, with perhaps a little bit of magnification, can easily differentiate many of them. That alone may be sufficient to differentiate pens that have the same color ink but have a different way of delivering the ink. The writing from a ballpoint pen, razor point, or a nylon tip can look quite different when carefully examined. Similarly a fountain pen produces a different looking line than any of the abovementioned types of pen. Examination of the lines produced by an inexpensive ballpoint pen compared to those from a high-quality pen (Figure 7.6), will usually show differences in both line quality and ink application. Further, each different mechanical device, such as a typewriter, computer printer, copying machine, or printing press, produces a characteristic type of printing.

Finally, the transfer medium may allow the differentiation of writing that is made with types of devices and that produce very similar appearing writing, when the ink, paint, wax, toner, or other transfer medium is examined.

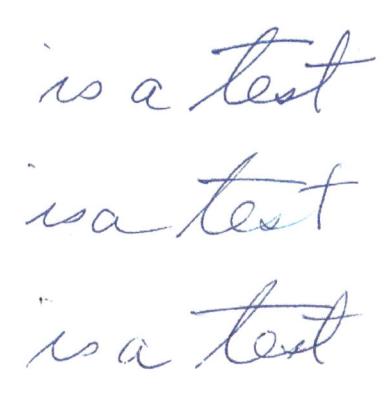

Figure 7.6 Magnified portion of writing by three different ballpoint pens showing differences in line quality.

Manufacturers of different writing instruments often use different ink compositions. Evidential writing may be made using spray paint, a crayon, a colored pencil, but, most commonly, it is from a pen of some sort. Today, ballpoint pens are by far the most common. One can examine the writing from two pens that appear to produce exactly the same color and, by extracting some ink from the writing and analyzing it, determine that the inks have a quite different chemical composition. Inks are made from dyes or pigments that are mixed to produce a particular color. One will discover that to achieve exactly the right shade of blue, the ink manufacturer may mix a little red, some blue, some black, and maybe a little yellow. Document examiners can remove a tiny sample of ink form the page and chemically separate the color components of the ink. Finding a different mixture of dyes can conclusively prove that two writings were not written with the same pen. For example, if someone tries to alter a document, they will usually try to find a pen with the exact same color ink. There is a good chance that they did not find the exact same pen, and, although the color matches well, the ink composition may be quite different. Examination of a document surface, writing instrument, and transfer medium can give valuable information in many questioned document examination contexts and needs to be considered as an integral part of each questioned document examination.

HANDWRITING COMPARISON—IDENTIFICATION OF WRITER OR DETECTION OF FORGERY

Development of an individual's handwriting

There is a good deal of science and technology underlying the examination and comparison of handwriting. It starts with one's basic handwriting style, which initially is determined during early schooling by the copybook (Figure 7.7) from which one learned to make the different letters. The teacher or a copybook would provide examples, and one would try and follow the example. This process begins with printing, and, when that is learned, moves on to cursive writing.

Everyone who attended a particular school at the same time learned the same basic style. Probably everybody in that school district, and maybe most of the school districts in that state, also learned that style. Nonetheless, there have always been a number of different handwriting copybooks being used by schools across the country. They have slight variations in style so that different individuals may have learned slightly different styles.

Because of the persistence of these class features in people's writing, it is often possible to recognize writing as being from a person who went to school in another country, because they likely learned from a different copybook. However, 25 years later, would one expect even individuals who went to the same school in the same year to still make their letters exactly the same way? For most individuals, writing does not remain static between grammar school and adulthood. As people mature and become more independent, their handwriting changes. Particularly during the teenage years, most individuals experiment with their writing style, and especially with their signatures. This eventually results in a style with which they are comfortable, and one that is usually quite distinctive. Again, this is particularly true of one's signature. Our handwriting evolves with time, but the major characteristics are usually set by the time one finishes one's formal education.

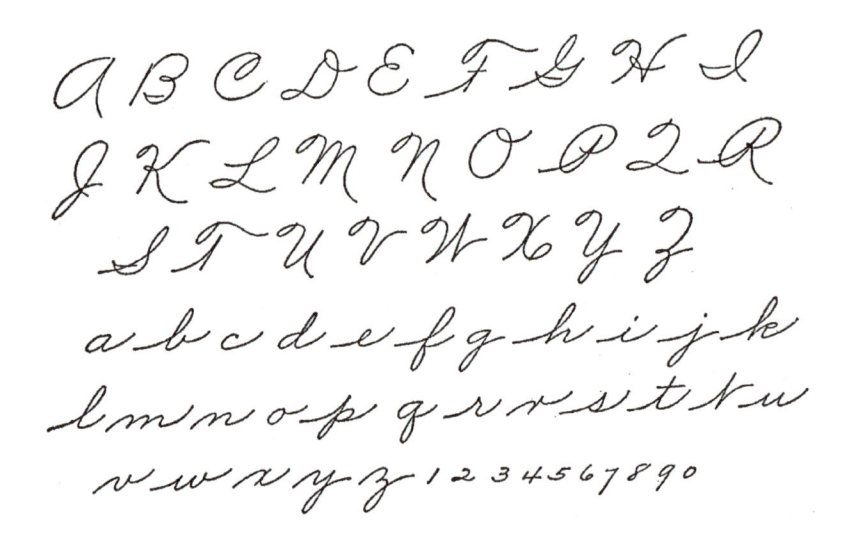

Figure 7.7 A page from the Palmer Method copybook used in many American schools to teach children to write.

The important point, from a document examiner's point of view, is that most individuals develop individual characteristics in their handwriting. By the time that one is in their late teens or early twenties, handwriting has largely stabilized. It will continue to evolve slowly with time, but basic characteristics are set for most people at young adulthood. If one were to look at a sample of the handwriting of the same individual in middle age, one would see that there have been some changes in the way it looks; however, careful examination, the way a document examiner looks at writing, would show that there is a great deal of consistency. Because of this drift, it is important, when obtaining known control handwriting samples, to try to obtain samples from about the same time period as the evidence samples were written.

There are other things that can affect one's handwriting in addition to aging effects. When people are sick or become infirm, their handwriting often deteriorates (Figure 7.8). This arises in the classic "deathbed will" situation where someone is holding the testator's hand while they sign, and they are barely writing at all. This often causes challenges to the will. Medical effects can come and go. Someone could have a stroke and their handwriting might deteriorate seriously, but, if their recovery is good, slowly come back to something like they had before they had the stroke. Alcohol or some drugs (therapeutic or illicit) also have the potential of influencing the features of a person's handwriting.

In document examination, as in most of forensic science, one is usually faced with comparisons between a known control and an evidence sample. A crucial point in these examinations of handwriting is the amount of normal variation in an individual's handwriting.

The variation within the known writing sample is usually significantly less than the variation between samples from different individuals. As discussed above, although handwriting tends to stabilize at adulthood, there is some day-to-day handwriting variation. Particularly when one is tired, or the conditions are less than ideal. Most students can attest that writing done while seated at a desk at home, is different than that done when seated in a classroom in a chair with a small arm on it that is not particularly stable. The key observation is that the variation within an individual's writing, caused by minor discomforts, is normally much less than the variation between that individual's writing and that of most other individuals. If that were not the case, one could not validly distinguish handwriting.

Figure 7.8 End of life aging can strongly effect the appearance and particularly steadiness of an individual's writing resulting in shaky character formation.

Mechanical and pictorial characteristics of handwriting

Superimposed on the class characteristics we learned from our copybooks are individual characteristics we have developed as we matured. The combination of class and individual characteristics is often sufficient that a handwriting examiner, with an adequate amount of written material and good known exemplar writing, may say, "In my expert opinion this writing was done by this person." Not everyone develops sufficient character to their handwriting, but the vast majority do. Because our writing has normal variation, the condition above; that is, sufficient evidential writing and sufficient standard writing is critical to successful comparison. The more writing available, the better the chance that the examiner will be able to come to a decision. If one has nothing but a single questioned signature, a large number of known control signatures would be needed. With other writings, a much larger sample of writing would be required since one's other signature is almost always much more distinctive than one's normal handwriting. Were one to have just three words of normal handwriting, and nothing else, an identification or exclusion would be unlikely, unless the writing showed a highly unusual writing style. On the other hand, if one had a full page of writing and if one obtained a set of high-quality standard writing written roughly contemporaneously and hopefully with the same general type of writing instrument, then there is a very good chance that an examiner can make an identification or an exclusion.

Handwriting comparison encompasses many more things than how the writing looks. When comparing an evidentiary document to known handwriting standards, the first area of interest is the appearance of the writing on the page. Such things include where the writing is located on the page and line, word and letter spacing, word usage, spelling, grammar, margins, whether the lines run parallel to the top of the page, and many other appearance characteristics. After looking at the general appearance, the examiner is concerned with the more detailed aspects of the writing. They are concerned with how the letters are made (letter forms), where pressure is applied, connections between letters, word and letter spacing, and many other subtle characteristics of the writing.

Handwriting comparison is strongly concerned with how letters are formed individually and into words. A well-trained examiner will look at each letter and how it is made (Figure 7.9), where it starts and ends, where pressure is heavy, and the general flow of the writing. Letter and word spacing are important characteristics as well as letter connection within words and diacritical marks, such as "I" dots and "T" crosses. The size ratios of the large and small letters and between the upper and lower portions of the same letter are also important. Slant of the writing is a useful characteristic. The ability to distinguish between different writers is dependent on the examiner's ability to fully

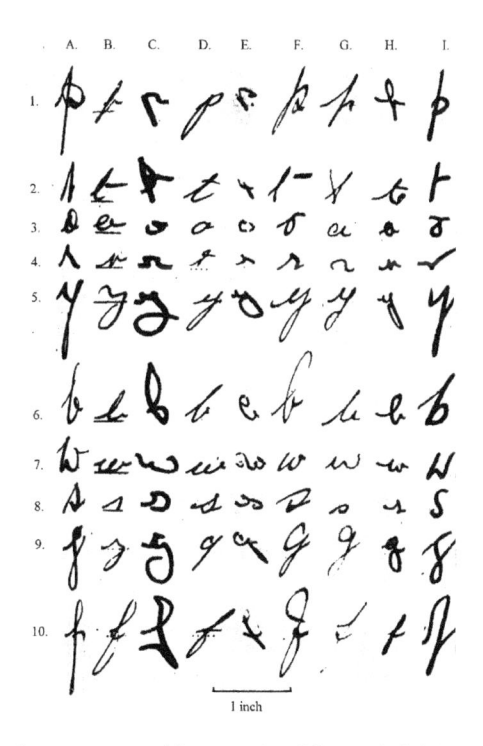

Figure 7.9 Cursive writing of the same letter compared from twelve different individuals.

appreciate the subtle differences in the way writers write. That ability is developed only after the careful study of writing mechanics and hundreds of hours of practice, usually under the tutelage of a seasoned examiner. Other factors are important in the detailed examination of how a document was written. The pictorial and contextual elements, such as style, spacing, grammar, and spelling are all important.

Some writers tend to cram the letters and/or words together, and others spread their writing out. Some writers have the letters close together with the words well-spaced, and some have the words close together and the letters well-spaced. There are myriad variations.

Importance of known standards to successful examination

The variability of handwriting, both in terms of day to day variation and variation caused by other physical changes, as indicated above, complicates the handwriting comparison process. A document examiner has the difficult problem of assessing whether the differences between a known sample and evidential writing are due to this natural variation in the author's handwriting or to the writing having been done by a different writer (forged). The more comprehensive the set of known standard writings, the more accurately the examiner can define and judge the normal variation level for that writer. Further, knowledge of when the evidence writing was supposed to have been written and the availability of standard writing from the same period or, at least within a reasonable time range, can further help in defining natural variation. Thus, both an appropriate quantity and a well-documented known writing sample are critical to the success of handwriting comparison. This may be frustrating for an investigator, who must usually gather these samples, but failure to do so can often lead to an inconclusive result. Handwriting standards fall into two general categories: authenticated, collected writings; and requested (ordered) samples.

Authenticated collected writing

Collected writings are the most useful, but often the most difficult to obtain. These samples are obtained from the subject's life experience and history. They can be letters written to friends, job applications, business papers, school records, loan applications, insurance claims, and a myriad of other sources. Table 7.1 shows some more of the possible sources of such standards. Although there are numerous possibilities, samples are often difficult to obtain. Original writing is highly desirable, and many of the sources can provide only microfilm or imaged copies of the original record. Further, many of the sources may simply not exist for a particular person. The investigator must be persistent and often patient when trying to obtain sufficient desired materials. These collected samples can provide an accurate record of an individual's writing and can often provide important information on how the writing looked in a particular period of interest, or how it changed with time. Further, each sample must be carefully authenticated to ensure that the sample is from the individual in question and not a spouse or someone else with the same name.

Requested (ordered) writings

Requested writings are those obtained either voluntarily, or through a court order, directly from the individual. Although this might sound ideal, because one can tell the individual exactly what to write, there are at least two problems. First, the individual now knows that he or she is a target of an investigation and may try to distort or disguise the handwriting. Second, this is a contemporaneous sample. If the evidence writing was done sometime before or if there has been a change in the individual's medical condition, the writing may have changed, making comparison more complex. To address the first problem, the sample must be taken under carefully controlled conditions. The material should always be dictated, and the dictation should be repeated several times. An individual trying to disguise their writing will tend to lose concentration, and the writing on the second or third dictation will tend to drift toward their normal writing habits. There are a number of other tricks that professional document examiners can apply in drafting a dictation that will help to provide a more truly characteristic standard. As a result, whenever one is applying for a court ordered sample, one should seek assistance from a well-qualified document examiner in drafting and obtaining the requested or ordered sample.

Dictation is critical since it allows one to gain information on important characteristics, such as spelling, page layout, and a variety of other important aspects, of how an individual writes, in addition to the mechanics of their handwriting. If someone is asked to copy a typed or printed exemplar, much of that information is lost. Spelling, word spacing, page layout, and line spacing will all tend to follow that on the template and not necessarily be characteristic of the individual's writing. Another result of the automatic nature of writing is that it is quite difficult for most people

Table 7.1 Where to find authentic handwriting samples

1. City records	Building Department permits City Auditor: cancelled checks, office records City Clerk: licenses (peddler, tavern, special permits, etc.); voter registration lists Personnel Department: Civil Service applications Permits: Dogs; building, parks, and other city facilities; applications
2. County records	County Clerk: Civil Service applications, claims for services or merchandise, fishing, hunting, marriage licenses Department of Taxation: State income tax returns Purchasing Department: bids and contracts Register of Deeds: deeds, birth certificates, public assistance applications, ID card applications Selective Service (local board): registrations, appeals
3. Department store records	Complaints and correspondence Credit applications Receipts for merchandise Signed sales checks Merchandise delivery records
4. Drug store records	Register for exempt narcotics, poisons Signed prescriptions
5. Hospital records	Admission and release forms Consent forms Checks and payment records Communications and letters
6. Library records	Applications for cards Check out and reserve slips
7. Education documents	Applications for entrance Athletic contests Daily assignments Examination and research papers Fraternity and sorority records Receipt for school supplies (laboratory, athletic gear) Registration cards and forms Federal and State Loan and Grant applications Scholarship applications

to disguise their writing. When trying to write in a way much different from one's normal writing, it is difficult to consistently maintain the disguise. Those automatic writing circuits want to take over. The importance of this phenomenon is discussed below.

Examination and comparison of handprinting

It seems that in recent years, with the replacement of the handwritten letter by computer email and social media contacts, more people are using handprinting instead of cursive writing. Much of what has been said about handwriting is also true concerning handprinting. However, the examination and comparison of handprinting is handicapped by the loss of the connecting strokes and several other characteristics useful in handwriting comparison. For individuals who avoid cursive writing and use handprinting almost exclusively, their handprinting will become more individual than those who use it occasionally to supplement cursive. Emphasis on handprinting appears to be a significant trend among younger writers for whom the computer and other electronic media have replaced cursive writing. As a result, examination and comparison of handprinting is much more important than it once was. Letter forms are useful as well as most of the other characteristics used with cursive writing. As a result, handprinting can be successfully compared and identified in many cases if sufficient printing evidence is available and good handprinted knowns are obtained.

Signatures

As indicated above, signatures are particularly distinctive, much more so than other writing. The distinctive nature of signatures (Figure 7.10) arises because one does not have to think about forming each letter of each word when one signs one's name. That is because one's brain has been "programmed" over time. Someone can write largely without

With more modern types of mechanical printing devices, there is still useful data that can be obtained. Any printer that is an impact device can develop misalignments and can suffer damage to its typing elements because of the potentially damaging effect of rapid and forceful physical movement. But with laser-type printers and other non-impact printers, like inkjets, the possibility of mechanical defects is virtually eliminated. Laser printers and most copying machines work by causing a light beam to trace out letters on an electrostatic drum. The toner (transfer medium) is attracted to the drum and then fused onto the paper from the drum. There are no writing elements and little opportunity for damage. Some misalignments are possible, and minor damage to the highly polished drum is not uncommon. With these machines, though, it is very hard to individualize a particular machine. The information available from examination of documents produced on such machines is quite limited, unless they develop some kind of a serious problem or suffer damage not related to their printing function.

Copying machines

Copy machines and the documents they produce can provide some useful investigative information. As indicated earlier, copies of handwritten documents are not usually suitable for handwriting comparison because one loses the important information concerning pressure points, lifts, retraces, and critical information about letter formation. One can develop useful information through analysis of the toner. In a copying machine and laser printers, the toner takes the place of the ink, particularly copiers using a xerographic or similar process. It is the toner that is attracted to the charged drum, transferred to the copy paper, then seared onto the paper using heat. Different copier or laser printer manufacturers tend to use different toner formulations. In addition, not only the printer manufacturers make toner. Analysis of the toner composition may provide the manufacturer of the copier or at least the company who compounded the toner. This can be useful when searching for a machine that might have been used to make a copy and for comparing copies as to source. One can often determine the make and the model of the copier used or at least the type of copy process the machine used. There is also the possibility of accidental marks on the drum. If the drum is scratched or otherwise damaged, this will usually show up on the copies produced. On some Xerox copiers; for example, a full revolution of the drum contains images of three pages, which results in every third page having a mark corresponding to the position of the damage to the drum. The spacing between repeating marks will vary depending on the copier model. When there are a sufficient number of different repeating marks, such marks are potentially individualizing characteristics. For example, marks on the platen of a copying machine, the area where the document to be copied is placed, will also be reproduced on the copy. Undried, opaquing liquid on a document can be transferred to the platen. This will effect copies until it is detected and wiped off. Older fax machines used a thermal printing process that can develop some class characteristics, but few individual characteristics. Most modern "plain paper" fax machines use a process similar to laser printers.

SOME OTHER USEFUL DOCUMENT EXAMINATIONS

Detection of alterations and erasures—usefulness of paper examination

Document examiners are often asked whether a document is the authentic original, or has there been an alteration? Detecting alterations and erasures can be extremely important. For example, if one has a lottery ticket that is just one digit different from a big winner, the temptation to alter that ticket (Figure 7.12a and b) can be fueled by millions of dollars. If only the 3 could be changed to an 8, it could be a big winner. It has been tried many times. One would not try this on a fifty-million-dollar winner, because the top winners are closely scrutinized. However, very substantial amounts are given out to winners upon presentation of a ticket and with little scrutiny of that ticket.

Most lotteries are now quite sophisticated and have many built in protections against alteration, microprinting that will not copy, or incorporation of a hologram, which is very hard to reproduce. Counterfeiting lottery tickets can be as tempting as altering authentic ones. With the quality of color copiers and digital scanners, simply making copies of money can produce quite authentic looking fakes. This is the impetus behind the U.S. Treasury Department redesigning the currency to add many new features. We now have off-center presidential portraits; a little line that gives the amount of the bill, which is only visible by holding the bill up to a bright light; there is microprinting that is so small that it is beyond the resolution of copiers; and other security features. The point is that people have over the years found it an attractive challenge to devise new ways to alter documents for their benefit.

Figure 7.12 (a) Lottery ticket as photographer under normal illumination, (b) Same lottery ticket photographed with infrared luminescence clearly showing the alteration.

Let us think about a simple document, such as a contract for the sale of pork bellies. That does not sound very exciting, but bacon is a major commodity. One signs a contract for a million pounds of pork bellies that represents a great deal of money. If the original contact is for a price of $1.16 per pound and the other party who is supplying the pork bellies subtly changes the amount to $1.76 per pound, even that difference is a lot of money. For a million pounds, adding $0.60 per pound is a $600,000 dollars difference. Unfortunately, there are many people who would be tempted to try document alteration for $600,000. There are many commercial and other agreements where circumstances can change, and one party is put at a considerable disadvantage. Just changing a few words or a number on an agreement could relieve that disadvantage.

People alter documents in their favor for economic reasons and other reasons as well. One of the simplest ways to determine if a document has been altered is to examine it carefully using good lighting and a stereomicroscope. Any mechanical process that will remove the desired words, will probably disturb the fine coating on the paper and make it more transparent to light or disturb the fiber mat of the paper. Both are detectible with careful visual examination. If this fails to disclose any alterations, there are many other examinations possible. For example, chemical or spectroscopic techniques may show changes that cannot easily be seen even with magnification. When one alters a document using an ink eradicator fluid, one must remove the original writing before replacing it with the desired change.

There are two processes: some sort of erasure and some addition of new writing. There are cases where one only needs to add something, but most often something must be erased so that something else can be added. A good document examiner will look for both and neither is easy to disguise from a knowledgeable examiner. Alterations are almost always done some time after the original document was executed. It may be years later, or in our pork belly illustration, it may only be 6 months or less. Perhaps there has been a sudden change in the market that means a large loss on the original contract. The point is that when the alteration is made, even if one can do an almost undetectable erasure, whatever must be added often cannot be done with the exact instrument that was originally used to create the document. It is also rather difficult to achieve perfect alignment, if it is being mechanically printed. If handwritten, obtaining the same or an identical pen, which was used when the document was originally created, could pose a significant problem.

One can certainly match the color of the ink. One can go to an office supply and buy thirty-six different blue pens or twenty-four black pens and select the one that appears exactly the same color on the document that one wishes to alter. In the world of inks, having the same color does not necessarily mean same ink. The technology for making virtually all inks involves mixing different dyes or pigments to get the desired color. There are many different combinations of colored dyes and pigments that will produce essentially the same color. One can use chemical analysis to look at the mixture of dyes or pigments in a particular ink sample. This kind of analysis will often show that inks, which look to be the same color, have different compositions and, therefore, must be from different writing instruments.

Although two inks look to be the same color to our eye (visible part of the electromagnetic spectrum), their interaction with other parts of the electromagnetic spectrum may still be quite different. Therefore, one might be able to use non-visible light examination to tell two identical appearing inks apart. Taking a normal visible light scan will show about the same thing that one would see looking through a stereomicroscope. On the other hand, illuminating the document with strong ultraviolet or infrared light and using a filter that filters out the visible light, but allows either ultra-violet or infrared light to pass, may allow one to see something that cannot be seen with normal vision. What one is doing is extending the range of human vision. Our eyes are only sensitive to a relatively small portion of the spectrum, called the "visible" portion. There are many areas of the spectrum our eyes cannot perceive, such as the ultraviolet and the infrared (Appendix A). Scanners or video cameras can be capable of "seeing" in these other regions and, in that way, can visualize differences that are not visible to the eye between two inks.

Many document examiners now have access to an instrument called a spectral comparator.

These instruments operate using a specialized video camera and a variety of light sources and filters to make quite of number of non-visible light examinations and comparisons much simpler and more effective.

One can also use mechanical and physical methods to detect alteration. One of the simplest is to look for changes in the paper brought about by the alteration. It was mentioned earlier that paper is matted cellulose fibers. In addition, it has a thin layer on top of the matted fibers to make the surface smoother and better able to accept the transfer medium (ink). This is normally either glue or clay, and is called sizing material. No matter how carefully one tries to erase writing or printing, the process of erasing will disturb that very thin sizing layer. Some of that material will be removed. This may not be visible to the naked eye or perhaps even with a microscope, but, if the document is examined by back lighting on a light box, the erased area will show up. It will show a lighter area because some of the opaque sizing material has been removed. The erased area will allow more light to penetrate the paper than through the area where the glue or clay sizing is undisturbed. Just a simple light box, such as a doctor would use to view an X-ray, is frequently all that is needed.

One can also look for alterations mechanically. One of the classic ways that document examiners look for erasures is using a very fine powder, such as lycopodium powder (pollen from a plant) or even fingerprint powder. One spreads a little bit of the fine powder all over the document that is suspected of having an erasure, shakes it around a bit, and then taps it all off. Where the erasure has occurred, again the surface coating layer will be disturbed. If the sizing is partially removed the paper, fibers are exposed, and a much rougher surface is present. The very fine powder is trapped in this roughened, fibers surface, and, since the powder is usually colored, the disturbed surface is made clearly visible. Usually this technique will work with chemical ink eradicator as well, since the chemical will also affect the sizing layer. Erasing fluids normally work by decolorizing the ink so it can no longer be seen on the document. The ink is still there but no longer visible to the eye. Non-visible light techniques will often disclose this writing that has just been made almost "invisible."

Another way that is useful for detecting alteration of a document, is a chemical separation of inks. If a particular section of the document is suspect, samples of the ink in that section can be compared to ink samples from an area not being suspected of tampering. A sample of the ink is taken by punching out tiny circles of writing from a few letters using a syringe needle. A number of these tiny circles of paper with ink on them are placed in a small tube, and a solvent is added to dissolve the ink. This ink solution is then separated into its components using a chromatographic method (TLC or HPLC) (Figure 7.13). The exact same procedure is used for the control samples from the unaltered portion.

The pattern of the dye components observed from suspect portion can be compared to that from the control portion to detect if a different ink was used.

Examination of charred documents or indented writing

Indented writing and charred documents present a different type of document problem. In one case, there is no transfer medium only mechanical indentation in the paper, and, in the other, the color of the paper has been changed to black or near black, so the ink can no longer be easily seen. Both situations are fairly common. For example, if someone involved in a white-collar crime wants to obscure the paper trail, they might take all the files and attempt to burn them. It turns out that paper does not burn as easily as one might think, particularly

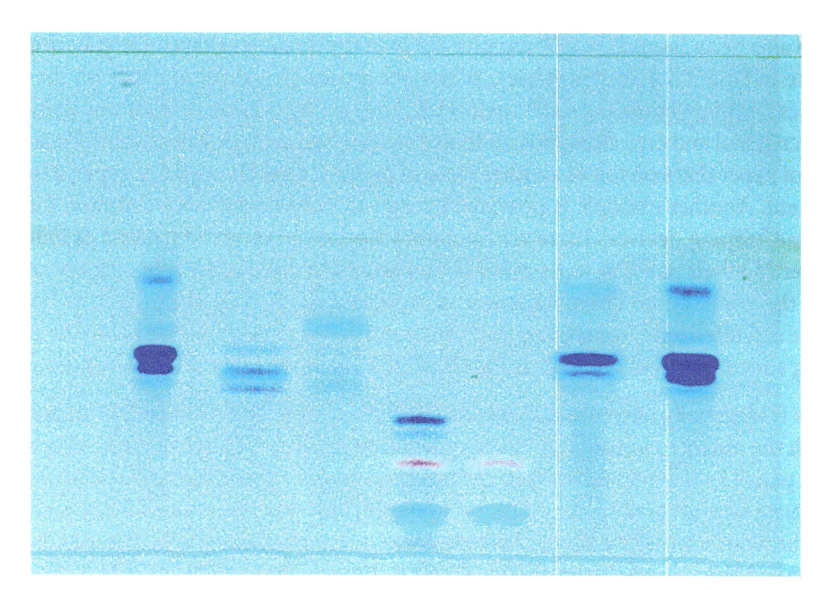

Figure 7.13 Thinlayer chromatographic separation of the pigments in several ballpoint pen inks.

Figure 7.14 Fragment of a charred document with the writing contrast enhanced by infrared luminescence.

when there is a stack consisting of many sheets. Although paper burns quite easily when a single sheet or when crumpled up sheets are ignited, when one has a stack of paper or a ledger book, it becomes quite difficult to provide sufficient air for rapid combustion. Unless one can keep stirring and separating the sheets, they will tend to char around the edges and not be consumed. Even if the burning has been more successful, one may recover the charred paper. It will be difficult to read because the writing has become black writing on black paper (Figure 7.14). However, the ink residue may still be there. There are several ways solve this problem through careful examination using transmitted light or using non-visible light photography to extend our human vision. This can be done most easily with a spectral comparator as mentioned above. If the original writing absorbs or reflects light outside the visible part of the electromagnetic spectrum and one uses the correct combination of light source and filters, one may make the ink appear as bright on a dark background and thereby visible again. In fact, the most difficult part is often finding a way to handle the charred sheets of paper without destroying this very delicate material. There are chemicals that one can lightly spray on the charred paper that will soften it and make it pliable and, thereby, much easier to handle.

handprinting
High Performance Liquid Chromatography (HPLC)
indented writing
ink composition
ink eradicator
light box
normal variation
oblique lighting
pigment
Platen
questioned document
requested writings
spectral comparator
surface coating (sizing)
threatening letter
Thin Layer Chromatography (TLC)
typewriter
Watermark
writing instrument
writing mechanics

Review questions

1. Discuss how questioned document examiners can provide information useful to many types of investigations.
2. What are the major factors in the development of an individual's handwriting?
3. List the two major types of handwriting control samples and their particular advantages and disadvantages.
4. List several important types of examinations that questioned document examiners provide other than handwriting comparison.
5. What are the main approaches for estimation of when a document may actually have been written?
6. What are the major methods used to decipher indented writing?
7. What are the major methods used to look for erasures on documents?
8. Discuss the health issues that can affect the appearance of an individual's handwriting.
9. What types of examinations do document examiners perform on typewritten, computer printed, and copied documents?
10. Discuss differences between handwriting and hand printing and how they can affect the examination of such documents.

Fill in/multiple choice

1. The variation in characteristics in known writing from a single individual is usually _____ (insignificant, less than, more than, equal to) the variation observed between different individuals' writing.
2. The production of a document in the broadest sense required three components, (1) _____, (2) writing instrument, and (3) _____, all of which may be examined to produce forensically useful information.

3. A known writing sample _____ (must, always, should never, should) contain the words and letter combinations present in the questioned document.
4. When collecting document evidence, it is important to carefully protect the document to avoid _____.
    a. a loss of trace evidence
    b. changing handwriting images
    c. loss of fingerprints
    d. a and c
    e. a, b, and c
5. The success of a questioned document examination is highly dependent on the quality and quantity of _____ (control writing, copies of the evidence, ink used in the writing, fingerprints developed) provided by the investigator submitting the evidence.

References and further readings

American Society of Questioned Document Examiners; www.asqde.org.

Baden, M. M. "Introduction to Plenary Session on the Lindbergh Kidnapping Revisited: Forensic Sciences Then and Now." *Journal of Forensic Sciences* 28, no. 4 (1983): 1035–1037.

Ellen, D. "*Scientific Examination of Documents: Methods and Techniques.*" 2nd ed. Abington, UK: Taylor & Francis Group, 2002.

Eye-Tracking Study Validates Handwriting Analysis; https://www.forensicmag.com/news/2015/05/eye-tracking-study-validates-handwriting-analysis.

Kam, M., P. Abichandani, and T. Hewett. "Simulation Detection in Handwritten Documents by Forensic Document Examiners." *Journal of Forensic Science* 60, no. 4 (2015): 936–941. doi:10.1111/1556-4029.12801.

Haag, L. C. "A Brief Chronology of the Lindbergh Kidnapping." *Journal of Forensic Sciences* 28, no. 4 (1983): 1038–1039.

Mnookin, J. L. "Scripting Expertise: The History of Handwriting Identification Evidence and the Judicial Construction of Reliability." 87 Virginia L. Rev. 1723 (2001).

Moenssens, A. A. "Handwriting Identification in the Post-Daubert World." 66 U. of Missouri at Kansas City Law. Rev. 251 (1997).

Mokrzycki, G. M. "Advances in Document Examination: The Video Spectral Comparator 2000." *Forensic Science Communications* I, no. 3 (1999).

Osborn, P. A. "Excerpts and Comments on Testimony by the Document Examiners in Regard to *State of New Jersey v. Bruno Richard Hauptmann*." *Journal of Forensic Sciences* 28, no. 4 (1983): 1049–1070.

"Questioned Document Examination", www.facultyncwc.edu/toconnor/425/425lectO5.htm.

Standards for Examining Documents for Alterations; SWGDOC; www.swgdoc.org.

CHAPTER 8

Toolmarks and firearms

Lead Case: "44 Caliber Killer" (Son of Sam case)

It began in July of 1976, and gripped the city of New York for over a year. There was a series of random shootings of young people, primarily in parked cars in lover's lane-type locations. It was almost 6 months before it was discovered that there was a serial killer involved. In New York City, each borough is a separate county with its own District Attorney and some County Government. Because the first incidents occurred in three different boroughs of New York City, the connection was not easily made. However, there is only one Police Department and one forensic laboratory for the entire city of New York. Firearms examination was to play an unusually large role in the investigation and solution of this case.

The initial discovery that several shooting incidents in different boroughs were connected was made in the firearms unit of the forensic laboratory. A firearms examiner was looking at the recovered bullet evidence from a recent shooting and remarked that the recovered evidence bullets had been fired from a .44-caliber weapon. He mentioned this because it was not a commonly encountered caliber. One of his colleagues, sitting a few feet away at another comparison microscope, mentioned that he, too, had recently encountered some .44-caliber evidence. The older case was found in the files and bullets from both were compared. It was soon evident that the bullet had been fired from the same type of weapon, and further examination disclosed that it was the exact same weapon. Further checking of fairly recent cases with .44-caliber evidence disclosed a third shooting in which the exact same weapon had been used. The .44-caliber serial killer investigation was begun when the submitting detectives were notified of the connection between their cases.

The popular press renamed it the "Son of Sam Case" a little later when the killer began taunting the Police with notes signed "Son of Sam." The shootings continued until August of 1978, in spite of one of the most extensive investigation in NYCPD history. The toll had risen to six dead and a number of others either permanently disabled or seriously injured—a terrible toll of young lives.

The somewhat unusual markings on the recovered bullets caused the firearms examiners to feel that it was highly likely that they were fired from a .44-caliber weapon manufactured by Charter Arms Company and, specifically, their model called "Bulldog." Pictures of this weapon were circulated to the investigators, and it became a significant focus of the investigation.

Because witnesses had seen the killer shooting from a two-handed military stance it was theorized that he might be a current or retired law enforcement officer. Files of the weapons possessed by current and past law enforcement officers were searched for any possessing Charter Arms Bulldog revolvers. As a result, hundreds of these weapons were brought in for testing in the firearms unit in hopes of finding the killer. Many samples were sent in by nearby law enforcement agencies for comparison, as well. The killer was not found, but the firearms examiners were even more convinced that this was the correct make and model weapon from examining hundreds of bullets from this make and model of weapon.

The case was finally broken when a systematic search of all parking tickets given near the location and during the period of any of the shootings was conducted. The officers checking out a Ford, which had receive a ticket fairly near the scene of a recent fatal shooting, went to Yonkers, a suburban community bordering the Bronx. The Bronx had been the scene of a number of the shootings. The car was parked on the street close to the address of the registered owner. When the investigators approached the car, a gun that resembled a Charter arms Bulldog was observed in the car. Backup was called, a search warrant for the car and dwelling were applied for, and the vehicle staked out. David Berkowitz left his apartment and approached his car and was stopped. He offered no resistance and freely admitted he was "Sam." The firearms examiners were relieved to find that the weapon seized from Mr. Berkowitz was indeed a Charter Arms Bulldog, and the test fires it produced were matched to many of the evidence bullets.

As with most high profile cases, there were those who said that Mr. Berkowitz had been set-up to take the blame for a more sinister individual. In addition to the firearms evidence, he had kept extensive notes on his activities that were discovered during the search of his apartment after his arrest. These writings included information that he intended another attack, very shortly, outside New York City on Long Island. These highly incriminating notes, as well as the taunting notes that had been sent to the newspapers and police, were unambiguously tied to Mr. Berkowitz by comparison to known samples of his hand printing. The identification of him as the writer was particularly strong because of the large writing sample available and the individuality of his hand printing.

After the arrest of Mr. Berkowitz, this rather bazaar story unfolded fairly quickly and, after being ruled mentally competent to stand trial, he pleaded guilty to six charges of murder and was sentenced to 365 years in prison.

> **LEARNING OBJECTIVES**
> - Understand the nature of toolmarks
> - The different types of toolmarks
> - The importance of looking for trace evidence associated with toolmarks
> - The proper ways to collect and preserve toolmarks
> - The examination and comparison process
> - General nature of firearms
> - The function and importance of the cartridge in the operation of a firearm
> - The importance of rifling to firearm performance and forensic examination
> - Learn the major important types of firearms
> - Proper procedures for collecting and preserving firearms and firearms evidence
> - Major steps in the examination of a firearm and firearms evidence
> - Growing importance of firearms data banks to investigation and prosecution
> - Potential utility of examination of even highly damaged firearms evidence
> - Uses of firearms evidence in reconstructing shooting incidents
> - How and why firearms serial numbers are defaced
> - Major techniques for restoring defaced serial numbers

TOOLMARKS—INTRODUCTION AND TOOLMARK DEFINITION

Toolmarks and firearms examinations are usually grouped together because, although firearms examination is a significantly larger volume of work in forensic science laboratories, intellectually the most important aspect of firearms examination is basically a toolmark examination. In addition, for many years the firearms section of most forensic laboratories did the toolmark cases as well, since many of the same skills are involved.

One can look at toolmark examination as the general category, and bullet and cartridge case comparisons as a subcategory. One can define a toolmark as a pattern resulting from a harder marking device; that is, the tool being forced against the surface of a softer object. Most toolmarks of forensic interest are marks left by screwdrivers (Figure 8.1), pry bars, wire and bolt cutters, and a wide variety of other tools often used to gain forced entry into a dwelling or even a simple cash box. The comparison of the toolmarks left on a bullet or cartridge case will be covered separately in the firearms section to follow.

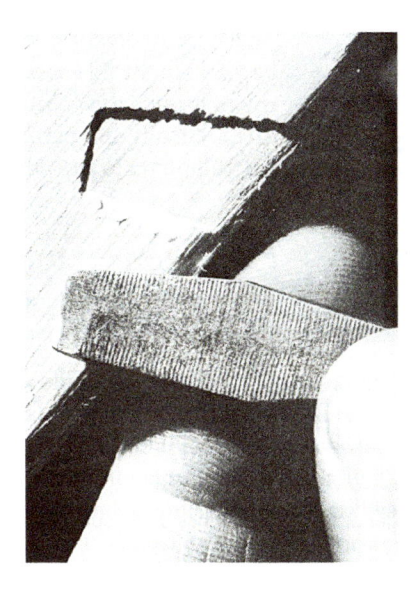

Figure 8.1 Screw driver and indented impression of blade in wood.

Toolmarks are of three types

The three types of toolmarks are indented, striated, or a combination of the two. Striated toolmarks are what we call scratches left in the surface of a softer medium by a harder tool. These striations are normally the result of the tool moving across the surface leaving behind a pattern of scratches caused by tiny defects in the face or edge of the tool that is in contact with the surface. They are the direct result of the tool moving across the surface.

On the other hand, impressions are formed by forcing the contact portion of the impressing tool into the surface. There is little motion involved. The third category is produced by a combination if impressing and moving the tool across the surface. Most commonly this occurs when a tool is pressed into a surface to separate two pieces of the surface and then slips out of the indentation and slides across the surface. One can imagine trying to pry open the lid on a cash box to gain entry, and the force being applied causing the tool to come out of the junction of the lid and box and slipping across the box or lid.

Residue from softer object is often deposited on the tool

Very often as a tool indents or slides across a surface, material is gouged from the surface and is left on the tool. Residue on the evidence tool is evidence that is often overlooked but potentially quite useful. Where this material is shown by chemical analysis to be consistent with the surface that was scratched, the connection of the tool to the marked surface is certainly strengthened. Such residue could be paint, oil, blood, iron, aluminum, or plastic. Although the material is usually left behind in microscopic amounts, and may not be visible to the naked eye, there may be sufficient material that careful microanalysis may provide a useful connection between the tool and the surface. As a result, the first step in examination of an evidence tool is to examine the tool under a stereomicroscope to see if there is anything foreign on the surface of the tool. Anything that is observed can then be removed and analyzed to see if it is consistent with the surface on which the toolmark was found. That is why it is so important that a control sample of the softer material from the scene be submitted as well as the tool and the toolmark, or even a cast of the toolmark. One can occasionally find material characteristics of the tool left behind in the toolmark as well.

Class and individual characteristics

Toolmarks can have both class and individual characteristics. The shape and general measurements of a toolmark are class characteristic which are best used to screen a suspected tools and indicate if it is possible that they could have caused the toolmark (see Figure 8.1). Sometimes the general class characteristics can be useful in indication the type of tool which made the mark. For example a toolmark made by a nail puller (Figure 8.2) will have different characteristics than one made by a pry bar.

FIREARMS EXAMINATION—BACKGROUND

Firearms examination can be one of the busiest and most important sections of a forensic science laboratory. In urban areas, especially large cities, many firearms cases are regularly submitted. Homicide is not one of the leading causes of death in the U.S., overall, with deaths due to firearms injuries ranging around 1% of all deaths. The rate (number per 100,000 of population) of firearms related deaths has declined with the overall crime rate for some time. Although firearms examiners were often not educated in science, the training and experience needed to become an expert firearms examiner is extensive. Today, more entry-level firearms examiners have four-year college science degrees than ever before, and a growing number have forensic science degrees.

Basic operation and important parts of a firearm

Before one can discuss the many aspects of firearms examination, one must have a basic appreciation for how firearms work and the vast range of firearms that exist. There are many possible definitions for a firearm and a simple one is: A firearm is a device that, using the rapid combustion of an energetic chemical, is designed to accelerate a projectile to a high velocity while directing it towards a target. This is quite general because, in addition to handguns and long guns, there are a great many other objects that can be considered firearms. Under the law in many states, persons can be charged with possession of firearm for possessing many things that most people do not usually consider guns or even weapons.

Major components of a firearm

A firearm has three major components: the frame, the action, and the barrel. The frame holds everything together, the action is the key working portion of the firearm, and the barrel directs the projectile toward the target. The major components of the action are the trigger, the storage area, the firing pin assembly, the breach, and the firing pin. The trigger, in most modern firearms, cocks the weapon by moving the firing pin assembly away from the breach against spring pressure. It also moves the cartridge from storage (cylinder or magazine) into the breach. The trigger then releases the firing pin assembly allowing it to strike the cartridge primer and fire the bullet down the barrel. The action then clears the empty cartridge from the breach. This is done in several ways depending on the type of weapon (see below).

The function of the cartridge in a firearm

The energy that provides the driving force for the projectile is contained in the cartridge.

Cartridges came into general use after the Civil War to replace the powder, rag (patch), and ball combination of the muzzle-loading firearm (Figure 8.5).

In muzzleloaders, a spark ignited the powder and the expanding gases from the powder forced the ball out of the barrel at a reasonably high speed. It is clear that taking all those pieces and putting them together in one nice convenient package, called a cartridge, was a good idea. No more need to pour powder down the muzzle, tamp down the powder,

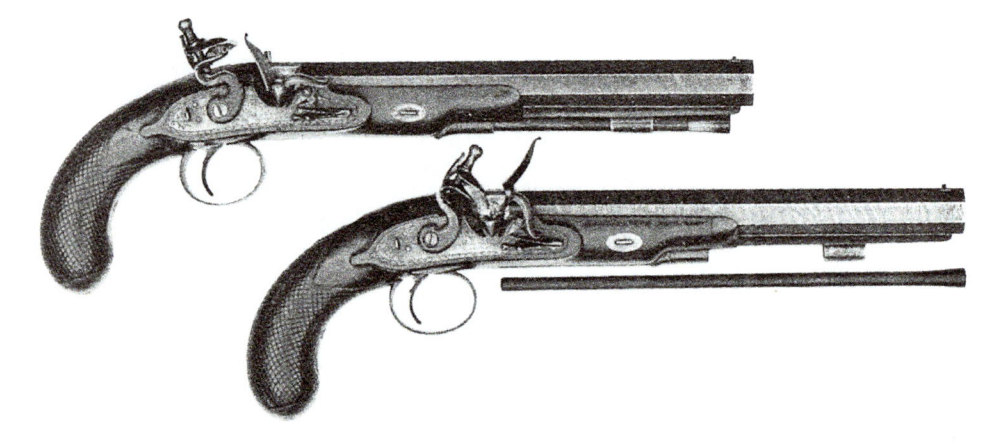

Figure 8.5 Example of muzzle loading firearm—dueling pistols.

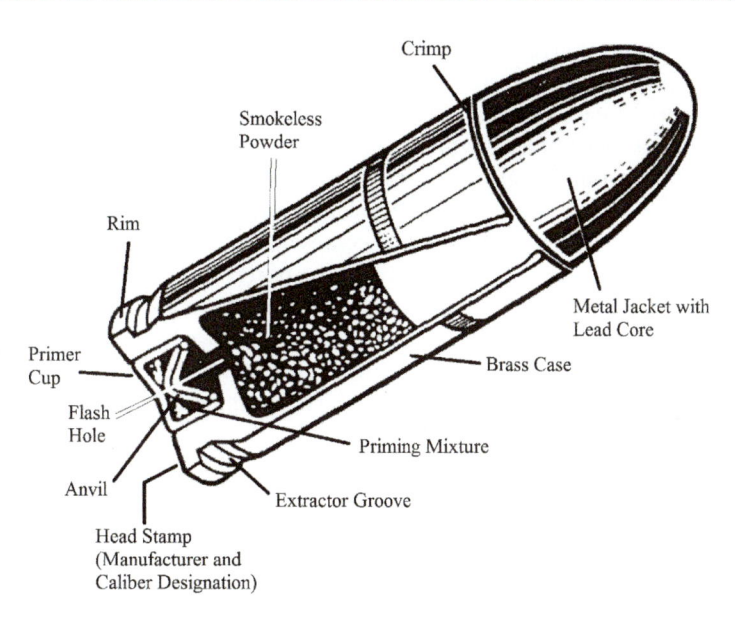

Figure 8.6 Diagram of the component parts of a cartridge case.

put in a ball and cloth, and tamp the materials together. Instead a neat little package was designed with the necessary components, which could fire from only the force of a firing pin striking it on the primer cup. There is a tendency to call the unit (cartridge) a bullet, but the bullet is just one component of a cartridge. A cartridge has four basic components (Figure 8.6). It starts with a case, which is the container of the unit.

A handgun or rifle cartridge is usually a cylindrical piece of brass, or nickel-plated brass, which holds the other three components together. The case is filled with small disks, cylinders, or balls of smokeless powder. Smokeless powder is made from cotton that has been treated, very carefully, with nitric and sulfuric acid to convert the cotton (cellulose) into cellulose nitrate. Once ignited, it burns very rapidly and very cleanly producing a great deal of heat and gas. The third component of the cartridge is the projectile (bullet), which is put into the case, and the case is squeezed down tightly around the bullet to make a tight seal. The bullet is usually made from lead or copper-coated lead.

The final component, the primer, is very important because it was the key to the development of the cartridge. It takes the form of a little soft metal cup that is part of the case at the opposite end to the bullet. The primer contains a tiny amount of shock sensitive material. This little cup has a small hole on the inside in contact with the smokeless powder. When the firing pin of the firearm hits that cup, it squashes the cup and causes the shock sensitive chemical in the primer cup to ignite and transfer a spark or a tiny flame through the hole into the powder charge to ignite the powder inside the case. In some .22 caliber cartridges, the primer material is in the rim of the cartridge case rather than a separate cup and the firing pin hits the rim of the cartridge. The burning powder gives off heat and creates gasses that develop high pressure inside the case. Because the bullet is swaged tightly into that case, these gasses are contained, and the pressure rapidly builds up until it forces the bullet out of the case and down the barrel at a high speed. This entire process is sometimes called the "firing train." Thus, a cartridge resembles a small explosive device (see Chapter 12), which directs the explosive energy to drive the projectile down the barrel. The single-unit cartridge provides the powder, containment, and the projectile, which characterized the old muzzle loading rifles and even cannons. Cartridges with their four major components, primer, powder charge, projectile, and case, are the heart of modern firearms. Whether they are tiny for small pistols or enormous artillery shells, the basic principle is the same.

One should not fall into the habit of calling cartridges "bullets." Bullets may do the ultimate damage, but cartridges provide the ultimate power of the firearm.

Rifling of the barrel makes modern firearms accurate

A key aspect of firearms identification for the forensic scientist is the fact that modern handguns and rifles have rifled barrels. This feature enables bullet to barrel individualization. When one is looking down the barrel of a handgun or rifle, one sees not a smooth cylindrical surface but a series of grooves spiraling down the inside of the barrel.

These grooves that spiral down the barrel are called rifling. Rifling is used because it was discovered many years ago a spinning projectile proceeds more accurately toward the target. When the projectile enters the barrel the grooves grip the projectile because the bullet is actually very slightly larger than the narrowest inside diameter of the barrel. Because the bullet is made of lead, which is fairly soft, it is compressed by the high areas (lands) of the rifling and actually flows slightly into the grooves. As a result, the bullet is forced to follow the helical path of the lands and grooves (Figure 8.7) as it passes down the barrel and out the front of the firearm. Since all this all happens almost instantaneously, the effect is that the bullet comes out spinning rapidly. That is the reason for cutting grooves (rifling) in the barrel. A projectile emerging from the barrel with a rapid spin travels a flatter trajectory and flies a truer path toward the target. Before the rifling of barrels, the projectile traveled down the smooth barrel and emerged with no significant spinning motion. The resulting path would not be as true and accurate as when the projectile is spinning. One can occasionally encounter a homemade weapon, usually referred to as a zip gun, which does not have a rifled barrel. But, generally, all modern, commercially manufactured handguns and rifles have rifled barrels.

The barrels of handguns or rifles are traditionally rifled by gouging the helical grooves in the steel of the barrel using a device called a broach (Figure 8.8).

Figure 8.7 Expended round showing the rifling—helical lands and grooves impressed on bullet by inside of a modern rifled barrel.

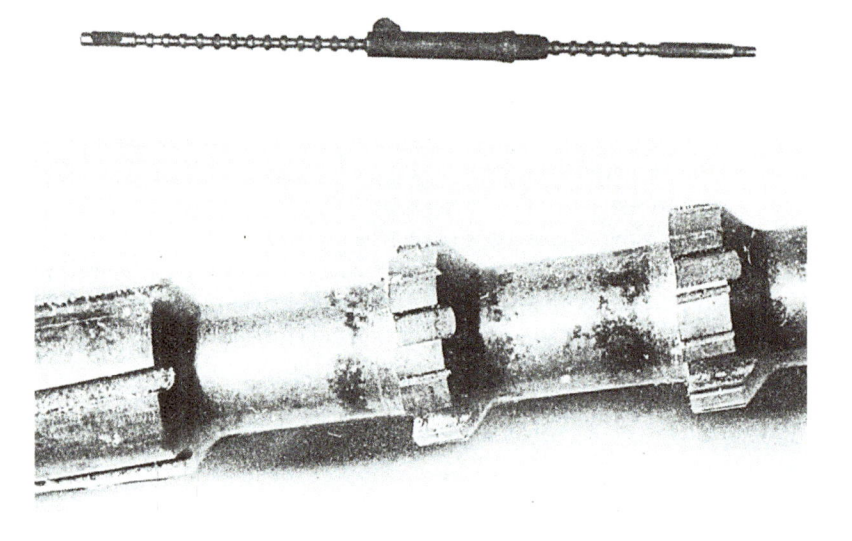

Figure 8.8 Images of a broach tool used to cut the rifling, helical grooves, into a gun barrel.

A broach is a hard steel device that has a series of sharp raised areas (teeth) circling a central shaft and can have as few as five or as many as 26 of these cutting surfaces. The helical grooves are cut by pulling this shaft through the tube, that is to become the barrel, and turning it at the same time. The outside diameter of these teeth is just slightly larger than the inside diameter of the barrel. There are actually several sets of these teeth along the central shaft, each one just thousandths of an inch larger than the one in front of it. Thus, each set of teeth gouges a little deeper until the grooves are of the desired depth. Nowadays, there are some other techniques to accomplish this as well, but the final result is the same; a series of grooves spiraling down the barrel to engage the bullet as it travels down the barrel and cause it to emerge spinning rapidly. At least partially because of the rifling in the barrel, modern firearms can be highly accurate. The raised areas between the grooves are called lands. The number of lands and grooves in a firearm barrel is a class characteristic of that firearm. The lands and grooves are not necessarily the same width, but there are usually the same number of each. Each barrel manufacturer believes that they use the ideal rifling and, therefore, make the best and most accurate firearm. In addition, the broach can be turned either to the right or to the left. There appears to be no strong advantage of one direction over the other, but some manufacturers prefer right twist and some manufacturer prefer left twist (Figure 8.9).

For example, Colt, which is one of the large manufacturers of firearms in the U.S., uses predominantly left twist, and Smith and Wesson, which is a rival firearms manufacturer, prefers right twist. The number, diameter, and direction of twist of the lands and grooves are class characteristics of a particular firearm and because their impressions are left behind on the projectile (bullet) they provide the firearms examiner with quite a bit of information about the firearm that fired a recovered bullet, without ever seeing that firearm. The lands on the barrel (high areas) are impressed into the softer bullet and these indentations are called *land impressions* and the softer lead flows into the barrel grooves leaving high areas on the bullet called bullet lands. These features do not run parallel to the length of the bullet, but appear at an angle because the bullet has been following a helical path through the barrel.

The most important class characteristic used in initial description of a firearm is the inside diameter of the barrel, which is called the *caliber*. It is the distance between two opposing lands and usually measured in hundredths of an inch in the U.S. and millimeters in the rest of the world.

Since the bullet is designed to fit tightly in the barrel, it is also the approximate diameter of the fired bullet. Thus, when a handgun or rifle is described as being a "22" (.22-caliber) that means the inside diameter of the barrel is twenty-two hundredths of an inch or a little under a quarter of an inch. The famous U.S. military side arm is called a "45," which means that bullet diameter is almost one half of an inch (0.45), and the very common semiautomatic pistol called a "9 millimeter" fires a bullet that is a little over one third (0.356) of an inch (there are 25.4 mm in an inch).

Figure 8.9 Two fired bullets one fired from a firearm with right twist rifled barrel and one from a left twist showing clear land and groove impressions and the angle caused by the helical twist of the rifling.

More commonly, the action is more complex, and cartridges are fed to the breach in a number of other ways. Many children learn about firearms by having a single-shot .22-caliber rifle as their first firearm. In semiautomatic firearms, the cartridges are fed by some mechanical action from a magazine or tube under the barrel by a spring action, fired by pulling the trigger, then the case is ejected by the action as a new cartridge is fed into the chamber. "Semiautomatic" means, and this is an important distinction from automatic, that the firearm discharges one cartridge sending the bullet down the barrel, then reloads a fresh cartridge each time the trigger is pulled. The trigger must be released and pulled again to fire another bullet. Reloading may be accomplished in several ways, including using the pressure of the exiting gasses, spring action, or a mechanical action like a pump or a lever. Semiautomatics are to be distinguished from automatic firearms, because the above series of steps is repeated when one pulls the trigger and holds it. The automatic firearm will continue firing either until the trigger is released or all of the cartridges have been fired.

Many rifles and handguns are designed for semiautomatic operation, and, in general, only military firearms are designed for fully automatic operation. The military firearms are usually capable of both semiautomatic and automatic operation. In military firearms, there is a selector lever that allows one to change from semiautomatic, to short burst, or to automatic operation. In the U.S., it is illegal for anyone, who is not in the military or law enforcement, to possess an automatic firearm without a special permit. The federal penalty for illegal possession of a fully automatic firearm is a minimum of a $10,000 fine and a year in jail. Infantry troops in the U.S. military are usually provided with firearms, such as the M16, the Russians and many others use the AK47, and the Israelis use the Uzi. These are very reliable firearms, which are also sold in many other countries. A civilian version of each of these firearms, which has the selector lever removed, is sold by the manufacturer and often copied by other manufacturers. Therefore, the civilian version is designed to fire only in the semiautomatic mode. It is possible to buy an illegal kit that will convert such firearms from a semiautomatic civilian unit into a military type firearm automatic firearm. The high incidence of automatic firearms in the movies and television crime shows, not surprisingly, does not correspond with reality. Most forensic laboratories encounter fully automatic firearms only rarely.

Classification of firearms by their functions

Another way to classify firearms divides them into rifles, handguns, and shotguns. Rifles are firearms, which are designed to be fired from the shoulder, and normally have relatively long barrels. As indicated earlier, they can be single-shot, semiautomatic, or fully automatic. The semiautomatics can have a great variety of different mechanisms for feeding cartridges into the breach for firing. One popular form uses a magazine, which is a spring-loaded device that can hold from five to 50 cartridges. The magazine is a device that pushes a cartridge into the breach as the fired case is expelled. It usually slips into the firearm just below the breech, in a rifle, and into the grip of a handgun. Some rifles have a tube that sits just below the barrel that acts as a magazine.

The advantage of the rifle is that it, generally, provides a projectile exiting at a high velocity. This is because it can accommodate a larger cartridge with more powder and has a long barrel that contains the gases and allows the bullet to build up speed as it is forced down the barrel. The burning of the powder produces the vast amount of gases that will force the bullet down the barrel, providing energy as long as the projectile is traveling down the barrel. The more energy and momentum the bullet has, the further it can go, the truer it will fly, and the more damage it can do.

The legal definition of a handgun is a firearm with a barrel less than a particular length.

The statutory length can vary from state to state but is usually about 9 or 10 inches. Handguns can be divided into three categories based on their mechanical action. There are single-shot handguns that are used almost exclusively for target shooting. Because they have few moving parts, they can be made with high precision and are very accurate and reliable. The other two types of handguns are revolvers and semiautomatic pistols, which are much more common. Revolvers (Figure 8.12) were developed in the mid-nineteenth century and are sometimes referred to as the "guns that won the west."

Revolvers have a rotating cylinder that holds anywhere from five to nine cartridges depending on the caliber. Only 0.22 caliber cartridges are small enough for the cylinder to hold as many as nine cartridges, while most 0.38 caliber revolvers hold five or six cartridges. When one pulls the trigger causing the firearm to fire, a lever turns the cylinder so that a new cartridge is in position to be fired. This mechanism allows the firearm to fire each time the trigger is pulled until all the cartridges in the cylinder are fired. The mechanism is much simpler, than a semiautomatic,

Figure 8.12 A typical revolver handgun with cartridges in revolving cylinder; in this case, a .45 caliber Colt made in 1965.

because the cartridge cases simply stay in the cylinder and are manually extracted later. The concept was developed by Samuel Colt and was the standard for handguns for many years. Because of this simple mechanism revolvers are generally considered more reliable than semiautomatic firearms. When one pulls the trigger, they fire, and they seldom jam or misfire.

A third type of handgun is the semiautomatic pistol (Figure 8.13). It is mechanically analogous to the semiautomatic rifle. Cartridges are usually placed in a magazine, which fits into the grip of the firearm and feeds a new cartridge each time the firearm is fired. Normally, a combination of spring action and expelled gasses is used to extract and eject the fired case and move the new cartridge into the chamber.

The empty cartridge cases are ejected with considerable force to the left or right (depending on the firearm) and are often found at the scene. A hurried exit by the shooter will usually result in at least some of the empty cases being left behind. Because of their mechanical complexity, only recently have improvements in design and reliability made them the weapon of choice for most police departments. In recent years, semiautomatic pistols have become the most popular handgun, because of their superior firing speed and the larger number of cartridges they hold. The magazine holds anywhere from 8 to 10 cartridges. It is now illegal for a handgun to hold more than 10 cartridges. Not many

Figure 8.13 Semiautomatic handgun (pistol) with magazine removed.

years ago there was no regulation of magazine size and fifteen or sixteen cartridge magazines were quite common. Since the law changed such magazines can no longer legally be imported or manufactured in the United States, but the existing magazines were not outlawed, and they sometimes may still be obtained.

Revolvers have the disadvantage, for forensic investigation, that the cartridge cases are retained in the cylinder. As a result, cartridge cases are often not left behind at the shooting scene—unless so many shots are fired that the shooter has to reload. Cartridge cases can have significant value in connecting a firearm to a crime scene. If they are not left at the scene, but taken away in the cylinder, this value is lost. As the popularity of semi-automatic firearms has increased, the occurrence of cartridge cases at shooting scenes has increased and their value as forensic evidence has been fully utilized.

With a semiautomatic pistol or rifle the cartridge case is ejected with considerable force and usually lands several feet from the shooter (Figure 8.14a and b). Because of its cylindrical shape, the case can then roll in any direction for a considerable distance. In most settings this may result in the case being difficult to find for the shooter.

There are no truly automatic handguns, although there are several firearms, that were designed to be very compact, automatic military rifles, which are not much larger than a handgun.

As indicated earlier, these cannot be legally sold to those other than military or law enforcement, except in semiautomatic form. Military weapons such as the AK47 (Figure 8.15) and others mentioned earlier sometimes are stolen or otherwise fall into the hands of criminals and thereby are submitted as evidence to forensic labs. Generally, rifles and handguns encountered in forensic labs are semiautomatic, even though the media regularly and incorrectly call them "automatic."

Finally, the third type of firearm is the shotgun (Figure 8.16).

Figure 8.14 (a and b) Expelled cartridge cases ejected from a semiautomatic weapon. This photo was taken at the scene from the Sandy Hook Elementary School shooting in Newtown, Connecticut 2012.

Soviet 7.62mm AK-47 Assault Rifle, right side view.

Figure 8.15 AK-47's are one of the most ubiquitous, popular military-style rifles.

Figure 8.16 Shotguns are generally designed for hunting birds or skeet shooting.

Shotguns are long guns, like rifles, designed to be fired from the shoulder, but they do not, generally, have rifled barrels. Shotguns were initially designed for hunting birds. They are designed to shoot a charge of steel or lead pellets. Although lead was the standard for many years, steel pellets must be used in many places to avoid lead pollution of the waters and woods where most hunting is done. If one is shooting at something flying fairly rapidly and at a sizeable distance, hitting it with a single projectile is very difficult. On the other hand if one sends up a "cloud" of lead shot, some of those pellets are more likely to hit that bird. As soon as the pellets emerge from a shotgun, they begin to spread and anything within the spreading pattern of pellets is going to be hit with a quantity of high-energy pellets. Shotguns are very damaging firearms, particularly at close range.

Shotgun ammunition (shells) can have a small number of fairly sizeable pellets or a large number of very small pellets, and a whole variety of combinations in between. These pellets are packed into that shotgun cartridge along with a large powder charge and a primer (Figure 8.17), exactly as handgun and rifle cartridges.

The primary difference is that because shotgun barrels are considerably larger in diameter than most handguns or rifles, there is room for more of everything. Even the smallest shotgun has a barrel diameter (bore) significantly larger than the average handgun or rifle.

As a result, because of the size of the shotgun shell, when a shotgun is fired there is a great deal of lead (or steel) emerging with a lot of force behind it. In addition to a variety of different sized pellet loads, shotgun shells are made with a single slug. This slug is the diameter of the shotgun bore and has helical grooves molded into it. Since the shotgun barrel is smooth, they do not lock in and cause the slug to spin rapidly, but are intended to create some spin because of air resistance. Therefore, shotguns are not nearly as accurate as a rifle or a handgun, but for relatively short distances or for causing a large amount of destruction, they are extremely effective.

The ownership of rifles and shotguns is not as legally restricted as that of handguns. Because of their availability, the illegal use of such weapons is quite common, particularly because they can be made concealable by cutting the barrel

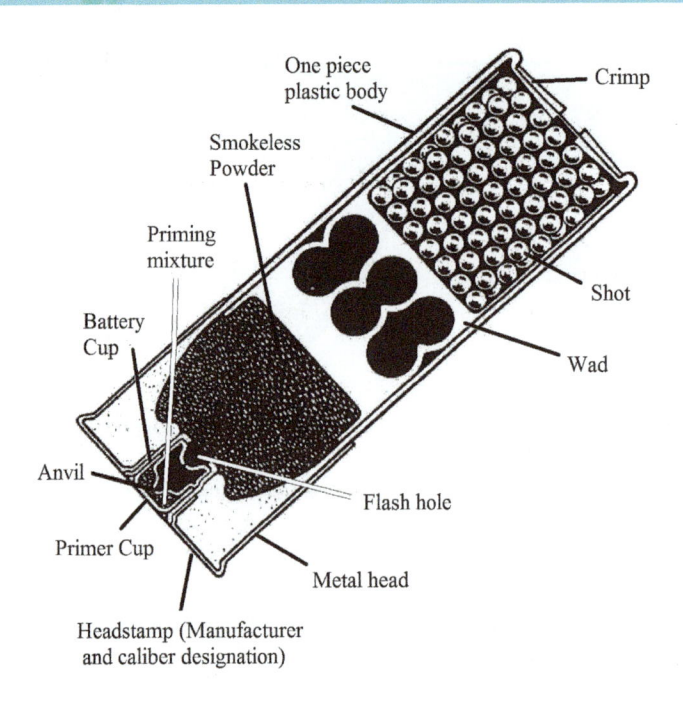

Figure 8.17 Cut away view of a shotgun shell showing the various components.

down. Long guns are hard to conceal when entering a store or confronting a person for the purpose of robbing them, but a shortened version removes that problem. A sawed-off shotgun is an extremely intimidating weapon; the large bore and the knowledge that at reasonably close range the shooter virtually cannot miss is a strong deterrent to resistance.

COLLECTION, EXAMINATION, AND COMPARISON OF FIREARMS EVIDENCE

Careful handling of firearm, bullets, and cartridge cases

Firearms evidence has great potential forensic utility, and it is particularly important that it be collected with care and properly protected and documented. Firearms, projectiles (bullets), cartridge cases, and objects that have been in their path must be handled carefully. Each can be of critical value to an investigation. In the United States, there are a great many crimes committed using firearms. In addition, accidental shootings, hunting accidents and variety of other serious incidents that involve firearms can produce evidence that must be carefully examined. The primary concern in handling and collecting firearms is safety. A firearm must be rendered safe (not able to fire) before collection and packaging. At times, this process will require manipulation to remove a cartridge (live round) from the chamber.

It is important to mark the cylinder position on a revolver and to mark recovered bullets and cases for identification. When a revolver is collected, one needs to note which cylinder position is lined up with the barrel of the firearm. The cartridge that is lined up with the barrel would be the next one to be fired. The empty cases and live cartridges in the rest of the cylinder will indicate which cartridges were fired and in what order. One often sees different brands or types of cartridges in the cylinder of a revolver, so this information may be important in reconstructing the events in a shooting incident.

The bullets and cases recovered at the scene must be marked for identification or placed in sealed containers that are properly marked. The evidence should be marked in a place where it is not likely to obscure any markings left by the firearm. For cases, this is on the side and away from the base or just inside the open end and for bullets it is usually on the base.

At one time, the nose was also considered a good place to mark a bullet since one wants to avoid damaging the lands and grooves. It is now recognized that a bullet may pick-up material from an object it hits or passes through, which may turn out to be important trace evidence for reconstruction, so it is best not to mark the nose.

Many times, firearms evidence may be collected under circumstances where it would seem to have little forensic value. It may be found property or there may be little initial controversy about what happened in a particular incident. Perhaps the firearm belongs to the victim rather than a suspect. Nevertheless, it is important that the firearm or firearms evidence be sent for forensic laboratory examination. The firearm, the projectiles, or the cases may be found to match evidence recovered in another, earlier crime. If the evidence is never submitted for laboratory examination, these important connections will never be made. As will be discussed shortly, the use of information from firearms evidence databases is one of the most rapidly evolving areas of forensic science. Sadly, in some jurisdictions it is not convenient to get such materials to the experts for examination and even if functional testing is required, the firearm may be test fired at the local training range and not fully examined.

Firearms evidence examination and comparison

Firearms examiners mostly examine firearms, projectiles, and cartridge or shot shell cases. Projectiles and cases are called "fired evidence." They bear marks unique to the firearm that produced them. Firearms examination is a multi-step procedure. The first step is a careful physical examination of the firearm for functionality, safety features, and physical condition.

Safety issues are extremely important for several reasons. Firearms must always be handled with respect and by someone with knowledge of their function. Poorly maintained or abused firearms can explode when being test fired, endangering the examiner and others in the immediate area. On occasion, there can be a bullet that didn't make it out of the end of the barrel for various reasons.

Firing a firearm with a bullet lodged in the barrel can cause it to explode in the shooter's hand, causing serious injury. It is also possible that a firearm's integrity was compromised by improper use, or disassembly and improper reassembly. The great amount of energy released by the explosion of the cartridge can be misdirected to the destruction of the firearm rather than the forcing of the bullet down the barrel and toward the intended target. This may result in injury to the individual who is firing the firearm or those nearby. Therefore, two key parts of the initial physical examination are: insuring that there are no cartridges left in the firearm and sighting down the barrel to make sure that it is not obstructed by a jammed bullet or other material.

Ordinarily, whoever collects the firearm and submits it for examination should remove all cartridges. This may not be quite as simple as it sounds. Someone unfamiliar with a particular firearm (and there are thousands of different ones) could miss a cartridge still in the firearm. Arguably, it may not be necessary to remove cartridges from a revolver, as long as the firearm is safe (hammer not cocked). With semiautomatic firearms, the magazine should be carefully removed and separately packaged. However, removing the magazine from a semiautomatic firearm does NOT remove a round from the chamber. Thus, a live round could inadvertently be left in the chamber of the firearm.

If a firearm has a light trigger pull or is dropped it may discharge. Usually firearms will not fire if dropped, but some may. Further, evidentiary firearms may have had safeties disabled or just not be in proper working order. It is important to carefully address all the safety concerns. The gun is normally dry fired (fired without a cartridge in the chamber) and cycled, if it's a semiautomatic by pulling back the slide, and cycling a cartridge through the firearm without pulling the trigger to insure that all the different parts are working.

Next, the basic class characteristics including the caliber, number, width of the lands and groves, direction of twist, make, model, and serial number are all noted. Interestingly, one of the largest single problems in firearms examination is proper identification of a firearm. Because there are literally thousands of different types of firearms in circulation, this is not a trivial task. Most firearms examiners, who examine a sizable number of firearms in a year, see a few each year that they have never seen before. There are new firearms being made all the time, and there are old ones that just are not very commonly seen. When an officer finds or seizes a firearm, it is their responsibility to identify it and locate the serial number. This information is necessary so that the firearm can be checked against the stolen firearms files kept by all states.

Forensic firearms examiners must check that the collecting officer has correctly recorded the identifying characteristics. Forensic examiners are much more knowledgeable about the myriad makes and models of firearms and which numbers are the primary serial numbers. Many firearms have numerous numbers stamped on them and determining which is actually **the** serial number can be confusing. That is a concern, not only to return the firearm to its rightful owner, but it can also be important for successful prosecution of an individual charged with theft of the firearm. If any of the descriptive information is not correct, the search will fail to find the proper firearm.

Test for functionality and to obtain control bullets and cases

Another important step in the examination of a firearm is to determine if it can be fired, and whether its safety features are functional. Firearms are also test fired to obtain known specimens of bullets and cartridge cases from that firearm. In most states, someone cannot be charged with a possession of a deadly weapon unless the weapon has been shown to be functional; that is, when the trigger it pulled, the gun fires. A significant number of firearms received in forensic laboratories are in poor condition and many will not fire for a variety of reasons. At one time, test firing was done into a long box filled with cotton (Figure 8.18), but it has been shown that water tanks (Figure 8.19) provide a better quality control specimen.

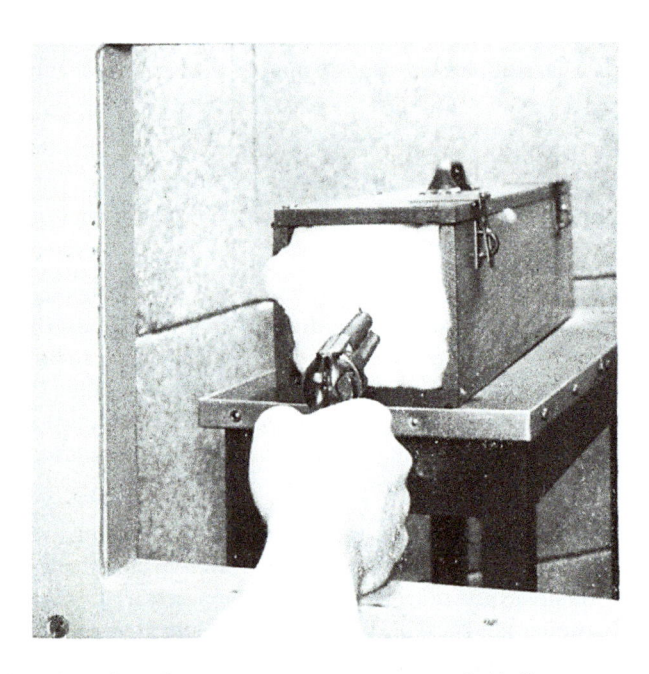

Figure 8.18 Test firing a hand gun into a box of cotton waste to recover expended bullets.

Figure 8.19 Test firing a hand gun into a water tank to recover expended bullets. (From ATF video.)

Therefore, in most modern firearms examination facilities, test firing is done into a large bullet recovery water tank. The recovery tank is usually 6–8 feet long and can be horizontal or vertical. For most handgun cartridges, the bullet has lost its energy after a few feet of travel in the water and settles to the bottom of the tank. Even bullets fired from high-powered rifles will lose their energy in less than 8 feet of water. The bullet is recovered from the bottom of the tank using a little net that can be raised (bringing the bullet up with it) or with a long stick that has putty on one end so that the bullet will adhere to the end.

Normally, two or three test fires are taken from each firearm. This allows the examiner to have known bullets to compare to each other in order to insure that the firearm is marking the bullets in a consistent manner. In addition, known cartridge cases for comparison are also obtained. Although the majority of firearms produce consistent markings on the bullets and cases, this may not always be the case. Obviously, if control bullets fired under ideal conditions cannot be matched to each other, it will usually not be possible to match evidence bullets to that firearm.

SIDEBAR 8.1
A very famous example of inconsistent marking is the firearm seized in the shooting of Dr. Martin Luther King. That firearm has been tested, over the years, by at least three different agencies and found not to produce consistent markings on fired bullets. As a result, it is not possible to say that the evidence bullets that killed Dr. King were fired from that gun.

Comparison of bullets, cartridge cases, or shot shell cases—the comparison microscope

The comparison of both bullets and cases are done using a *comparison microscope*. Firearms and toolmark comparisons use reflected light with the bullet comparison microscope (Figure 8.20) instead of the transmitted light generally used for transparent or translucent objects, such as hairs and fibers.

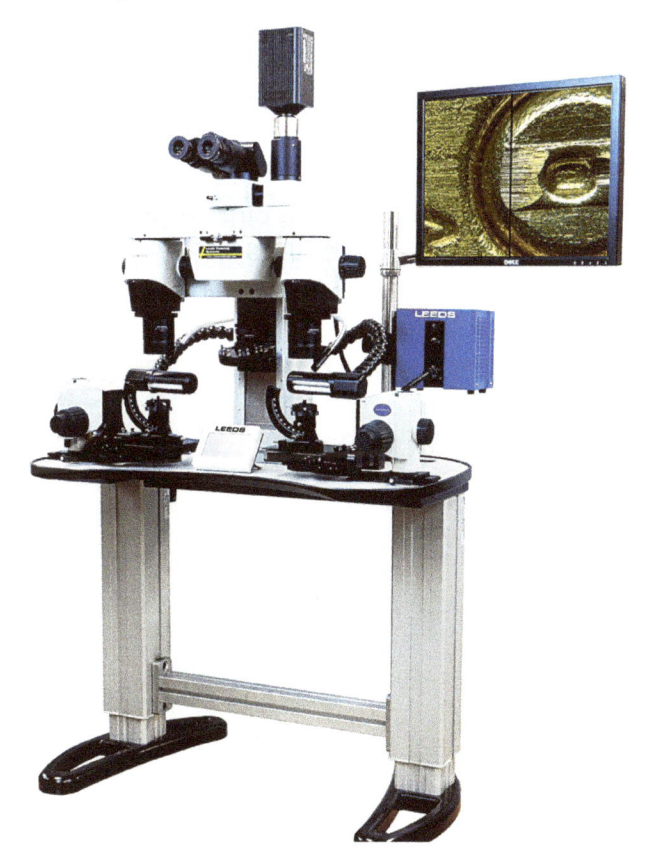

Figure 8.20 A reflected light comparison microscope used for comparing evidence and known bullet and cartridge case samples. The insert shows a comparison of breach face marks on two cartridge cases. (From Leeds Forensic Systems, Inc. www.leedsmicro.com. With permission.)

> **CASE—SIDEBAR 8.2**
>
> The so-called CBS murder case arose when a hired killer executed a victim as she approached her car in the middle of the afternoon on a rooftop parking garage in New York City. As he dragged her toward his van and shot her, three employees of CBS observed the struggle. They started towards him to see what was happening, and he turned on them and shot and killed all three and then fled the scene in his van. Eventually, the investigation narrowed to Donald Nash as the prime suspect, and he was put under surveillance at his home. When he left that area, he was followed, eventually stopped, arrested, and his van searched. Two of the most critical pieces of evidence in the CBS murder case were two cartridge cases. An exhaustive search of the defendant's van, which included taking out the seats, disclosed a cartridge case wedged under the driver's seat. When his home was carefully searched, some wooden steps from the backyard into the house were torn up and a cartridge case that had fallen between the two boards of the step was recovered. Although the gun used in the murders was never recovered, the firearms examiner was able to say that these two cases found in different places under the control of the defendant were fired from the same gun as those recovered from the murder site (parking garage). These results provided the strongest connection between the defendant and the murders.

Cases fired from semiautomatic firearms also may have additional markings from the extractor and/or ejector. These are the parts of the firearm that remove the case from the breech of the firearm and eject it from the firearm, so that a fresh cartridge can be loaded from the magazine. Extractor and ejector marks, of course, are not found on cases fired in a revolver. Sometimes, theses extractor or ejector markings on the case can provide enough information to allow individualization of the case when compared against test fired cases from a particular firearm. Sometimes cases fired from a semiautomatic firearm may also show magazine marks. When one loads the cartridges into the magazine or when they leave the magazine, roughened areas on the magazine may leave characteristic scratches on a cartridge.

Shot shell cases can also have markings that are useful for associating or individualizing a fired shell with a particular shotgun. So, even though the projectiles (pellets or rifled slug) exit through a smooth barrel and, thus, are not marked, the shell case can acquire a firing pin impression, breech face, and potentially extractor and/or ejector marks; just as rifle and pistol cartridges do.

The firing mechanism of a shotgun is quite analogous to that of a rifle or pistol. The same basic principles govern the operation. When the firearm fires, the pellets go out the front and the shot shell is pushed back against the breech face. When the firing pin hits the primer area, it too leaves a mark. If it's a semiautomatic shotgun there is a lever that can make marks on the fired and ejected shell case as the new shell is loaded in the breech. Shotgun shells were traditionally largely made of paper, but that has been replaced by plastic. Plastic or paper are not usually useful for preserving markings. However, the breech end of almost all shotgun shells is still brass with a soft metal primer cup and, therefore, can record marks in exactly the same way as handgun or rifle cartridges. Thus, shotgun shells can be individualized to a shotgun just as handgun and rifle cartridge cases can be individualized to a particular firearm.

Association of cartridges or bullets to firearm or manufacturer using databases

As mentioned earlier, bullets can often be associated with a particular ammunition manufacturer or to possible makers of firearms, even down to a make and model from the class characteristics left on the bullet by the rifling. Often this is true of a recovered cartridge case as well. Frequently the manufacturers name or a well-known trademark and the caliber are impressed on the base of the case, which makes identification of the ammunition manufacturer quite simple. Even when this is not the case and the markings on the case appear quite obscure, many of these so-called head stamps have been collected in reference books that associate head stamps with the particular manufacturer of that case.

As noted earlier, firearms evidence now makes up one of three major evidence databases in forensic laboratories. Chapter 6 describes the AFIS databases containing fingerprint images. In Chapter 11, the CODIS databases containing DNA profiles will be discussed. For fired evidence, NIBIN (Figure 8.24) contains images of bullets and cartridge cases recovered from scenes of crimes or test fired from firearms submitted to a laboratory after seizure by law enforcement. NIBIN started as local or regional databases, which were combined into a fully national database. They have become a very useful tool in law enforcement, since bullets and cartridge cases can now be associated with a firearm, even though the firearm has not been found. If the evidence comes from a firearm that has been recovered and test fired, the source of the fired evidence is thereby identified. These databases enable law enforcement to make connections between cases (Figure 8.25), even from jurisdictions that are far apart, through the bullet or cartridge

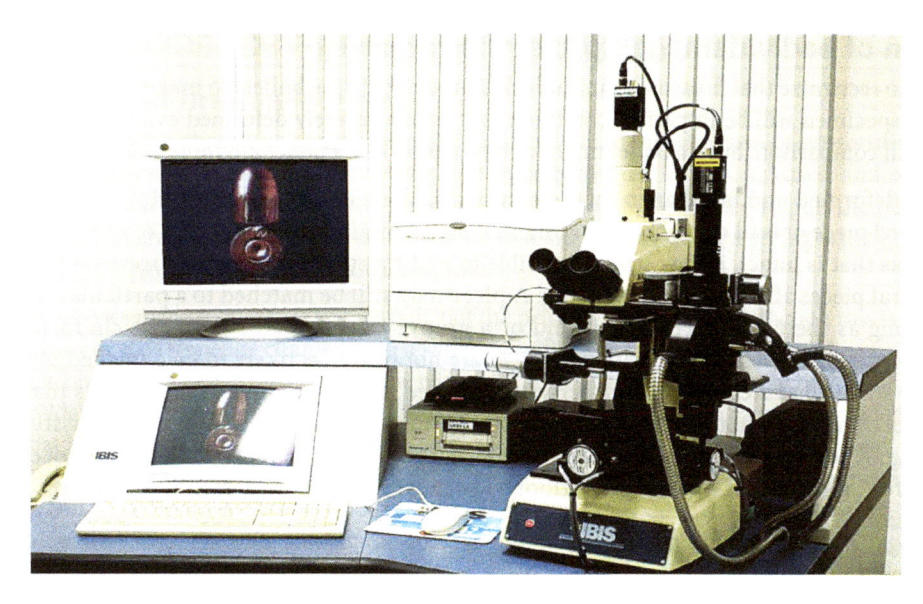

Figure 8.24 Equipment station to collect and search firearms images for and from the national NIBIN database.

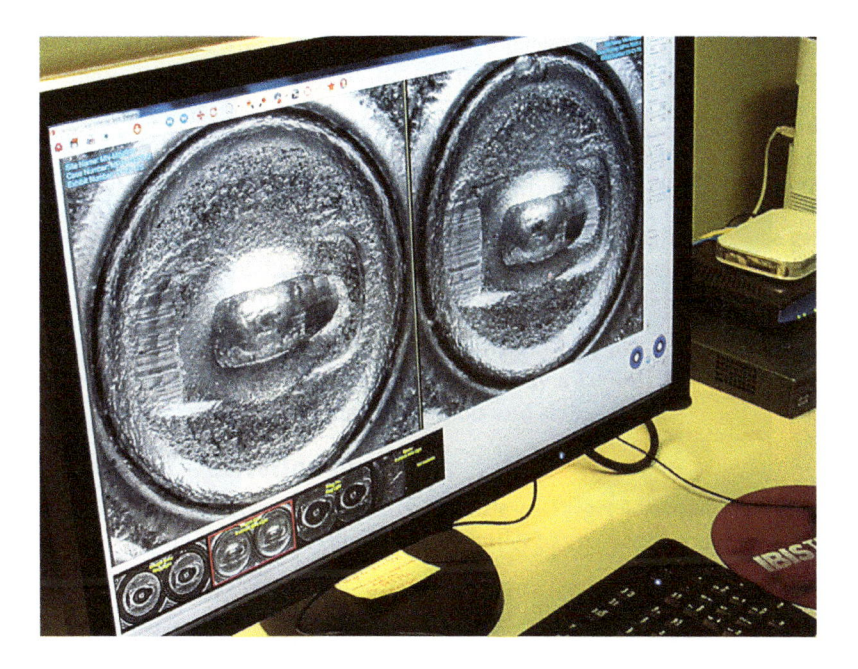

Figure 8.25 Comparison of breach face marks on evidence casing to cartridge case image in NIBIN database.

case evidence. Such case associations can be significant in moving an investigation forward by allowing the investigators on the associated cases to work together and look for possible connections.

Firearms examination and comparison has been the subject of considerable research in recent years and significant progress has been made. One exciting area of research is the three-dimensional capture of the markings on bullets and cartridge cases. The field is now beginning to use this technology, and it holds great promise in making such evidence even more useful.

The General Rifling Characteristics (GRC) File, mentioned earlier, has now been incorporated into NIBIN so one can search the class characteristics (caliber, twist, lands, groove diameters, etc.) on bullets and casings in that database to identify types of weapons and manufacturers as well from pre-comparison data or on evidence not suitable for comparison.

SERIAL NUMBER RESTORATION

Another important aspect of firearms examination is serial number restoration. Firearms manufactured by a legitimate manufacturer are serialized for identification. The manufacturer stamps a unique serial number on the frame of each firearm that identifies that particular firearm. This is done with an automatic machine that has a counter that goes up by one as each a frame is stamped. Because each manufacturer must, by law, keep track of the model and serial number of each firearm they produce, the serial number is the key to identifying a firearm.

Those who come into possession of a firearm illegally or who use a registered firearm to commit a crime would often like to make it impossible to uniquely identify the firearm. There are a number of ways to try to remove or obscure the serial number. The Federal Bureau of Alcohol, Tobacco, and Firearms (ATF) has the responsibility to keep track of legally manufactured and registered firearms. They have quite an effective system of tracing the movement of firearms from manufacturer to first seller to initial owner. In addition, subsequent sales through licensed-firearms dealers are also recorded. If one recovers a firearm at a crime scene or from someone who is not its rightful owner, it is useful to know the previous owner or owners of that firearm. Further, each state keeps records of firearms reported stolen, and a recovered firearm can be checked against those records as well. If someone has come into possession of a firearm illegally or has stolen it from somebody who obtained it illegally, there will be a break in the recorded chain of possession. In addition, it should be mentioned that many private transactions are often not recorded at all, and, even if they are, the information is seldom recorded in a way that can be easily retrieved.

CASE—SIDEBAR 8.3

Tracing a shotgun provided an important lead in solving a tragic murder case. A young couple and their daughter were returning home from a short trip late in the evening. As their car pulled into their driveway in a quiet residential suburb, their headlights picked up an individual standing in the driveway holding a firearm. The exact motivation of the assailant is not known, but it was discovered later that he had forced his way into the home of an elderly couple, only a short distance away, and robbed them a few days before. Either he was concerned that the driver was going to hit him, or he accidentally discharged the shotgun he was holding, as a result he killed the husband. In his rush to get away, he left the shotgun behind. The investigators were able to trace the firearm, through several owners to a pawnshop in a nearby city. The pawnshop owner was persuaded to disclose to whom he had sold the shotgun, and this lead the police to the killer.

There are a number of common methods used to obliterate (deface) serial numbers. Because of the importance of serial numbers in proper identification of firearms and the need for "untraceable" firearms by criminals, many of the firearms that are seized show evidence that someone has tried to remove the serial number.

In many cases, the serial number is so badly defaced that it is largely gone or impossible to read. The defacing (rendering it unreadable) of a serial number on a firearm is a felony, in itself, in most states. Most firearms examiners develop some skill at serial number restoration. Those who deface serial numbers use a variety of techniques from scratching, to gouging, to even grinding them off. It is not uncommon for as many as 20% of the firearms received by a firearms unit to have a defaced serial number, which can vary in readability from having one or two numbers unreadable to being virtually undetectable. This problem can be addressed in several ways. The easiest and best is if there is a hidden serial number. A number of manufacturers place a second copy of the serial number in a very inconspicuous place. The individual removing the serial number is unlikely to know of this hidden serial number and therefore fails to remove it as well. A knowledgeable firearms examiner will know where to look for the hidden serial number and the problem is solved. Unfortunately, most firearms do not have hidden serial numbers.

The approach then must be to try to chemically or physically restore the defaced serial number. When the serial number is stamped into the frame of the firearm, the pressure of the die that impresses the number deforms the crystal structure of the metal to a considerable depth into the metal. This disruption of the crystal structure goes well below the actual indentations that form the visible serial number. When the crystal structure is deformed, it changes the metals susceptibility to attack by chemicals.

To take advantage of this fact, the firearms examiner smooths out the area where the serial number was originally stamped using emery cloth or a small hand grinder. If there are gouges or deep scratches, they are smoothed out, if possible. The examiner then takes one of several etching solutions that have been found to work well on the type of metal of which the gun is made. These are generally strongly acidic solutions that will attack the metal and dissolve it slowly. One usually dips a cotton-tipped swab (Q-tip) in the etching solution and paints it on the area where the serial number was stamped. The acid will preferentially eat away at the metal where the crystal structure has been deformed; that is, where the numbers were, and the numbers may begin to reappear. This is a tedious process where the area is cleaned, examined, and acid is reapplied until one can read some of the numbers. The numbers do not always all come up at once, so one has to go slowly and if a number becomes visible it must be photographed or recorded.

The first numbers that appear may be etched away before some of the other numbers have become visible, so the process must be done in steps. If one is patient and the individual doing the defacing did not go deep enough, acid etching can reveal the serial number in many cases.

> ### CASE—SIDEBAR 8.4 OFFICER PIAGENTINI SHOOTING
>
> During the civil unrest of the 1960s, a radical group lured two New York City police officers into an ambush using a bogus 911 call and killed them. They stole their firearms and made much of their success in striking a blow against their oppressors. Many years later, information was developed that one of the officer's guns has been buried in a field in Mississippi. A team was sent to recover the firearm to confirm the accuracy of the information. A gun was recovered and submitted to the forensic laboratory covered with mud (Figure 8.28). It was consistent with the type of firearm carried by police at the time, but was so badly rusted that there was no discernable serial number. The firearm was cleaned and acid etched, and it was possible to restore the entire serial number (Figure 8.29), which proved that this was the firearm carried by the murdered police officer.

It seems that the experience of most laboratories is that etching is effective in more than half the cases. Even if one is not able to read all the digits of the serial number, one can search the computerized databases with the information one has and often determine if the firearm is in the database or narrow it to several possible firearms. For example, one knows the make and model and caliber of the firearm from examining it. It may be a .44 caliber Charter Arms Bulldog revolver and one can read six of the eight digits of the serial number. When the database is searched, only one or two firearms that fit the available data are likely to be found that have been reported stolen. As a result, even the partial serial number restoration will allow determination if this firearm had been reported stolen.

The same techniques are often used on engine blocks from recovered stolen automobiles, where it is also common to encounter defaced serial numbers. A number of other things are also serialized by stamping a number in metal and can be processed similarly. In the "Crafts" case (Chapter 1), restoration of the serial number on a chain saw recovered from the Lake Zoar reservoir permitted investigators to establish that the suspect was the originally purchaser the chain saw.

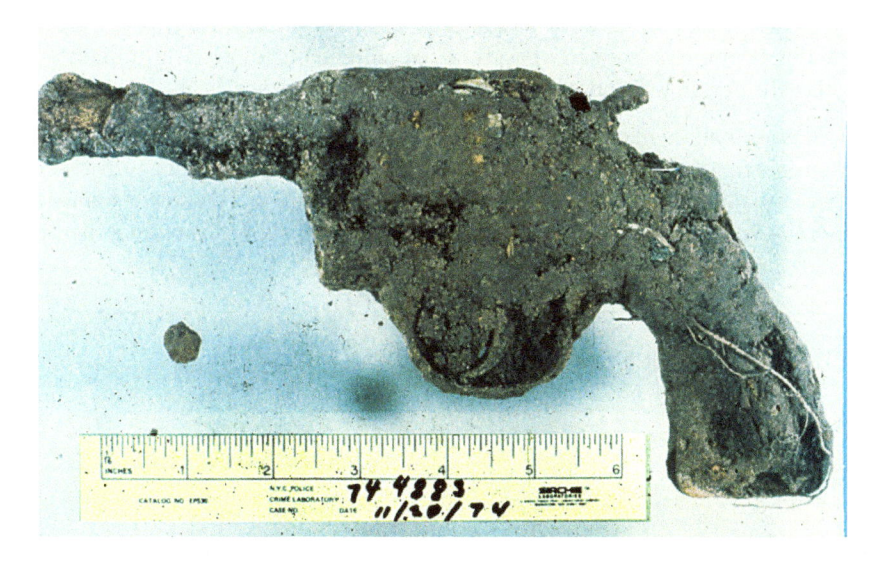

Figure 8.28 A handgun recovered from a field after having been buried for many years.

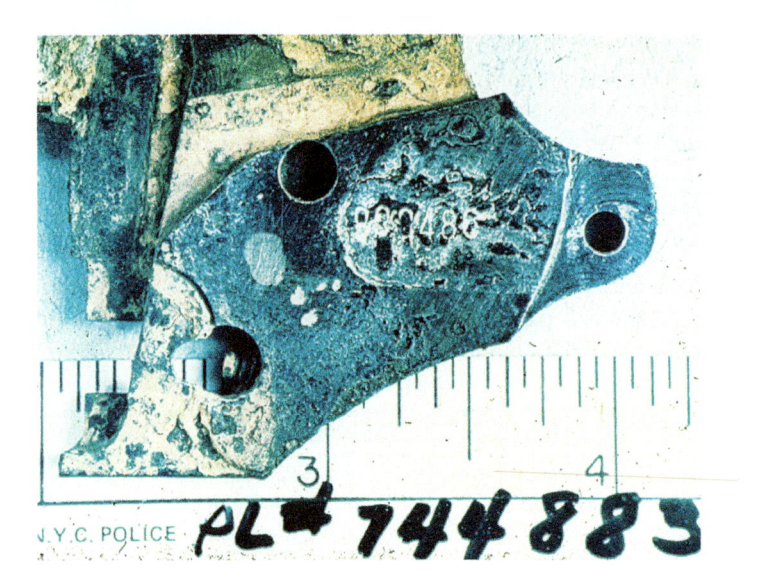

Figure 8.29 The serial number revealed after acid etching of long buried gun. The serial number indicated that this was the gun taken from murdered NYC police officer Piagentini.

There are several other techniques that can be used for restoring serial numbers, such as magnetic particles, electrolytic etching, and others. They all are based on the fact that stamping deforms the crystal structure of the metal and thereby changes the way the metal behaves. A majority of forensic laboratories use the acid-etching technique since it works quite adequately and is simple and inexpensive.

THE FIREARMS AND TOOLMARK EXAMINER PROFESSION

We have noted in earlier chapters that each major pattern evidence area is specialized, and each has its own professional societies, sometimes its own journals, and active peer group organizations that set the practice standards. The primary professional organization of firearms and toolmarks examiners is the Association of Firearms and Toolmark Examiners (AFTE). Almost all examiners in public agencies and many in private practice belong to AFTE. For years, AFTE published a newsletter, which has now evolved into a peer-reviewed journal. AFTE has its own annual training seminar and has an examiner certification program that was established a number of years ago.

In recent years, with the evolution of the various Technician and Scientific Working Groups, SWGGUN was formed to examine and recommend standards of practice in the firearms and toolmark field. Until perhaps a decade ago, most firearms examiners were not necessarily trained or educated in one of the natural sciences. However, that picture is changing. Many laboratories now have the same basic entry requirements for every criminalist—including a natural science degree—whether the person will end up in the firearms, trace evidence, or the DNA section of the lab.

Firearms and toolmark examination, like the other specialized pattern evidence areas (fingerprints, questioned documents), requires extensive training beyond formal education. The initial training for a fully casework-ready firearms and toolmark examiner usually takes about 2 years. As with all professions, competent forensic examiners make a commitment to "life-long-learning" and continue to try to improve their skills throughout their professional careers.

Key terms

- automatic weapon
- breach face
- bullet recovery tank
- caliber
- cartridge
- defacing
- etch

firearm
firearms comparison microscope
firing pin
firing pin impression
magazine
groove
gunshot residue
handgun
indented mark
land
land impression
NIBIN
powder pattern
primer
projectile (bullet)
revolver
rifle
rifling
safety
Scanning Electron Microscope (SEM)
SEM/EDX
semiautomatic weapon
serial number restoration
shotgun
shotgun shell
slug
smokeless powder
striated mark
toolmark
trajectory
trigger pull
twist

Review questions

1. What role does trace evidence play in the examination of tools and tookmarks?
2. How are toolmarks properly documented, protected, and collected?
3. What is the function and importance of the cartridge in the operation of a firearm?
4. List the three types of handguns and how they differ.
5. What is the process for comparing bullets and firearms evidence to a suspected weapon?
6. What is the role of the General Rifling Characteristics database in the investigation and prosecution of firearms cases?
7. List some of the many ways that firearms evidence may be used in the reconstruction of firearms-related incidents.
8. Discuss the importance of and methods for restoration of defaced serial numbers.
9. Discuss the role of NIBIN in the investigation of cases involving firearms.
10. Why have cartridge cases become such an important type of firearm evidence in recent years?

Fill in/multiple choice

1. Where possible, the best method of submitting an evidentiary toolmark for comparison is to send _____.
   a. a photograph
   b. the mark itself
   c. a casting
   d. a tape lift
2. The inside diameter of a gun barrel is known as its _____.
3. The four components of a handgun or rifle cartridge are the primer, _____, _____, and case.
4. Handgun and rifle barrels have helical grooves on the interior of their barrels to _____.
   a. Improve the accuracy of the weapon
   b. Cause the bullet to spin
   c. Grip the bullet as it enters the barrel
   d. All of the above
5. When the serial number on a weapon has been defaced to make it unreadable, _____.
   a. The weapon cannot be properly identified
   b. The weapon can be identified by other part numbers
   c. The serial number can often be chemically restored
   d. The serial number can often be read by using image enhancement

References and further readings

Association of Firearms and Toolmarks Examiners; Web site, www.afte.org.

FirearmsID.com/new-index.htm.

Cristina Cadevall, and Niels Schwartz; Forensic Inspection for Firearms: Comparison Techniques in Optical Profilometry; Forensic Magazine; October/November; 2013; www.forensicmag.com

Cowan, M. A., and P. L. Purdon. A study of the "'Paraffin Test". *Journal of Forensic Sciences* 12 (1967), 19.

Heard, B. *Firearms and Ballistics: Handbook, Examining and Interpreting Forensic Evidence*. Chichester, UK: John Wiley & Sons, 1996.

Katterwe, K. "Modern Approaches for the Examination of Toolmarks and Other Surface Marks." *Forensic Science Review* 8, no. I (1996): 46–72.

Moenssens, A. A., J. E. Starrs, C. E. Henderson, and F. E. Imbau. "Firearms Identification and Comparative Micrography." In *Scientific Evidence in Civil and Criminal Cases*, eds. A. Moenssens, F. E. Imbau, J. E. Starrs, and C. E. Henderson. Westbury, New York: Foundation Press, 1995.

O'Mahony, A. M., I. A Samek, S. Sattayasamitsathit, and J. Wang. "Orthogonal Identification of Gunshot Residue with Complementary Detection Principles of Voltammetry, Scanning Electron Microscopy and Energy-Dispersive X-ray Spectroscopy: Sample, Screen, and Confirm." *Analytical Chemistry* 86 (2014): 8031–8036.

Springer, E. "Toolmark Examinations: A Review of Its Development in the Literature." *Journal of Forensic Sciences* 40, no. 6 (1995): 964–968.

Validity of Toolmark identification in Forensics; Published on *Forensic Magazine* (http://www.forensicmag.com)

Zeichner, A., and N. Levin. "Casework Experience of GSR detection in Israel, on Samples from Hands, Hair, and Clothing Using an Autosearch SEM/EDX System." *Journal of Forensic Sciences* 40, no. 6 (1995): 1082–1085.

CHAPTER 9

Digital evidence and computer forensics

Raymond J. Hsieh

Lead Case: Actual Cases Involved Digital Evidence BTK Killer

The BTK investigation in Wichita, Kansas was a serial homicide investigation spanning decades and causing a great deal of local fear. The investigation began in the mid-1970s, spanned 30 years, and concluded with the arrest of a 59-year-old employee of Park City, Kansas, a small community adjacent to Wichita. Dennis Lynn Rader born March 9, 1945, is an American serial killer who murdered 10 people in Sedgwick County (in and around Wichita, Kansas), between 1974 and 1991. The serial killer first struck in January 1974, with the murder of four members of the Otero family. He killed again just 3 months later but waited nearly 3 years before striking for the third time.

Rader was particularly known for sending taunting letters to the police and newspapers. There were several communications from the BTK from 1974 to 1979. The first was a letter that had been stashed in an engineering book in the Wichita Public Library in October 1974, which described in detail the killing of the Otero family in January of that year. In early 1978, he sent another letter to KAKE, a television station in Wichita, claiming responsibility for the murder of the Otero's. In this letter, he suggested a number of possible names for himself, including the one that stuck: BTK (Bind, Torture, and Kill). He demanded media attention in this second letter, and it was finally announced that Wichita did indeed have a serial killer at large.

By 2004, the investigation of the BTK Killer had gone cold. He committed his last murder in January 1991. It was 30 years after the first murders, between January 1974 and February 2005, that the BTK Killer resurfaced with great media attention, triggering an intensive 11 months investigation by local, state, and federal law enforcement agencies that finally brought the case to a successful close. Because the BTK serial killer had been silent for so long, many in the community had believe that he was dead, had moved to another state, or was incarcerated somewhere. The case riveted the community when the murders first began occurring, again when the BTK resurfaced in 2004, and two facets of the investigation proved to be unusual. The first is the way the Wichita Police Department was able to stay in communication with the killer through the media, and the second involves the application of DNA and Computer Forensic Science (digital forensics) in the investigation.

On March 19, 2004, the killer sent a letter to the Wichita Eagle newsroom. In its report on the letter, the newspaper reported that BTK was claiming responsibility for the September 16, 1986, murder of a young mother of two who was found inside her Wichita home. The author of the letter claimed that he had murdered Vicki Wegerlie on September 16, 1986, and enclosed photographs of the crime scene and a photocopy of her driver's license, which had been stolen at the time of the crime. Prior to this, it had not been definitively established that Wegerlie was killed by BTK. DNA collected from under the fingernails of that victim provided police with previously unknown evidence. They then began DNA testing hundreds of men in an effort to find the serial killer. Altogether, some 1100 DNA samples would be taken.

Wichita police held a news conference on March 25, 2004, confirming the reemergence of the BTK Killer and asked citizens with any information to contact the police department. An e-mail account and a post office box were set up to accommodate tips. In the first 24 hours following the news conference, almost 400 tips were received. By mid-May, the tips received by the police exceeded 2000.

As a new chapter in the story of the serial killer began to unfold, the Wichita Police Department started adapting new investigative techniques. The police corresponded with the BTK Killer in an effort to gain his confidence. Then in one of his communications with the police, Rader asked them if was possible to trace information from floppy disks. The police department replied that there was no way of knowing what computer such a disk had been used on, when, in fact, digital forensic methods actually did exist. Rader then sent his next message on a floppy to the police department. The digital forensic examiners for the police found metadata embedded in a deleted Microsoft Word document that was, unbeknownst to Rader, still on the disk. The metadata, recovered using the forensic software EnCase, contained the name Christ Lutheran Church, and the document was marked as last modified by "Dennis." When the police searched on the Internet for Lutheran Church Wichita and "Dennis," they found his family name and were able to identify a suspect: Dennis Rader, a Lutheran deacon. The police also knew BTK owned a black Jeep Cherokee. When investigators drove by Rader's house, they noticed a black Jeep Cherokee parked outside. The police now had strong circumstantial evidence against Rader, but they needed more direct evidence to detain him.

They somewhat controversially obtained a warrant to test the DNA of a Pap smear Rader's daughter had given at the University of Kansas medical clinic while she was a student there. The DNA of the Pap smear was a near match to the DNA of the sample taken from under the victim's fingernails, indicating that the killer was closely related to Rader's daughter. This was the evidence the police needed to make an arrest.

On February 25, 2005, Rader was detained near his home in Park City and accused of the BTK killings. At a press conference the next morning, Wichita police chief Norman Williams announced, "The bottom line is BTK is arrested."

After his arrest, Rader talked to the police for several hours. He stated he chose to resurface in 2004 for various reasons, including the release of the book *Nightmare in Wichita: The hunt for the BTK Strangler* by Robert Beattie. He wanted the opportunity to tell his story his own way.

On February 28, 2005, Rader was formally charged with 10 counts of first-degree murder. He made his first appearance via videoconference from jail. He was represented by a public defender. Bail was continued at $10 million.

On June 27, the scheduled trial date, Rader changed his plea to guilty. He unemotionally described the murders in detail and made no apologies.

On August 18, Rader faced sentencing. Victims' families made statements followed by Rader, who apologized for the crimes. He was sentenced to 10 consecutive life terms, which requires a minimum of 175 years without possibility of parole. Because Kansas had no death penalty at the time the murders were committed, life imprisonment was a maximum penalty allowed by law.

LEARNING OBJECTIVES
- Outline computer crime, digital evidence, and digital media evidence
- Present the various crimes in which digital evidence can play a part
- Distinguish between data integrity and data authentication
- Detail the various digital-related devices
- Familiarize readers with computer hardware components
- Explain the steps involved in the collection and processing of digital evidence
- Outline hashing, digital fingerprinting, and how they factor into digital investigations
- Detail chain of custody issues and consideration in digital evidence preservation and analysis
- Explain unallocated file space and the concept of slack
- Outline various network components and hardware
- Acquaint the reader with various gaming consoles, media, handheld, phone, mobile, thumb drives, and other such data transmission and storage devices
- Offer a brief overview and summary of off-the-counter tools available to digital forensic investigators
- Provide an overview of IP addresses, e-mail, Internet cache, cookies, and browsing history

INTRODUCTION

Digital devices are everywhere in the current world, assisting people to communicate locally and globally with ease. But, unfortunately, this new technology can be used in a negative way. Conventional crimes with digital evidence have been increasing rapidly. Computer Forensics are not only helpful for solving **computer crimes** like child pornography, child exploitation, or hacking, but are also being used to solve many other crimes, such as terrorist attacks, drug trafficking, robbery, organized crime, tax evasion, and homicides.

WHAT IS DIGITAL EVIDENCE?

The National Institute of Justice (NIJ) defines **digital evidence** as information and data of value to an investigation that is stored on, received, or sent by a digital-related device or attachment. This evidence can be collected when digital-related devices or attachments are seized and secured for examination.

Digital evidence has the following features:

- Is latent (hidden), like fingerprints or DNA evidence
- Crosses jurisdictional borders quickly and easily
- Can be altered, damaged, or deleted with little effort
- Can be time-sensitive.

Digital evidence is also referred to as any electronic data saved or sent using a computer-related device that supports or refutes a theory of how an offense occurred or that addresses critical components of the offense, such as intent or alibi (Casey, 2004). Digital evidence involves data on computers, video recordings, audio files, and digital photos. This evidence is essential in computer and cyber crimes or even traditional crimes, but it is also valuable for facial recognition, crime scene photos, and surveillance tapes/hard drives.

The details of digital-related devices or attachments will be discussed in the following section. Since crimes are trending toward those which involve evidence such as; video clips, audio recordings shot by smart phones, or video footages from surveillance DVR (Digital Video Recorder) system, we may narrow down the evidence into **digital multimedia evidence (DME)**.

What is digital multimedia evidence?

The Scientific Working Group on Digital Evidence and the Scientific Working Group on Imaging Technology (SWGDE/SWGIT) indicate Digital Multimedia Evidence (DME) is any information of probative value that is either stored or transmitted in a digital form including, but not limited to; film, tape, magnetic and optical media, and/or the information contained therein (LEVA, 2016).

When Digital Multimedia Evidence is needed in the legal process, protection of data integrity and data authentication are significant requirements during the forensic procedure. These two terms are usually interchangeable and can be confusing; we will need to clarify them in this section. SWGDE/SWGIT defines those terms as follows.

Data Integrity: The process of confirming that the data presented is accurate and untampered with since time of collection.

What must be determined with data integrity is whether the DME has been tampered with or changed since it was originally made. Changing the data is not necessarily contraindicated for preserving evidence reliability. In some cases, DME, especially for video evidence, will need to have the image aspect ratio calibrated or have the colors/brightness of the images corrected or sharpened. Each procedure could change the evidence data, but not the **validity** of the data representation (LEVA, 2016).

Data Authentication: The process of substantiating that the data is a precise representation of what it purports to be.

DME displays virtually, not physically. For instance, in some video evidence, there may be some aspect ratio errors or color errors. Data Authentication needs to verify the following possible questions:

- Does DME precisely represent what it purports to be?
- Pertaining to video evidence, is what happened in front of the surveillance camera the same visual representation in time and space as what was recorded on the DVR?

RAM (Random Access Memory): Stores programs and data currently being used by the CPU. The maximum amount of data that RAM could store is measured by bytes. The modern computer usually is measured in millions of bytes called, Megabytes (MB). RAM is volatile memory where the data is lost when the computer power is turned off. RAM allows data to be stored and retrieved on a computer. It is usually associated with DRAM, which is a type of memory module.

ROM (Read-Only Memory): Mainly used to store firmware. ROM is one kind of non-volatile memory used in computers and other digital devices. ROM is a significant part of the computer BIOS (Basic Input/ Output System). ROM is computer memory on which data has been prerecorded.

HD (Hard Drive): Hard Drives keep data or computer programs that are not currently being used by the CPU. HD is the essential part of the computer for storage. Internal hard disks install in a drive bay, attach to the motherboard using an IDE (or ATA), SCSI, or SATA cable, and are powered by a connection to the Power Supply Unit (PSU). The transfer of data drive-to-drive is relatively fast. It is not recommended to boot via a suspect computer. If you are booting to your computer forensics workstation, install the suspect's hard drive on your workstation. You will need to configure your new hard drives, so that both your HD (master) and suspected HD (slave) are each cabled and pinned as masters (if they are IDE HDs). Slave-to-Master data transfer should provide a safe acquisition performance.

HOW IS DIGITAL EVIDENCE PROCESSED?

The Technical Working Group for the Examination of Digital Evidence (TWGEDE) suggests that there are four steps in processing digital evidence.

Evaluation: Computer forensic scientists should assess digital evidence completely with respect to the scope of the case to determine the course of action to take.

Acquisition: Digital evidence, by its very nature, is fragile and can be tempered, damaged, or ruined by improper handling or examination. Examination is best conducted on a copy of the original evidence. The original evidence should be acquired in a manner that preserves and guards the integrity of the evidence.

Examination: The purpose of the examination process is to extract and analyze digital evidence. Extraction refers to the recovery of data from its media. Analysis refers to the interpretation of the recovered data and putting it in a logical and useful format.

Documentation and presentation: Actions and observations should be documented throughout the forensic processing of evidence. This will conclude with the preparation of a written report of the findings.

COLLECTION OF DIGITAL (MULTIMEDIA) EVIDENCE

Digital evidence may come into play in any serious criminal investigation like homicide, aggravated assault, rape, stalking, carjacking, burglary, child abuse or exploitation, counterfeiting, extortion, gambling, piracy, property crimes, drug trafficking, and terrorist acts. Since digital (multimedia) evidence is hidden data, the most important thing is to know how to properly retrieve and interpret it.

Under current technology, handheld consoles can carry encoded messages between criminals. Even newer household appliances, such as a refrigerator with a built-in TV, could be used to store, view, and share illegal photos. It is **vital** that crime scene investigators be able to recognize and properly collect **potential** digital evidence.

Originals versus copies

In the analog world, the determination of originals versus copies is generally clear cut. Analog signal is a continuous signal which contains time-varying quantities. We actually live in the world of analog. There are infinite colors to see; unlimited tones to hear; infinite smells to sense. The whole analog system could build up noise or distortion during duplication; it creates **degradation,** which leads to the loss of quality of an electronic signal during the duplicating process. In other words, there is original data, which owns the very best quality of signal, and the first generation copy of data, which has slightly lower quality than the original. The loss of quality increases with each newly generated copy.

A digital signal is a discrete value at each sampling point. It could be represented by computer sample data, which would be established with a series of bits that are either the number 1 (refer as ON) or the number 0 (as OFF), so that there are NO fractional values. A digital signal is good for computer processing meaning it is more reliable. Technology now uses digital signals to quickly replace a large amount of analog applications and devices. During the process of duplication, it simply copies the computer data onto a new digital media. Therefore, regardless of how many generations of copies, each copy of data is exactly the same as the original one, so that even the thousandth copy will be as good as the original. There is no differentiation between originals and copies.

Performing an analog-to-digital conversion is called, **sampling** or **digitization**. **Nyquist's Theorem** indicates, to reconstruct the original analog signal to digital data without any information loss, the sampling rate must be at least twice the bandwidth of the signal. If not, then **aliasing** will happen. What is aliasing? When the system is measured with insufficient sampling rate, it leads to a distortion when the signal is reconstructed from samples. The recreated original (discrete) signal will create some artifacts that are different from the original analog (continuous) signal. For audio files, aliasing makes a buzzing sound, and photos aliasing causes jagged edges or stair-step artifacts on the images.

Hashing of digital files

Everyone has unique fingerprints by which one can be identified. But in the virtual world, how do people differentiate digital files? There is a way called **Hashing**. Hash is able to validate digital evidence. What is Hash? It is a document's identifying fingerprint. Basically, hashing is the process of using a mathematical algorithm, which calculates a value from a stream of data that will be used to identify the file. Hashing is used to determine whether a digital file has been tampered with or not.

When we deal with digital evidence duplication, all effort should be taken to assure that the working copy of the original does not insert any new data or reduce information that could alter the meaning of the original evidence. Hash is a method of ensuring that the working copy is identical to the original.

Hashing algorithms—MD5 and SHA1

Validating digital evidence requires the use of a hashing algorithm tool, which is able to generate a unique hexadecimal number to represent a digital file. We may refer to that unique hash value as **digital fingerprint**. Normally, if two documents have the same hash values, we may say they are statistically identical; even when those two files have two different filenames. Making even a minor alteration in one of the files (such as to add or replace a single bit of data), will lead to a different hash value. There are two computer forensic hashing algorithms that are recognized in the forensic community. The first one was named, Message Digest 5 (MD5), which is a mathematical formula that processes file data into a hexadecimal code or a hash value. If even a single bit in the data has been updated, the hash value will also be changed. MD5 Hashing function has a 128-bit value that is unique to the stream of data. The odds of any two different documents having the same MD5 is one in 2^{128}.

The second hashing algorithm is a newer one, called Secure Hash Algorithm version 1 (SHA-1), which was created by the National Institute of Standards and Technology (NIST). The SHA-1 Hashing function has a 160-bit value that is unique to the file being hashed. The odds of any two different files having the same SHA-1 is one in 2^{160}. SHA-1 is gradually replacing MD5, even though MD5 is still widely used (Nelson et al., 2010).

Digital evidence chain of custody

In order for any digital evidence to be accepted by a court, the proponent must authenticate the evidence. That means, the party must show that the evidence truly is what it purports to be. One of the more important rules, which needs to be highlighted when dealing with digital evidence, is the chain of custody. In simple words, the definition of digital evidence chain of custody is a timeline that depicts how digital evidence was seized, analyzed, and preserved, in order for it to be shown as evidence in court. For any further forensic analysis, work should not be done on the original evidence but on a copy version, so that the duplication of digital evidence maintains its untampered content in an unbroken chain of custody, which shows electronic file preservation and its reliability and accuracy. For digital evidence legal admissibility, hash values of seized digital evidence are no different than photographs of a crime scene. The entire hard drive and relevant documents or any other storage media should be hashed as soon as possible.

Building a sound chain of custody is imperative since digital evidence can be easily changed. Maintaining a chain of custody for digital evidence, at a minimum, must ensure that the following procedures were followed:

- A complete copy was made with hash verification
- All digital media was secured
- No data has been altered

Chain of custody documentation should be preserved for all digital evidence, so that a record of the relevant hash values could be referred to the court. By comparing two hash values, the digital evidence will be authenticated and become admissible so a clear chain of custody documentation can be presented to the court.

Forensically sterile conditions for suspect digital devices

Prior to any further forensics analysis on suspect digital devices, we are required to sterilize all media that will be used in forensic imaging (duplication). Since we are only allowed to work on a copy version of suspect digital evidence (not on original evidence), the condition of any new media should be forensically wiped and verified prior to analysis. This procedure should obstruct data corruption from previous analysis or a possible data contamination from destructive computer programs.

ANALYSIS OF ELECTRONIC DATA

Any analysis applied to digital evidence must take place on a copy of the original. If the original evidence is being processed, there will be no way to reproduce the results. The original evidence furnishes the function of control, much the same as any control sample used in forensic scientific examination. Without any effective control, any conclusions drawn from the digital evidence will be suspect (Berg, 2000).

Unallocated file space and file slack

Under Windows-based computer systems, when files are deleted, the content of the data is not truly erased. Data from the deleted file remains intact in an area called unallocated space. Unallocated file space and file slack are both significant sources in computer forensics. Unallocated file space potentially includes intact files, remnants of files and subdirectories, and temporary files that were made and erased by computer applications and operating system (OS). All such files and data fragments could be sources of digital evidence. This also poses a security risk for sensitive information for those who thought the data was erased.

The file system utilizes fixed, seized containers or blocks named clusters. In Windows-based systems, they normally write in 512-byte blocks called sectors. **Clusters** are groups of sectors that are used to allocate the disk storage space in Windows-based operating systems. Briefly, Windows systems allocate disk space for files by clusters. If a new file is assigned a number of clusters, then the total file size will be equal to or smaller than the number of clusters multiplied by the size of one cluster. Since file sizes rarely match the size of multiple clusters perfectly, there is a tiny space in which some data could reside between the end of a logical file and the end of a cluster. This is called drive slack.

The drive slack could divide into RAM slack and file slack. We are going to use a simple two-sector cluster to illustrate slack space. A simple example is to use Notepad to type "CSI" and save the file. The file size is 3 bytes, since there are only three letters in the document.

RAM slack

As indicated above, Windows-based systems usually write in 512 bytes per sector (like a container). That means, whenever the OS wants to save a file into the file system, it would write in blocks of 512 bytes with a minimum of at least one container size of 512 bytes. So, if there is not enough information to fill the last sector of the file in the last cluster, the OS will write random data from the computer's RAM to the unfilled area in the last sector of the file. Login account ID, file fragments, deleted photos, e-mails, notes, and passwords are often discovered in RAM. That sensitive information could possibly be transferred to and reside in RAM slack.

The file shown in Figure 9.2 contains only one cluster which has two sectors for simplicity.

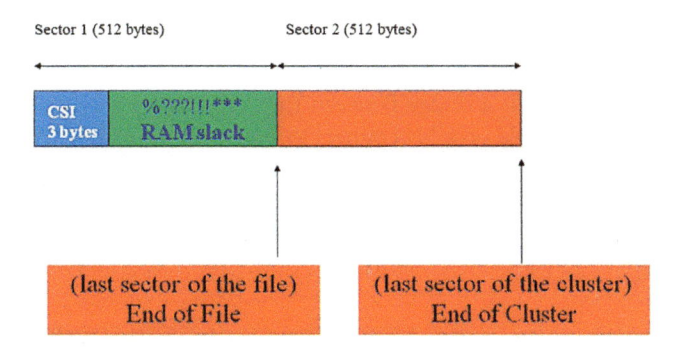

Figure 9.2 Illustration of RAM slack. The file shown contains only one cluster which has two sectors.

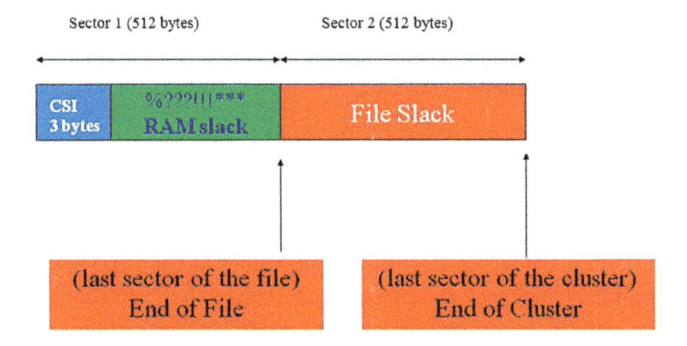

Figure 9.3 Illustration of file slack. The file shown contains only one cluster which has two sectors.

File slack

As in the example above, RAM slack covers to the last sector of a file only. Notice the remaining sectors which are still affiliated with the last cluster assigned to the file, but have no data placed in them. Since there is no new data to rewrite into those remaining sectors, it could preserve remnants of previously erased files or the data which existed before. This is the space we call file slack.

The file shown in Figure 9.3 contains only one cluster, which has two sectors for simplicity.

Deleted/un-deleted process

Computer forensics can recover deleted files. We may say that this is an undeleting process. When a computer OS deletes a document, it does not erase the data. It simply updates the pointer to the document and informs the file system that the document no longer is available, and the space is now open for any new data for storage. For instance, in the FAT (File Allocation Table) file system, when a file is being deleted, the computer OS will replace the first character of file name to hex code E5, which displays as a lowercase Greek letter sigma (σ). The sigma symbol is a marker, which advises the OS and file system that the document is no longer available, and the space is now open for any new file to be written into the same cluster location. The area of the space where the deleted files are located becomes **unallocated file space**. You may say, the unallocated space is like a junk yard and waiting for recycling. The unallocated space is now available to accept new data from newly-made documents or other files needing more space.

Usually deleted files can be used to show the culpability of perpetrators by showing their willful behavior of deletion, which was used to try to cover their negative intention.

Forensic tools

Write-Blockers safeguard digital evidence to prevent it from being written upon. There are software and hardware write-blockers that provide the same feature, but in a different fashion.

Software Write-Blockers: It usually updates forensic examination computer BIOS interrupt 13 to hinder writing to the specified hard drives or media.

Hardware Write-Blockers: They serve as a bridge between the suspect's hard drive and the forensic examination computer. Write-Blocker Devices could possibly connect to a forensic workstation via FireWire, USB, SATA, and SCSI controllers. Write-Blockers usually have a feature to remove and reconnect drives without having to turn down to examination computer.

Commercial tools are known as "All-in-One" Tools and are software used to aid with extracting the information from the computer or any digital evidence.

Using National Institute of Standards and Technology (NIST) computer forensics tools

The National Institute of Standards and Technology supports the Computer Forensics Tool Testing (CFTT) project to test and validate the Computer Forensics software to improve digital evidence admissibility in legal procedures and supervise research (such as to construct the categories of CF, and to determine CF category requirement) on the tools of Computer Forensics. For further information on CFTT project, visit http://www.cftt.nist.gov

We have also listed a number of computer forensic tools in the Appendix.

INTERNET CRIME

Our world has changed tremendously because of the development of the Internet. The Internet has caused the appearance of new crime patterns and previously unheard crimes that transcend all continents and continue to challenge law enforcement and prosecutorial measures, which have traditionally been used to investigate, prosecute, and punish criminal actions (Pittaro, 2010). The term of cyber crime could also be referred to as Internet crime.

Computer forensics analysis of Internet data

The existence of data in digital format, that is representative of people activities in a virtual world, creates digital footprints and trails of their daily lives. Computer forensics is able to show readily one's history of internet activities.

Web Cache: Is also called **HTTP cache** or **Internet cache** that is a facility for web browsing temporary storage. It is designed to improve loading speed while browsing the Internet. Basically when you browse a web site, the page and all its photos, & files are sent to the browser's temporary cache on the hard drive. If the web page and files contained on that web page need to be re-visited again (the web pages have not been modified since your last visit), the browser opens the page directly from Internet cache instead of downloading the page again. This is technology to speed up internet browsing. Web cache could be a valuable area to examine when investigating a cyber crime. When those web cache files that have been erased it is often possible to undelete those cached files that can be very useful for forensic investigation.

Web Cookies: Is also called **HTTP Cookies** or **Internet Cookies** are a reliable way for websites to remember the user's browsing activity by storing a small piece of text onto a user's hard drive. When a user revisits the website, the browser sends the cookie back to the server to notify the website of the user's previous browsing activity. Obviously, web cookies can be vital evidences for cyber crime investigation.

Internet Browsing History: Web browser software (such as Chrome, Firefox, and Internet Explorer) automatically stores a list of web sites which a user has surfed lately, so that one is able to go back to web pages they have visited. The Internet browsing history is there for users' convenience. The Internet browsing history could be a valuable resource for digital evidence and it can be reviewed through computer forensics software.

E-mail forensics

E-mail is one of the most important applications on the Internet for our daily communication including messages with documents or photos delivery. It is used not only from computers, but also from smartphones or mobile devices. However, with the increase in e-mail phishing, scams, and fraud attempts, forensic investigators need to understand how to review and analyze the unique content of e-mail messages.

E-mail messages examination

E-mails are made of two main components; they are the **message header** and **message body**. The header part includes routing information about the e-mail and other information such as the source and destination of the e-mail, the IP address of the sending computer and time related information. The message body contains the actual message of the e-mail; that is, the message subject and text. That's the part one usually reads. Sometimes, the body could also include attachments in the form of MIME (Multipurpose Internet Mail Extensions) or SMIME (Secure/Multipurpose Internet Mail Extensions). However, message headers are often the most significant part for examining e-mail messages for forensic identification.

How to review e-mail headers

Use Microsoft Outlook Office 365 e-mail header, then log into the account of interest, and highlight the message you would like to review the header of, right click to choose "View Message Details" (See Figure 9.4).

View Google mail e-mail header

To view e-mail headers in Google Mail, log into the Google account, and click into individual messages. There is an arrow down key on the right-hand upper corner, click "Show Original" (see Figure 9.5).

Figure 9.6 identifies the e-mail header, attachment, and e-mail body.

Figure 9.7 illustrates how to identify a sender's IP address from the e-mail header.

IP Address: Each computer used on the Internet must be assigned a unique address. The number called Internet Protocol Address (IP Address), which is a 32-bit number divided into 4 parts. It is usually written as a series of four numbers in the range 0–255, with a single dot (.) separating each number or set of digits, such as 100.6.9.86.

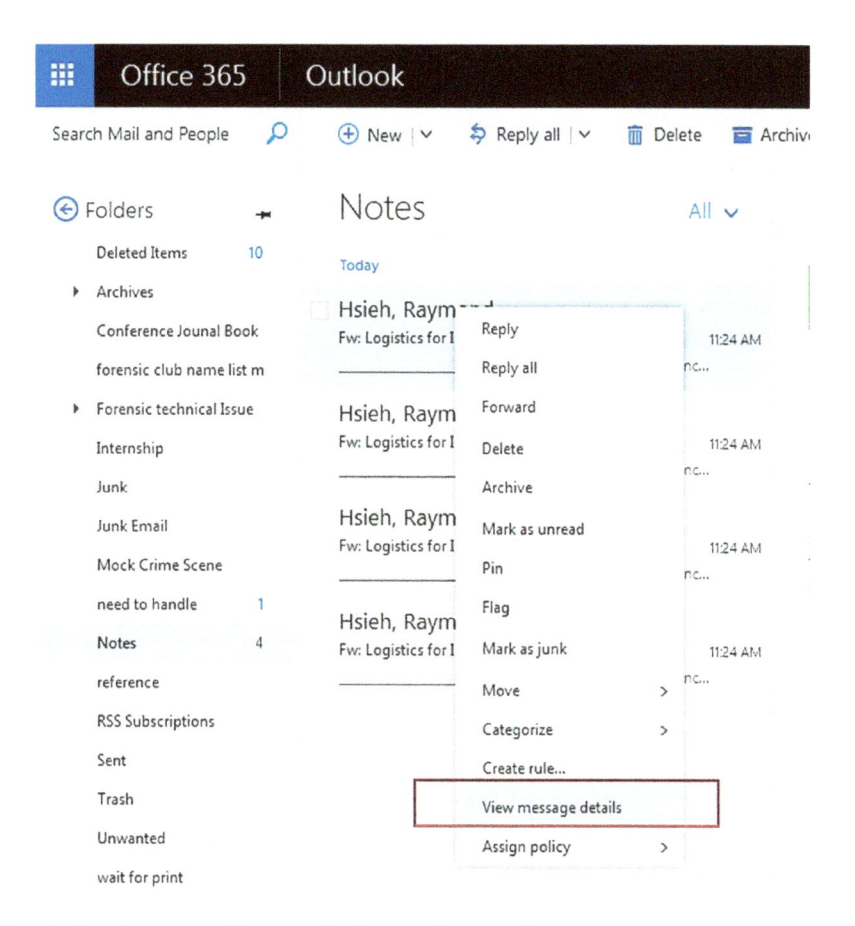

Figure 9.4 Reviewing the header can provide more details regarding e-mail messages.

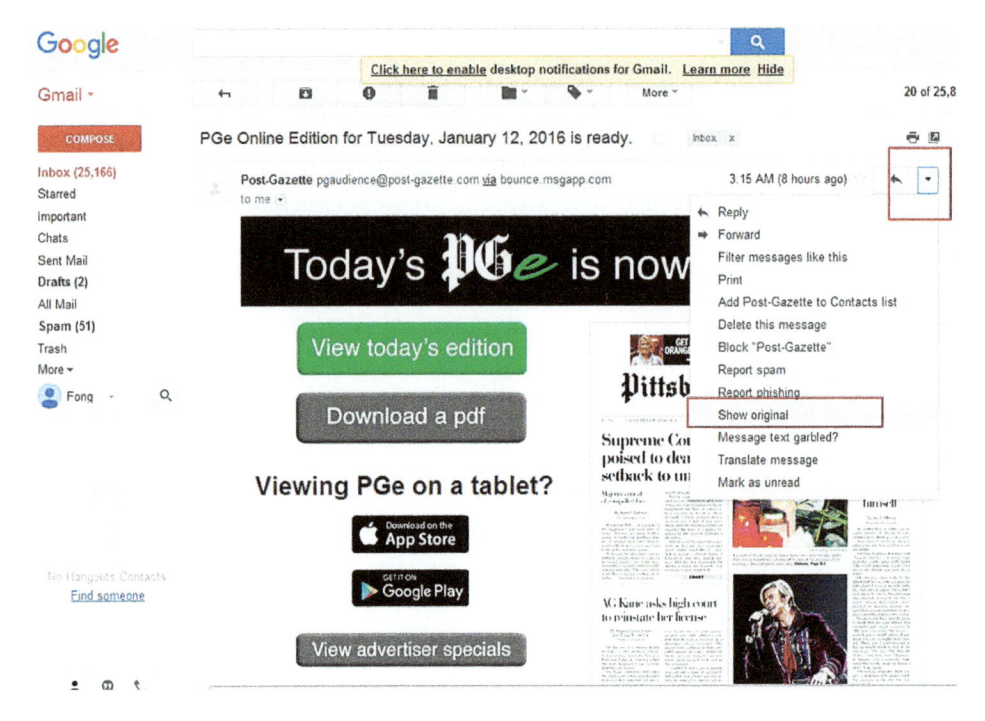

Figure 9.5 Headers can be reviewed in Web mail platforms, including Google Mail illustrated here.

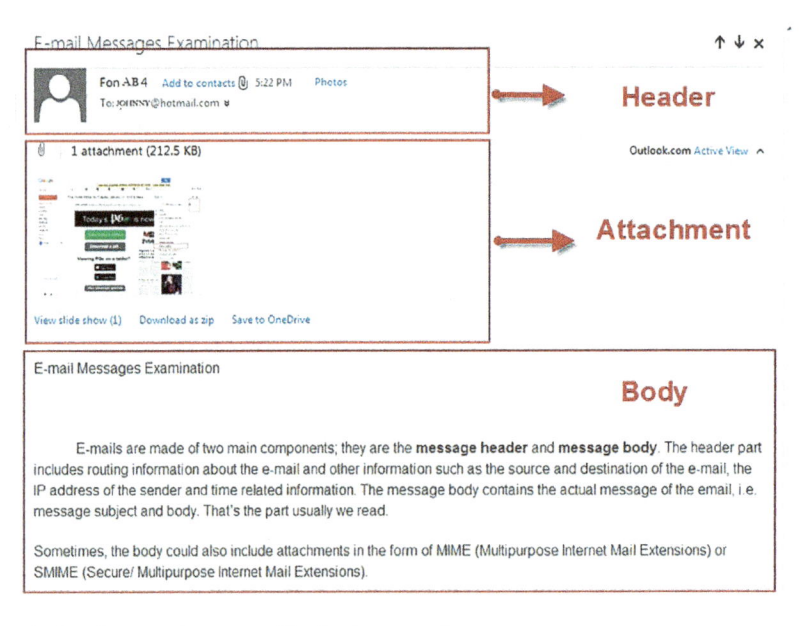

Figure 9.6 Identifying the header, attachment, and body of an e-mail.

```
x-store-info:sbevk12QZR7OXo7WID5ZcVBK1Phj2jX/
Authentication-Results: hotmail.com; spf=pass (sender IP is 209.85.220.174; identity alignment result is pass a:
X-SID-PRA: sale3_@gmail.com
X-AUTH-Result: PASS
X-SID-Result: PASS
X-Message-Status: n:n
X-Message-Delivery: Vj0xLjE7dXM9MDtsPTA7YT0xO0Q9MTtHRD0xO1NDTD0w
X-Message-Info: gamVN+8Ez8V+RHg+F+brAQuxypb5rSmunG2ydpzgta/jeYM6a6dzLXtLOOvCcoYA7HEc+ILvZ8TkIRu7ygjm8ccSEXUMTvXl
Received: from mail-qk0-f174.google.com ([209.85.220.174]) by BAY004-MC2F2.hotmail.com over TLS secured channel
         Tue, 12 Jan 2016 09:22:14 -0800
Received: by mail-qk0-f174.google.com with SMTP id n135so258057559qka.2
         for <rjh_@hotmail.com>; Tue, 12 Jan 2016 09:22:14 -0800 (PST)
DKIM-Signature: v=1; a=rsa-sha256; c=relaxed/relaxed;
         d=gmail.com; s=20120113;
         h=from:message-id:date:user-agent:mime-version:to:subject
          :content-type;
         bh=vuEGgsHvjmMSvXpikZUqJUnuVXbZF6K/1dhwOSC+Jc4=;
         b=wUg7TrjjjI2+bCLa+kly+I+bFVW8+haFCQEo/FPmoXr8ToVEb/uRYDgnKOfzNty+Hr
          qOYVpMNut1fqAaDS6xey2W11ctY/m6TjgtePiAK4yDFowZnVBZ92pyTdQ71y4HnGiddv
          xF8QMZ5IITmQVuRPf1aIVWHXZJNZci6Dag30LEI880b+nuiNvhQV9DRYYHMJ3nMT1OWEn
          OTFwbfkbiYEJ7eNxktUaiRrpM6eY8BBw27mpQCtrcDz0cUdxkRJSijy+Z1EtyZSbVBHn
          EE4mh16y5+2o3fPB73KQrSVEdDY8wHD1NbnwO3R5E5P9/zfBdO+vomMphQgk645o06Yu
          naaA==
X-Received: by 10.55.27.98 with SMTP id b95mr30873080qkb.51.1452619334425;
         Tue, 12 Jan 2016 09:22:14 -0800 (PST)
Return-Path: <sale3_@gmail.com>
Received: from [192.168.1.162] (pool-100-6-90-86.pitbpa.fios.verizon.net. [100.6.90.86])
         by smtp.gmail.com with ESMTPSA id u16sm39639992qka.22.2016.01.12.09.22.13
         for <rjh_@hotmail.com>
         (version=TLS1 cipher=ECDHE-RSA-AES128-SHA bits=128/128);
         Tue, 12 Jan 2016 09:22:13 -0800 (PST)
From: Fong384 <sale3_@gmail.com>
X-Google-Original-From: Fong     <fong: :@gmail.com>
Message-ID: <56953645.7010506@gmail.com>
Date: Tue, 12 Jan 2016 12:22:13 -0500
User-Agent: Mozilla/5.0 (Windows NT 6.0; WOW64; rv:17.0) Gecko/20130509 Thunderbird/17.0.6
MIME-Version: 1.0
To: rjh2677@hotmail.com
Subject: E-mail Messages Examination
Content-Type: multipart/mixed;
         boundary="------------090505030509040305020309"
X-OriginalArrivalTime: 12 Jan 2016 17:22:14.0670 (UTC) FILETIME=[C79CE2E0:01D14D5D]

This is a multi-part message in MIME format.
```

Sender IP → (pool-100-6-90-86.pitbpa.fios.verizon.net. [100.6.90.86])

Figure 9.7 Identifying a sender's IP address from the e-mail header.

E-mail tracing

This actually examines e-mail header information to search for routing information to locate where a message has been sent from. To audit or develop paper trails of e-mail routing information can be important as evidence in legal a presentation. One can use EmailTrackerPro (http://www.emailtrackerpro.com/index.html) or Visual Route (http://visualroute.visualware.com/) by Visualware to visually monitor e-mail traffic information.

Mobile device/smartphone forensics

Mobile forensics is a relatively new discipline. Smart phones are ubiquitous in today's world, and many crimes have some involvement of phones. Therefore, cell phones have emerged as a vital resource for both criminal investigators and defense lawyers. The biggest hurdle is dealing with constantly changing models of smartphones.

What is a smartphone? Basically, a smartphone is a cell phone with an advanced, mobile operating system (such as Mac iOS, Black Berry OS or Google Android OS) that not only allows one to place phone calls, but also adds in features once only found on computers (such as edit photos/video clips or send and receive e-mail). People put a lot of personal related information on smartphones. If a smartphone is involved in a legal case, you may tap deeper into it to discover a hidden gold mine of personal information.

The following crucial information could be possibly collected from a smartphone.

- Call logs: Incoming, outgoing, and missed calls
- Text and Short Message Service (SMS) messages

- E-mail
- Social Media Messaging (Facebook, Twitter, What's App, Web Chat & Line, Instagram...etc.)
- Instant-messaging (IM) logs
- Web pages content
- Photos
- Personal calendars
- Address books
- Voice recordings (such as W4A)
- Music files (such as MP3)

Smartphone basics

There are two major radio systems in cell phones technologies, **CDMA (Code Division Multiple Access)** and **GSM (Global System for Mobiles)** that establish a gap you can't cross. That's the reason you can't use Verizon's phones on an AT&T network and vice versa. Five of the top seven carriers in the U.S. use CDMA: Verizon Wireless, Sprint, MetroPCS, Cricket, and U.S. Cellular. Only two carriers in the US use GSM: AT&T and T-Mobile. The majority of the world uses GSM (especially in Europe and Asia.) and the U.S. is basically a CDMA country. Details regarding the two systems are outlined in Table 9.1.

Acquisition procedures for mobile devices/smartphones

All smartphones or mobile devices have volatile memory, so it is significant to collect data prior to a power failure. Proper search and seizures for smartphones are vital, if it is off, do not turn it on. One also needs to locate any extra power supply or recharger as quickly as possible. If the smartphone is left on; however, it could receive calls and text messages during transport to the lab and data could be overwritten or deleted, so that you must isolate the phone/device from incoming signals with one of the following alternatives:

- Place the device in a paint can
- Use the Paraben Wireless StrongHold Bag (https://www.paraben.com/), which compiles with Faraday wire cage standards
- Use at least 8 layers of antistatic bags to block the signal

Mobile forensics tools

Paraben Software (https://www.paraben.com/): Paraben developed the first mobile forensic tool on the market. The tool provides logical and physical acquisitions as well as password bypassing and file system extractions in one easy-to-use tool. There are several tools like DS (Device Seizure), Mobile Field Kit, StrongHold Faraday Protection, Project-A-Phone, and JTAG Analysis Tool.

Table 9.1 Comparing GSM and CDMA

Digital Network	Description
GSM	SIM (subscriber identity module) card, the onboard memory device that identifies a user and saves all his information on the phone.
	You can swap GSM SIM cards between phones when a new one is necessary, which enables you to transfer all of your contact and calendar information over to a new phone with little effort.
	GSM internally based on Time Division Multiple Access (TDMA) techniques and GPRS (General Packet Radio Service).
CDMA	CDMA carriers store user information; includes phone book and scheduler information, on the carrier's database.
	This service makes it possible to not only change to a new phone with little trouble, but it also gives users the ability to recover contact data even if their phone is lost or stolen.
	CDMA replies on 1xRTT or CDMA 2000, which has the capability of providing ISDN (Integrated Services Digital Network).

BitPim software (http://bitpim.org/): It allows you to view and manipulate data on many CDMA phones from LG, Samsung, Sanyo, and other manufacturers. This includes the PhoneBook, Calendar, WallPapers, RingTones (functionality varies by phone), and the File system for most Qualcomm CDMA chipset-based phones.

Cellebrite UFED Forensic System (http://www.cellebrite.com/Mobile-Forensics): The system works with Smartphones/mobile devices. They include, UFED Touch, UFED 4PC, UFED TK, UFED InField Kiosk, and UFED Cloud Analyzer, which is able to analyze cloud data sources that represent a virtual goldmine of potential evidence for forensic investigators.

Other free resources for iPhone forensics

Investigator Ryan Kubasiak's Web site (http://appleexaminer.com/): It offers a variety of useful Mac tools for evidence collection.

Forensic Magazine (http://www.forensicmag.com/) and **SANS** (http://www.sans.org/): They are helpful resources for mobile forensics news, tips, and trends.

ACTUAL CASE EXAMPLES INVOLVING DIGITAL EVIDENCE

BTK

Known as the "BTK Killer"—Which stands for "bind, torture, and kill"

Dennis Rader is one of the most notorious serial killers in the United States. Dennis Rader, the self-proclaimed "BTK: Bind, Torture, and Kill" serial killer was born in 1945, and was raised in Wichita, Kansas along with his three brothers. He is a father, husband, Boy Scout leader, Air Force veteran, and active church member/president; many would say that Rader was a typical nice guy. With the reputation of being an easy-going and caring man, he wasn't really a noticeable person. Ultimately, it would be his ability to blend in that would keep him from being caught for so many years.

Between the years 1974–1991, Rader single-handedly tortured and murdered a total of 10 people beginning with Joseph Otero, his wife Julie, son Joseph Jr., and daughter Josephine. Dennis Rader was arrested in February 2005 and charged with committing 10 counts of murders since 1974 in the Wichita, Kansas area. After 14 years of silence from 1991, he reappeared in 2004 by mailing a letter to a local newspaper taking credit for a 1986 killing. The BTK Killer dropped 11 packages, in total, to news stations at various locations. The packages included cards, letters, photos, and sketches for the murders he committed. He sent his last package on February 16, 2004. A package arrived at the studios of KSAS-TV containing a letter, a piece of jewelry, and a purple diskette referred to as "Test Floppy for WPD review." A Computer Forensic Examiner analyzed the diskette and found that metadata of the Microsoft Word document on the diskette was originally from the Christ Lutheran Church in Wichita and the name "Dennis." The BTK Killer had apparently thought he had erased the original contents of the diskette. However, he did not know the file could be undeleted and restored by a computer forensic scientist.

A quick Internet search brought up a website for the church mentioning its current council president, Dennis Rader. He was arrested, tried, and sentenced to 10 consecutive life-sentences. The people of Kansas were finally able to breathe easier and no longer worry about the evil predator that had caused terror in the state for almost 2 decades.

Death of Caylee Anthony

On July 15, 2008, 2-year-old Caylee Anthony's grandmother (Cindy Anthony) reported her missing. After months of the investigation focusing on Casey Anthony, Caylee's mother, Caylee's skeletal remains were found near her house. Throughout that time, Casey Anthony lied repeatedly regarding her daughter's whereabouts.

She was tried for murder and her trial lasted 6 weeks, from May to July 2011. The police sought the death penalty and alleged Casey wanted to free herself from parental responsibilities and killed her own daughter by administering chloroform and applying duct tape. The digital forensic report related to the Anthony investigation, included computer evidences of Google searches of the terms "neck breaking" and "how to make chloroform" shortly before Caylee disappeared in 2008. The Prosecution showed that Casey Anthony had conducted computer searches on the term "chloroform" 84 times. Furthermore, among some other photo evidences collected from the computer of an ex-boyfriend of Casey Anthony, Ricardo Morales, a poster was found with the caption "Win her over with Chloroform."

Due to the complication of the case, a jury acquitted Casey Anthony of murder charges, but found her guilty on lesser charges, four counts of giving False Information to Law Enforcement.

Appendix

This appendix contains descriptions of most of the widely used forensic software packages available to those doing digital forensic investigation.

ILooKIX (http://www.perlustro.com/) is an all-in-one computer forensic suite created by Perlustro L.P. With the world's first Law Enforcement Windows forensics toolset, ILookIX, now redefines the future of computer forensic investigations. ILooKIX includes the ability to find and mark files that appear to be human images. These images may include pornographic and contraband, such as abusive images or illicit images otherwise.

IXImager, from **ILooKIX**, stands alone in the field of computer forensics imaging products. It stands in that position for a very simple reason—it has more breadth of use, court validation, standards validation, and testing in hostile environments than any other imaging systems that exist.

Maresware (http://www.dmares.com/) is software produced by Mares and Company and is used for Linux and, henceforth, is named Linux Forensics, which provides tools for investigating computer records while running under the LINUX operating system on Intel processors. It is useful to all types of investigations, including law enforcement, intelligence agency, private investigator, and corporate internal investigator.

Used within a forensic paradigm, the software enables discovery of evidence for use in criminal or civil legal proceedings. Internal investigators can develop documentation to support disciplinary actions, yet do so non-invasively, to preserve evidence that could end up in court. Mares and Company also offer free downloads as well as training on their website.

Encase Forensic (https://www.guidancesoftware.com/) from Guidance Software, is one of the standard technology solutions for capturing, analyzing, and reporting on digital evidence. Powerful filters and scripts enable investigators to build a case on forensically sound evidence. From the simplest requirements to the most complex, EnCase is the premier computer forensic application on the market. It gives investigators the ability to image a drive and preserve it in a forensic manner using the EnCase evidence file format (LEF or E01), a digital evidence container validated and approved by courts worldwide.

Encase Forensic contains a full suite of analysis, bookmarking, and reporting features. Guidance Software and third-party vendors provide support for expanded capabilities to ensure that forensic examiners have the most comprehensive set of utilities. EnScripts and customizable filters allow examiners of all experience levels to quickly parse out relevant data for further review with pre-built EnScripts or by developing their own EnScript tools. EnCase Forensic also offers powerful hidden volume detection and volume rebuilding capabilities, allowing investigators to review evidence that would have been irretrievable with other computer forensics applications.

It is also worth mentioning that Guidance Software sells hardware disk-write blockers called, FastBloc, which provides the extra measure of assurance that no writes occur on the device.

DriveSpy (https://www.digitalintelligence.com/) is a DOS-based forensic tool developed by Digital Intelligence, Inc.

Unlike Byteback, DriveSpy is an extended-DOS forensic shell. DriveSpy provides an interface that is similar to the MS-DOS command line, along with new and extended commands. The entire program is only 110KB and easily fits on a DOS-boot floppy disk.

DriveSpy provides many of the functions necessary to copy and examine drive contents. All activities are logged, optionally down to each keystroke. If desired, logging can be disabled at will. You can examine DOS and non-DOS partitions and retrieve extensive architectural information for hard drives or partitions. DriveSpy does not use operating system calls to access files and it does not change file access dates.

Additional functionality includes:

- Create disk-to-disk copy (supports large disk drives).
- Create MD5 hash for a drive, partition, or selected files.
- Copy a range of sectors from a source to a target, and where the source and target can span drives or reside on the same drive.
- Select files based on name, extension, or attributes.
- Unerase files.
- Search a drive, partition, or selected files for text strings.
- Collect slack and unallocated space.
- Wipe a disk, partition, unallocated or slack space.

DriveSpy helps you to select files based on name, extension, or attributes, It also allows you to review the sectors or clusters in its built-in hex viewers. Another useful DriveSpy feature is a search engine that allows you to search a partition or drive for specific text strings. DriveSpy provides basic command-line functionality that is portable enough to carry on a single floppy disk and use at the scene. After you create an image of a drive, DriveSpy can assist you in examining the images content. (Solomon et al., 2004)

Forensic Replicator, (https://www.paraben.com/index.html) from Paraben Forensic tools, is another disk imaging tool that can acquire many different types of electronic media. It provides an easy-to-use interface, to select and copy entire drives or portions of a drives. It also handles most removable media, including Universal Serial Bus (USB) micro drives. Replicated media images are stored in a format that can be read by most popular forensic programs. Forensic Replicator also provides the ability to compress and split drive images for efficient storage.

The ISO CD Rom option allows you to create CDs from evidence drives that can be browsed for analysis. This option can make drive analysis much easier and more accessible for general computers. You don't need to mount a copy of the suspect drive on a forensic computer. You can use searching utilities on a standard CD-ROM drive. Forensic Replicator also offers the option of encrypting duplicated images for secure storage. Paraben also sells a Firewire or USB-to-IDE, write blocker, called Paraben's Lockdown, as a companion product. Forensic Replicator requires a Windows operating system to run. ("XBox Forensics," n.d.)

FTK (Forensic Toolkit) or FTK Imager (http://accessdata.com/) from AccessData Corporation is a set of forensic tools that includes powerful media duplication features. FTK can create media images from many different source formats, including:

- NTFS and NTFS compressed
- FAT12, FAT16, and FAT32
- Linux ext2 and ext3

FTK generates CRC or MD5 hash values, as do most products in this category, and for disk copy verification, FTK provides full searching capability for media and images created from other disk imaging programs. Image formats that FTK can read include:

- EnCase
- SMART
- SnapBack
- SafeBack
- SafeBack (not V3.0)
- Linux dd

FTK is a Windows-based utility and, therefore, requires that the user boot into a Windows operating system and provides a very powerful tool set to acquire and examine electronic media. FTK provides an easy-to-use file viewer that recognizes over 270 types of files.

There are four different types of controls that might be important in forensic testing of evidence: knowns, alibi knowns, substratum, and blank. All of these are important in evidence collection and testing.

Known (may be called **Exemplar** or **Reference**) **Control**. Many types of forensic tests discussed in this book involve comparisons between a questioned (evidentiary) specimen and a known specimen (whose origin is known with certainty). In those situations, the known (which may also be called the exemplar or the reference) is essential for the comparison. Examples are known blood, or a buccal (cheek) swabbing, from a person (for comparison with DNA types from evidentiary blood or body fluid specimens), known fibers from a carpet (for comparison with evidentiary fibers from a suspect's clothes), and known paint from an automobile suspected in a hit-and-run case (for comparison with a paint smear transferred onto a bicycle).

Alibi (may be called **Alternative**) **Known Control**. An alibi or alternative known is the same as a known control but from a different source. Police might suspect, for example, that bloodstains on a suspect's clothing are from an assault victim. The assault victim's blood would therefore be the known control. But the suspect says the bloodstains came from a fight with someone in a bar (and he names the person). Blood from that person would then be an alibi known in the case.

Substratum Control. As explained above, the term "substratum" refers to the underlying material or surface on which evidence is found or has been deposited. The term "substrate" is sometimes used for the same purpose.

With most blood and physiological fluid evidence, suspected ignitable liquid residues in fire debris (Chapter 12), and various types of materials or "trace" evidence (Chapter 14), investigators must remember to collect a sample of the substratum separately from the evidence in order to permit the analyst to interpret the scientific tests properly.

Generally, the substratum control is subjected to the same testing as the evidence (which was on the substratum already) to make sure it is the evidence giving the test result and not the underlying surface material. Most of the time, substratum controls are not necessary for interpretation of DNA profiles.

Blank Control. A blank control refers to a specimen known to be free of the item or substance being tested. Scientists use blank controls as negative controls (generally along with positive controls) to be sure tests and test chemicals are working properly. Generally, investigators do not have to worry about blank controls.

The discussion of substratum controls brings up the issue of potential contamination. Contamination can be an important consideration with crime scene evidence because investigators have no control over evidence items until those items have been recognized, documented, and properly collected. In the biological evidence arena, forensic scientists have to be concerned about the possibility that human biological material, which has nothing to do with the case events, might get deposited onto, or be admixed with, human biological evidence that is case related. This problem has always existed but is perhaps of greater concern today in some circumstances because DNA-typing methods are so sensitive; that is, they are capable of generating DNA types from tiny traces of material.

Criminal justice practitioners often infer that contamination is the result of mishandling by crime scene or laboratory personnel. In theory, evidence might become contaminated in several ways. There could be biological material on an item or surface before the biological **evidence** was ever deposited. Biological material might be deposited onto evidence during scene searching and/or processing activities. Or, biological material might be deposited onto evidence during laboratory examinations and/or manipulations. Neither investigators nor forensic scientists have any control over the history or circumstances of evidence prior to its recognition and collection. A major purpose of collecting and testing the substratum—and sometimes alternative known—controls discussed above is to check for the possibility of a contaminant. DNA that was on a substratum before the biological evidence was deposited is often referred to as "background" DNA to avoid negative implications of the word "contaminant." Investigators and scientists do have control over what happens to evidence following its recognition and during their activities at scenes and subsequently.

It is important to emphasize that contamination is not an important issue in many cases—those that have large quantities of biological evidence (Figure 10.6) in comparison with the quantities of any contaminant that might be present, even in theory. There is usually so much biological evidence compared with any contaminant that only the results from the evidence are observed. Generally, only when the quantities of the contaminant start to approach the quantities of the evidence in a mixture does the possibility of contaminant interference become a significant consideration. Studies have shown that typically a contaminant present in ratios of 1:20 or more do not produce conclusive DNA results for the contaminant using standard PCR-STR protocols. As the amount of contaminant approaches the quantity of the evidence, the analyst would obtain results indicating a mixture. However, when techniques are used for mitochondrial DNA or low copy number ("touch") DNA, these small amounts of contaminant may be detected. It is also important to emphasize that contamination does not necessarily imply error nor accidental or intentional wrongdoing on the part of investigators or scientists. Clothing that has been worn virtually always has trace contamination from saliva spray or sweat from the wearer.

Thus, there can be unavoidable contamination, but most contamination of evidence by investigators and scientists is avoidable. Controls help forensic scientists detect the presence of contaminants and correctly interpret the results of their tests. Good crime scene and evidence collection practices on the part of investigators, and good laboratory techniques on the part of evidence technicians and scientists, are all designed to prevent contamination of any evidence items. A number of crime scene and laboratory practices are specifically aimed at avoiding contamination of the evidence specimens. These are discussed nearby.

SIDEBAR Good crime scene and laboratory practices sidebar

The "good practices" discussed here are those designed to protect evidence from any contamination by the investigators or scientists, or any other avoidable source, and to protect investigators and scientists from anything harmful in the evidence.

Perhaps the simplest and most basic precaution is protective gloves. Often, latex gloves are used, but some people are allergic to latex. Nitrile gloves may also be used, and they afford some additional protection against chemical hazards. Gloves prevent investigators' or scientists' hands from coming into contact with biological evidence that might have some infectious agent present. Viruses that cause AIDS, hepatitis B and C, and herpes viruses can be present in blood or body fluids. Although there is probably very little risk for someone to get these viruses into their systems from dried biological evidence, the risk may be greater if the biological evidence is still wet. Wearing protective gloves constitutes a normal and widely recommended precaution in handling biological material that is potentially biohazardous.

Gloves also protect the evidence from the evidence handlers—investigators and scientists. Depositing skin cells from the hands or fingerprints is avoided. Evidence handlers should also take precautions to avoid their hair getting into evidence. And they should realize that coughing or sneezing may contaminate evidence.

Sterile solutions and sterile gauze for evidence collection avoid the problem of potential bacterial contamination. Some bacteria may have the ability to destroy biological evidence. They can also contribute unwanted, background DNA. Bacteria are everywhere and probably cannot be completely avoided. As a rule, they do not prevent successful biological evidence analysis. But taking simple precautions to avoid contamination is sensible.

There are strategies to prevent contamination in laboratories as well. Analysts always wear protective gloves. Face masks are recommended to prevent the addition of small amounts of DNA and body fluids from taking over the evidence. Additional protective equipment—booties, hair covers—may also be necessary depending on the nature of the scene or evidence. In some DNA laboratories, the DNA profiles of all the employees are kept on file so that the lab can immediately spot contamination of a specimen from a lab person. All DNA analysis relies on PCR (discussed in detail in Chapter 11), and PCR makes large numbers of copies of certain regions of DNA that are important in forensic DNA profiling. The large quantity of copies creates a potential for contamination within the laboratory that must be avoided. Separate rooms in the lab are devoted to pre- and post-PCR activities for these reasons.

As we have noted in the main text, investigators and scientists must try to avoid contaminating evidence themselves. They have no control over pre-existing contamination, however. Furthermore, if the quantity of evidence is very large in comparison to the contaminant, the contaminant will have no effect on the DNA profile. The signal from the evidence DNA will overwhelm the signal from any contaminant. For example, coughing on a large bloodstain is almost certainly not going to change the DNA profile. There is so much bloodstain DNA compared with the small amount that may be deposited from the cough that only the bloodstain profile will be seen.

Initial examination of and for biological evidence

Blood or other physiological fluid stains or residues may be found on almost anything. Part of crime scene processing involves recognizing the presence or possible presence of such evidence, and then properly collecting and preserving the item. Items submitted to the laboratory often have obvious stains (Figure 10.5) on them, but sometimes they do not. Further, an item can have some obvious stains as well as other stains that are subtle and more difficult to find. The initial examination of items by a criminalist in the laboratory is designed to evaluate them for possible evidentiary value. This includes but is not limited to searching for biological stains. With visible stains or with areas where analysts perceive subtle staining or the possibility of staining, preliminary blood (and physiological fluid) tests are used to determine whether there is really anything to examine further. Positive preliminary tests indicate possible presence of blood or body fluids. The item will then be subject to confirmatory tests. Initially examiners may cut out or swab stains or areas that are of interest and/or that have given positive preliminary tests. Subsequent examinations are then done on the swabs or cuttings.

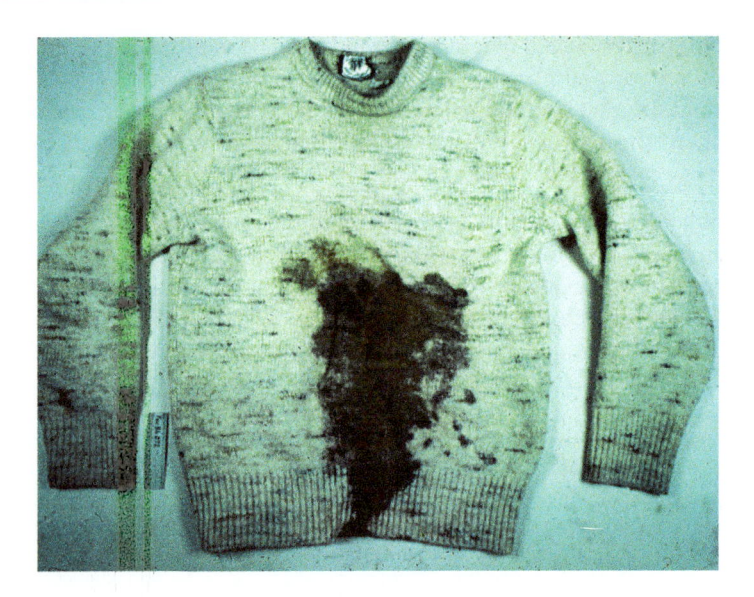

Figure 10.5 Heavily blood-stained sweater.

Because the first examiner may be the only person in the chain of analysis who actually sees the evidence in context, it is crucial to make good lab notes and, if necessary, sketches or photographs. Done properly, this strategy allows the evidence (and any results obtained) to be put into proper case context. At this preliminary stage of examination, perhaps more so than anywhere else, it is essential to remember that criminalistics involves the evaluation of physical evidence in the context of a case, not just the detailed, serial analysis of submitted items. In large busy laboratories, there may be considerable division of labor in handling the volume of evidence items. Under those conditions, the role of the initial examining criminalist becomes critical.

To some extent, logic dictates that the search for evidence at a scene and on items submitted to the laboratory, will be conditioned by the type of case and its circumstances and the type of evidence. But crime scene investigators and criminalists must always be open to finding something unusual or unexpected. It is also important for both scene investigators and criminalists to be thorough. Getting too focused on or caught up in the obvious may cause one to overlook something important. There is considerable truth to the old saying: "One sees mainly what one is looking for." This cognitive bias has been the basis of much criticism of forensic science in recent years.

It may also be noted that presumptive testing can be used to evaluate evidence items at scenes for possible collection, packaging, and submission. Often, these activities are entirely the responsibility of crime scene investigators, but in some places on some occasions, laboratory examiners may go to a scene to provide assistance or advice. Regardless, the only reason to do presumptive testing at a scene is to make a decision about collection and submission of the item. If the item is going to be collected and submitted, there is often no point in doing field tests because the laboratory will do the testing.

FORENSIC IDENTIFICATION OF BLOOD

Just as identification, individualization, and reconstruction may be seen as the three principal aspects of a criminalistics investigation, they may also be seen as the three main objectives of biological evidence analysis. *Identification* (which can also be called *classification*—see Chapter 1) means showing what the biological evidence is and whether it is blood, semen, saliva, or other. *Individualization* is the DNA typing of biological evidence to try to show that it almost certainly came from a particular individual, or that it did not come from that individual. *Reconstruction* in this context is primarily the interpretation of blood patterns (Chapter 4). It could also involve interpreting physiological fluid stain patterns, such as whether a semen stain seems to have resulted from drainage following intercourse.

Species determination in bloodstains is part of "identification." This chapter, then, discusses the identification aspect of blood and physiological fluids. DNA typing is covered in Chapter 11.

With blood and physiological fluid stains and with several other types of physical evidence, there are two categories of identification tests: presumptive or preliminary, and confirmatory.

Preliminary (or **presumptive**) tests are used for screening specimens that might contain a particular substance or material. Usually the tests are sensitive, but not necessarily specific for that substance or material. They are also faster and less expensive than confirmatory tests. Non-specificity means that there may be false positives and false negatives. In a "false positive," other substances or materials give positive results. In a "false negative," the test does not register when the substance or material is actually present. Ideally, preliminary tests should give a minimum of false negatives. The reason for minimizing false negatives is to maximize the test's utility as a screening tool. If the test gives a low rate of false negatives, specimens that test negative can be assumed to be negative and not to require any further testing. If a screening test does not meet this criterion, it does not serve its primary purpose—which is to avoid having to do confirmatory tests on every specimen. A good screening test prevents laboratories from having to do confirmatory testing on every item and, thus, increases efficiency and evidence throughput.

Confirmatory tests are generally more complicated, often more expensive, and may require more analyst time than preliminary tests. Confirmatory tests must by definition be entirely specific for the substance or material for which they are intended. A positive confirmatory test is interpreted as an unequivocal demonstration that the specimen contains the substance or material. For some types of biological evidence, no confirmatory tests exist. That is, the only tests available are presumptive. Under those circumstances, the most an analyst can ever say is that a specimen "might have," "could have," or "is consistent with containing" the substance.

Preliminary or presumptive tests for blood

Preliminary (presumptive) tests for blood have been around for more than a century. Most of them are color tests; that is, there is a distinctive color, or color change, when the test is positive. The tests that are in fairly widespread use today include: phenolphthalein (Kastle-Meyer), tetramethylbenzidine, leucomalachite green, and luminol. All the tests, except luminol, work on the same principle—they take advantage of the peroxidase-like properties of hemoglobin.

Peroxidases are enzymes that occur widely throughout nature, but are especially prevalent in some plants and bacteria. Horseradish is an especially rich source of peroxidase. An enzyme is a protein that can speed up a chemical reaction without being changed in the process. And a peroxidase is an enzyme that specifically catalyzes the reduction of peroxide:

$$\text{Reduced Dye} + \text{Peroxide} \xrightarrow{\text{peroxidase}} \text{Oxidized Dye} + \text{Water}$$
$$(\text{color 1}) \qquad\qquad\qquad\qquad (\text{color 2})$$

"Reduced" and "Oxidized" are chemical concepts that have to do with a change in chemical structure. The point here is that the reduced and oxidized forms of the dye are different colors. The reaction works as a blood test because hemoglobin acts like a peroxidase. Luminol, under certain conditions and in the presence of peroxide, produces light as a reaction product; that is, some of the energy released in the reaction is emitted as light. Substances like luminol that can produce light by undergoing chemical reactions are called *chemiluminescent*.

Presumptive blood tests are routinely used in the laboratory to screen evidence items for the possible presence of blood. The tests may also sometimes be used in the field at crime scenes. They are very sensitive and can give positive results with small amounts of blood that are invisible to the naked eye. This is particularly useful if any blood at a scene has been cleaned up or diluted before the scene is actually processed.

These tests can be done directly—by applying the chemicals directly to the specimen—or by a transfer technique. Direct application of the chemicals to a stain is not recommended for most cases, but can be especially useful to develop patterns in blood. In the latter method, which is always recommended over direct testing, a sterile cotton swab (like a Q-tip) is lightly moistened using sterile saline or water. The moist end is used to lightly swab the specimen one wants to test, in order to transfer a small quantity of the suspected substance to the swab. The test chemicals are then added, in sequence, to the swab, and any color changes observed. Prior to testing the sample, chemicals should be added to a second swab (blank) as a *blank control*. Any color changes that occur on the blank swab when

Sperm cells can be thought of as DNA delivery vehicles. Their function is to find a receptive female egg, attach to its cell membrane, and deliver their DNA to form a fertilized egg or zygote. Mature, fertile males may have from about 15–80 million sperm cells per mL of semen.

As with bloodstains above, there are presumptive and confirmatory tests for semen. Positive presumptive tests show that semen may be present and that additional, confirmatory testing is indicated. In addition, light of certain wavelengths may cause semen stains to fluoresce, and these light sources can be used as searching aids.

People have known for a long time that semen stains often fluoresce brightly when exposed to ultraviolet (UV) illumination. Fluorescence is the emission of light caused by the excitation of the molecules in a material by higher energy light. Accordingly, a UV light, sometimes called a Woods lamp, can be used in examining evidence to help locate stains. Semen stains may also fluoresce under illumination with certain lasers and alternate light sources of the kind that are commonly used in latent fingerprint work (Chapter 6). Typically, light at a wavelength of 450–475 (blue) viewed through an orange filter is used to locate body fluids (Figure 10.11). Most body fluids are readily seen as fluorescent stains using these parameters. Using these lights to help find stains is merely a finding aid since the body fluids cannot be distinguished with this procedure. Just because a stain fluoresces under one of these light sources doesn't necessarily mean it's a seminal stain, just that it should be further tested.

Historically, there have been a number of different preliminary tests for semen used in forensic science laboratories. Most of them are no longer used. The only currently important one is called the "acid phosphatase" test. Essentially it is a color test, and can be done using a cotton swab the same way that was described above for presumptive blood testing, except, of course, the chemicals are different. Acid phosphatase is an enzyme manufactured in the male prostate gland, and semen usually contains a lot of it. Acid phosphatase is quite robust in dried stains, and the test is generally positive with actual semen stains unless there has been bacterial degradation of the stain or it has been damaged by exposure to heat or some other environmental extreme. The acid phosphatase test is sometimes also called the "ACP" or "AP" test.

Enzymes, as noted earlier, are nature's catalysts—a *catalyst* can speed up a chemical reaction without itself entering into that reaction or being changed by it. Further, the catalyst is only required in small quantities compared with the compounds actually reacting. Many of the reactions taking place in living cells and tissues that are necessary to maintain life would be too slow if there were nothing to speed them up. The catalysts for all these reactions are proteins, and they are called *enzymes*. Enzymes are generally very specific for the particular reaction they catalyze. As a result, the reaction can be set up in the laboratory to serve as a "test" for the enzyme in a biological specimen. Enzymes can be useful in forensic science as "identification markers." Acid phosphatase is an identification marker for semen. Because other cells and biological materials in the natural world, besides semen, contain acid phosphatase, the acid phosphatase test is a presumptive, not a confirmatory test.

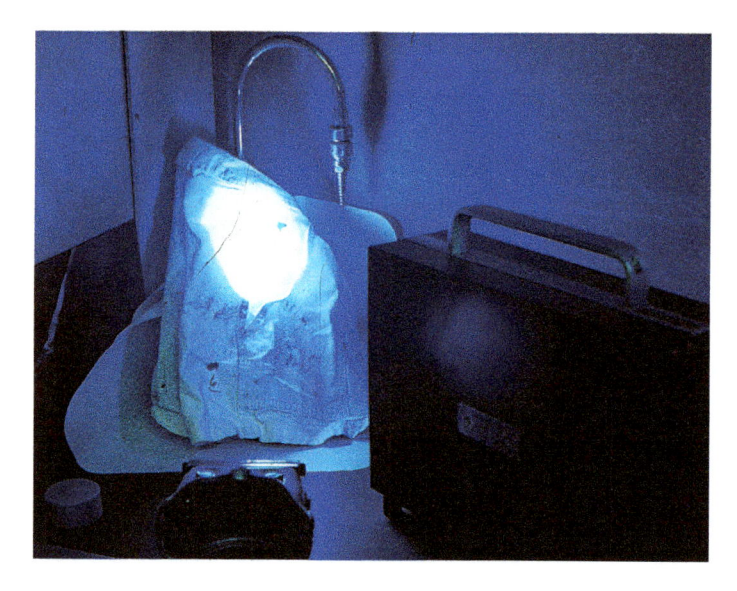

Figure 10.11 Use of an intense alternate light source to help make semen stains more visible.

The oldest *confirmatory* test for semen in dried stains, and still the best test, is finding spermatozoa, the male reproductive cell, in a smear extracted from the dried stain and viewed through a microscope. Forensic scientists have known to look for sperm in suspected semen stains as a means of identification since the 1840s, and the procedure is still common today. If sperm are seen, there is no doubt that the specimen had to come from a semen stain. Histological dyes (compounds that are used to color tissue preparations on microscope slides, so the components will contrast with one another and be visible) are often used to "color" the cells during preparation of the smear. In a vaginal swab preparation that contains sperm, a large number of vaginal epithelial cells (cells shed from the tissues lining the inside of the vagina) are also visible along with the sperm cells. For many years, finding sperm was the only confirmatory test for semen in a stain. If sperm could not be found, an analyst could not say for sure that the stain was semen. There are several reasons why sperm might not be found in an actual semen stain. A man might be pathologically azoospermic (a medical condition that prevents normal sperm manufacture), or he might have had a vasectomy—a surgical contraceptive procedure that prevents sperm from ever getting into the semen. In addition, sperm may not always be easy to find in some dried stains for various reasons. Identification of the seminal proteins called "p30" by most forensic scientists ("PSA" for "prostate specific antigen" or "PA" for "prostatic antigen" by most medical people) and semenogelin have solved this problem.

Identifying semen in forensic specimens where there are no sperm cells is an old problem. The protein called "p30," "prostatic specific antigen," or "prostatic antigen," appears to be *almost* unique to human semen. In recent years, this same protein has taken on a role as a "marker" in the screening of middle-aged men for prostatic cancer. The p30 is made in the prostate, and when there is prostatic cancer present, it spills out into the bloodstream. Doctors usually call the protein "PSA," so testing for it in blood as a means of screening men for prostatic carcinomas is usually called a "PSA test." The lab techniques used by forensic scientists to identify the protein for semen identification, and by clinicians for cancer screening, are essentially identical. P30 is identified using immunological methods—similar to those described earlier for species testing. Specific antibodies for p30 are available commercially for use in the various test protocols. The formats in use by forensic laboratories are the same as when testing for human blood: double immunodiffusion, immunoelectrophoresis, ELISA, and the commercial immunochromatographic test cards. As with the test for human blood, laboratories have all but abandoned other procedures for the (simpler) immunochromatographic test strip (Figure 10.12).

The P30 test card method is based on a test in a kit that has antibodies specific for p30 immobilized on a membrane strip. There is a color change in a positive test. A control portion on the strip indicates that the test is working properly if no color change is noted from the sample. Researchers have found that p30 is also present in other body fluids such as urine, fecal matter, sweat, and mother's milk, but in much lower levels than in semen. Because p30 has been found in these body fluids in detectable levels, most laboratories now consider the detection of p30 as only a "strong indicator" that semen is present. A test for the presence of semenogelin, seminal vesicle-specific antigen, is now available in a test strip from Independent Forensics as a rapid semen identification test. Semenogelin appears to be specific to semen and shows no cross-reactivity with other body fluids or with semen from other animals. Since there is no cross-reactivity, a reaction with other body fluids, a positive reaction is considered confirmation of the presence of semen.

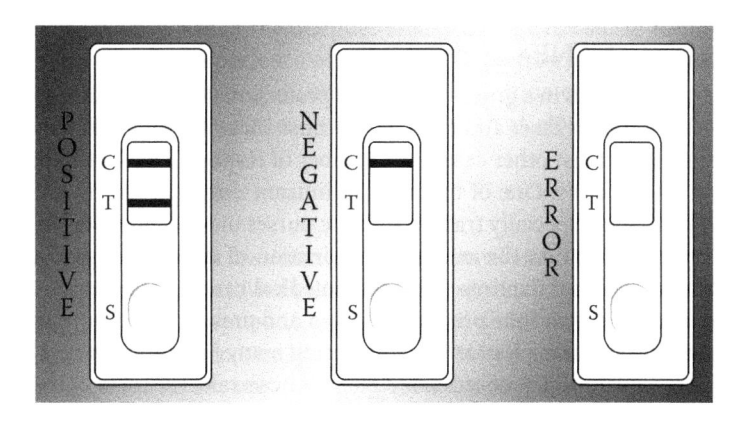

Figure 10.12 The immunochromatographic test strips can show a positive, negative or error where it did not function correctly.

Table 10.1 Sexual assault evidence collection

Example of evidence kit item or action

Item	Reason for including in kit/collection of evidence item
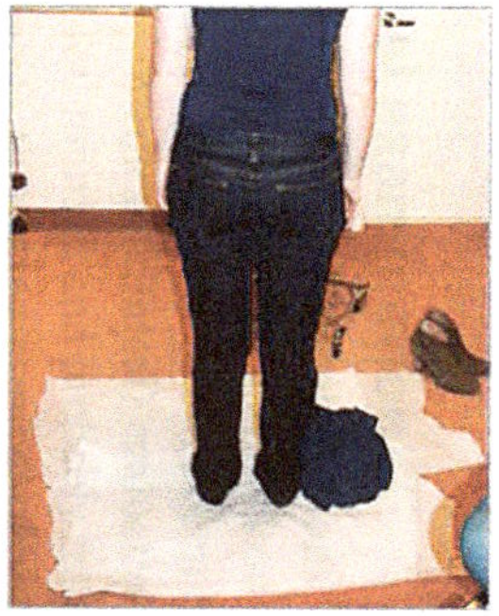 Large (approx. 4 ft. square) piece of clean white paper	Victim may stand on paper to disrobe, allowing any trace materials to fall onto the paper. The paper can be re-folded and packaged for possible later examination of contents
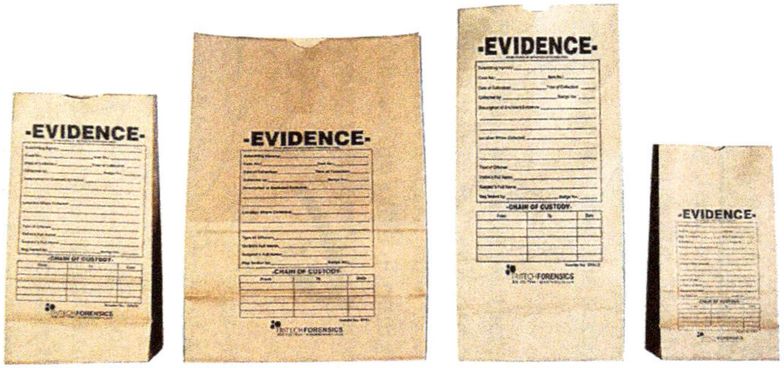 Various paper bags	For clothing items, which should be air dried, then packaged separately. Use additional bags or clean paper to wrap items.
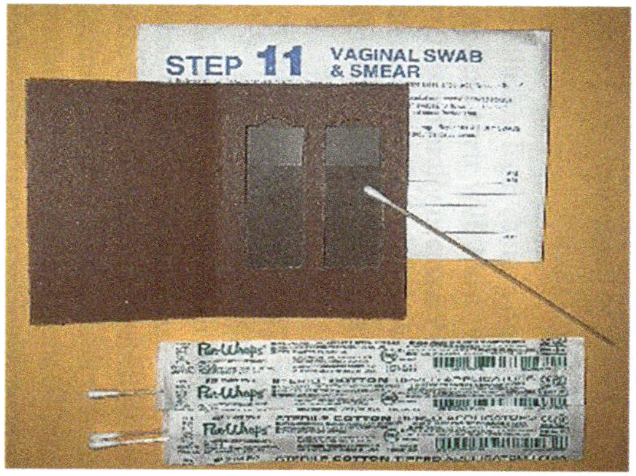 Vaginal Oral Anal Swabs & Smears	Swab up any body fluid residues present in the vaginal vault, or the oral, or anal cavities. Smears prepared from the swabs. Used by lab examiners or SANEs (often after histological staining) to look for spermatozoa, confirming the presence of semen

(Continued)

Blood and physiological fluid evidence 245

Table 10.1 (*Continued*) Sexual assault evidence collection

Example of evidence kit item or action	Item	Reason for including in kit/collection of evidence item
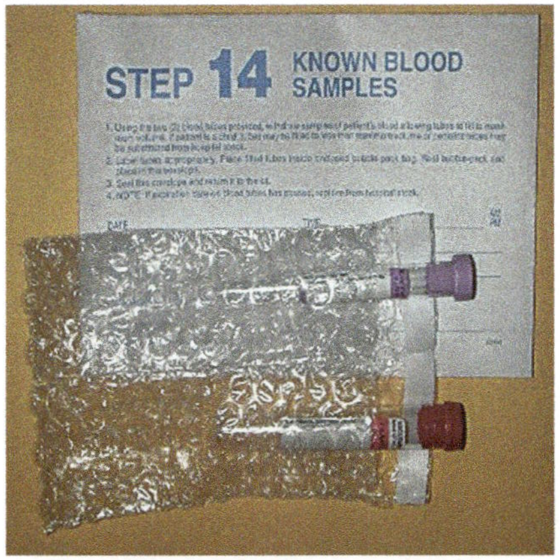	Vacutainer blood tube or buccal swab	Victim "standard" or exemplar (known) for DNA typing
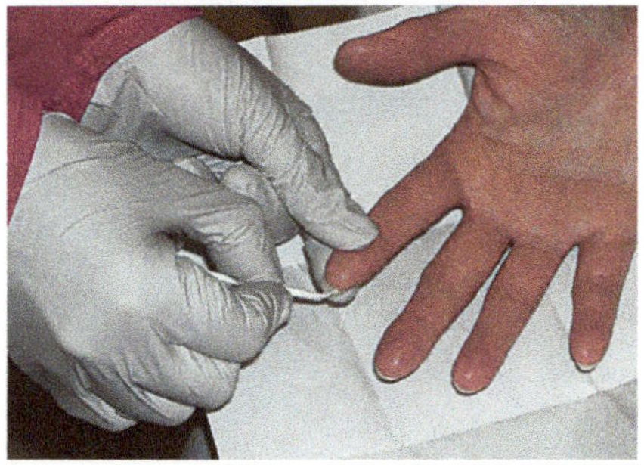	Wooden stick or fingernail clippers	Fingernail undersides may be scraped; or fingernails may be cut and saved intact over clean paper. Fold and place in outer envelope.
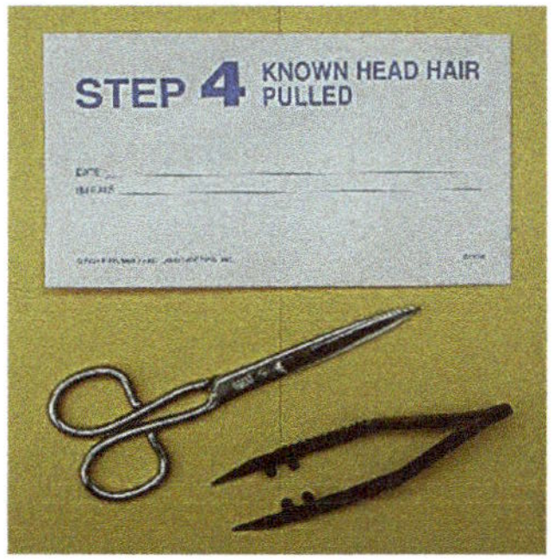	Container for head hair standards (knowns)	Head hair collected for possible later comparison, on the chance that some victim head hair transferred to suspect or his clothing or belongings

(*Continued*)

Table 10.1 (*Continued*) Sexual assault evidence collection

Example of evidence kit item or action	Item	Reason for including in kit/collection of evidence item
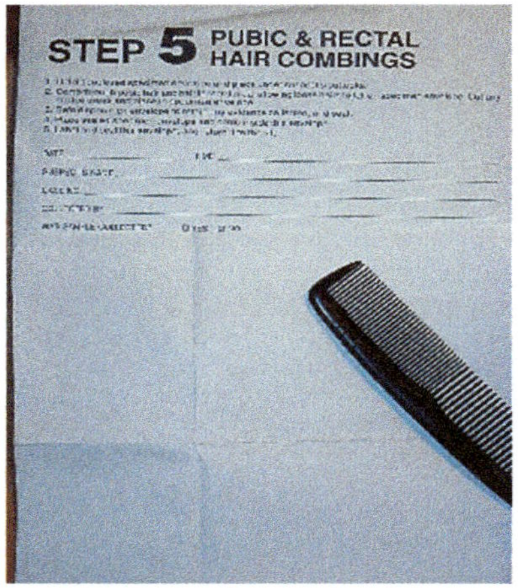	Container for pubic hair combings (and sometimes comb)	Pubic hair combings may contain perpetrator's transferred pubic hair or trace materials that corroborate aspects of the assault history
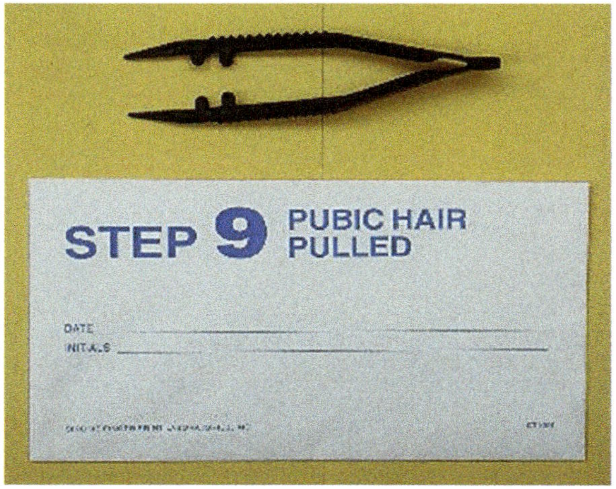	Container for pubic hair standards (knowns)	Known pubic hairs from victim are required (for elimination) to properly examine the public hair combings

Evidence from a complainant's person is generally collected and packaged by means of a so-called sexual assault evidence-collection kit (sometimes called a "rape kit"). These are essentially collections of containers and instructions for what to collect and how to collect it (see Figure 10.14). Both the kits themselves and the instructions have become more complex and sophisticated in recent years as more effort has been put into coordinating the response to sexual assault investigations and giving more careful consideration to what is to be collected. The National Protocol for the Medical Examination of Victims of Sexual Assault issued by the DOJ Office of Violence Against Women has outlined the recommended medical protocols and evidence collection that reflect "state of the art" forensic practices. Kits represent a consensus view of what evidence should be collected, and how it should be packaged. Many jurisdictions (which in this context can often mean a state) have their own sexual assault evidence collection kits. State law may specify the composition of a coordinating group for sexual assault complaint response, and among its responsibilities can be the design and periodic re-design of the sexual assault evidence-collection kit. State law may also require hospitals to take and keep sexual assault evidence until law enforcement picks it up or for a proscribed period. It may be noted that, although we say "she" and "her" often in this chapter in referring to sexual assault complainants, there are also male victims. Some places have kits designed to be useable for either females or males. Other places have separate kits for males. The male kits have often been designed more for use with suspects in sexual assault cases than for use with the adult male complainant.

One issue that has come to light in recent years is the problem of "backlogged" sexual assault evidence kits. A case is backlogged at the laboratory when the evidence is not analyzed in a timely manner. The actual number of kits that remained untested in 2015 in the United States is not known, but estimates range from 50,000 to hundreds of thousands of kits. When this evidence sits unexamined, a perpetrator may be unidentified and still in the community, or there may be an innocent person incarcerated. Both of these outcomes are devastating to criminal justice. Local jurisdictions are now making a concerted effort with the help of the Department of Justice to analyze all of the backlogged kits by identifying any semen or other biological evidence samples, and conducting DNA analysis on those samples.

Types of sexual assault cases and their investigation

From an investigative and forensic-science point of view, there are three (primary) types of sexual assault cases. The first two involve adult victims and consist of those in which the identity of the offender or suspect is unknown (identification cases); and those in which the identity of the offender or suspect is known, but where he claims the sexual relations were consensual (consent cases). The third type of case involves children.

DNA profiling is expected to be informative and helpful in identification cases, but not in consent cases. If the suspect does not deny sexual contact with the complainant, DNA profiling adds nothing to the case. These cases often succeed or fall on the credibility of the parties, since there are rarely other witnesses or other items of physical evidence that can settle the matter. Injuries to a complainant that have been properly documented at the time of the complaint may be helpful in some consent cases.

Cases involving children, especially younger children, are often investigated by a specialized child protective services agency and specially trained medical personnel. Only occasionally do these cases have physical evidence of the kind normally collected and submitted to forensic science labs, so forensic scientists are rarely involved.

It may be noted that the distinction between "children" and "adults" is a legal one, in the sense that state law defines the age at which a person may consent to sexual relations. A male who has sexual relations with anyone younger than the age of consent, as defined by state law, is guilty of sexual assault under the law—this is often called "statutory rape." These situations can come up with teenagers. Many teenagers are sexually mature before they reach the age of consent. In those circumstances, sexual assault evidence kits may be collected just as they would be for an adult complainant.

It should be noted too that the absence of physical evidence does not prove that an assault did not occur. Absence of semen on a vaginal swab could be the result of condom use, failure of a perpetrator to ejaculate, or too much time having elapsed between sexual contact and evidence collection.

Finally, it is worth noting that sexual assault complainants must provide a series of consents for different procedures. They must first agree to be transported to an emergency department or SANE–SART facility. Next, they must give consent for medical treatment. Separate consent is often required to take specimens from their person (that is, to use the sexual assault evidence-collection kit) and forward those specimens to a forensic science laboratory. Failure to obtain written consent at any stage of the process effectively ends it. Parents or guardians must give the consent if the victim is younger than the age of consent in that state. And, even if a victim is treated, and a kit taken and sent to the laboratory, she may withdraw her complaint or decide not to further press the case at any time.

Drug-facilitated sexual assault—"Date-Rape" drugs

During the 1990s, a new twist on the sexual assault problem began to gain attention. Cases started surfacing where a victim had been drugged—often surreptitiously. Several drugs are commonly encountered in this context: Rohypnol® (flunitrazepam), sometimes other benzodiazepines (such as clonazepam), gamma-hydroxybutyric acid (GHB), and ketamine (Figure 10.15). Generally speaking, drugs of the benzodiazepine class are depressants, and different ones are approved for the management of a variety of conditions, from anxiety to sleeplessness. Rohypnol is perhaps the most potent of the class. Although it is available in most of the world, its use and importation are banned in the U.S. because it was never approved for sale by the FDA and it has become associated with drug-facilitated sexual assault. GHB is approved for very limited use as a human drug. Ketamine is a human and veterinary anesthetic. All these drugs and others have been implicated in sexual assaults They have in common that they are all depressants, tending to make a person less in control, and they have amnestic effects (particularly Ketamine); that is, the person does not to remember much about what happened while under the drug's influence. The drugs are frequently used along with alcohol, which tends to make their effects more pronounced. One of the drugs may be added to an alcoholic drink as way of getting it into the intended victim. Alcohol itself has long been associated with sexual assaults, in that drinking alcohol is one of the biggest risk factors for sexual assault among teenaged and young adult women. But the association of the drugs mentioned above with sexual assault is relatively recent.

One generally thinks of drug-facilitated sexual assault (DFSA) in terms of a victim being surreptitiously drugged by a sexual predator perpetrator. And many cases do happen that way. However, there is also evidence that these drugs are sometimes abused for recreational purposes. Women may thus be sexually assaulted when they are under the influence of a drug they ingested knowingly. This is still drug-facilitated sexual assault.

Drug analysis and forensic toxicology will be discussed in Chapter 13. It is important to note here that the drug-facilitated sexual assault problem has required forensic toxicologists to become involved in examining evidence from sexual assault complainants. Now generally sexual evidence-collection kits include urine samples. While many hospitals test specimens for drugs and poisons for medical treatment purposes, many of the hospital protocols will not detect the drugs commonly used in these cases. This is why specific procedures for the collection

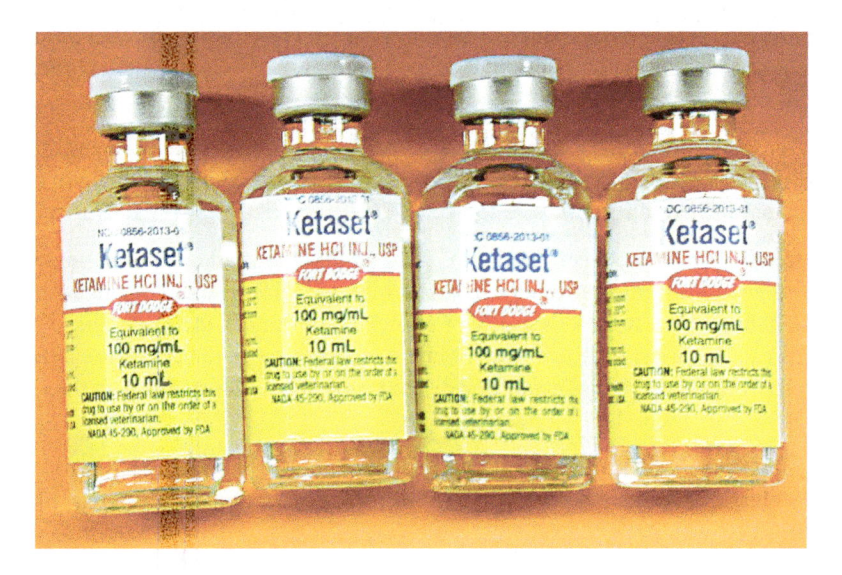

Figure 10.15 Vials of Ketamine a veterinary anesthetic often misused as a date rape drug. Photo credit: user *Psychonaught*; Retrieved from Wikipedia and used per Creative Commons0 1.0 Universal https://creativecommons.org/publicdomain/zero/1.0/legalcode).

of biological samples and submission of these to a forensic toxicology laboratory is so important in cases of suspected drug facilitated sexual assault. Special consent that clearly outlines what this testing is for and may detect is also required in most jurisdictions to obtain samples for DFSA agents. It should finally be noted that the full extent of this problem is not yet clear. For various reasons, it is difficult to gather good epidemiological data on the problem.

BLOOD AND BODY FLUID INDIVIDUALITY—TRADITIONAL SEROLOGICAL (PRE-DNA) APPROACHES

It is the genetic (inherited) characteristics that can be found in cells in both blood and other physiological fluids that are used to try to determine origin; that is, from which person did the blood or body fluid come. Before DNA typing, blood groups (blood types), red cell isoenzymes, certain plasma (serum) proteins, and hemoglobin variants could be used to partially individualize biological evidence. Since the late 1980s, DNA typing has become virtually the exclusive method for the individualization of biological evidence. DNA typing is the subject of the next chapter.

In recent years, a number of individuals have been exonerated after DNA testing was conducted on biological evidence associated with the crime committed. To provide some understanding of the limitations and value of serological testing that was conducted before the DNA era, a brief discussion of the pre-DNA genetic typing characteristics and methods concludes this chapter.

The classical or conventional (pre-DNA) genetic markers

In the 1990s, DNA typing replaced all genetic analysis that had previously been used for biological evidence. But for most of the twentieth century, forensic scientists used genetic markers other than DNA to try to associate or disassociate blood and body fluid evidence (from a particular individual). Today, those pre-DNA systems are often referred to as "classical" or "conventional" genetic markers.

There are five categories of conventional genetic markers: (1) blood groups; (2) isoenzymes; (3) plasma (serum) proteins; (4) hemoglobin variants; and (5) the HLA system. Different systems in the first four categories were used in the forensic analysis of biological analysis. The HLA system consists of different "types" on white blood cells or tissues. HLA typing is an important part of tissue matching before a tissue or organ transplant in clinical medicine. HLA typing was also widely used for parentage testing until DNA typing replaced it.

The first blood group, the ABO system, was discovered in 1901 by Karl Landsteiner (1868–1943). Because ABO typing is so important in the field of blood transfusion, Dr. Landsteiner received the Nobel Prize for Physiology or Medicine in 1930. ABO blood typing was first applied to bloodstains in criminal cases by Dr. Leone Lattes (1887–1954) of Turin, Italy, in 1913. From that time until about 1950, many more blood groups were discovered, and some of them were used in the typing of bloodstains in forensic cases.

The isoenzymes and plasma proteins were discovered on the 1940s and 1950s. Isoenzymes are enzymes that occur in multiple molecular forms. That is, there are several similar but distinguishable protein molecules that all perform the same enzymatic activity. The differences between the forms is usually very slight. These different forms of the enzyme exist because of genetic variation. The plasma proteins are different for the same reasons as the isoenzymes—the DNA segment responsible for making them has a few different forms. And similarly, there are a couple of common variants of the protein hemoglobin, which carries oxygen to all the cells in the body. One of the most widely recognized hemoglobin variants is "sickle." Commonly but not exclusively found in Africans and African-Americans, this gene can cause mild anemia problems in people who have one of the genes, but it causes serious illness in persons both of whose hemoglobin beta-chain genes are "sickle." Beginning in the late 1960s, forensic scientists at the then Metropolitan Police Forensic Science Laboratory began applying isoenzyme, plasma protein, and hemoglobin variant typing to bloodstains in criminal cases. Use of these genetic-marker systems spread quickly to forensic labs in other countries, including the U.S.

How does typing genetic markers help "Individualize" a biological specimen?

The whole idea behind genetic typing of biological evidence is to be able to tell who the specimen came from. Before DNA typing, a bloodstain or semen stain could never be attributed to only one person just because all the types between the stain and a person matched. The most we could say is that the person was part of a fraction of the population of persons, any of whom could have been the depositor. Of course if the types did not match, we could say absolutely that the person was *not* the depositor. DNA typing works essentially the same way, except that in "match" cases (where the evidence and the person have the same types or profile) the chance that someone other than the person deposited the evidentiary stain is usually very low. Thus, it should be noted that if traditional genetic testing *excluded* an individual, subsequent DNA typing will not include that suspect. This is a question that often arises during cold case evaluations if investigators are not familiar with traditional genetic testing.

To understand how this whole process works, a little background in **population genetics** is necessary. More is said about genetics and population genetics in Chapter 11. An important concept in genetics is that of a *gene*. A gene can be thought of in one sense as a region, or segment, of the DNA sequence that tells the cell how to make a particular protein or enzyme. Thus, we can talk about the "gene for hemoglobin" or the "gene for blood type B." The location on the DNA molecule (the location on the chromosome) where the sequence for a particular trait or characteristic is determined is called a *gene locus*, or simply a *locus*. "Locus" is Latin for "place." At any particular *locus* in one person's DNA, there are two genes, one on the maternal chromosome and the other on the paternal chromosome. The genes making up this pair of corresponding genes at a given locus are called *alleles*. Any person could have the alleles at a locus be the same (homozygous), or the two alleles could be different (heterozygous). At many different human genetic loci (*loci* is the plural of *locus*), there are quite a few different genes (alleles) that a person can have. Population genetics looks at how often alleles at some locus occur in a population. For example, a population geneticist might ask: "How many blood type B people are there in the population of the U.S.?" An answer to that question tells the population geneticist how frequently the gene for blood type B occurs in the population. The way one finds out the answer to a question like this is by blood typing a lot of people. As a practical matter, it is impossible to type everyone. So, a *sample* of the population is typed, usually at least several hundred people should be typed. The frequency of blood type B in the sample provides an *estimate* of the frequency in the entire population. *Statistics* provide mathematical rules for figuring out how large the samples have to be in relation to the level of uncertainty in the resulting measurements. Use of genetic marker frequencies in populations to "individualize" evidence is illustrated in the box nearby.

SIDEBAR Population genetics/individualization of evidence

This example illustrates how was used population genetics to estimate the frequency of two genetic traits, blood type B and isoenzyme PGM type 2 + 1+, helps "individualize" a specimen from a person with these two types. Two important concepts in this example are the *independence* of the genetic loci, permitting the use of the *product rule*. These principles are still important in calculating frequencies after DNA profiles are developed.

Let's say we find out that the frequency of blood type B in the U.S. White population is about 10%. (It turns out that the frequencies of the alleles or types are not always the same in populations of different racial/ethnic groups, so forensic scientists often separate the frequency estimates for the different groups). Knowing that, we can say that a bloodstain that is type B can only have come from someone in about 10% of the population. We have excluded 90% of the population as possible depositors. Still, 10% of the population is a lot of people. Taking this line of thinking a step further, let's say we find out that the frequency of the isoenzyme PGM type 2 + 1+ is about 20% in the same population for which the frequency of blood type B was 10%. Let's also say that we can establish that the inheritance of blood type B and the inheritance of PGM are *independent*; that is, the inheritance of one of them does not influence the inheritance of the other one. Now, if we knew that a bloodstain was blood type B *and* PGM 2 + 1+, the combination is expected to be found in only 2% of the population (10% times 20%). That is, if the loci are independent, the frequencies can be multiplied to get an estimate of the frequency of people who have both types—the product rule. So in this illustration, 98% of the population has been excluded as possible depositors. If a person in the case has the same types as were found in the stain, we can say that he or she is *included* as a possible depositor, along with 2% of the population. The 2010 U.S. census of population reported that there were 308,745,538 people in the U.S. Of those, 12,830,632 lived in Illinois. Within Illinois there were estimated to be 9,109,749 White people. Based on our genetic marker population studies, we could then say that 182,194 (2% of the total) of them are expected to be *both* blood type B and PGM 2 + 1+. If a person involved in the case is different in blood type or in PGM type from what is found in the bloodstain, he/she is *excluded* as a possible depositor. If we typed more genetic characteristics in the same specimen and they were all independent, the number of people who would be expected to have that same profile of types keeps getting smaller. We will develop this same reasoning for DNA types in the next chapter.

CASE ILLUSTRATIONS—THE POWER OF DNA VERSUS CONVENTIONAL GENETIC SYSTEMS

One of the remarkable, although fully expected, consequences of the widespread availability of DNA typing has been the exclusion of a number of people previously convicted of serious crimes, especially but not exclusively sexual assaults. Sometimes, conventional genetic testing was done but did not exclude the person at the time of the crime and the trial.

From the discussion in this chapter about the individualizing power of the conventional systems, it is clear that they could not even come close to pinning down a particular person as a depositor of blood or semen. This does not mean that the typing conducted was in error. It was state-of-the-art analysis before DNA testing. While some suspects were excluded using conventional systems, others were included because of the test limitations. In sexual assault cases, the vaginal swabs taken from complainants following the incident contained mixtures of blood group factors and isoenzymes from both the woman and the semen depositor. Analysts had to try to interpret the typing results in terms of the people involved in the case. We will shortly see in Chapter 11 that it is often possible to separate the male and female fractions of DNA in these semen-vaginal mixtures, and this ability is obviously a great help in interpreting the typing results.

In one illustrative case that happened before DNA typing was available, both the complainant and the suspect were type B. They were also both "secretors." We did not discuss the "secretor" characteristic in the main text, but suffice it to say, that in secretors the blood types could be detected in other body fluids. The evidence in the case, such as the vaginal swab on which semen was identified, showed "type B secretor" characteristics. From these results, an analyst could only say that the semen on the swab could have been deposited by a male who was a type B secretor, a type O secretor, or a nonsecretor of any ABO type. That is, approximately two-thirds of white males and three-quarters of black males could have been depositors. So, even though the suspect was included in the group of possible depositors, the group was very large. The suspect was convicted at trial largely on the strength of the testimony of the complainant in the case. Sometime later, however, DNA typing showed that the semen on the swab could not have come from this particular man.

In another case, a man who was a type O secretor was convicted of sexually assaulting two different women at different times. He could not be excluded as the depositor in either case based on ABO types nor on the PGM isoenzyme types. Several years later, however, the evidence was analyzed using DNA typing, and the man was excluded as the depositor of the semen in either victim.

There are many cases like these in different jurisdictions. The conventional genetic systems just did not have the individualizing power to exclude non-depositors, whereas DNA typing nearly always does have it.

Key terms

acid phosphatase

alibi known control

anticoagulant

antigens

blank control

blood cells

catalyst

confirmatory test

contamination

crystal tests

drug-facilitated sexual assault

enzyme

forensic biology

Chapter 11

DNA analysis and typing

Richard Li

Lead Case: Security Guard Guilty of Homicide

A week after Dr. Srb, a sociology professor at a community college in Connecticut, died of a heart attack, his widow went by herself to clean out her husband's office. It was a quiet Sunday afternoon. A campus security guard was on patrol, and suddenly gunshots were heard. The guard found a woman's body in the second-floor woman's restroom of the Social Sciences building. She was later identified as the widow of the professor. Police were alerted by a 911 call. Detectives responded to the scene and found Mrs. Srb's partially clad body on the floor of the restroom next to a sink. Her upper body was fully clothed including a jacket and scarf. Her lower body was exposed, and her skirt and panties were partially pulled down. There were no witnesses and no immediate leads.

When the state police detectives and the forensic laboratory who were contacted to assist in the investigation arrived at the campus crime scene, they carefully studied and documented the scene, and a detailed search was conducted. The following physical evidence was found:

1. A "forensic" light source was used to search the floor, and three spots of apparent fresh semen stains were found.
2. Three bullet wounds were located in the upper back and two bullet wounds in the front of the upper body. Autopsy results indicated the victim was shot five times—three times anteriorly and twice posteriorly.
3. Only one spent shell casing was found under the victim's scarf. The suspect must have picked up the other spent shell cases but overlooked the one beneath the scarf.
4. Apparent seminal stains were found on her scarf. These stains were fresh and moist.
5. Hairs were found on her stockings. They were consistent with human Caucasian pubic hairs.
6. Two bullet holes were found on the divider wall of the restroom. Two spent bullets were recovered from the wall.
7. Paper towels were found in the wastebasket in the restroom. These paper towels were wet and appeared to have been recently used.

Detectives interviewed the security guard after they arrived on the scene. The guard provided a statement and indicated that he was on patrol outside the Social Sciences building when he heard several gunshots. He said he rushed into the building, looked on every floor, and found the body in the second-floor women's restroom. He immediately called the police, and he also informed the investigator that he did not carry any weapon. Police searched the building and the guard's office and found no guns.

A search warrant was obtained to search the guard's body, clothing, and vehicle, and to collect known blood and hair samples. The following results were obtained:

1. A gunshot residue test kit was collected from his hands. The test results were negative.
2. Seminal stains were found on the front, lower left side of his uniform. These semen stains appeared to be fresh.

3. 9 mm fired cartridge casings were found inside of his right boot. The casings were the same brand as the spent casing found under victim's scarf.
4. Head and pubic hair samples were collected from the guard. Two pubic hairs found on the victim's stocking had similar microscopic characteristics to the known pubic hair sample from the suspect.
5. A known blood sample was also collected for DNA tests. DNA extracted from the semen stains has the same profile as the known DNA from the suspect. The profiles had extremely low probabilities of a match by chance.
6. The guard's car had heavy seat covers. After the seat covers were removed, a 9 mm automatic pistol was found under the right front seat.
7. A firearms examiner test fired the weapon and found that the bullets recovered from Mrs. Srb's body and from the restroom wall were fired from the same gun.
8. Bullet trajectories were reconstructed. These trajectories show that the victim was first shot twice in the front while she was in a standing position. Three additional shots were then fired at close range.

The security guard was arrested and charged with first degree murder. He subsequently pled guilty and is currently serving a life sentence without the possibility of parole.

> **LEARNING OBJECTIVES**
> - What DNA is, including its structure and its functions
> - How genetic marker typing helps individualize biological evidence—some concepts of population genetics
> - Where DNA is found in the body—nuclear and mitochondrial DNA (mtDNA)
> - How DNA technologies developed; RFLP, dot-blots, and STRs
> - What the polymerase chain reaction (PCR) is and its importance in biological research and forensic DNA analysis
> - Current DNA-typing methods: how they work and how DNA typing individualizes biological specimens
> - DNA databases—CODIS
> - The forensic applications of DNA typing: criminal, civil, human identification, parentage testing
> - Some of the newer DNA technologies: Y-chromosome and single nucleotide polymorphisms (SNPs)
> - The strengths and limitations of DNA technology, and how they relate to the media hype and the ultimate potential

Forensic DNA typing is based on the fact that DNA is the genetic material of all living organisms, including human beings. Through DNA, genetic information is passed from generation to generation, from parents to children. This process and its details form the basis of the science of genetics. The rules of inheritance—which are faithfully followed in all sexually reproducing, multicellular organisms—were worked out by an Augustinian monk named Gregor Mendel (1822–1884, Figure 11.1). He lived in a monastery in what was called Brünn, Austria (today called Brno, in the Czech Republic). Mendel figured out the basic rules of inheritance while working with pea plants in the monastery garden. He purposely interbred them and observed the specific characteristics of the plants in the next generation. He did publish his work, but in a relatively unknown journal. Around 1900, when scientists began actively researching genetics, they rediscovered Mendel's experiments. Today, the basic rules of inheritance are often called the Mendelian principles, as a tribute to Mendel.

Figure 11.1 Gregor Mendel is considered the father of modern genetics.

GENETICS, INHERITANCE, GENETIC MARKERS

Genetics is the science of inheritance—how parents pass their traits and characteristics to their offspring. Inheritance can be extremely simple—where a cell divides to form two identical daughter cells. It can also be more complicated—where human parents produce sperm and egg cells, each of which contains half the genetic material of the parent, and these cells recombine at fertilization to form a unique cell capable of developing into a new individual.

In higher animals, including humans, DNA, the genetic material, is organized into *chromosomes* that are found in the *nucleus* of most cells. Humans have 46 chromosomes. Forty-four of them are paired—that is, there are 22 pairs. The other two are called sex chromosomes: X and Y chromosomes. Women have two X chromosomes, and men have one X and one Y. Individuals inherit one member of each of the 22 pairs from their mothers and the other member from their fathers. They also inherit an X chromosome from their mothers. The fathers' sperm may provide either an X chromosome (producing a female) or a Y (producing a male).

When men make sperm or women make eggs, a modified cell division process takes place that randomly sorts the pairs of chromosomes so that each reproductive cell (egg or sperm) contains 23. At fertilization, the characteristic chromosome number, 46, is restored.

DNA regulates cell activity mostly by controlling how cells make proteins. Some of the proteins are structural, and others are enzymes. Often, the presence or absence of a particular enzyme, which in turn means that some chemical reaction occurs or does not occur, is the basis for a particular trait or characteristic. Before laboratory tools were available for the analysis or typing of DNA directly, geneticists and forensic scientists looked at the products (proteins, enzymes, and other characteristics like blood types) of genes encoded by DNA to do genetic analysis and to try to forensically "individualize" biological evidence. The proteins, enzymes, and so on that were used for these purposes were commonly called genetic markers—"genetic" because of their being inherited, and "markers" because they could serve as a means of narrowing down the number of people in a population who could have deposited the biological evidence. Genetic differences among people that enable them to be distinguished is called genetic polymorphism. It will be discussed in more detail below.

DNA—NATURE AND FUNCTIONS

DNA was first "discovered" in the nineteenth century, and by the 1940s it was clear that DNA is genetic material, the chemical blueprint of life itself. In 1953, James Watson and Francis Crick, in collaboration with Wilkins and Franklin, used X-ray crystallography to figure out the chemical structure of DNA; the double helix.

DNA is a very large molecule that consists of units called *nucleotides* (Figure 11.2a). The components of a nucleotide are an organic base, a five-carbon carbohydrate called ribose, and a phosphate moiety. In DNA, the base can be one of four compounds: guanine, cytosine, adenine, and thymine; commonly abbreviated G, C, A, and T. The complete DNA molecule consists of two strands. Each strand is a polymer of nucleotides, each of which has an attached base that can be A, T, C, or G. In the complete, double-stranded DNA, the bases are always paired in a specific way: A is always paired with T, and C is always paired with G. Figure 11.2b shows the double-stranded structure.

The base pairing rules establish that wherever there is an A in one strand, there will be a T in the other strand, and wherever there is a C in one strand, there will be a G in the other. The two strands are said to be complementary. This specific base pair bonding holds the two strands of the double helix together. The strands can be separated by heat or under certain chemical conditions. Because of the positioning of the terminal phosphate residues on the two strands, one strand is said to run 5' to 3' (5 prime to 3 prime—these designations refer to numbered positions in the ribose sugar structure), and the other strand runs 3' to 5'. Because of the base pairing rules, the sequence of one strand dictates the sequence of the complementary strand. Finally, the double-stranded DNA forms a stable helical structure, the now familiar double helix (see Figure 11.2b).

It is the *sequence* of bases that is the primary structure of DNA and that makes every individual's DNA unique, thus enabling it to be a chemical repository of information specific to each individual. The entire complement of DNA is referred to as the genome. Because intact DNA is double-stranded, each position is occupied by a base pair, either A-T, T-A, C-G, or G-C. For this reason, it is common to state the length of a double-stranded DNA molecule in

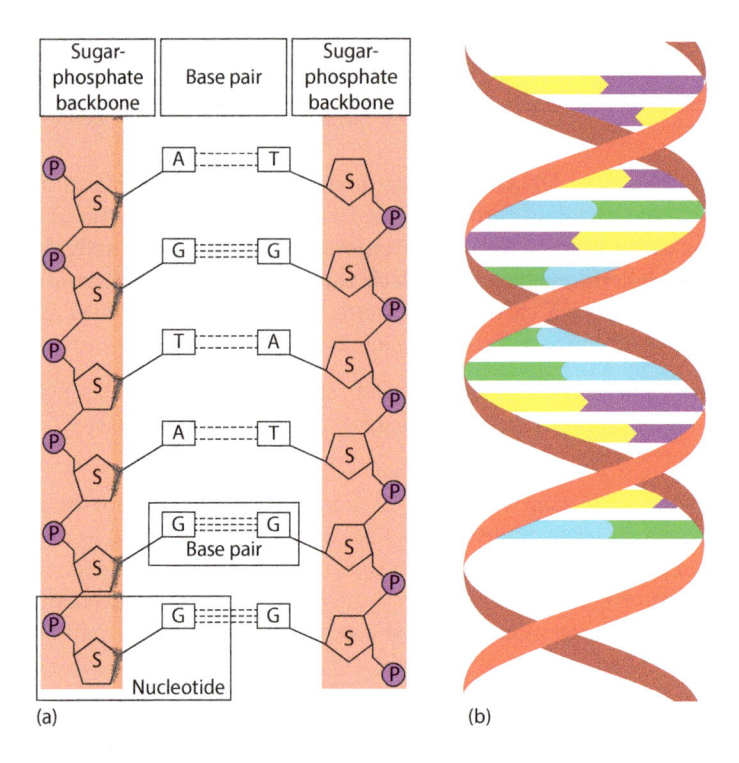

Figure 11.2 (a) Diagram of double-stranded DNA. One nucleotide is boxed. The sugar-phosphate backbone of the strands is apparent, along with the bases A, T, C and G. The strands sit next to one another, and the base pairings A-T and C-G are always followed. This structure twists to form the double helix. (b) Double helix structure of DNA. The two strands are held together by weak bonds between the paired bases, represented by the "ladder rungs" in the diagram. (Adapted from a drawing at Access Excellence @ the National Health Museum [www.accessexcellence.org/].)

base pairs (bp). Human DNA has a total of around 3.5 billion bp. The sequence is not just a random assortment of base pairs in linear order. The base sequences in the coding portions of DNA conform to what is called the genetic code. The majority of human (and other mammalian) DNA is noncoding. However, the function of this noncoding DNA is largely unknown. Quite a bit of the noncoding DNA has repetitive sequences. There are long and short repetitive sequences scattered throughout the genome. That is, the same sequence will be found in hundreds or thousands of different places throughout human DNA. A particular type of repetitive sequence is very important for forensic DNA typing: tandemly repeated sequences, or tandem repeats. The sequence that is repeated can be as short as two bases (e.g., ATATATATAT…) or as long as 100 bases. In a tandemly repeated sequence region, the repeated sequence occurs in a string of units, set head to tail. In any particular repeated region, there can be several to several hundred tandem repeats.

As mentioned, the two strands of DNA can be separated in the laboratory by heat or certain concentrations of salt solution. During the polymerase chain reaction, an important process to forensic DNA analysis (described later), strand separation must occur so that the *primers* can anneal to a specific sequence on the individual strands. A primer is a short single strand of DNA that is complementary (A for T and C for G) to a short sequence in the larger strand of DNA (see later). Separation of the two DNA strands is called "melting," or denaturation. However, DNA strands also have to separate in the cell to permit certain functions. During DNA synthesis, one strand is used as a template to make a new complementary strand. Furthermore, during transcription—the process of transcribing the DNA sequence into single stranded RNA—the strands must separate enough to permit synthesis of messenger ribonucleic acid (mRNA) from the coding strand. This mRNA is later used as a template to synthesize proteins.

You may have read or heard about the human genome project. For most of the 1990s, government and private laboratories worked to determine the complete sequence of human DNA. The project has been completed, and now genetic scientists have to try to figure out the roles and functions for the sequences. As this is accomplished, the possibilities of designing real cures for genetic disorders and of designing drugs for very specific therapeutic tasks become more feasible.

DNA has two principal functions. One function is that DNA can make exact copies of itself. This function is essential every time a cell divides, meaning that the daughter cells are to be identical copies of the parent cell. Thus, although humans consist of billions of cells (many of which are constantly being replaced), every one of them has the identical content of DNA. Every time a cell divides to produce genetically identical daughter cells (a process called *mitosis*), a complete, identical copy of DNA must be made so that each daughter cell can have a copy. Mitosis is the way single-celled organisms reproduce, and it is the way multicellular organisms grow, develop, and repair damaged cells.

DNA replication is catalyzed by enzymes called *DNA polymerases*. Quite a few different kinds of this enzyme are present in various cells and organisms throughout nature, but, in effect, they all do the same thing. A particular type of DNA polymerase is necessary for the polymerase chain reaction (see later). DNA polymerases catalyze the synthesis of DNA using one of the strands as a template. During the DNA synthesis, the polymerases fill in the appropriate nucleotides with the complementary bases of the second strand and bond them together into the complementary strand (Figure 11.3). DNA polymerase cannot attach to a single strand of DNA. There has to be at least a primer, which is a small segment of the complementary strand that is going to be extended by the synthesis, already in place. The polymerase enzyme attaches to the terminal nucleotide of the incomplete strand and catalyzes the bonding process of the complementary nucleotides one at a time, essentially sewing up the complementary strand, thus producing the double stranded DNA. The second principal function of DNA is more complicated. DNA, through its chemical structure, controls all cell functions because the DNA base sequence determines the chemical structure of all the proteins. Some years ago, geneticists figured out that DNA can specify protein structure because there is a correspondence between certain DNA base sequences and certain amino acids, which are the building blocks of proteins. This correspondence is called the *genetic code*. It is thus sometimes said that "DNA codes proteins."

Proteins are made up of units called amino acids that are chemically hooked together in a sequence. There are about 20 naturally occurring amino acids found in proteins. The DNA sequence must therefore be able to specify 20 different amino acids so that the sequence of bases in DNA will dictate the sequence of amino acids in proteins. Since there are only four bases, the correspondence can't be one to one. If a two-base sequence specified each amino acid, only 16 different ones could be accommodated. So, the fewest number of bases in a sequence that can "code for" each amino acid, and have enough sequences to be able to do all the amino acids, is three. Since there are 64 possible different three-base sequences, there is some degeneracy in the code sequences; more than one DNA triplet sequence

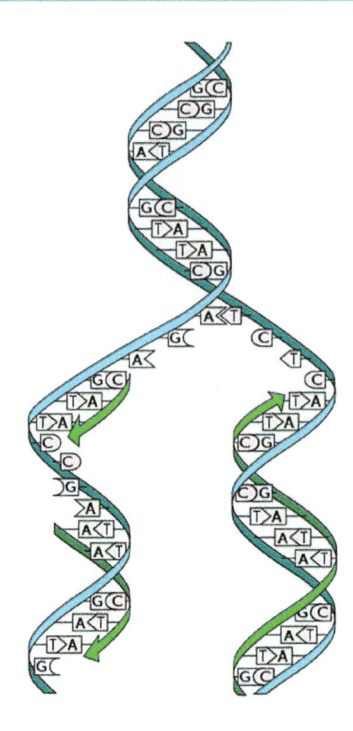

Figure 11.3 DNA replication. The double helix is "unzipped and unwound with one strand (turquoise) serving as a template for the synthesis of the complementary partner strand (green)." Nucleotides are matched to synthesize the new partner strands into two new double helices. (From Madeleine Price Ball used per Creativ Commons Attribution-ShareAlike 3.0 https://creativecommons.org/licenses/by-sa/3.0/legalcode.)

can code for the same amino acid. As mentioned, this set of triplet sequences that code for amino acids is the genetic code. Working out the "words" in the genetic code and how the coding takes place is one of the major milestones of twentieth-century molecular biology. DNA does not do the actual coding directly. An intermediary molecule, mRNA, is the actual template for protein synthesis. The process of making mRNA from DNA is called *transcription*. The mRNA is single stranded and is a replicate of one strand (the coding strand) of the DNA from which it was made. Again, the process of making protein from mRNA is called *translation*. This process is illustrated in Figure 11.4.

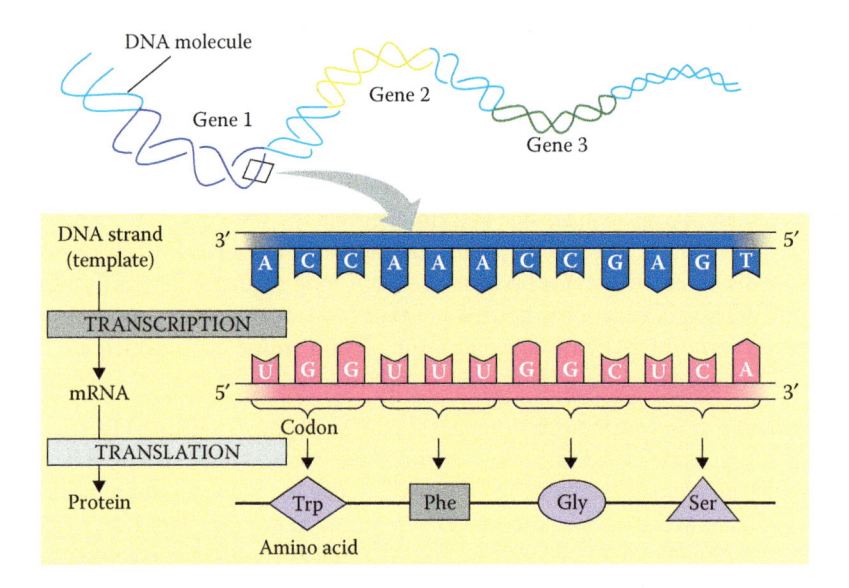

Figure 11.4 Diagram showing DNA-determined protein synthesis. Transcription of DNA to mRNA takes place in the nucleus while the translation of mRNA to protein takes place in the cytoplasm.

WHERE DNA IS FOUND IN THE BODY—NUCLEAR (GENOMIC) AND MITOCHONDRIAL DNA (mtDNA)

Every person starts life as a single cell. As that cell and its progeny divide and differentiate to form the millions of cells that make up different organs and tissues, DNA is faithfully replicated during each cell division. Thus, every cell winds up with a complete copy of DNA. This DNA is always found in the nucleus (a structure bounded by its own membrane and containing the nuclear DNA within the chromosomes) of the cell and, thus, often called *nuclear DNA*. It may also be called *genomic DNA*. Most of the time, when forensic scientists are talking about DNA it is nuclear DNA that they refer to. Later we will introduce and describe another kind of DNA in the cell (mitochondrial DNA), which can also be used in forensic cases.

There are two important exceptions to the general proposition that every cell in the body has a complete, identical copy of the person's DNA. The first is red blood cells (see Chapter 10).

Mature red cells have no nucleus and thus no nuclear DNA. The second exception is the germ cells: spermatozoa and ova (sperm and egg). These cells have undergone a type of cell division that results in their having 23 chromosomes instead of 46. They have one member of each of the 22 pairs and one additional of either an X or a Y.

Because most body cells have a copy of the set of chromosomal DNA, almost any tissue or organ can be used for DNA typing—at least in theory. Isolating DNA from some tissues (such as bones) is not as simple as from blood, tissues, or other bodily fluids. Thus, not every forensic laboratory routinely performs DNA typing from sources other than blood, semen, or saliva.

In addition to a nucleus, cells also have structures within them called *mitochondria*. At a simple level, mitochondria can be thought of as the cell's power plants. They have the components and enzymes necessary for cells to make the most of the energy they need. This energy is stored in a chemical form called adenosine triphosphate (ATP) until it is needed. There are hundreds to thousands of mitochondria in every cell. Mitochondria have a small quantity of their own DNA. The mitochondrial genome, as it is sometimes called, consists of about 16,000 base pairs, as opposed to around 3 billion base pairs for nuclear DNA, making it on the order of 100,000 times smaller in size than nuclear DNA.

Mitochondrial DNA (mtDNA) has two regions in its sequence, called HV1 and HV2 (HV stands for hypervariable region), which show quite a bit of variation between individuals in populations. This variation is not random throughout the sequence but tends to be characteristic of certain base pairs along its length (mtDNA has a universally agreed upon numbering system of its base sequence). Forensic scientists take advantage of this variation in some specific situations and with specific types of evidence. These are developed later in the chapter.

MtDNA has another feature that makes it different from nuclear DNA—its mode of inheritance. It was noted that every person inherits half of his or her nuclear DNA from the father and the other half from the mother. The maternal half is contributed from the nucleus of the egg cell, and the paternal half is contributed from the nucleus of the sperm cell at the time of fertilization. However, mtDNA is inherited *only* from one's mother, with no paternal contribution to it (Figure 11.5). Because the sequence of mtDNA is quite stable from generation to generation, a sequence can often

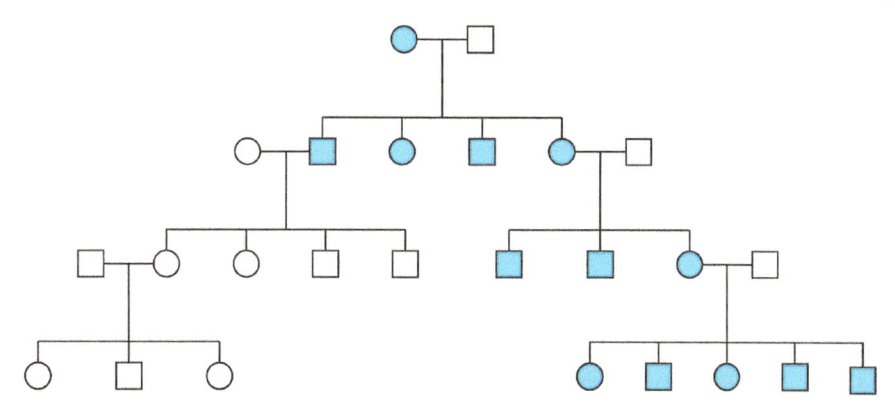

Figure 11.5 Human inheritance of mitochondrial DNA (mtDNA). Individuals who inherited the same mtDNA are indicated by blue symbols. Females and males are indicated by circles and squares, respectively.

be traced through the maternal lineage for many generations. This strictly maternal inheritance has implications for the application of mtDNA typing in human identification cases, to be developed later in the chapter.

Another feature of mtDNA that makes it a good candidate for human identification cases (Case Study 11.1), especially if the DNA must be recovered from old, decomposed, or skeletonized remains, is its much higher number of copies in each cell. Each cell has hundreds to thousands of mtDNA molecules and, thus, hundreds to thousands of times as much mtDNA as nuclear DNA. Because DNA can be destroyed by ambient conditions, the likelihood of isolating a usable mtDNA sequence is greatly increased by the fact that so many copies are present to begin with.

CASE STUDY 11.1 IDENTIFICATION OF UNKNOWN SOLDIER FROM THE VIETNAM WAR

Mitochondrial DNA (mtDNA) typing (sometimes called "mitotyping") is the method of choice among DNA-typing technologies for the identification of old, highly decomposed, and/or skeletal human remains. The mtDNA is more robust in these specimens, and the laboratory is more likely to obtain results. MtDNA typing is actually sequencing—the analyst determines the sequence of a small segment of the mtDNA. This segment tends to show variation at certain base positions among different individuals.

MtDNA is inherited exclusively from one's mother. Unless a mutation occurs, the sequence of the variable segment in mtDNA passes unchanged from mother to offspring, generation after generation. Mothers pass their mtDNA to their sons, but the sons cannot pass it on to anyone else. Daughters, on the other hand, pass it along to their offspring, and so forth. For this reason, the reference specimen that is needed to use mtDNA for human identification is one from the suspected person's mother or another person who would share her sequence (a sibling, maternal aunt, etc.). If the reference and questioned sequences match, a probability of a match by chance can be computed. With mtDNA, these probabilities are high in comparison with those obtained from 13-locus nuclear DNA comparisons. That is, a mitotype may occur in every couple thousand persons in a population. However, the identification of remains is not based on the mitotype alone. There is other circumstantial evidence. In the case of remains of military personnel from theaters of battle, the remains may be associated with a particular airplane or ship or location. There may be articles or fragments of clothing or belongings associated with the remains that suggest a certain person. The mtDNA typing is used to help confirm these suspected identifications. MtDNA typing of remains is of no value unless there is some indication of who the person might be. Those clues allow the laboratory to determine who in the suspected person's maternal line may be available to provide an appropriate reference specimen for comparison.

For nearly 80 years, the United States has maintained the Tomb of the Unknowns at Arlington National Cemetery in Arlington, Virginia. It is also sometimes known as the Tomb of the Unknown Soldier. On March 4, 1921, Congress approved the burial of an unidentified American soldier from World War I in the plaza of the new Memorial Amphitheater. The white marble sarcophagus has a flat-faced form and is relieved at the corners and along the sides by neoclassic pilasters, or columns, set into the surface. Sculpted into the east panel, which faces Washington, D.C., are three Greek figures representing Peace, Victory, and Valor. Inscribed on the back of the Tomb are the words:

> HERE RESTS IN
> HONORED GLORY
> AN AMERICAN
> SOLDIER
> KNOWN BUT TO GOD

The Tomb sarcophagus was placed above the grave of the Unknown Soldier of World War I. Just to the side are the crypts of unknowns from World War II, Korea, and Vietnam. Those three graves are marked with white marble slabs flush with the plaza. For each war, a service member whose identity could not be determined was selected to lie in these crypts.

The unknown soldier of the Vietnam conflict was designated by Medal of Honor recipient U.S. Marine Corps Sergeant Major Allan Jay Kellogg Jr. during a ceremony at Pearl Harbor, Hawaii, May 17, 1984. The Vietnam Unknown was transported aboard the *USS Brewton* to Alameda Naval Base, California. The remains were sent to Travis Air Force Base, California, on May 24. The Vietnam Unknown arrived at Andrews Air Force Base, Maryland the next day.

(Continued)

CASE STUDY 11.1 (Continued)   IDENTIFICATION OF UNKNOWN SOLDIER FROM THE VIETNAM WAR

Many Vietnam veterans and President and Mrs. Ronald Reagan visited the Vietnam Unknown in the U.S. Capitol. An Army caisson carried the Vietnam Unknown from the Capitol to the Memorial Amphitheater at Arlington National Cemetery on Memorial Day, May 28, 1984.

Officials had stated that the identity of the Vietnam Unknown Soldier was likely either Lieutenant Michael J. Blassie or Captain Rodney Strobridge, of the Air Force. The two were shot down on May 11, 1972, near An Loc, about 60 miles north of Saigon, where the remains of the Unknown Soldier were found. Blassie's wallet and identification gear from his A-37 fighter were found near the remains, leading to speculation that the remains might be those of Blassie. But initial medical tests could not confirm identification.

The Blassie family, convinced that the remains belonged to their relative, pressed the Pentagon for exhumation, which finally took place in May, 1998. MtDNA testing at the Armed Forces DNA Identification Laboratory (AFDIL) in Rockville, Maryland, confirmed that the remains in the crypt were those of Lieutenant Blassie. They were handed over to his family for reburial in a cemetery near the family home in St. Louis, Missouri.

The identification was announced by President Bill Clinton himself. These new methods of DNA identification make it likely that other unknowns may be identified in the future. Also, it is unlikely that any service members from future conflicts will remain unidentified. "It may be that forensic science has reached the point where there will be no other unknowns any more. So we have to look very carefully about where we go from here…," as former, Secretary of Defense William Cohen noted at the time.

COLLECTION AND PRESERVATION OF BIOLOGICAL EVIDENCE FOR DNA TYPING

The principles and techniques described in Chapter 10 for collecting, packaging, and preserving biological evidence fully apply here. Biological evidence should be thoroughly dried before packaging—that is the most important factor in preserving DNA to obtain DNA profiles.

Substratum comparison specimens, discussed in Chapter 10, are important when biological evidence is collected for DNA analysis. These specimens will be more important in some circumstances than in others. With large and/or concentrated blood or semen stains, so much DNA will be obtained from the cells in those stains that the presence of background biological material will simply not be an issue, because there will be so little of it in comparison to the primary evidence DNA. In the case of smaller, diluted, less concentrated or trace-type stains, the amount of DNA from the primary evidence might be less. In that case, the comparison specimen is more important, because processing it in parallel with the primary stain helps to establish whether or not there is "background" DNA that may interfere with or complicate the interpretation of the DNA profiles from the evidentiary stain.

Compared with the other genetic markers that were used before DNA methods became available (described in Chapter 10), forensic DNA typing is more robust. That doesn't mean one can always get DNA and/or acceptable DNA profiles from evidence no matter what its condition has been. But it is known from experience that DNA profiles can sometimes be obtained from evidentiary stains that are old or have been subject to adverse environmental influences. Age of stain tends to make biological stains insoluble, or "fixed" to the substratum, so that it is very difficult to extract anything from them. Additionally, heat tends to mimic the effect of aging in that respect. DNA tends to be degraded in biological traces or stains that are damp or warm. Enzymes called DNases, which degrade DNA by chemically cutting it into small pieces, can be released from cells during the putrefactive and autolytic (a type of chemical self-destruction) processes that occur after death. DNases are also present in some bacteria that may infect and grow in biological stains that are not dry. It is important to note; however, that no one can predict in advance whether a biological trace or stain will yield DNA, nor whether it will yield a suitable profile. The only way to find that out is to try typing it.

DEVELOPMENT AND METHODS OF DNA ANALYSIS

The groundwork for developing DNA typing methods for forensic casework was laid down by genetic scientists throughout the 1970s and 1980s. Many techniques for manipulating DNA were developed in research laboratories to obtain DNA sequence information, to map the human and other organism's genomes, and to learn how genes work.

Over the years, there has been special interest in learning about genes that are directly or indirectly involved in genetic defects or diseases. In the process, much was learned about DNA functionality.

As mentioned earlier, a great majority of human DNA is not "functional;" that is, it does not code for protein structure. In humans, nonfunctional DNA represents about 80% of the total. This nonfunctional DNA has been called anonymous DNA or "junk DNA," but those characterizations could be premature. It could be that the functions of some of that DNA have just not been discovered yet. Because the non-coding portion of the nuclear DNA is much more varied between individuals, it is more useful for forensic identification than the coding portion. The nonfunctional DNA has repetitive sequences. There are several different types of repetitive DNA sequences, but almost all forensic applications of DNA typing have exploited what is called *tandemly repeated sequences*.

A tandemly repeated sequence is a head-to-tail repeat of the same sequence of bases. Some repetitive sequences are as short as two bases, and others may be dozens of bases long. The number of repeats can vary, and it is the *number of repeats* that constitutes the polymorphism (the occurrence of different forms of DNA sequence variations in human population). People differ in how many repeats of a sequence they have at a particular locus. For this reason, some of these regions in DNA are called Variable Number of Tandem Repeat (VNTR) loci.

There have been three "generations" of DNA-typing technology. The first was called RFLP typing; the second was based on polymerase chain reaction and mainly involved dot-blot techniques; the third is called STR typing, and is the current method. Before describing the typing methods; however, a discussion on DNA isolation from a biological specimen is in order.

Isolation of DNA

An important aspect of DNA analysis is the isolation, or extraction, of DNA from biological evidence such as blood, semen, and saliva stains. This process is usually done before any of the DNA typing steps, no matter what method or system is being used. The most common method for DNA isolation involves digesting the evidentiary material with a broad spectrum proteinase (an enzyme that hydrolyzes and thus breaks down the proteins). This digestion breaks down the proteins and assists in disrupting cellular and nuclear membranes to release DNA. Next, DNA can be separated from other cellular components and extracted using ion exchange resin, organic solvents, or silica-based devices (Figure 11.6). Subsequently, DNA quantitation tests are performed to determine how much human DNA has been obtained (Figure 11.7a–c).

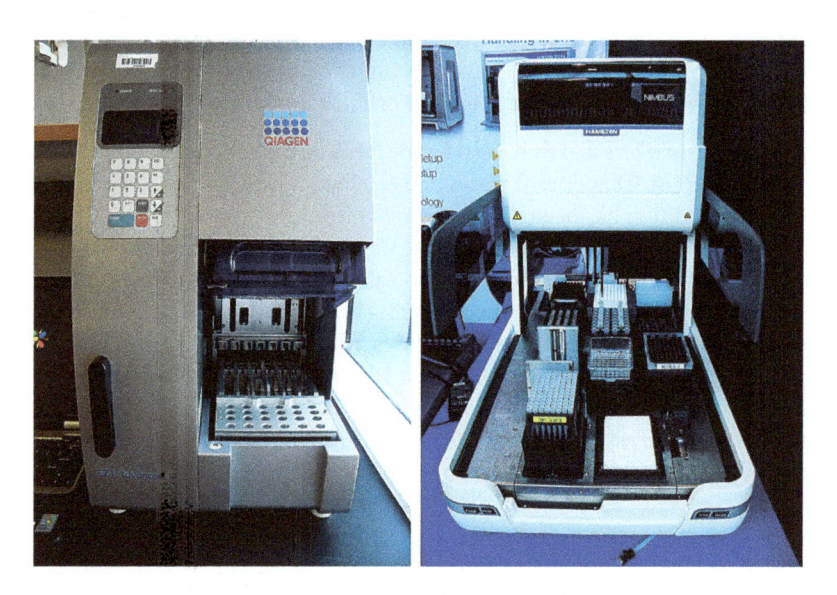

Figure 11.6 Automated bench-top DNA purification and liquid handling systems. On the left is an automated bench-top DNA purification device that enables the isolation of DNA from a wide variety of forensic samples. On the right is an automated bench-top liquid handling workstation for setting up tests of DNA analysis.

DNA analysis and typing   265

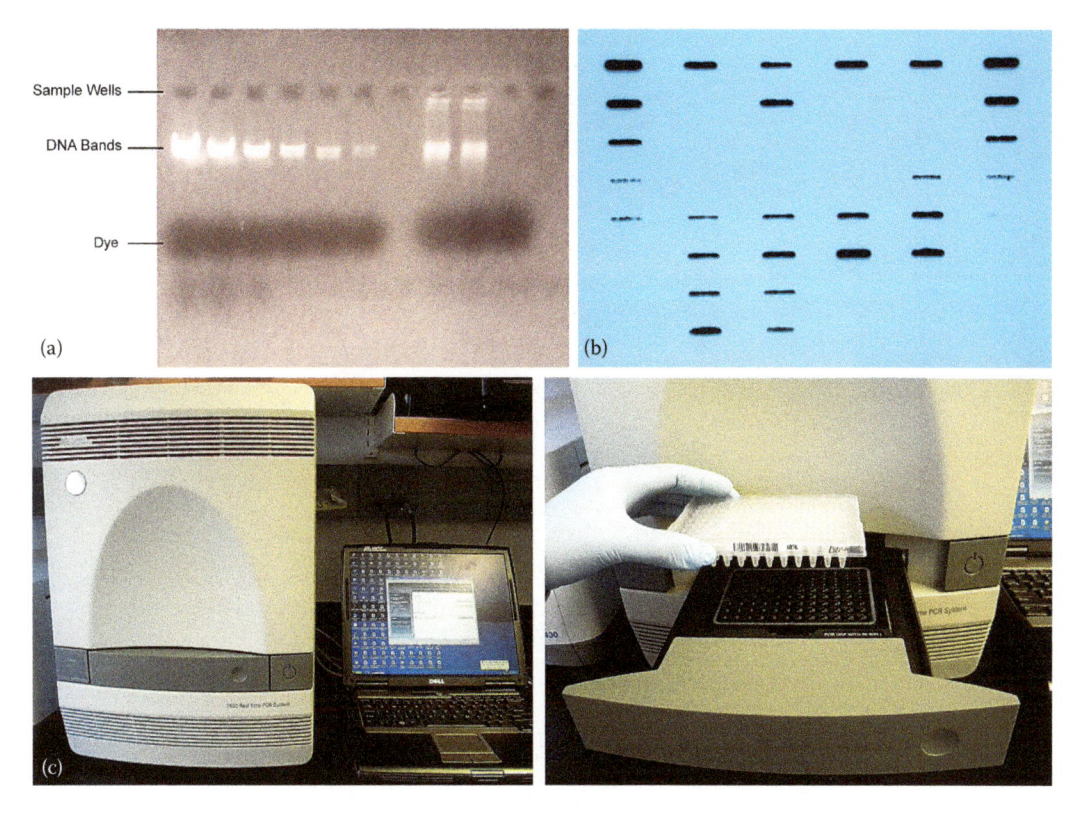

Figure 11.7 (a) A minigel for assessing DNA quantity and quality. This gel, made of agarose, is run in a special electrophoresis chamber. The current causes the DNA to migrate out of the sample wells and down the gel. The dye helps the analyst know how long to leave the electrophoresis running. The brightness of the DNA band is proportional to the quantity of DNA. Additionally, a minigel can assess the quality (degree of DNA degradation) of DNA. (b) Quantiblot film for estimation of the amount of human DNA present in samples. DNA specimens are applied via a slot template to a special nylon membrane. The membrane is treated with a special human DNA probe. The probe can emit light. If the membrane is placed together with X-ray film for a time, the emitted light will expose the film. The darkness of a band is proportional to the amount of human DNA. Note that the minigel provides an estimate of the quantity of total DNA in the specimen, while the quantiblot provides an estimate of the quantity of human DNA. This technique for human DNA quantitation is now obsolete, as most laboratories have switched to a new technique called real-time PCR. (c) On the left is a real-time PCR instrument. On the right is a 96-well plate, which can be used to quantitate multiple samples simultaneously.

A variation of the standard extraction procedure is used for isolating DNA from mixtures of sperm cells and vaginal epithelial cells, such as those commonly found in vaginal swabs or drainage stains in panties or other items from sexual assault cases. This process is called *differential extraction*. In this extraction method, the more easily broken epithelial cells have their DNA released without disrupting the sperm cells. When used on a mixture stain, this first fraction is almost exclusively epithelial cell DNA (usually female DNA). Then, the intact sperm cells are separated, they are disrupted, and a fraction that is almost exclusively male DNA is obtained (Figure 11.8). When this process is successful, the ability to separate the male and female DNA during the extraction process is very helpful in obtaining individual DNA profiles from sexual assault evidence.

The beginning—restriction fragment length polymorphism (RFLP)

In the 1980s, geneticists found a number of VNTR loci (a singular locus is the specific location of a DNA sequence on a chromosome) in the human genome that could be typed using *restriction fragment length polymorphism* (RFLP) methods. The typing method, now largely obsolete in forensic science, involved cutting the DNA into smaller pieces with specific enzymes (called restriction endonucleases), separating the resulting fragments by electrophoresis, transferring the separated fragments to a nylon membrane by a process known as Southern blotting, and finally detecting the specific alleles using a radioactively labeled or chemiluminescence-labeled DNA probe. This intensive process is not very automatable and, therefore, time-consuming and requires a lot of effort (Figure 11.9).

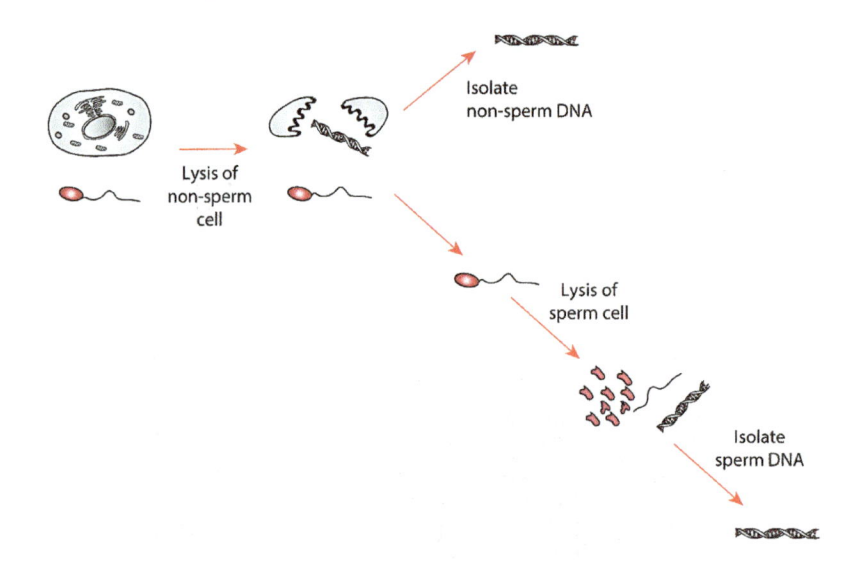

Figure 11.8 Differential extraction process used to separate sperm cells from nonsperm cells. Nonsperm cells are lysed first; sperm cells are resistant to such treatment. The nonsperm cell DNA is extracted. The sperm cells are then lysed to extract the sperm DNA. (Adapted from Li, R. *Forensic Biology*, (2nd Ed.), CRC Press, Boca Raton, FL, 2015.)

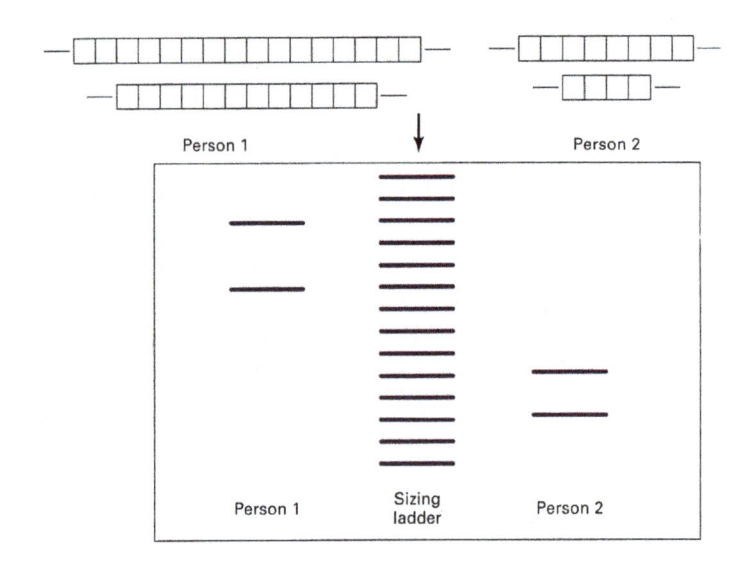

Figure 11.9 Diagram illustrates restriction length polymorphism (RFLP). Each repeat unit of VNTR is represented as a little box. People have two copies of a particular chromosome, and each copy can have a different number of repeats at each particular VNTR locus. Using RFLP, the electrophoresis will cause the different-sized fragments to be separated. Smaller fragments migrate farther than bigger ones. A sizing ladder is a special mixture of DNA fragments of known sizes, which analysts use to estimate the sizes of the fragments from the person's DNA.

In 1985, Sir Alec Jeffreys, a geneticist at the University of Leicester, England, reported in the scientific journal *Nature* that the RFLP DNA-typing method was extremely powerful in individualizing people, and further that he had used the techniques to help solve some alleged blood relationship cases in connection with British immigration law (Figure 11.10). Around that same time, there was a double sexual assault/homicide in an English village. This new DNA-typing method was applied to analyze evidence from the case (Case Study 11.2).

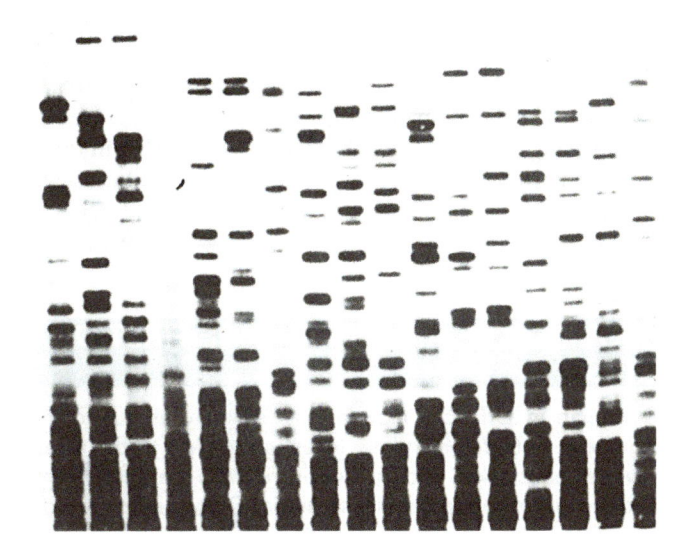

Figure 11.10 RFLP performed using multilocus probes. In the diagram of Figure 11.9, each person can have a maximum of two bands (at a locus) representing DNA fragments. That is in fact how forensic RFLP typing was done. RFLP could also be done using so-called multilocus probes, and then each person would produce multiple bands that looked like a bar code. Sir Alec Jeffreys' earliest cases were done using multilocus probes, as shown here.

CASE STUDY 11.2   THE NARBOROUGH RAPE MURDERS—ALEC JEFFREYS AND THE FIRST USE OF DNA TYPING IN A CRIMINAL CASE

This case, which was the catalyst for the development of DNA-typing technologies by many forensic laboratories throughout the world, centered in a village called Narborough, Leicestershire, England. In 1983, a 15-year-old girl was raped and murdered as she was walking home along a country lane. The initial investigation yielded no suspects. Three years later, another young girl turned up dead in Narborough, sexually assaulted, and murdered in a similar manner. Police arrested a seventeen-year-old named Rodney Buckland, a worker in a local mental hospital, who made statements incriminating himself in the second murder but proclaiming his innocence in the earlier one. Police, certain that Buckland had raped and killed both girls, sent semen samples from the two attacks, along with a blood sample from the suspect, to Leicester University for Sir Alec Jeffreys to examine with his new RFLP DNA-typing method. The DNA analysis confirmed that the same offender had committed both crimes, but it also showed that Buckland could not have been the perpetrator. His incriminating statements were false. Thus, Buckland was the first man ever exonerated by DNA typing. Without it, he might well have gone to jail for life.

Police then began a massive manhunt for the real perpetrator, conducting what would today be called a "biological evidence dragnet." All the sexually mature males in the village were requested to provide reference blood specimens. Most did so—over 5,000 men in all. DNA typing was very new at this point. Forensic science laboratories didn't even do DNA typing as yet. Sir Jeffreys agreed to perform DNA typing in his lab at Leicester. But because this form of DNA typing is so complex and time-consuming, the Home Office forensic science laboratory first "screened" all the specimens using the conventional genetic marker systems we described earlier. That is, all the men whose specimens allowed them to be excluded as potential semen donors were excluded. The remaining specimens were then DNA typed. In the first round of DNA typing, none of the specimens matched the profiles of the semen donor. The perpetrator, a 27-year-old named Colin Pitchfork, was ultimately found, arrested, and convicted of the offenses because he had paid someone to give the voluntary blood specimen to the police using his name. He bragged about this while drinking in a pub but was overheard and eventually turned in. The surrogate admitted to the police what he had done. Pitchfork's DNA profiles matched those found in the semen recovered from both victims.

The crime writer, Joseph Wambaugh, describes this interesting case in its entirety in a book called *The Blooding*.

It was the first time that DNA typing had been used in a legal case in the United States, pre-dating the Andrews case (Case Study 11.3) which represented the first use of DNA in the United States in a criminal matter.

Later, Cetus scientists devised an additional typing system officially called PM, but widely called "Polymarker" among forensic scientists. The same design was used for PM as had been used for HLA-DQA1. PM actually consisted of five additional genetic loci. Once PM was available, the HLA-DQA1 and PM loci were combined into a typing system having a total of six genetic loci.

For a number of years, the HLA-DQA1 and PM typing systems were the only PCR-based methods available and were widely used in many laboratories. Some laboratories attempted RFLP typing on specimens first. If there was insufficient DNA or if RFLP did not yield satisfactory results, the PCR-based methods were then used. Other laboratories never developed the ability to do RFLP, and they did PCR-based techniques instead. In the years when the DNA-typing options were RFLP or PCR-based reverse dot-blot systems, RFLP was preferable where possible, because it usually provided considerably better discrimination of individuals in the population (near individualization) than HLA-DQA1 and PM methods, and because only RFLP data were databased (see later).

CURRENT DNA-TYPING METHODS—SHORT TANDEM REPEATS (STRs)

From the beginning of forensic DNA-typing development, it was recognized that RFLP was too slow and not sufficiently sensitive for widespread forensic casework and database demands. But, at the time there were no alternatives. As PCR-based methods were developed, it was clear that ideal forensic methods would be based on the PCR, and that they would have to provide a level of individualization as good as or better than the RFLP methods.

It was noted previously that RFLP typing utilized several VNTR genetic loci. Different VNTR loci can differ in repeat unit length—the number of nucleotides in a single repeat unit. The loci used in RFLP typing can have their repeat unit length as long as 50 or more base pairs. They are sometimes called *minisatellites*. Another category of VNTR has a repeat unit length of 15-20 base pairs. An example of this is called D1S80. In the time period when HLA-DQA1 and PM were the primary PCR-based typing systems, Cetus (Roche) also developed and marketed a typing kit based on D1S80 typing using the PCR method. Some laboratories used this typing kit in conjunction with HLA-DQA1 and PM in their casework. The category of tandemly repeated DNA sequence that forms the basis of current DNA-typing methods has repeat unit length of 4–6 base pairs. These loci are called *short tandem repeats (STRs)* or *microsatellites*. There are hundreds of these STR loci in the human genome. Over the years, 13-standard loci were chosen (the required STR regions are recently increased from 13 to 20 in the U.S., see later). Typing the 13 chosen STR loci provides a high level of individuality in the population, thus minimizing the probabilities of a match that would occur by chance in the databases. Exact correspondence between two DNA samples has been widely accepted to allow a qualified examiner to testify that, "In my expert opinion these two samples come from the same individual."

Several things had to be considered in choosing the STR loci that would be typed using this newer, faster DNA technology. First, a big advantage of DNA typing is that a person's DNA profile can be placed into a database. If that profile shows up in some subsequent situation, the database can be searched, and the individual can be identified. Databases will be discussed in more detail later. From the beginning, the intention was to have all the forensic laboratories in the United States participate in entering profiles into the databases and then being able to later search the databases for potential matches to unknown profiles. To make this possible, every laboratory must utilize the same standard STR loci and methods that are reproducible from laboratory to laboratory. In other words, a profile generated for a person or specimen in any one laboratory must be the same as the profile that would be generated in any other laboratory. For that reason, a set of STR loci had to be universally agreed upon by participating laboratories.

Another consideration was the degree of individualization that the profile would have in the population. This matter is further discussed later. It should be clear that the more loci that were typed and included in the profile, the fewer the people who would be expected to have that profile—that is, the greater the individualization. There were also some technical reasons underlying the choice of the 13 STR loci.

Most laboratories in the United States type the STR loci using instruments from Applied Biosystems (now Thermo Fisher Scientific) based on a technique called capillary electrophoresis (Figure 11.13). Capillary electrophoresis is just a variant of the electrophoresis process described in Chapter 10. These instruments also use laser light sources

DNA analysis and typing 271

Figure 11.13 Photograph of multi-capillary electrophoresis instrument (ABI PRISM 3500® Genetic Analyzer).

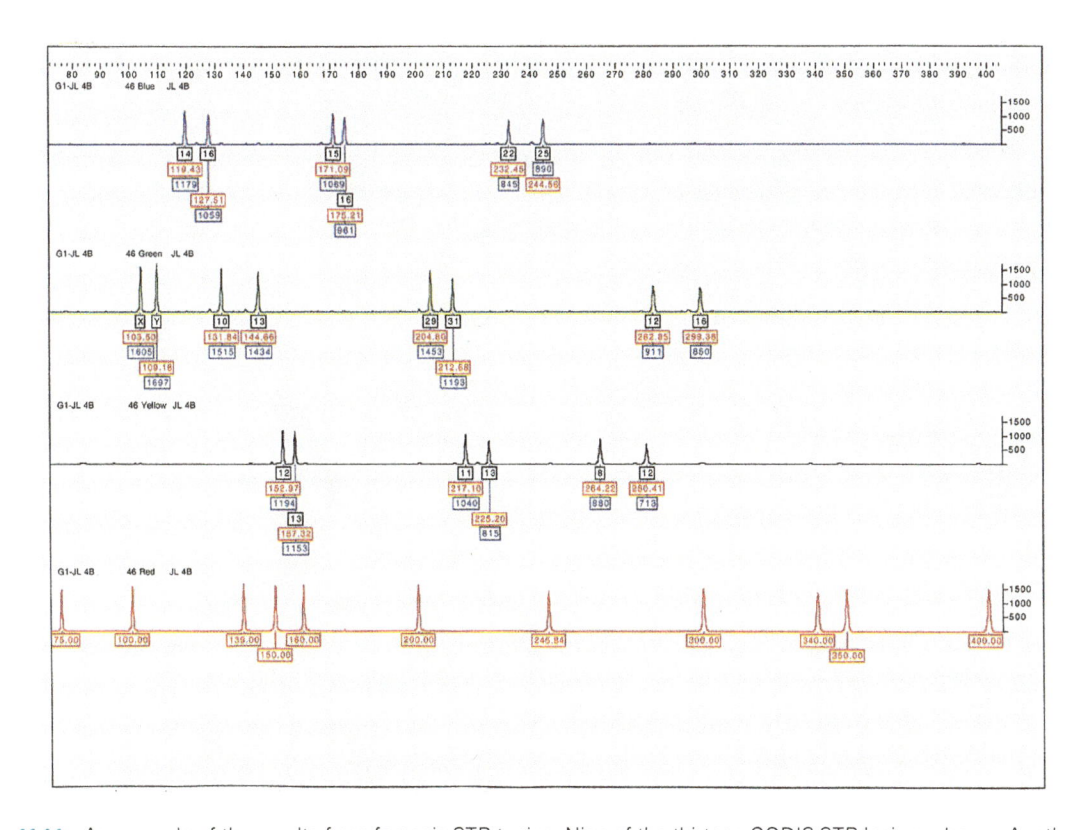

Figure 11.14 An example of the results from forensic STR typing. Nine of the thirteen CODIS STR loci are shown. Another typing kit can be used to profile the other four CODIS loci. Computer software assigns genotypes to each locus based on its programmed algorithms. The profile also provides gender information (XY for males, XX for females). Forensic DNA scientists analyze results like this to solve cases by comparing the DNA profiles of evidence and known reference samples.

to induce fluorescence in the amplified STR products and measure it. Software supplied with the computer that controls the instrument analyzes all the data and translates it into STR profiles (Figure 11.14). See the box "More on the Science: DNA Electrophoresis, Capillary Electrophoresis, and Genetic Analyzers" for further discussion on DNA electrophoresis and genetic analyzers. There were also laboratories that used polyacrylamide gels to separate the amplified STR fragments and determine the profiles.

> **MORE ON THE SCIENCE: DNA ELECTROPHORESIS, CAPILLARY ELECTROPHORESIS, AND GENETIC ANALYSIS**
>
> In Chapter 10 in a "More on the Science" box, we described electrophoresis and isoelectric focusing. Electrophoresis is the separation of charged molecules (proteins or nucleic acids usually) in an electric field. A solution called a buffer solution is used to hold the acidity of the solution constant. Scientists measure a property called pH to determine acidity, and buffer solutions have constant pH.
>
> In a buffer solution, a protein or nucleic acid molecule has a constant net charge, because its net charge is determined by the acidity of the solution in which it is dissolved. As a result, it will move in a particular direction in an electric field.
>
> Electrophoresis is used all the time in DNA analysis. At pH values around neutrality, DNA is very negatively charged and will migrate toward the + pole in an electrophoresis setup. DNA scientists use electrophoresis to evaluate the quality and the amount of DNA isolated from cells, and to look at the products of the PCR reaction. This kind of electrophoresis is regularly done in gels made from a substance called agarose.
>
> Many laboratories, as noted, use genetic analyzers from Applied Biosystems for their DNA typing and profiling. These genetic analyzers can also be used for DNA sequencing. They employ a special kind of electrophoresis known as capillary electrophoresis (CE). CE is not different in principle from the other kinds of electrophoresis we have talked about. But CE is done in very finely drawn glass capillary tubes, and the molecules that are being separated (DNA fragments in this case) move in a viscous polymer solution.
>
> The "analyzer" portion of the instrument consists of a laser and a detector system capable of seeing fluoresced light of various wavelengths (see in the Appendix—Scientific Tools of the Trade). In forensic DNA analysis, certain dyes that can be excited by the wavelength of the laser are used to tag the PCR primers. Since the primer is incorporated into the PCR product, every molecule has the dye. The capillary electrophoresis separates the molecules on the basis of size to some extent (in addition to the separation based on net charge). Smaller molecules move through faster than larger ones. DNA scientists have control over the size of PCR products because they can position the primers anywhere they want (within certain limits). When a molecule with a dye passes by the laser beam, it is excited, and fluoresces at a characteristic wavelength that is picked up by the detector.
>
> The whole system can be made more efficient by using different dyes on different primer molecules. Now it is possible for molecules, even if they are the same size, to be distinguished by the detector because the fluorescence of each dye used is of a different wavelength. In this way, many STR regions can be analyzed simultaneously.

THE POWER OF DNA TO INDIVIDUALIZE BIOLOGICAL EVIDENCE

In Chapter 10, it was discussed that a combination of the classical (conventional) genetic types could be used to give increasingly greater probabilities of individualization and that DNA typing works the same way.

The level of individuality from DNA typing depends on the population genetics of the *genotypes* of the loci that are chosen for the overall DNA profiling. Typing one locus might give a genotype shared by many people in a population. Typing another locus, and considering both loci together, gives a profile that fewer people will have. Adding a third and fourth locus, and so on, to the profile continues to reduce the number of people who share it.

The box "More on the Science: DNA Population Genetics/Individualization" shows the level of individuality obtained from the profiles generated by all three generations of DNA-typing technologies: RFLP, HLA-DQA1, PM, D1S80, and STR. One can see a couple of things from reading through the science box. One is that the estimates of the frequency of each genotype in the population can be multiplied to give an estimate of the *random match probability*; that is, the probability that a randomly selected person in the population will have the same DNA profile as that shown by the evidence. For many complete DNA profiles, this probability is extremely low. But one of the common misunderstandings about these numbers is the assumption that they are actual distributions of the genotype in the population; In fact, they are not—rather they are probabilities. Thus, even though the numbers can be astronomical, it does not mean that there cannot be another person with the same profile—it just means that the probability is very low. A second thing to realize is that there are different estimates of random match probability according to ethnic/racial group. Because the frequencies of genotypes within loci can and do vary in different populations, the different estimates for each of them are commonly given so as not to mislead the courts. Most laboratories can compute random match probability for Caucasians (as in the science box table), African Americans, and certain Hispanic populations. There are also data available for populations, such as Native Americans belonging to certain tribes. The application here is that a defendant whose DNA matches evidence in a case might claim that he is a member of a minority population in which the genotypes observed are more common than in the larger population. It is important to remember that in a forensic case, the race or ethnic group of the actual depositor of evidence is not known with certainty. Even in a DNA match case with very low probabilities of a match that would occur by chance, there is still a small chance that another person has the same profile. Also since close blood relatives will usually share more common loci than unrelated individuals the generalized common match probabilities will be affected.

MORE ON THE SCIENCE: DNA POPULATION GENETICS / INDIVIDUALIZATION

As we have noted in the main text, there have been three generations of DNA-typing technologies. In the following table are shown the probabilities of a match that would occur by chance for the RFLP, HLA-DQA1, PM and D1S80 loci, and then for the 13 so-called core CODIS STR loci.

The information is presented for Caucasians in U.S. populations, based on population studies that have been done and published in the literature. Similar calculations can be performed for other populations and racial/ethnic groups that have been sampled. We present the frequencies as decimal numbers (that is, 0.03 is the same as 3%, etc.). In the final combined profile for each major category, the reciprocal probability of a match by chance is also given (i.e., 1 out of however many total people).

To give an idea of the range of numbers possible, we present the extreme cases at each end of the spectrum of possibilities—one where a person had the most common type at every locus, and the second where a person had the rarest type at every locus. In real casework, the numbers should always be somewhere in between these two extremes.

Note that in the STR example, a person who had the most common type at every one of the 13 loci would be expected to occur only once in 1.66×10^{11} people. That is a very large number: 166,000,000,000 or 166 billion. Right now, there are about 7×10^9, or 7 billion people on earth. You could say, therefore, that this DNA profile should not occur more than once in the population of earth. Remember, that the calculated number is a probability based on estimates of the population frequencies. Nonetheless, there is discussion among forensic scientists now about stating that a profile like this "originated from the person it matches to a reasonable degree of scientific certainty." For various reasons, many laboratories do not yet state DNA results in such simple terms. They continue to quote numbers. However, the FBI Laboratory has adopted a policy that if certain statistical genetic conditions are met, they will testify as to "origin."

DNA population genetics table

Technology	RFLP	
Locus	Most common	Least common
D1S7	0.0127	0.0013
D2S44	0.0257	0.00014
D4S139	0.0459	0.00194
D10S28	0.0139	0.00032
D17S79	0.1207	0.0008
Combined	2.542×10^{-8}	9.039×10^{-17}
Combined: One in	39.3×10^6	1.1×10^{16}

(Continued)

MORE ON THE SCIENCE: DNA POPULATION GENETICS / INDIVIDUALIZATION (Continued)

Technology	Dot-blot and D1S80	
Locus	Most common	Least common
HLA-DQA1	0.1429	0.007
Five PM[a]	0.0217	1.3×10^{-6}
D1S80	0.2	0.000001
Combined	0.0006	16^{-16}
Combined: One in	1800	10^{16}

[a] LDLR, GYPA, HBGG, D7S8 and GC

Technology	STR	
Locus	Most common	Least common
D3S1358	0.114	6.05×10^{-6}
VWA	0.1166	2.98×10^{-5}
FGA	0.0655	1.04×10^{-4}
D8S1179	0.0137	1.04×10^{-4}
D21S11	0.0841	6.50×10^{-6}
D18S51	0.0651	4.11×10^{-5}
D5S818	0.2903	6.55×10^{-6}
D13S317	0.1969	1.28×10^{-3}
D7S820	0.1174	6.05×10^{-6}
CSF1PO	0.1954	6.05×10^{-6}
TPOX	0.2762	6.05×10^{-6}
THO1	0.1384	6.05×10^{-6}
D16S539	0.1847	6.15×10^{-6}
Combined	6.027×10^{-12}	3.597×10^{-62}
Combined: One in	1.66×10^{11}	2.78×10^{61}

DATABASING AND THE CODIS SYSTEM

DNA genotypes at individual locus are designated by numbers, letters, or combinations of numbers and letters. A combination of DNA genotypes from one individual (such as a combination of genotypes at five RFLP loci, the combination of HLA-DQA1 and PM genotypes, or a combination of genotypes at 13 STR loci) is called a DNA profile. Over the years, 13 STR loci are required in the U.S. The number of required STR loci will soon be increased from 13 to 20 (also called expanded CODIS Core STR Loci) in the U.S. Since a DNA profile can be written as a set of numbers or numbers and letters, it is easy with computer technology to store DNA profiles in searchable *databases*.

Early in DNA technology development, people realized that from a law enforcement viewpoint, it would be very helpful to store the DNA profiles of convicted offenders in a database. Then, if those individuals committed other crimes and biological evidence were recovered, they could be readily identified through their DNA profiles. Since sex offenders typically leave semen evidence behind, having the DNA profiles of convicted sex offenders in a database would help identify these individuals if they were to commit another crime after being released from prison. A fairly high rate of recidivism occurs among felons released from prison, and among sex offenders in particular. All the states have now passed legislation enabling storage of DNA profiles from those convicted of felony sex crimes. Many states also collect and store DNA profiles from persons convicted of any felony. There are some variations in the laws of individual states as to which offenders are DNA typed and stored.

Some law enforcement and other public officials have suggested putting criminal suspects in a database, but that has not happened in the United States on a broad scale because of privacy and civil liberties concerns. Recently, however, several states have passed legislation allowing the inclusion of DNA profiles of arrestees and/or suspects into the database. Criminal suspects are included in the database in the United Kingdom. In the U.S., DNA database laws are based on a balance between usefulness to law enforcement and concerns about the protection of individual rights and privacy. Arguments for enlarging the databases rely to a certain extent on the fact that forensic DNA profiles are anonymous identifiers; that is, they do not tell you anything about an individual. They can only be used for comparison with a DNA profile generated from a case specimen. Those opposed to enlarging the databases argue that while these "anonymous" DNA sequences have no known function now, a function could be discovered later that would then give information about a person. Those same people also argue that once someone's biological specimen or DNA is in the possession of law enforcement, it could be taken out of storage at a later date when the technology might allow the determination of information about someone's appearance or health status from his or her DNA.

In the U.S., there are several tiers to the forensic DNA profile databases: the first is the Convicted Offender Index, containing the profiles of convicted offenders. The second is the Forensic Index, often called the "forensic file," containing the profiles from biological evidence (such as semen evidence) from unsolved cases. There is also a third tier consisting of DNA profiles of missing persons which are obtained by analyzing cells from their personal items (e.g., a toothbrush). This tier may contain parental DNA profiles that can be helpful in identifying the person or the person's remains. Recently, the fourth tier, the Rapid DNA Index System (RDIS), has been proposed; it shall store the DNA profiles of reference samples, such as buccal swabs, processed at police booking stations (Figure 11.15).

There are also three "levels" of the databases: national, state, and local. The national file is called *Combined DNA Indexing System (CODIS)*, and is maintained by the FBI Laboratory. Many states and localities have their own databases, called SDIS and LDIS, respectively. This system of interconnected databases enables cross-jurisdiction searches for DNA profiles developed in new cases. You will hear people use the term CODIS for all levels of the databases.

As noted, each state, through its own laws, controls which convicted offenders are put in the database in that state. Every state allows inclusion of DNA profiles of offenders convicted of sex crimes into the database. Some states allow the inclusion of offenders convicted of other felonies as well as arrestees. State laws also regulate how far back in time the convicted offender database goes, whether juveniles are included, and whether offenders convicted but placed on probation are included, among other matters. Profiles in local databases that meet appropriate criteria can be uploaded to state databases, and profiles in state databases meeting appropriate federal criteria can be uploaded to CODIS.

As of September 2015, there were over 11,962,222 convicted offender profiles, 2,120,729 arrestee profiles, and 657,298 forensic profiles in CODIS. There had been over 296,490 "hits," or matches of case specimen profiles, to profiles in the Forensic Index or in the Convicted Offender Index. This information is updated regularly on the FBI website at https://www.fbi.gov/about-us/lab/biometric-analysis/codis/ndis-statistics. Hits to the Forensic Index help connect cases that police might not otherwise realize were connected. Hits to the Convicted Offender Index tentatively identify the depositor of the biological evidence. The usual procedure is that when a suspect is identified by a

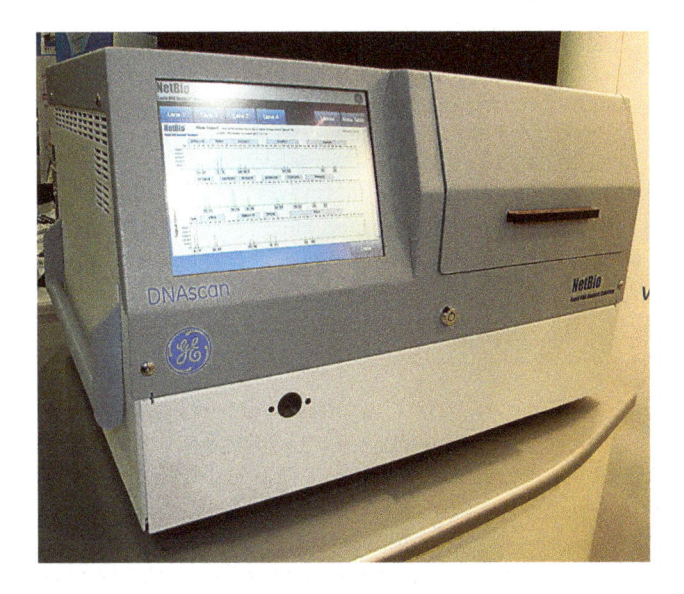

Figure 11.15 Compact Rapid DNA device for processing DNA evidence at booking stations.

database profile match, a court order is requested to obtain a new DNA sample from that individual. That sample is profiled and then entered for comparison to the evidence sample. This avoids the possibility of an administrative error or sample mix-up causing the wrong individual to be prosecuted. Database matches that have helped solve several cases are described in Case Study 11.4.

> ### CASES STUDY 11.4 DEVELOPING SUSPECTS THROUGH DATABASE MATCHES
>
> Many years ago, when DNA profiling was still new and the databases were in the process of being built, CODIS "hits" (matches) were something of a novelty. Today, with the technology well-established and the databases much larger, they are an almost everyday occurrence. A few representative cases will illustrate the point.
>
> There are a couple types of potential CODIS matches. A new, unknown profile is searched against the convicted offender database. A match here putatively identifies the profile, because the offender database profile belongs to a specific individual. We say "putatively" identifies the profile, because the laboratory does not rely on this "hit" or to make the match. A new specimen from the suspected depositor is taken, re-profiled, and compared with the unknown profile. If these are identical, the match is made. We would expect them to be the same, of course, unless some kind of specimen mix-up or labeling failure had occurred. The laboratory would then calculate the probability of a match by chance. Another possibility is that the new, unknown profile matches a profile in the forensic file. This links together the two cases from which the profiles came as having the same DNA depositor—but there is no connection to a specific person. Still another possibility is that a new convicted offender profile matches a profile in the forensic file. That connects this offender to a previously unsolved crime.
>
> #### CASE A
> In June of 2005, a case in Illinois was solved by a match between a profile in the forensic file and a convicted offender profile. A 2001 sexual assault, in Des Plaines, Illinois, was linked to a similar case in Northlake. The suspect's DNA profile matched the evidence DNA profiles. His DNA profile had been entered into the database in 2003, following a conviction for driving under the influence. The suspect confessed to the unsolved Des Plaines rape after he was confronted with the DNA evidence.
>
> #### CASE B
> A Pennsylvania man pleaded guilty to the rape of two teenagers after his DNA profile, entered into the convicted offender database following a conviction for robbery, matched the semen profiles from the sexual assault cases. Following the CODIS match, witnesses picked the suspect out of a photo array. A new specimen was next obtained, profiled, and matched the male profiles from the sexual assaults.
>
> #### CASE C
> In June of 2005, in North Carolina, the State Bureau of Investigation Laboratory used a technique called "familial searching" to identify a suspect from CODIS. The laboratory was searching for the profile of an offender from a case involving the rape and murder of a newspaper editor. The profile was not in CODIS, but another similar one was—one from a person who would have the same parents as the offender being sought. It turned out that the CODIS profile belonged to a convicted offender who had a brother. Police followed the brother and retrieved his discarded cigarette butts for DNA comparison. A match was obtained, and the case was solved.
>
> This case illustrates two points. The first is that there is nothing illegal about the police seizing a discarded item (like a cigarette butt or an empty soda pop can) that contains enough biological material to get a DNA profile. Second, familial searching is still somewhat controversial in the U.S., though legal. It is more common in Europe and the United Kingdom.

All the states, the United States Army Criminal Investigation Laboratory, and the FBI are participating in CODIS. However, the states are at all different levels of development in their DNA database programs. Some states started early and have extensive programs (e.g., Florida, Virginia, and California). Their CODIS databases are typically larger than those in states whose programs are less developed. CODIS is one of three major databases containing forensic data that are available to law enforcement. The other two were discussed previously. The AFIS databases contain fingerprint images (Chapter 6), and the NIBIN databases contain bullet striation and cartridge case marking images for firearms identification (Chapter 8).

APPLICATIONS OF FORENSIC DNA TYPING

There has been so much hype in the popular media surrounding DNA typing that a person could easily think that DNA technology is the solution to every criminal justice problem in the nation. In fact, DNA typing is applicable only in circumstances where biological evidence has been deposited and recovered, and yields DNA that can be typed. It should also be pointed out that even though DNA-typing technology has come a long way in terms of automation and throughput, there are still serious backlogs in processing both casework samples and convicted offender database (CODIS) samples in many places. Until these backlogs are cleared up, and there are sufficient resources both financial and human available to keep up with the demand, the potential of forensic DNA typing cannot be fully realized. Two recent reports from the National Institute of Justice documented these persistent backlog problems and made recommendations to provide substantial additional funding to help alleviate them. Further details are described in the box "More on the Science: DNA Casework and Database Backlogs."

> **MORE ON THE SCIENCE: DNA CASEWORK AND DATABASE BACKLOGS**
>
> In March 2003, the National Institute of Justice issued its report to the U.S. Attorney General on DNA casework and database backlogs. The Attorney General had requested the report. The full report is available as document NCJ 199425 from the NIJ.
>
> There were six major recommendations:
>
> 1. Improve the DNA analysis capacity of public crime laboratories.
> 2. Help state and local crime labs eliminate casework backlogs.
> 3. Eliminate existing convicted offender DNA backlogs.
> 4. Support training and education for forensic scientists.
> 5. Provide training and education to police officers, prosecutors, defense attorneys, judges, victim service providers, medical personnel, and other criminal justice personnel.
> 6. Support DNA research and development.
>
> A substantial appropriation was requested in the president's budget to help address these recommendations and the details contained within them.
>
> In December 2003, a final report to NIJ entitled *National Forensic DNA Study* was published. This document was the final report on Grant No. 2002-LT-BX-K 003 by researchers from Washington State University and Smith, Alling, Lane, P.S. The Executive Summary of Report Findings concludes the backlog of unsolved rapes and homicides in the United States is massive. Through the data collected from a large, representative sampling of local law enforcement agencies in the United States, the study arrives at the following pertinent estimates:
>
> - The number of rape and homicide cases with possible biological evidence that local law enforcement agencies have not submitted to a laboratory for analysis is over 221,000.
>   Homicide cases—52,000 (approximate)
>   Rape cases—169,000 (approximate)
> - The number of property crime cases with possible biological evidence that local law enforcement agencies have not submitted to a laboratory for analysis is over 264,000.
> - The number of unanalyzed DNA cases reported by state and local crime laboratories is more than 57,000.
>   State laboratories—34,700 cases (approximate)
>   Local laboratories—22,600 cases (approximate)
> - Total crime cases with possible biological evidence either still in the possession of local law enforcement, or backlogged at forensic laboratories, is over one-half million (542,700).
>
> A significant proportion of law enforcement agencies continue to misunderstand the potential benefits of DNA testing.
>
> *(Continued)*

> **MORE ON THE SCIENCE: DNA CASEWORK AND DATABASE BACKLOGS (Continued)**
>
> Next, while these figures address the first question as to the approximate size of the case backlog, a second question remains regarding how the backlog became so large. The answer to this question is quite complicated, and the phenomenon of a substantial growth in criminal case backlogs involves a variety of factors—some of which are vexing and difficult to manage. However, a series of questions posed to local law enforcement agencies and forensic laboratories revealed several interesting patterns of response that, when considered collectively, begin to provide an accurate picture of how the backlogs tend to develop and why they continue to exist. The following provides some of these responses regarding why cases with DNA evidence have not been submitted to the laboratory:
>
> - 50.8% of responding local law enforcement agencies indicated that forensic DNA was not considered a tool for crime investigations.
> - 31.4% responded that no suspect had been identified.
> - 9.2% indicated that the prosecution had not requested testing.
> - 10.2% responded that a suspect had been identified, but not yet charged.
> - 23.6% of responding agencies suggested that DNA evidence from unsolved cases was not submitted for reasons relating to poor funding.
> - 9.4% indicated a lack of funding for DNA analysis.
> - 10.4% indicated inability of laboratories to produce timely results.
> - 3.8% indicate crime laboratory is not processing requests for DNA testing.
>
> Both state and local crime laboratories are overworked, understaffed, and insufficiently funded. Processing times at crime laboratories pose significant delays in many jurisdictions. State laboratories take an average of 23.9 weeks to process an unnamed suspect rape kit, and local laboratories average 30.0 weeks for such tests. The cost for testing these rape kits was estimated at $1,100 per case, a number that does not account for many overhead costs. Both state and local laboratories indicated that personnel needs were among the most significant concerns for their DNA programs. Specifically, most crime laboratories expressed the need for supplemental funding for additional DNA staff; several laboratories indicated that their priority concern was for additional funding to augment current salaries to avoid the loss of skilled personnel to other prospective employers. A strong need was also reported for reagents (chemicals used in DNA analysis) and for technical equipment used for DNA analysis.
>
> Finally, most laboratories also reported that, while federal funding has played an important role in assisting with backlogged DNA cases, the proportion of their overall DNA budgets funded through federal sources is minimal. Only 20.5% of state crime laboratories receive 50% or more of their funding from federal sources; that figure is only 4.5% for the local laboratories.
>
> There is additional discussion in the full report on the potential benefits of maximizing DNA analysis in criminal cases.

The major applications of forensic DNA typing are criminal cases, civil cases (primarily disputed parentage), and identification of persons when other simpler methods fail or cannot be used, such as in mass disasters (Chapter 6). Most evidence coming into public forensic science laboratories is from criminal cases. For the DNA section of the laboratory, these will be primarily evidence from sexual assault cases and blood stains. Sexual assault cases significantly outnumber homicide cases in most jurisdictions. There are also the less common saliva or other bodily fluid evidence cases, and situations involving DNA analysis from less common biological materials like dandruff, fingernails, and so on. One example is to analyze "touch DNA." "Touch DNA" refers to the DNA resulting from the transfer of skin cells when an individual makes contact with an object at crime scenes. For example, "touch DNA" testing is often needed to analyze samples derived from evidence, such as fingerprints, and tools or weapons handled by perpetrators. The testing involves the analysis of samples with very small amounts of DNA. This can usually be carried out by using more specialized STR analysis procedures, achieving high sensitivity of the testing to identify perpetrators. However, this approach also increases the appearance of artifacts, which may arise from the contamination of environment-born DNA. As a result, the interpretation of "touch DNA" testing results can sometimes be challenging. Case studies 11.5 and 11.6 show how DNA typing can include or exclude suspects in criminal matters. Remember, that the obverse of the very high degree of individualization provided by the technology is the equally high probability of excluding a person (nondepositor) who was not the depositor of the evidence. DNA technology in the criminal case arena has received as much attention for excluding suspects as for including them, especially in some of the cases where individuals were convicted prior to the availability of DNA typing.

CASE STUDY 11.5 DNA INCLUSION IN CRIMINAL CASES

Recently, an Illinois man being held in a June 2003 sexual assault was linked by DNA to an earlier case from February 2002. This is a case where a DNA profile from the suspect matched the semen from the case under investigation. But when the CODIS forensic file was queried, the profile showed up in the earlier case. This same man is linked to a third sexual assault through a database hit.

If these cases go to trial, an examiner will testify as to the DNA profile matches, and that the probability of a match by chance in the population is very low, probably less than one in many billion. Since there is no consensual connection between this man and the women, there is no possible innocent explanation for the DNA match.

Today, there are hundreds of cases like this one every week. Many of the cases that reach the DNA sections of laboratories are sexual assaults. As noted in the previous chapter, there are a greater number of sexual assault cases than murder cases. Also, as noted there, DNA profiling is informative (probative) if the perpetrator is a stranger. If the people are acquainted, and the suspect does not deny sexual contact but claims it was consensual, DNA profiling is not very probative.

Similarly, in cases involving blood transfers, DNA profile matches are informative if there is no innocent explanation for the blood evidence. If there is, the profile match carries much less weight in helping prove a person committed a crime.

It is a good idea to keep in mind, too, that a DNA match establishes a connection between a person and recovered biological evidence. This connection may be highly incriminating or it may have an innocent explanation. By the same token, a DNA exclusion shows that a person was not the depositor of the evidence tested. Under some circumstances, the exclusion may strongly suggest or almost establish innocence, but in other circumstances it may not. Popular media often say things like "DNA testing proves so-and-so is innocent …," or "DNA testing proves so-and-so is guilty …." DNA should be thought of as a way of linking or disassociating biological evidence from individuals, not as a mechanism for establishing guilt or innocence.

A project begun at the Cardozo Law School in New York by attorneys Barry Scheck and Peter Neufeld, and known as the "Innocence Project," has worked to use DNA evidence to exclude people who were convicted prior to the availability of DNA analysis. Up to early 2016, the project had established post-conviction DNA exonerations in 337 cases. Cases are screened to ensure that excluding the person convicted as a depositor of the case evidence will essentially establish the individual's innocence. Many of the cases were sexual assaults, and the semen in the evidence was not deposited by the man who was originally convicted. Some states have established their own versions of The Innocence Project. Further information has been included in the chapter references.

CASE STUDY 11.6 DNA EXCLUSIONS IN CRIMINAL CASES

As we have noted in the main text, DNA is as valuable as an exclusionary tool as it is as an inclusionary one. The lead case in Chapter 9 (Dotson) is a DNA exclusion case. Two cases will illustrate the exclusionary value of DNA. One involves an exclusion in an active criminal case. The other involves a post-conviction exclusion.

CASE A: EXCLUSION OF A SUSPECT IN AN ACTIVE CRIMINAL CASE

In June of 2005, a little girl named Riley Fox was reported missing in Joliet, Illinois. She had been in her home with her brother and father. Her mother was in Chicago taking part in a charity walk. The father, Kevin Fox, told police he had gone to a street festival the night before. He said he had left the two children in the care of their grandparents, and, after picking them up around midnight, he had put them to bed. In the morning, the front door to the home was open, but Kevin Fox said he did not know whether his daughter had opened it and wandered off.

Hundreds of volunteers helped search for the child, and her body was found later in Forked Creek, 4 miles from the family's home. An autopsy determined that Riley Fox had been drowned. She had also been sexually assaulted. Suspicion fell on the father, and he was questioned and eventually arrested after the police reported that he had given them a confession. Mr. Fox's defense attorney later alleged that the confession had been obtained after lengthy questioning and improper suggestions of what would happen next. The police denied the allegations.

(Continued)

CASE STUDY 11.6 (Continued)   DNA EXCLUSIONS IN CRIMINAL CASES

Kevin Fox was held in jail, unable to make bail, for 8 months. No DNA testing was immediately performed on the vaginal swab taken from Riley Fox. The state police laboratory indicated that they were never given a chance to do the testing before the specimen was sent to the FBI Laboratory. The FBI Laboratory scientists indicated that the prosecutors had told them to stop testing once the confession was obtained. Mr. Fox's defense counsel challenged the state police laboratory's explanation, and she pressed to have the specimen tested at a private laboratory. This testing excluded Kevin Fox as the source of the semen, and effectively established his innocence and the falsity of the confession he had given. There was considerable discussion and finger-pointing about why it had taken such a long time to do the tests and obtain the results. This case illustrates a DNA exclusion in an active criminal case. Because of the severity of the crime, that it was committed against a child, and because the suspect was the victim's father, the importance of the DNA testing in this case is difficult to overstate.

In the investigation, the DNA profile from the swab did not match any convicted offender profiles, nor did it match any profiles in the forensic file; thus, the case was unsolved. Later in 2010, police finally tracked down the suspect, Scott Eby, believed to be Riley's killer. After DNA evidence linked him to the case, Eby was charged on five counts of first-degree murder and one count of predatory sexual assault and received a life sentence without the possibility of parole.

CASE B: POST CONVICTION EXCLUSION OF RONALD JONES

On March 10, 1985, a 28-year-old woman was raped and murdered in an abandoned motel on the South Side of Chicago. Seven months later, Chicago Police detectives obtained a signed confession to the crime from Ronald Jones. He was 34 years old at the time, an alcoholic, and he lived in the neighborhood where the crime had occurred.

Jones would later claim that he signed the confession because he had been beaten by the detectives. The detectives claimed the confession was voluntary. The confession stated that the victim was a prostitute, when in fact she had no history of prostitution. It went on to say that Jones had agreed to engage in sex for money with the victim, but that she was killed in a subsequent dispute over payment. No physical evidence linked Jones to the scene or to the victim. The confession stated that he had ejaculated, but the state had contended there was an insufficient quantity of semen for testing. The confession was held admissible at Jones' trial in 1989. He was convicted by the trial jury and sentenced to death. The Illinois appellate defender eventually convinced the Illinois Supreme Court to permit DNA testing on the semen recovered from the victim at the time of the crime. The testing was performed in 1997, and showed that Jones could not have been the semen depositor. For a time, prosecutors considered retrying Jones on the murder charge (now decoupled from the sexual assault because of the DNA exclusion), but eventually, in 1999, all the charges against him were dropped, and he was freed.

The Jones case is an example of several hundred cases around the country in which post-conviction DNA testing has excluded the convict of being the source of semen or blood associated with the original crime. Around a dozen of the cases in Illinois involved men who were, like Jones, awaiting execution. Former Illinois governor George Ryan placed a moratorium on executions when these cases began to surface, and he eventually commuted the sentences of all the inmates on death row to life imprisonment just before he left office.

DNA typing is useful in many civil case situations, as well. The most prominent of these is *disputed parentage* (usually disputed paternity). In the U.S., most public forensic science laboratories do not do, and have not traditionally done, disputed parentage cases. Primarily for historical reasons, parentage cases tend to be done by clinical or other laboratories dedicated to that type of work. Those laboratories rarely do any criminal casework. Since maternity is rarely in doubt, most parentage cases involve doubtful paternity. The cases usually arise when a mother seeks support payments for her child, and the accused man denies paternity. The cases may also arise through the intervention of the Aid to Families with Dependent Children program in a state. The states have an interest in trying to determine the paternity of children and enforcing court orders for support, because this can result in savings of the maternal and child welfare program funds. In a typical case, a mother, child, and putative father are tested. The laboratory examines the alleles inherited by the child. Then, knowing which ones of those were contributed by the mother, the laboratory expert looks to see if the putative father could have contributed all the non-maternal alleles. Exclusion shows that the putative father is not the biological father. Failure to exclude the putative father proves that the putative father is the biological father because the chance of an unrelated person with even half of the same DNA profile in the population is remote. In addition, remember that the burden of proof in civil proceedings, which is preponderance of evidence, is lower than in criminal cases, where the burden is beyond a reasonable doubt (Chapter 2).

Once in a while, an unusual case arises where a sexual assault results in pregnancy. In those circumstances, parentage testing can be used to help support a criminal charge against a defendant. Case Study 11.7 illustrates this application.

CASE STUDY 11.7 PARENTAGE TESTING IN SUPPORT OF CRIMINAL SEXUAL ASSAULT

A number of years ago, investigators from the state's Special Prosecutor for Nursing Homes came to a local laboratory for help in a particularly troubling case. They were investigating a sexual assault on a young woman who had been in a coma for nearly ten years and was discovered to be pregnant during a routine physical examination. Through some interviews, they had developed a possible suspect, who had worked at the nursing home at about the right time and who had been seen leaving the victim's room several times when there was no reason for him to be there.

The investigators felt that they did not have enough evidence to obtain a search warrant to obtain a DNA sample from the suspect. The family of the victim was quite cooperative and allowed a sample of tissue from the fetus to be obtained. This was in the early days of forensic DNA, but some limited work had been done on obtaining DNA from stamps and envelopes and amplifying it using PCR. It was suggested that if they could track down a mailing from the suspect that could be authenticated as being from him, it might provide a partial DNA profile. This was done, and when the possibility of paternity was explored he was included as a possible father. This information along with the other circumstantial evidence was taken to a judge who then issued a search warrant to obtain a blood sample from the suspect.

The victim's parents elected to allow carrying the pregnancy and when the baby was born, samples of its genetic material were taken. A full paternity trio (mother, child, and possible father) was run, and the suspect was found to be included and the father with a high degree of probability. He was convicted and sentenced to a long term in prison.

The third major application of forensic DNA typing is identification of decedents in criminal cases or mass disaster situations; forensic DNA typing is used when the traditional identification methods, such as witness testimony, fingerprints, and dental records, cannot be obtained or used for any reason. With human remains that are fairly recent, nuclear DNA typing may be used for identification. There are two ways this might be done. First, the DNA profile of the remains might be directly compared with that of some reference specimen, such as the suspected person's used toothbrush or a tissue biopsy specimen retained in a pathology laboratory. Second, parentage testing methods can be used if parents or descendants of the putative decedent are available to provide specimens. With skeletal and/or older remains, mtDNA is the method of choice for identification because it is more robust in the older biological materials. It should also be noted that mtDNA is present in hair shafts. Forensic hair examination and comparison is discussed in Chapter 14. The important points in the DNA-typing context are that hair roots are made up of cells, and nuclear DNA can be typed if enough of it can be extracted. The hair shaft does not contain any nuclear DNA, but mtDNA typing is possible. The Armed Forces DNA Identification Lab (AFDIL) in Dover, Delaware, has extensive experience using mtDNA typing to identify the remains of United States military personnel recovered from war zones, sometimes decades after death (see Case Study 11.1 earlier). It should be noted that DNA typing is the most complicated and costly of all human identification methods, and it should be considered a last resort. Further, DNA typing for this purpose is not useful unless investigators already have a putative or suspected identity. The suspected identity enables collection of appropriate reference specimens for comparison. It is remotely possible that a DNA profile from one of these cases could be in a database, but it is rather unlikely. In the past decade, DNA-typing methods have been used extensively to help identify victims in several plane crashes. The most overwhelming application of this kind, thus far, has been the attempted identifications of all the remains recovered from the World Trade Center terrorist attacks even though unidentified remains are mostly only a small piece of bone. Many identifications have been made, but the work is a slow process because there are so many specimens. Additionally, an accurate and complete list of the victims was not compiled, and there are likely to be many victims for whom no reference specimen is available.

Most operational forensic science laboratories that have biology/DNA sections can perform nuclear (STR) DNA typing in routine blood and other bodily fluid specimens. Some laboratories may be reluctant to tackle unusual specimens, such as bones or other tissues. However, most forensic laboratories will send these unusual specimens to specialized laboratories. The majority of forensic laboratories are not equipped to perform mtDNA typing. It is done in a relatively small number of specialized public and private laboratories, and it is not a routine method in forensic laboratories.

NEWER DNA TECHNOLOGIES

New DNA-typing methods are continually being developed. Some are only used in a few specialized laboratories or are becoming more widely practiced as time goes on.

We noted at the beginning of the chapter that human cells (other than spermatozoa, egg, and red blood cells) have 46 chromosomes; 44 of them are paired. The remaining two are called the sex chromosomes. Males are XY, and females are XX. A male inherits his Y chromosome from his fathers (Figure 11.16). As a result, Y-chromosomal typing can be useful in paternity testing. In sexual assault case evidence, which is typically a mixture of male and female cells, the only current method for "separating" male and female contributions to the mixture is differential extraction (see earlier). This procedure does not always achieve the desired separation. Further, it only works if the male contribution is in the form of sperm cells. In a mixture of female and male epithelial cells, the isolated DNA will be a combination of male and female DNA. In Chapter 10, we mentioned that men can be azoospermic—have no sperm cells because of a medical condition or vasectomy. Such men do have epithelial cells in their semen. In sexual assault cases where the men do not possess sperm cells, a mixture of male and female epithelial cells occurs in semen-vaginal mixtures. In such mixtures, the typing results are more complicated and harder to interpret than results from specimens originating from one person. There are a number of polymorphic genetic loci (such as Y-STRs; Figure 11.17) on the Y chromosome that can be typed, and many laboratories are now able to type Y chromosome loci. The advantage here is that only male cells in a mixture have Y chromosomes. Thus, the DNA profiles are, by definition, those of the male contributor. Y-chromosomal-typing procedures are coming into more widespread use because of the importance in analyzing mixture samples from sexual assault cases.

Now that the human genome has been completely sequenced, a great deal has been learned about DNA variability in the human population. One kind of variability is called the single nucleotide polymorphism (or SNP, sometimes pronounced as "snip"). This is the same kind of variability we described in mtDNA—one person differs from another at a particular location in DNA by one base. One person might have C while another has G. There are millions of SNPs in the human genome. Certain SNPs can be selected and used to essentially individualize biological evidence, just as STRs are used now. Changing from STRs to SNPs would require much change in forensic DNA technology. This shift would mandate DNA sequencing (Figure 11.18) in every forensic DNA analysis, whereas currently, DNA is not often sequenced. New equipment and technology such as next-generation sequencing techniques (Figure 11.19) would have to be obtained and installed in hundreds of forensic laboratories. SNP typing techniques would require dozens of validation studies to insure their utility for evidence analysis. Population studies would have to be done to determine the different SNP frequencies so the probabilities of a match that would occur by chance could be calculated. And, perhaps the biggest obstacle, all the databases would have to be redone. Every biological specimen that had been profiled for a database would have to be reanalyzed to get the SNP profile. Forensic DNA typing may eventually be done by SNP analysis, but right now it does not appear to offer any advantage over current technology sufficient to justify the effort and cost of making a change.

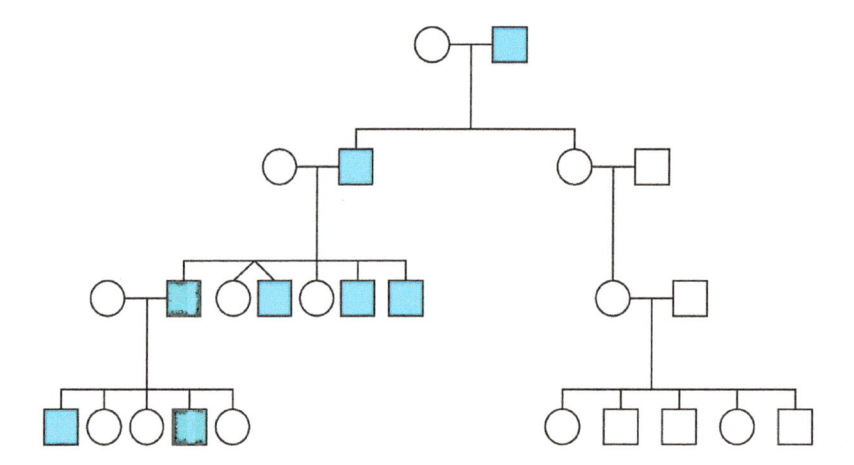

Figure 11.16 Human inheritance of the Y chromosome. Individuals who inherited the same Y chromosome are indicated by blue-symbols. Females and males are indicated by circles and squares, respectively.

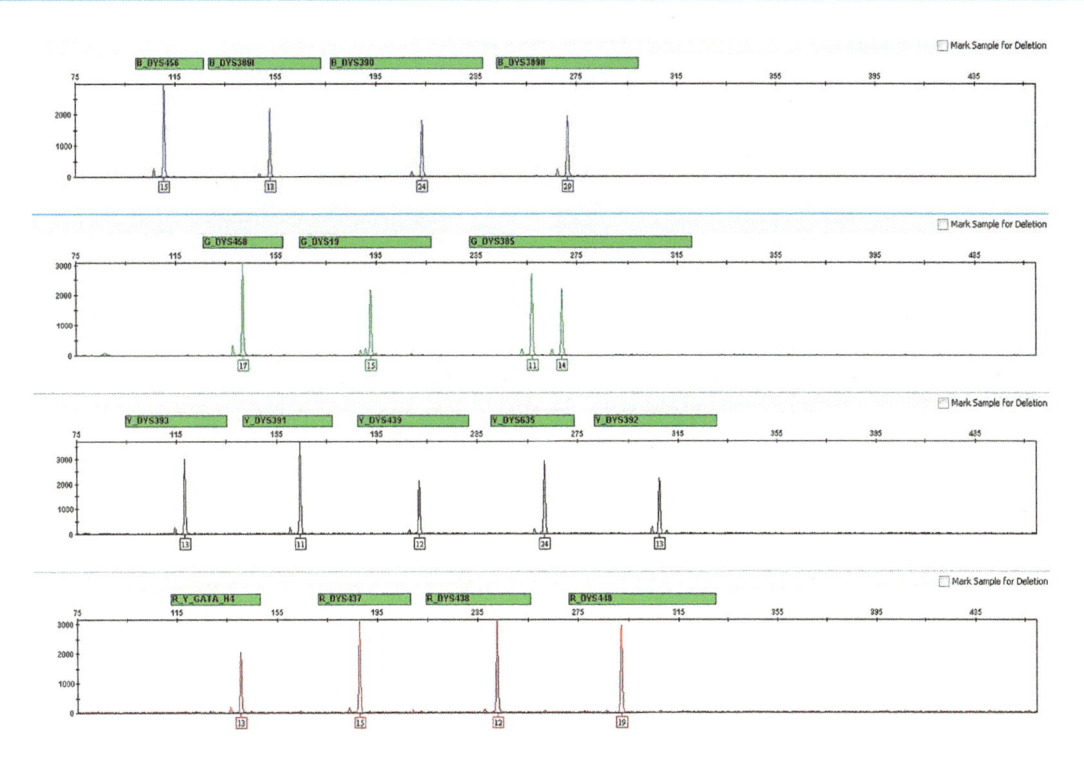

Figure 11.17 Y-STR profile (AmpFLSTR®Yfiler® PCR Amplification Kit). The genotype of the DNA profile is shown. DYS19: 15. DYS385 a/b: 11, 14. DYS389 I: 13. DYS389 II: 29. DYS390: 24. DYS391: 11. DYS392: 13. DYS393: 13. DYS437: 15. DYS438: 12. DYS439: 12. DYS448: 19. DYS456: 15. DYS458: 17. DYS635: 24. Y GATA H4: 13. (From Li, R., *Forensic Biology* (2nd Ed.), CRC Press, Boca Raton, FL, 2015.)

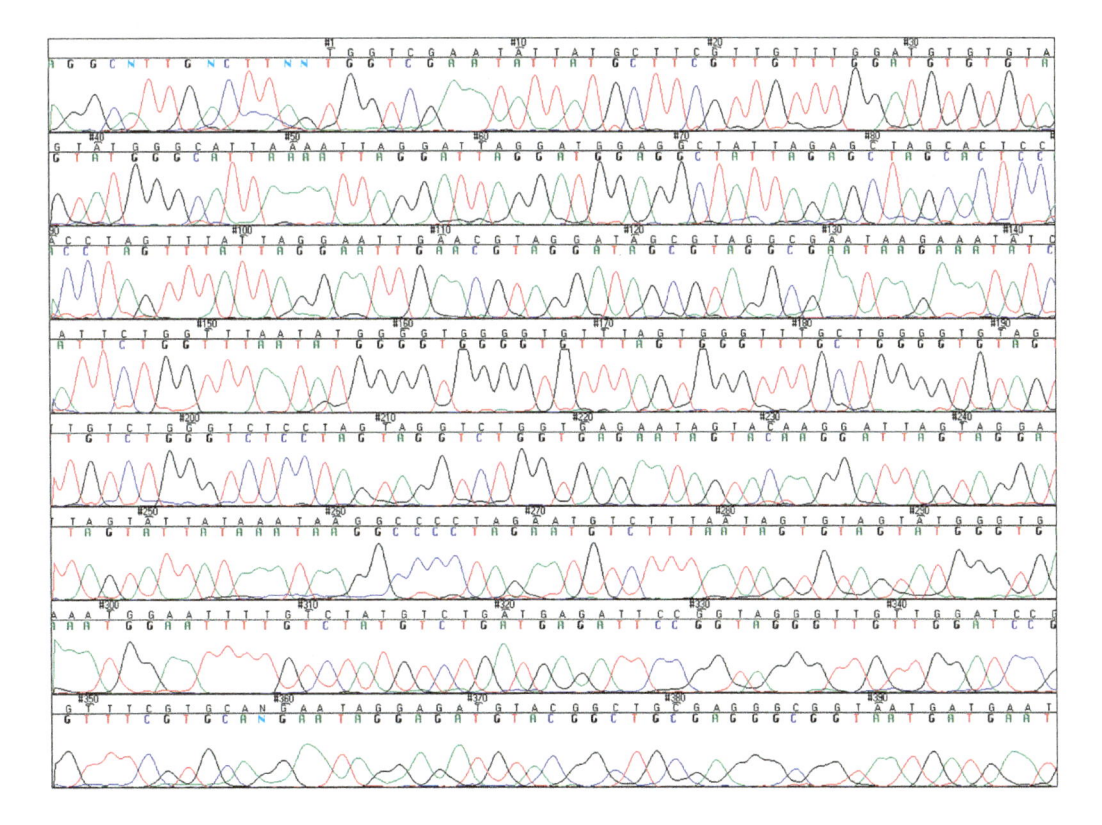

Figure 11.18 Electropherogram representing a DNA sequence. The sequences are presented as peaks with colors corresponding to each type of base (A, C, G and T). The sequence is read from left to right (5′ to 3′) and the bases are noted.

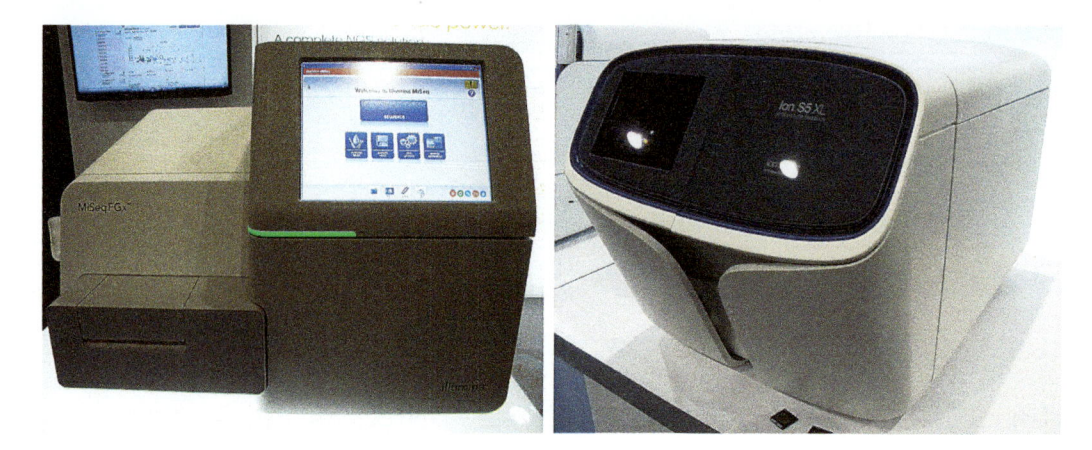

Figure 11.19 Photograph of NGS desktop sequencers. On the left, a MiSeqFGx manufactured by Illumna, and on the right, an Ion Torrent Ion S5 XL manufactured by Thermo Fisher Scientific are shown.

It is worth mentioning here, too, that forensic DNA analysis is not always restricted to humans. Animals and plants also have DNA, and efforts are underway to try and exploit the value of plant and animal DNA in forensic cases and a few successful applications have occurred.

As the genomes of animals are sequenced, and the individualizing segments of DNA identified, laboratories will be able to use technology very similar to what we have described for people to individualize animals, like dogs or cats, for example, if an appropriate specimen is available for testing.

The DNA sequences of plants are likewise being studied. It will take time to work out the sequences of many species, but already there is an interest in trying to see whether seized marijuana or cocaine, for example, can be DNA profiled to provide information about where the specimens came from. It might be that plant contaminants in these specimens, such as pollen or bits of leaf material, could also help in this respect. This information can be helpful to investigators who try to track the movements of shipments of illegal drugs.

DNA profiling also has a role to play in the efforts to detect and thwart bioterrorists. The DNA of many infectious disease agents, such as viruses and bacteria, has been sequenced, making it possible to detect these agents fairly quickly and unequivocally. Technologies are in the works to quickly take air samples, filter out any bioparticulates, and determine what they are. These techniques can help investigators detect potential bioterrorist threats. Ideally, they would be detected before deployment, but even if it were after the fact, it would help police, military, and emergency service response personnel to know what precautions were necessary, and possibly how to contain the agent or how to best treat the living victims.

STRENGTHS, LIMITATIONS, PROMISE, HYPE

DNA technology is the most revolutionary tool to become available to forensic scientists, perhaps in the whole history of the profession, since fingerprints. For the first time, biological specimens can be effectively individualized—shown to come from one person. Equally important, DNA typing will almost certainly exclude someone who did not deposit a biological specimen. DNA can be typed from many different kinds of biological evidence other than blood and semen stains, making it a more versatile tool in criminal investigations. Also, the ability to store profiles in a database and search for them at a subsequent time is revolutionary. The forensic DNA database represents another important tool (besides fingerprints) that can be used to reliably identify unknown persons.

There are, however, limitations to the technology. First, many crime scenes or criminal situations do not yield any useful biological evidence. In those cases, DNA profiling is not possible. Additionally, in many criminal cases where there is biological evidence, there can be a perfectly innocent explanation for the biological materials being there.

Although DNA technology is comparatively straightforward in many cases, complications can arise. Mixtures of DNA from two or more individuals; for example, can become difficult to sort out in terms of the depositors of biological evidence. DNA typing from very small quantities of biological materials (like dandruff flakes, partial fingerprints, or one hair root) is of questionable reliability at present. This type of specimen is often called "low copy number" DNA evidence. Interpreting the results from typing low copy number DNA evidence is probably not ready (reliable enough) for courtroom use as yet.

As we have said, the technology is revolutionary and exciting. And the media have made the most of it—not only in straight news reporting but in popular entertainment programs. The tendency by TV and radio to try to reduce everything to 15-second sound bites, and by all media to oversimplify scientific stories to the point where they are no longer even correct, has not helped the public really understand DNA typing.

One big problem that has received some, but perhaps not enough, attention in the public media is backlogs. Many laboratories have become overwhelmed with a combination of case materials submitted for DNA typing on the one hand, and specimens from convicted offenders to be typed and entered into CODIS on the other. In many places, there are both casework and CODIS backlogs. None of this is helpful in realizing the full promise of DNA technology. More recently, the federal government has put considerable resources in the hands of the states to help alleviate the backlogs. However, it is going to take time.

As backlogs are diminished, we can look forward to a time when cases will be worked quickly and efficiently. In addition, as more and more individual profiles are added to the databases, CODIS hits (matches between database profiles and specimen profiles) will become even more frequent and routine than they are now.

Key terms

- chromosomes
- Combined DNA Index System (CODIS)
- differential extraction
- disputed parentage
- DNA polymerases
- forensic DNA database
- forensic DNA typing
- genetic code
- genetic markers
- genotype
- microsatellite
- minisatellite
- mitochondria (sing., mitochondrion)
- mitochondrial DNA (mtDNA)
- mitosis
- nuclear/genomic DNA
- nucleotide
- nucleus
- polymerase chain reaction (PCR)
- primer
- restriction fragment length polymorphism (RFLP)
- sequence
- short tandem repeat (STR)
- tandemly repeated sequence
- transcription
- translation
- variable number of tandem repeat (VNTR)

Review questions

1. Describe chromosomal inheritance in humans.
2. Describe the structure of DNA.
3. What is the "genetic code?"
4. Where do you find DNA in the body? What is the forensic usefulness of the different types of DNA?
5. How do you collect and preserve biological evidence for DNA analysis?
6. What is RFLP DNA typing and how does it work?
7. What is PCR? How does it work? What are its advantages for forensic work?
8. What are STRs? How does STR DNA typing work?
9. What is CODIS? How does DNA databasing work?
10. Describe how DNA typing works in blood transfer cases; in sexual assault cases; in human remains identification cases.

Fill in/multiple choice

1. The first widely used method of forensic DNA typing is called restriction fragment length polymorphism (RFLP) analysis. The individuality detected in this type of DNA analysis, as it is used by different forensic labs, is due to
   a. differences in coding (functional) genes
   b. variable numbers of tandemly repeated sequences in different people
   c. slight differences in hemoglobin sequences detected by allele specific oligonucleotide probes
   d. all of the above
2. An individual inherits 23 _____ from each biological parent.
3. No two individuals, except for identical twins, can be expected to have exactly the same combination of
   a. blood types
   b. serum groups
   c. nuclear DNA markers
   d. mitochondrial DNA markers
4. The advantage of PCR-based methods is that they work well with both _____ DNA and require _____ amounts of DNA.
5. All state DNA databases are currently authorized to collect samples for DNA databasing from
   a. every child born in the state
   b. every male between 15 and 30 years old in the state
   c. all individuals convicted of specified crimes in the state
   d. all individuals arrested for specified crimes in the state

References and further readings

Breeze, R., B. Budowle, and S. Schutzer. *Microbial Forensics*, Burlington, MA: Elsevier Inc., 2005.

Cardozo Law School (Yeshiva University, New York) Innocence Project Web site:www.innocenceproject.org. (The Innocence Project was started by attorneys Barry Scheck and Peter Neufeld.)

Connors, E., N. Miller, T. Lundregan, and T. McEwen. Convicted by Juries, Exonerated by Science: Case Studies in the Use of DNA Evidence to Establish Innocence After Trial. Research Report, National Institute of Justice, NCJ 161258, June 1996.

Coyle, H. M. *Forensic Botany: Principles and Applications to Criminal Casework*. Boca Raton, FL: CRC Press, 2005.

Jeffreys, A. J. "Genetic Fingerprinting." *Nature Medicine* 11, no. 10 (2005): 1035–1039.

Jeffreys, A. J. "The Man Behind the DNA Fingerprints: An Interview with Professor Sir Alec Jeffreys." *Investigative Genetics* 4, no. 1 (2013).

Jobling, M. A., and P. Gill. "Encoded Evidence: DNA in Forensic Analysis." *Nature Reviews Genetics* 5, no. 10 (2004): 739–751.

Kobilinsky, L. "Deoxyribonucleic Acid Structure and Function—A Review." In *Forensic Science Handbook*, ed. R. Saferstein, vol. 3, pp. 287–357. Englewood Cliffs, NJ: Prentice Hall, 1993.

Northwestern University, Chicago, IL: Law School Innocence Project, Chicago, IL, www.law.northwestem.edu/wrongflilconvictions/; Medill School of Journalism Innocence Project, www.mediIl.northwestem.edu/specialprograms/innocence/.

Roewer, L. "DNA Fingerprinting in Forensics: Past, Present, Future." *Investigative Genetics* 4, no. 1 (2013).

Rudin, N., and K. Inman. *An Introduction to Forensic DNA Analysis*. 2nd ed. Boca Raton, FL: CRC Press, 2001.

Shaler, R. C. "Modem Forensic Biology." In *Forensic Science Handbook*, ed. R. Saferstein, vol. 1, 2nd ed., pp. 525–613. Upper Saddle River, NJ: Prentice Hall, 2002.

van Oorschot, R. A. H., K. N. Ballantyne, and R. J. Mitchell. "Forensic Trace DNA: A Review." *Investigative Genetics* 1, no. 1 (2010).

Westphal, S. P. *DNA Profiles Link Dope to its Source*. New Scientist Print Edition, June 9, 2006.

CHAPTER 12

Arson and explosives

Lead Case: The Unibomber

"For 17 years the FBI had been investigating a series of bombings in the case the Agency referred to by the acronym UNABOM" (UNiversity and Airline BOMber). On a very chilly day in April in a remote area of Montana, there were over 100 agents surrounding a tiny 10-foot by 12-foot cabin. When a local forest service officer hailed the inhabitant of the cabin, a rather disheveled Ted Kaczynski stepped out and agents served a search warrant and entered the cabin. Within a short time, word was radioed to the agent in charge of the massive investigation that the contents of the cabin clearly indicated that they had the correct man. Thus, ended one of the longest and most frustrating investigations in the history of the FBI.

However, the difficult job of proving beyond a reasonable doubt, in a court of law, that Kaczynski was guilty of this series of bombing was in many ways just beginning. This reclusive bomber had been extremely careful to minimize the clues that law enforcement could use to find him or prove that he was the bomber. For all those years, he had resisted the temptation to preach his cause, thereby making the tracing of this lone bomber almost impossible.

The story started on May 25, 1978, at Northwestern University where a security guard was injured when opening a package. In total, there were 16 bombings over the course of 17 years that resulted in three deaths and quite a number of serious injuries. In 1995, the bomber finally broke his silence with the publication of his manifesto in *The New York Times* and the *Washington Post*. This strange and rambling document concerning the evils of modern industrial society was published partially because the mysterious bomber promised to stop the series of bombings if they would publish the tract. This publication caused his brother to reluctantly come forward and indicate to the authorities that his reclusive brother, who lived in a small cabin in Montana, might well be the author. This was the first real break in the case and lead directly to the arrest of Dr. Theodore Kaczynski as the prime suspect of being the "Unabomber."

Forensic science had already played an important role in the difficult investigation by clearly establishing that these 16 bombings were all the work of the same individual. Kaczynski had been careful to use as little traceable material as possible in his devices and to remove markings from any commercial parts used. After all, he was an individual of great intellect having a PhD in mathematics and a short career as a college professor before becoming so estranged from society that, with his brother's financial help, he bought the tiny cabin in which he lived in seclusion and fashioned his bombs.

The initial role of forensic science and physical evidence was to convince the FBI that one individual was responsible for the construction of a continuing series of bombs. This belief was based on unmistakable similarities in the construction of the bombs sent to individuals with no obvious connection to each other and over an extended period of time. Just three of the many characteristics, which led to this conclusion, were similarly constructed electrical fusing systems, highly unusual loop switches, and the fact that all but one of the bombs were enclosed in a handmade wooden enclosures. In addition, he used end caps held in by retaining pins to seal the pipes containing the explosive material and this was a system, virtually, unique to the Unabomber's bombs. Although the series of bombs showed evolving sophistication of construction and increasing emphasis on causing serious damage and injury, the careful examination of materials recovered after each explosion showed these consistent and very unusual characteristics.

In addition, each of the devices after the first two were found to have a small metal plate with the letters "FC" stamped into it. There were many additional, highly technical characteristics that also strongly indicted a common source for this series of devices.

Once Dr. Kaczynski was identified and in custody, there remained the problem of proving in court that he was indeed the Unabomber. The fact that he was a recluse and a highly intelligent individual made this task much more difficult. There were no witnesses to any of the incidents since he either left the package or sent it through the mail. In fact, the famous Unabomber image that was circulated for years was based on a description given by one individual who had seen someone walking away from an area shortly before a bomb went off. Again the burden would fall on examination of physical evidence. Fortunately, his tiny cabin proved to be a treasure trove of evidence that could be used to link him to the series of bombings.

The listing of items removed from that cabin, which had no running water or electricity, runs to about 700 items and included all the necessary materials to construct explosive devices including some of the highly unusual items mentioned above, such as his unusual electrical fusing systems, loop switches, and many kinds of wood for construction the boxes. In addition, there were clearly labeled, and neatly stored samples of many different chemicals associated with homemade explosives, including the ones used in his devices. He has all the necessary tools, pipes cut to the size, which he used in his bombs, and many other important linkages to his style of improvised explosive device. Further, the investigators found a completed device, and another partially completed one. Other important items of physical evidence included numerous handwritten notebooks with meticulous descriptions of his experiments to develop his style of device and constant efforts to improve the devices. Several typewriters were also found that could be linked to his manifesto and some of the mailing labels, portions of which had been recovered after most of the explosions. He had other notes that listed the individuals to whom he had sent bombs and other individuals who appeared to be under consideration as possible recipients of future devices.

There was much legal maneuvering, but a jury was picked, and pretrial motions were all filed when, just before testimony was to start, Dr. Kaczynski's motion to defend himself was denied and he decided to plead guilty in exchange for an agreement that he would not receive the death penalty but would spend the rest of his life in prison with no chance of parole.

1. "Success Took Seventeen Years", Johnston, D., *New York Times*, May 5, 1998
2. www.crimelibrary.com/terrorists_spies/terrorists/kaczynski/1.html
3. Affidavit of Assistant Special Agent in Charge Terry D. Torchy U.S. District of Montana, www.courttv.com/archive/casefiles/unabomber/documents/affidavit.html
4. The Unabomber case, www.unabombertrial.com/archive/1998/01298.07.html

LEARNING OBJECTIVES
- The science underlying combustion (fire).
- Commonly encountered fuels.
- Importance of pyrolysis in the combustion of solid fuels.
- Useful investigative information available from careful examination of a fire scene.
- Primary reasons for individuals setting arson fires.
- Proper examination and processing of materials collected in the investigation of suspicious fires.
- The most commonly encountered accelerants used in arson fires.
- Laboratory analysis process for evidence from suspicious fires.
- The science underlying an explosion.
- Commonly used explosive materials.
- Necessary components of an explosive device
- Processing and sampling of an explosion scene
- Laboratory analysis of explosive devices and residues.
- Different approaches to examination of exploded and unexploded devices.
- How explosives are identified from the analysis of explosive residues.
- Nature of improvised explosive devices (IEDs)

FIRE AND ARSON

The nature of combustion—flaming and glowing combustion

To understand arson and explosions, it is very helpful to know a bit of the basic chemistry and physics of fire. This understanding is critical to recognizing what has happened at a fire or explosion scene. Some knowledge of the nature of combustion provides important insights needed for effective investigation of such incidents. Therefore, a look at the nature of combustion is a good starting point. To a chemist, combustion is simply a rapid oxidation reaction. It is the combination of oxygen, usually from air, with carbon from some sort of fuel. When oxygen and a carbon source combine with each other to form carbon dioxide and water, the resulting chemical reaction gives off heat. Chemical reactions that give off heat are called *exothermic* reactions and the combustion reaction is highly exothermic. When the reaction is complete, all the hydrogen in the fuel is converted to water (H_2O) and all the carbon is converted to carbon dioxide (CO_2). When one "flicks one's Bic," one sees a flame caused by butane lighter fuel reacting with oxygen from the air to form carbon dioxide and water. In real fires the reaction is not always complete, but the vast majority of the fuel is converted to carbon dioxide and water. Where combustion is not complete, usually due to lack of enough oxygen, an important byproduct is carbon monoxide. Because structure fires seldom have ideal conditions, carbon monoxide poisoning is a frequent cause of death for those trapped at fire scenes.

Successful combustion (fire) requires a balance of Fuel and Oxygen and some ignition source to get the reaction started. Once the reaction begins, under most conditions, it gives off enough heat to keep itself going and under the proper conditions to increase in intensity. There are three common mnemonic devices used to assist in remembering the requirements for a successful fire (oxidation reaction). The most common are the fire triangle and the fire pentagon. The fire triangle (Figure 12.1) shows diagrammatically that one needs fuel, usually something that has carbon in it, oxygen, normally air, and heat to get it started. The fire pentagon is a little more complete description of what one needs to have a *sustained* combustion; that is, a fire. Three of the five sides of the pentagon are identical to those in the triangle: fuel, oxygen, and heat; in addition, the pentagon represents two additional components: ignition, something to get the combustion started, and a sustainable chain reaction. This is the chemist's way of saying there is nothing present to keep the heat produced from sustaining the chemical reaction. Certain materials can interrupt the combustion reaction, either by reducing the heat or the chain reaction necessary to maintain the fire. These materials are good candidates for preventing or extinguishing fires. The most common and obvious one is water. Water is capable of "absorbing" considerable heat. It helps extinguish a fire by removing the heat necessary for the sustained reaction. Other chemicals, that interrupt the chain reaction, are used to make fire retardant materials (e.g., fabrics for children's sleepwear or upholstery in airplanes) or may be used in chemical fire extinguishers.

There are two major types of combustion: flaming combustion (visible flame) and glowing combustion (glowing coals). An important requirement for *flaming* combustion is the need to have both the fuel and the oxygen in the gas phase. This is not intuitively obvious because one can put a match to a piece of paper (a solid, and clearly not in the gas phase) and see a flame almost immediately. Nonetheless, to have flaming combustion (Figure 12.2), the general rule is that both the fuel and the oxygen must be in the vapor phase. For sufficient heat to be produced to sustain flaming, rapid combustion, the fuel, and the oxygen have to interact with each other rapidly, and can do so at the rate necessary when they are both in the vapor phase. Since the reaction is using the oxygen from the air, which is already the

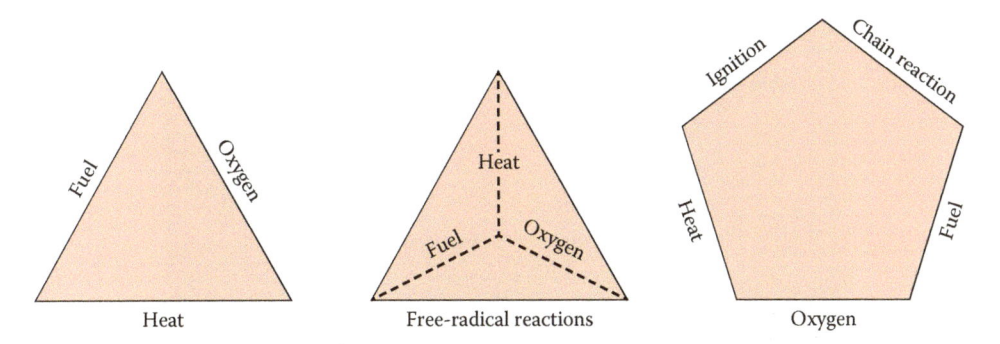

Figure 12.1 Mnemonic Devise used to remember the items required for sustained combustion: fire triangle, fire tetrahedron, and fire pentagon.

Figure 12.2 A large building fully enveloped in flames.

vapor phase, the issue is getting the fuel into the vapor state. A spark and/or heat is required to get the fuel into the vapor phase. As will be discussed later, one of the other roles for heat is breaking down most solid fuels to produce products that are vapors (in the gas phase) or can be easily vaporized. Most common solid fuels (wood, wax, paper) can either be readily vaporized or can be broken down (pyrolysed see section "Pyrolysis of solid fuels", page 296) by heat to produce materials that are readily vaporized.

However, there are a few solid fuels that cannot be easily vaporized. Charcoal is a common example. That is why charcoal lighter fluid, or an electrical heater is needed to get a charcoal fire started. Although the charcoal will burn and give off a considerable amount of heat once started, flames are typically not visible (once the lighter fluid burns off). The charcoal is still burning, and it will eventually be completely consumed—transformed from carbonaceous material to carbon dioxide, water, and a little bit of ash. What is occurring here is called *glowing* combustion (Figure 12.3), where the fuel is a solid and the oxygen is in the vapor phase. The combustion reaction is not rapid

Figure 12.3 A campfire with largely glowing combustion and little flaming combustion.

enough to produce a flame, although it is rapid enough to sustain itself and give off sufficient heat to cook one's burgers. Glowing combustion is too slow for an arson fire but is fine for cooking and may be important in accidental fires. A lit cigarette dropped into a couch cushion; for example, may smolder for a long time before building up enough heat and transforming enough of the cushion material to provide fuel in the vapor phase before it changes over to flaming combustion.

Common fuels used to produce flaming combustion

Common fuels start out as gas, liquid or solids. As indicated above, both fuel and oxygen must be in the gaseous phase to support flaming combustion. There are certain fuels that are already in the gaseous phase and, therefore, immediately suitable for flaming combustion. The most dangerous of these is hydrogen. Hydrogen not only burns readily and with a very hot flame, but is easily ignited.

SIDEBAR

One of the most famous aviation disasters of all time occurred in New Jersey when the Hindenburg Zeppelin exploded and burned violently in a thunderstorm, killing many on board and on the ground. That zeppelin (blimp) was filled with hydrogen and its violent destruction by the explosion stopped the commercial use of zeppelins (blimps) for many years. Blimps are now filled with helium, a non-flammable gas.

On the other hand, hydrogen is the most environmentally friendly possible fuel, because the product of its combustion is only water, no carbon dioxide. As a result, there is considerable interest in developing ways to use hydrogen as a clean fuel for automobiles. One difficulty is developing safe ways to store and transport hydrogen.

A much more common gaseous fuel is natural gas. Natural gas is used for home heating and cooking in much of the United States and in many industrial applications, as well. Where it is readily available, it is usually the fuel of choice. Natural gas is primarily methane (CH_4), which is a gas at room temperature and burns very cleanly. Similarly propane, another room temperature gas is widely used as fuel where natural gas is not readily available. Propane is supplied from tanks to provide readily transportable fuel for lanterns, home stoves, camp stoves, and many other items.

Liquid fuels are more common. Gasoline may be the most well-known fuel for transportation vehicles. Many millions of gallons of gasoline are consumed by the automobiles, buses, and trucks that move us and our goods, from point to point. Although gasoline is a liquid fuel, some of it is always in the gas phase, even at room temperature (as is obvious because we can smell it). Thus, gasoline is a liquid fuel that is easily vaporized by warming and, therefore, readily supports flaming combustion. Fuel oil, on the other hand, also has a strong odor, but at room temperature very little is in the gas phase. In fact, one can extinguish a flaming match in fuel oil. It is an excellent fuel; however, because once it is heated to vaporize enough to support flaming combustion the combustion produces heat that will then vaporize sufficient additional fuel to sustain the fire.

Another common liquid fuel is ethyl alcohol (ethanol). In flaming desserts, such as Bananas Foster, Crepe Suzettes, or any other flambé dish, the flame is from ethanol, from an alcoholic beverage. Any alcoholic beverage with over 40% alcohol, if warmed, will produce enough ethanol vapor to be readily ignited and provide the pretty blue flame associated with a flaming dish or drink (Figure 12.4). The ethanol quickly burns off, leaving behind that nice flavor associated with the beverage. In recent years, ethanol has been added to gasoline to reduce petroleum usages. Kerosene is another very common liquid fuel that falls between gasoline and fuel oil in terms of vapor above the liquid. It is the fuel of jet engines, lanterns, often camp stoves, and efficient space heaters.

Recent years have seen government sponsor research for renewable energy primarily from crop waste or inexpensive non-traditional crops. This is now beginning to produce some limited availability of a few renewable fuel and means that forensic laboratory are going to start seeing some non-petroleum combustible liquids used as arson accelerants.

There are also many very common solid fuels. Although wood has become a more popular fuel in recent years, it is not usually thought of as a heating fuel, but wood stoves are quite popular in rural areas. Wood and many other solid fuels must be pyrolyzed to provide enough volatile material to support flaming combustion. Coal is actually a much more widely used solid fuel, particularly in industrial and commercial heating. Coal is derived from vegetable matter that has been under high pressure and at high temperature deep in the earth for millions of years. During that time, it

Figure 12.4 Use of alcoholic beverage to flame a desert drink.

has been largely converted to carbon, but does contain other residues from its plant beginnings. It is an excellent fuel that has been used for centuries. Many polymeric materials, such as plastics, will also burn, but are not usually used as commercial fuels. Unfortunately, because some plastics produce carbon monoxide, hydrogen cyanide or other toxic gases when they burn, the combustion of plastics can kill people in accidental fires. Most commercial fuels burn largely to carbon dioxide and water, which are not toxic. Finally, a common solid fuel is charcoal. Charcoal is perhaps best known as the fuel for grilling food in the backyard. It is not hard to light, and it produces that nice smoky taste so desirable to many. Charcoal is wood that has been very slowly heated and burned in an oxygen deficient atmosphere, until it is essentially pure carbon. That is why charcoal will not vaporize, and will only burn slowly at its surface (glowing combustion) producing a nice even heat for a long time, but not producing a flame. The flames one sees on the barbeque grill are from fats and oils dripping from the food and then vaporized and ignited by the hot charcoal.

Measures of fuel characteristics—measures of combustibility

Two of the most common measures of combustibility are flash point and fire point (Table 12.1). Flash point is the lowest temperature at which a liquid produces enough vapor to be ignited by a small flame. This was originally done in what was called an open cup flash point tester. The flash point tester has a little cup in which the flammable liquid is placed so that it could be slowly heated, and a small flame can be swung over the cup. The cup was heated and tested with the flame until it ignited and lowest temperature that produces ignition is the flash point. Flash point is a very important characteristic of a fuel because it indicates the temperature below which it is safe to handle the fuel with little danger of accidental ignition. Simply, if a fuel has a very low flash point then it vaporizes easily and must be handled with great care and if it has a high flash point it is not very easy to light and safer to handle. Of course room temperature gaseous fuels have flash points below room temperature. The fire point is the temperature at which there is enough heat to cause combustion even in the absence of a source of ignition.

Another useful measurement used to characterize fuels is the flammable range.

Flammable range is a measure of the percentage of the fuel that, when mixed with air, is needed to sustain combustion. For sustained combustion, the correct ratio or amounts of fuel and oxygen must be present for the chemical reaction of combustion to be self-sustaining. If there is too much or too little of either fuel or oxygen, one does not have ideal conditions and combustion will not be sustained. There is a particular range of fuel to air ratios for each fuel that will sustain combustion. Hydrogen is particularly dangerous because it has such a wide flammable range, from 4% to 74%, much wider than almost any other fuel. This makes hydrogen a particularly serious fire threat. On the other hand, gasoline, which one thinks of as very flammable, is only flammable in the 1% to 6% range. If you have significantly more than 6% of gasoline vapor mixed with air, it will not burn. Most common liquid fuels have a similar flammable range, in the 2% to 10% range.

Table 12.1 Flash point, boiling point, and flammable range of a number of flammable materials

Material or fuel	Flash point, °F	Boiling point, °F	Flammable range	Ingnition temperature, °F
Hydrogen	Very low	−423	4.1–74	1075
Carbon monoxide	Very low	−313	12–75	1290
Methane	Very low	−258	5.5–14	1200
Ethylene (etherne)	Very low	−155	3–34	1010
Ethane	Very low	−126	3.2–12.5	986
Acetylene (ethyne)	Very low	−119	2.5–80	635
n-Propane	Very low	−44	2.4–9.5	874
n-Butane	−216	31	1.6–6.5	806
Ethyl ether (diethyl ether)	−49	94	1.8–50	355
n-Pentane	−40	97	1.4–8.0	588
Petroleum ether (benzine)	−40	95–175	1.4–5.9	475
Gasoline	−50	102–230	1.3–6	495
Acetone	−4	134	2.2–13	1000
Methyl alcohol (methanol)	54	148	6.3–6.5	880
n-Hexane	−7	156	1.2–6.9	477
Ethyl alcohol (ethanol)	48	173	3.3–19	800
Methylethyl ketone (2-butanone)	35	175	1.8–11.5	960
Benzene	50	176	1.4–8	1075
Isopropanol (2-propanol)	53	180	2–12	852
n-Heptane	30	209	1.1–6.7	~450
Toluene	50	231	1.3–7	1025
Butyl alcohol (n-butanol)	97	243	1.5–11	695
n-Octane	56	258	0.9–6	450
Cellosolve (2-ethoxyethanol)	111	275	2.6–16	460
Naphtha (Stoddard solvent)	100–110	300–400	1.1–6	450
Kerosene	110–185	300–600	1.2–6	490
Ethylene glycol	232	388	3.2–?	775

SCIENCE SIDEBAR

Another important property of both fuels and combustion products is called the relative vapor density. A rule in chemistry says that one mole of a chemical compound, which is its molecular weight expressed in grams, will occupy a volume of 22.4 L at standard temperature and pressure. In other words, when one vaporizes water, which has a molecular weight of 18, a little over one half of a shot glass (18 g) will produce 22.4 L of gas (steam) at about room temperature. That is roughly a volume the size of a 5-gallon can. Whatever the material, the molecular weight of it in grams it will occupy 22.4 L when it is in the vapor phase. Thus, if a fuel has a molecular weight of 400, there will be 400 g of that fuel in about five gallons of vaporized fuel. Therefore, the higher the molecular weight of a material, the more of that material will be needed to occupy 22.4 L when in the gas phase. Air is about 79% nitrogen, 21% oxygen, and has small amounts of several other gases. Nitrogen has a molecular weight of 28, and oxygen has a molecular weight is 32, so one could estimate the molecular weight of the gaseous mixture we call air at about 29. If you have a mixture of gases like air that has an average molecular weight of 29, then 29 g of it will also occupy 22.4 L. The net result is that most materials when vaporized are much heavier than air and a few are lighter. The important fact about vapor density is that any vapor composed of a material that has a molecular weight higher than 29 will tend to settle, because it is heavier than air. It will tend to push down on the air, forcing the air up. Any vapor composed of material that has a molecular weight less than 29 will tend to rise. So hydrogen, which has a molecular weight of 2, is much lighter than air and it would rapidly rise. On the other hand, carbon dioxide, which is the combustion product of almost all carbon-based fuels, has a molecular weight of 44.

Since carbon dioxide is denser (heavier) than air, it will tend to settle and push the air away from areas near the floor (see the sidebar on vapor density nearby). If there is limited ventilation, the carbon dioxide will displace the needed oxygen (in the air) and slow the burning process. That is why carbon dioxide fire extinguishers (Figure 12.5) are effective. If the carbon dioxide is directed to area of active burning, it tends to keep the oxygen away and slows down or even extinguishes the fire.

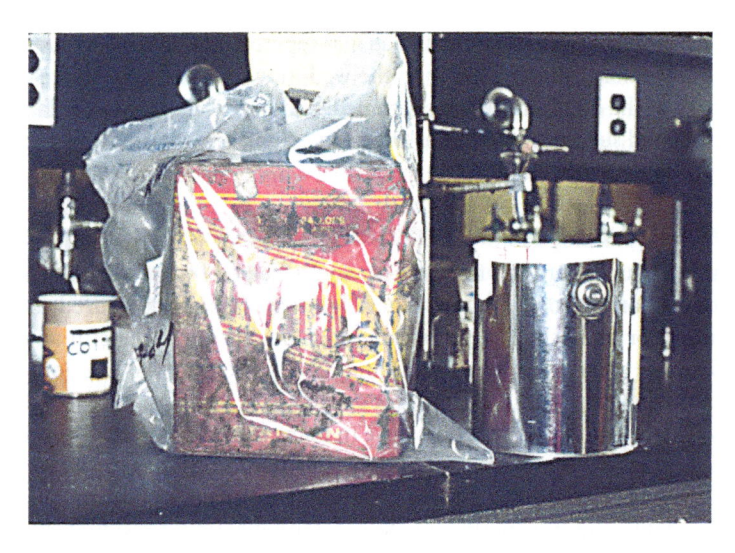

Figure 12.11 Cans recovered from near a fire scene to be screened for possible residual flammable liquid.

Figure 12.12 A shoe print left in soil close to the point of entry into an arson scene.

toolmarks, blood, or anything else that might be of value to the investigation, should not be overlooked because of the preoccupation with collection of debris to search for possible accelerant.

Remember that ignitable liquid residue evidence is useful in establishing that the fire was arson, but it is seldom helpful in identifying the arsonist. On the other hand, an arsonist may leave fingerprints on a can or other item, footwear impressions (Figure 12.12), even blood if cut by glass broken during forced entry. Those types of evidence could prove to be valuable. Recovered timing or ignition devices can sometimes provide a "signature" of a particular arsonist (see lead case).

LABORATORY ANALYSIS OF DEBRIS AND OTHER SAMPLES

Preparation of gas or liquid samples and processing of residue samples

Laboratory analysis of debris and other fire related evidence is a significant part of the work of virtually every forensic laboratory. After the initial examination and description of the evidence, one must prepare the samples for analysis.

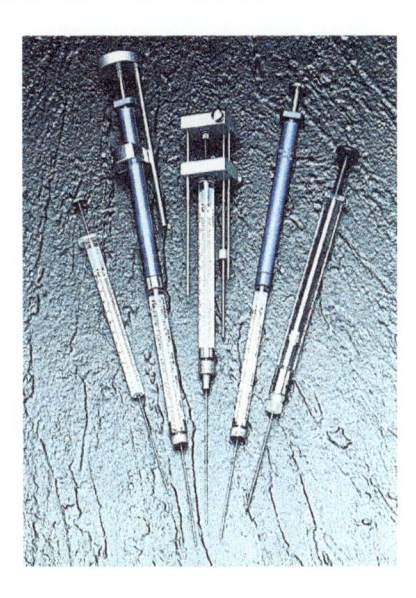

Figure 12.13 Typical microliter syringes used for injecting very small liquid samples into a gas chromatograph.

Gas samples are obtained, usually from the headspace above "empty" can by drawing vapor into a gas tight syringe and then directly injected into the instrument. Liquids samples are even easier since the sample is drawn into a special microsyringe (Figure 12.13), which is used to inject it into a gas chromatograph. See the sidebar on chromatography and GC nearby. These are the easiest to process.

Three primary techniques for preparation of debris samples

Most of the time, forensic laboratories receive burnt debris, soggy carpeting, or a variety of other things collected from a fire scene that may contain accelerant. If there is an ignitable liquid residue present, it is usually in trace amounts and mixed with the debris. There are three common methods of processing such material to try to separate or concentrate any ignitable liquid from the debris. Since fires are usually extinguished with large volumes of water, the ignitable liquid residue may be washed away. It is fortunate that the most commonly used accelerants are not soluble in water. Although they may still be mechanically removed by the flow of the water, if they are absorbed into debris, the water will not extract them efficiently.

The simplest of the three techniques is called heated headspace.

The laboratory receives a metal paint can with debris collect from the scene at a possible point of origin of the fire. The examiner will usually briefly lift the lid and take a little sniff. If there is an odor of gasoline or other recognizable ignitable liquid, then probably the best technique is heated headspace. It is not the most sensitive technique, but it is simple, avoids introduction of contaminants, and works well where there is adequate amounts of accelerant present. If, however, only a burnt odor is detected, then one of the other techniques may be best. Another reason for taking a quick look into the sealed can is to make sure there are no ignition devices present that could cause the closed can to explode when it is heated. The paint can lid is punctured with a nail, and the small hole is covered with a piece of tape. The can is then placed in an oven and set at a temperature that will vaporize any ignitable liquid residue, but not create enough steam to pop the lid of the can. Seventy or eighty degrees Celsius will usually work well. After allowing an hour or more for the contents to thoroughly equilibrate, the can is removed from the oven and the air space above the debris quickly sampled. The higher the vapor pressure of the material the more will be in the gas phase, and since most accelerants have fairly high vapor pressures, there will often be a significant amount in the gas phase. Sampling is done simply by inserting the needle of a gas-tight syringe into the can through the hole, and drawing a small sample of the vapor into the syringe. This vapor sample is injected immediately into the gas chromatograph.

Heated headspace is a very simple technique with little chance for contamination. Nothing is added to the sample and handling is kept to a minimum. The weakness of the technique is that it is not good for trace amounts of ignitable liquid residues or for high boiling or mixtures of ignitable liquids.

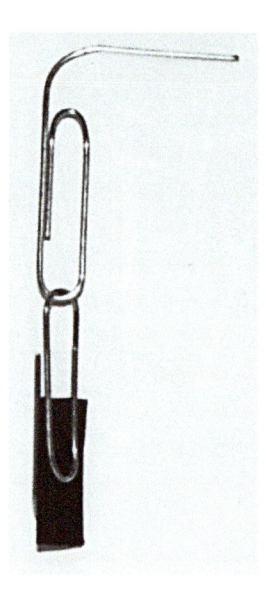

Figure 12.14 An activated carbon strip used for recovering flammable material from an enclosed debris samples.

Carbon strip or tube absorption is now the technique most used by forensic laboratories. The active component of the absorption strip or tube is activated charcoal, which is a very high surface area carbon material that strongly absorbs hydrocarbon vapors that come in contact with it. These strips (Figure 12.14) were developed for the badges that people who work in chemical plants wear to determine any exposure to chemicals in the workplace atmosphere. It was discovered that a piece of one of those strips could be used to sample hydrocarbon and other chemical vapors from the headspace in a can of debris from a fire. This ignitable liquid residue isolation technique involves suspending a small absorption strip with its thin coating of activated carbon, in a can of debris and heating the can as with the heated headspace technique. During heating, any volatile residues are vaporized and adsorbed onto the charcoal strip. The strip is then removed and any adsorbed hydrocarbon (ignitable liquid) residue washed off using a volatile solvent, such as diethyl ether, carbon disulfide, or pentane. The wash can then be analyzed directly, or gently concentrated by evaporation of the volatile solvent. This technique has proven to be efficient and sensitive for detecting any residual ignitable liquid in arson debris. There are some newer variations on the basic technique, such as the use of solid phase microextraction fibers, but the basic principle as the same. Because of the sensitivity of the absorption methods, one must be particularly concerned with avoiding contamination. The strips must be handled carefully and exposing the strips to the atmosphere in a laboratory area where arson evidence is stored has to be avoided.

Solvent wash is the last of the common technique for recovering possible accelerant material form arson evidence. Solvent wash works better for higher boiling ignitable liquids that cannot be easily vaporized, and for extracting ignitable liquid residue from larger pieces of debris. A low boiling solvent, which can be gently evaporated to concentrate any residue extracted, is used. The items to be treated are placed in a clean container and the extraction solvent added. The solvent is swirled around the debris (Figure 12.15) and allowed to make contacts for some time. For a number of years, fluorochlorocarbon solvents (Freons) that were available as refrigeration liquids were used. Unfortunately, because of concerns about their effect on the atmospheric ozone layer, they are no longer readily available. They were ideal because they were very volatile and could never be mistaken for ignitable liquids since they are not flammable.

There are several solvents available that can be used in place of the Freons. This is a more macro technique than the absorption tube or the strip and can be used with larger quantities of evidence. After the solvent wash, most of the solvent is removed to concentrate any ignitable liquid residue present, and then analyzed as with any of the other techniques.

Laboratory examination of specimens prepared from fire debris

Once any potential accelerant present in the debris is isolated and/or concentrated by one of the above techniques, it must be analyzed to determine if there is an ignitable liquid residue present. Chromatography, and particularly gas chromatography (GC), is a technique that uses the partition between a mobile and stationary phase to separate mixtures of chemical compounds (see the sidebar nearby). Since most of the common accelerants are complex mixtures of many compounds, this is an ideal way of analyzing them.

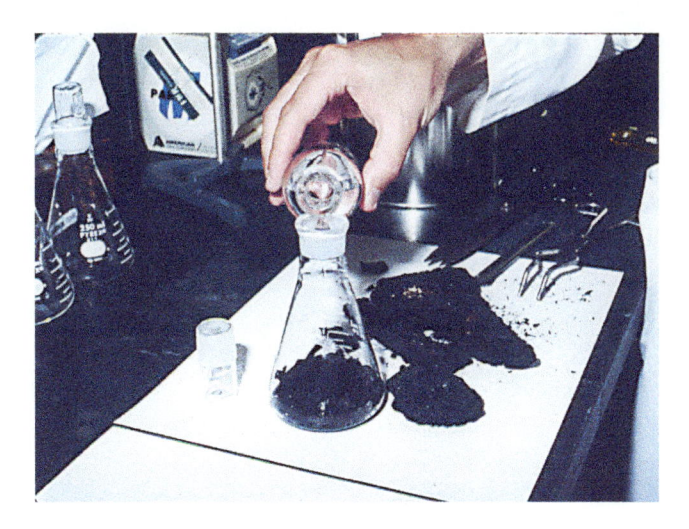

Figure 12.15 Debris submitted from a fire scene being washed with a low boiling solvent to recover any traces of flammable material left on the debris.

Gas chromatography

Gas chromatography is the technique of choice in searching for ignitable liquid residues in debris from suspicious fires. Forensic laboratories use it for examination of materials from suspicious fires and a great variety of other analyses as well.

SCIENCE SIDEBAR   Chromatography and gas chromatography

Chromatography is a name given to certain chemical separation methods. The word itself comes from a Greek word meaning "color," because the procedure was originally used to separate colored pigments from plants. There are several different kinds of chromatography that are commonly used by forensic scientists, these are: paper, thin-layer, gas (once called "gas-liquid"), and high-performance liquid. Paper chromatography is largely a thing of the past. The others are typically abbreviated TLC, GC, and HPLC, respectively. Chromatography brings about separation of substances based on their differing affinity for one or the other of the chromatographic "phases." For example, in thin-layer chromatography, there is a solid phase and a mobile liquid phase. A plate, coated with a layer of finely divided solid adsorbent, is placed vertically into a tank so that it just contacts a liquid (the mobile phase). By capillary action, the liquid "climbs up" the solid phase. Substances that have been placed near the bottom of the plate and absorbed by the solid phase ahead of time may have greater affinity for the liquid phase as it moves, and be carried along with it. Or, they may have more affinity for the solid phase, and move more slowly. Because different substances have differing affinities for the phases, they are separated.

In gas chromatography (GC), the mobile phase is a gas (usually helium). The "liquid" phase is a liquefied high molecular weight waxy absorbent that coats the inside of a long column. The substances one wants to separate are heated, normally dissolved in a solvent, and injected into the heated inlet to the Gas Chromatograph. This inlet, into which the injection is made, is held at a high temperature causing the material injected to vaporize. The vapors then pass onto the column, and are carried along by the gas. Over the years, chemists have designed many different liquid phases and discovered many different instrument conditions for separating a great variety of different compounds.

In this way a fire debris residue sample can be separated into its many components (Figure 12.16). The flowing gas moves the sample through the column, which has a thin coating of a high boiling liquid on the walls. The thin layer of the liquid on the walls causes the components of the sample to be partitioned between the liquid (stationary phase) and the flowing gas (mobile phase). This partitioning is a dynamic process in which the different chemical compounds dissolve in the liquid coating and then are re-vaporized many thousands of times during their passage through the column. The different components are moved along through the column while in the gas phase, but are stationary when dissolved in the liquid phase. Since each different chemical compound has a slightly different tendency to move from the gas to the liquid phase, they take different amounts of time to pass through the column. The more time a compound spends in the liquid layer the less time it is moving and the longer it will take to make the trip through the column. Conversely, the more time a compound spends in the gas phase the sooner it will emerge. The net result is that even very similar chemical compounds with only a slight difference in their partitioning between the mobile and stationary phase will take slightly different amounts of time to pass through the column. Any slight differences in their boiling points or their attraction to the stationary phase will cause them to emerge at different times. At the end of the column there is a detector that tells the operator when something other

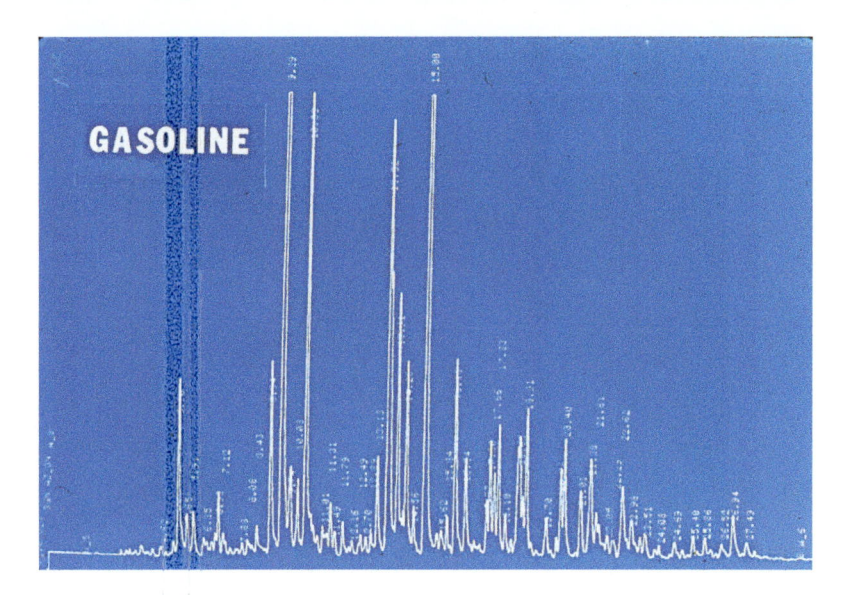

Figure 12.16 A gas chromatogram obtained from a known gasoline standard.

than the carrier gas is emerging from the column. Accordingly, the length of time each component of a mixture takes to traverse the column can be measured accurately. In addition, the sizes of the peaks produced by the detector are proportional to the quantity of the component represented by that peak in the mixture.

The concentrated solution of a possible accelerant isolated from the debris is injected into the column and the clock is started. If one has some gasoline residue in that sample, one will see many peaks on a display, one for each different chemical in the gasoline. Because there are over 200 different chemical compounds in gasoline, one obtains a very complex pattern of the emerging peaks. Some components are major components and will have large peaks, and some are minor components and will produce only tiny peaks. The gas chromatograph is designed to allow one to carefully control conditions so that each run with the same specimen is reproducible. This data is usually collected in a computer and can be examined to compare results from different samples. The pattern of when the peaks emerge, and their relative size is characteristic of the mixture of materials injected.

Accelerant classification by pattern recognition—importance of reference collections

Every time one injects gasoline, even different brands of gasoline, the same general pattern of peaks is observed in the GC. Different accelerant compounds will generally produce very different patterns. One can develop a library of known standards with examples of the things that someone might use as an accelerant. Specimens of any material that might be used as an accelerant, such as kerosene (Figure 12.17), paint thinner, lacquer thinner, lantern fuel, and many more, are obtained from vendors, then injected into a GC, and the chromatograms obtained under exactly the same conditions, to build the library. The National Center for Forensic Sciences has collected a very large database of chromatograms of different combustible liquids, computerized the data, and makes this database available to all forensic laboratories.

One complication with fire debris extracts is that the most volatile components of an accelerant tend to be lost and will not be seen in the GC patterns from residues samples. Therefore, chromatograms of pure samples that have not been collected from debris like gasoline, kerosene, or paint thinner do not look exactly the same as GC patterns from evidence samples (Figure 12.18). Thus, although chromatograms from fire debris can be compared with those of liquid accelerants to get some idea of what the accelerant was. GC is not and should not be considered an ignitable liquid identification technique. An analyst may be able to tell from the chromatogram that a low-, medium- or high-boiling

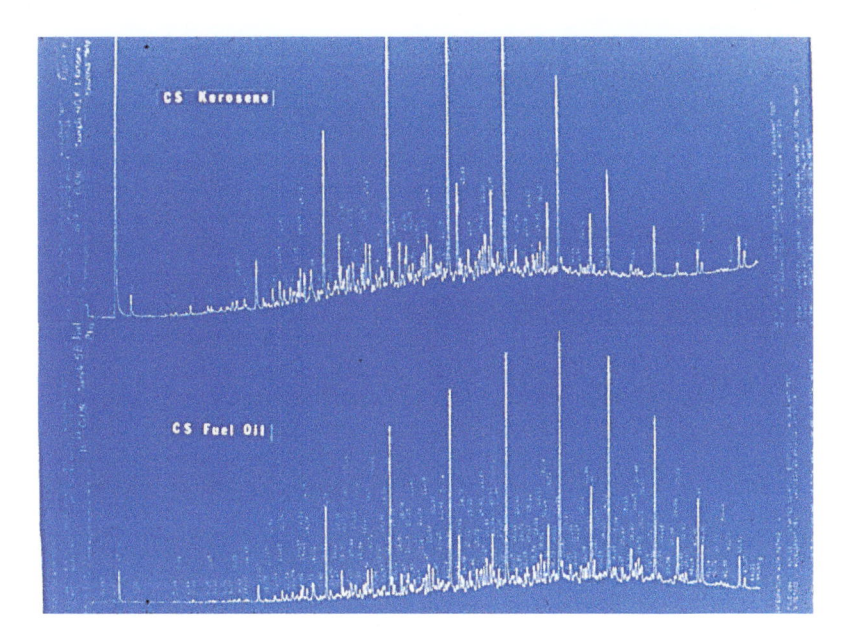

Figure 12.17 Gas chromatograms of kerosene and fuel oil standards.

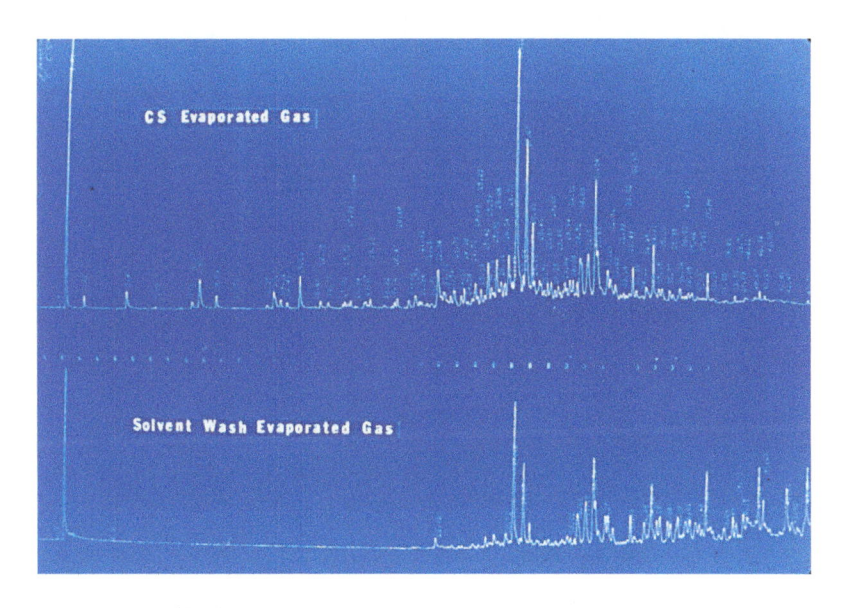

Figure 12.18 Gas chromatograms of highly evaporated gasoline.

ignitable liquid was present (i.e., an accelerant), but generally will not know its identity with certainty. Petroleum-based accelerants are usually divided into nine classes as shown in Table 12.2.

Identification of individual components, identification of ignitable liquid, and indirect identification

Most forensic laboratories now have in instrument called a gas chromatograph/mass spectrometer (Figure 12.19) (GC/MS for short), which allows the identification of each individual peak in a gas chromatogram. The mass spectrometer

Table 12.2 Classification of ignitable liquids according to current ASTM standards

Category[a]	Examples
Gasoline	Most of the compounds are in the C_4–C_{12} range
Petroleum distillates	Some cigarette lighter fluids, charcoal starters, paint thinners, dry cleaning fluids, and jet fuels; kerosene; diesel fuel
Isoparaffinic products	Aviation gas, specialty solvents, some charcoal lighters
Aromatic products	Some paint and varnish removers, xylenes, toluene-based products, some automotive parts cleaners, and industrial cleaning solvents
Naphthenic paraffinic products	Cyclohexane-based products, some lamp oils
n-Alkanes products	Pentane, hexane, heptane, some candle oils
De-aromatized distillates	Some camping fuels, charcoal starters, paint thinners, charcoal starters; odorless kerosene
Oxygenated solvents	Alcohols, ketones, some lacquer thinners, fuel additives, surface preparation solvents, lacquer thinners, industrial solvents
Others-miscellaneous	Single component products, some blended products, some enamel reducers; turpentine products, various specialty products

[a] All the categories except gasoline may be divided roughly into "light," "medium," and "heavy," according to boiling points.

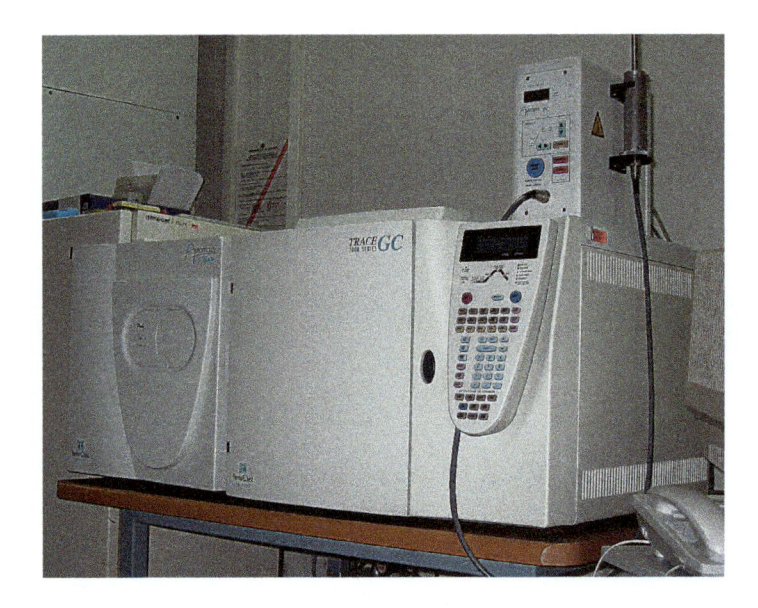

Figure 12.19 A Gas Chromatograph/Mass Spectrometer instrument used to help analyze complex samples where simple pattern recognition is not sufficient for identification. (From User Polimerek. Used per Creative Commons Attribution-ShareAlike 4.0 International https://creativecommons.org/licenses/by-sa/4.0/legalcode.)

is used as a sophisticated detector for the gas chromatograph. As each material comes to the end of the column in the gas chromatograph, it is passed directly into the mass spectrometer. As each compound enters the mass spectrometer, it is bombarded with high-energy electrons and, as a result, breaks into many fragments. The mass spectrometer measures the size of these fragments, and plots them out in a graph form (mass spectrum). Because virtually every different chemical compound will show its own characteristic mass spectrum (pattern), it is possible to identify that chemical entity from its characteristic mass spectrum. See the sidebar nearby. Once a particular chemical compound is identified, a chemist can say whether or not it is flammable and perhaps whether it is known to be a component of any commonly encountered ignitable liquid.

SCIENCE SIDEBAR
The combination of the gas chromatograph and mass spectrometers is a marriage made in heaven. The gas chromatograph vaporizes the components of a mixture, separates the mixture into its individual components and feeds each component, one at a time, into the mass spectrometer. The mass spectrometer produces a pattern for identification that usually allows each component to be chemically identified. This combination of separation of a mixture into its individual components (GC) and the ability to then unambiguously identify each of these components (MS) using one very versatile instrument has become one of the most useful scientific tool in a forensic laboratory. Not only is it important in arson and explosive analysis, but is one of the chief tools of the drug chemist and the toxicologist as well.

As we noted earlier, when a potential accelerant is identified in fire debris, it is termed "ignitable liquid" reside. The laboratory does not identify the residue as "accelerant," because even though the compound is a potential accelerant, the chemical identification does not demonstrate why the compound was present at the scene nor that it was used as an accelerant. This is an important point for fire investigators to understand. It is now recommended by the International Association of Arson Investigators and others that the laboratory identifies ignitable liquid residues in fire debris according to the categories in Table 12.2.

Although the primary emphasis in analysis of debris from suspicious fires is looking for traces of possible arson accelerants that may remain, one can sometimes obtain indirect evidence of the presence of an ignitable liquid even if none of the organic components remain. Sometimes, inorganic chemicals are used in petroleum products, particularly gasoline, to improve its properties as an automotive fuel. Because these materials are not volatile, they will be left behind after the organic components have all evaporated or been destroyed. For many years, gasoline manufacturers used tetraethyl lead, and some other very similar compounds as gasoline additives. More recently, manganese compounds have been used as gasoline additives. Detection of such materials in debris might be an indicator that gasoline was once present.

Comparison samples
In Chapter 3, we discussed different control specimens that were necessary for the comparison of various types of evidence. These were knowns, alibi knowns, substratum, and blanks. Many books on fire and arson investigation advised investigators to try and collect unburned areas of carpeting, flooring, etc. to serve as "controls" for the burned debris when the lab looked for ignitable liquid residues. The reason for this advice was that certain natural and synthetic materials contained volatile compounds that produced GC peaks when released from the substrata by heating or extraction techniques. Fire investigators are still advised to collect unburned or unaffected areas of burned objects if they can. Sometimes, there is no such area available. However, these samples are not to be considered "controls." Arson investigators now realize that a "control" specimen is one with a completely known history. Specimens from fire scenes, even though unburned, are still unknowns in terms of any possible ignitable liquid residue content. Accordingly, they are to be called "comparison samples." Although these intrinsic volatile compounds in certain materials are usually not a source of confusion in an ignitable liquid residue analysis, the comparison sample can sometimes be helpful in interpreting the results of an analysis.

Examination of the other physical evidence collected
In processing the scene of a suspicious fire, identification of a possible accelerant is very important in making the determination of whether a particular suspicious fire was the result of arson. However, it is particularly important to remember that identification of a possible accelerant is not usually, by itself, very helpful in identifying the arsonist. As mentioned in the section Collection of other physical evidence, it is important that fire scene processing include the search for all other types of physical evidence, such as bloodstains (Figure 12.20), fingerprints, toolmarks, and many other useful items of physical evidence. Laboratory analysis of such material can provide useful investigative information. Fingerprint searching through AFIS and DNA through CODIS can do a great deal to move an investigation along. Should a suspect be developed footprints or toolmarks may become useful. In forensic labs other sections than the Arson Section can be enlisted to deal with physical evidence collected. Such evidence can be critical in the investigation of the incident, and in the identification and prosecution of an offender.

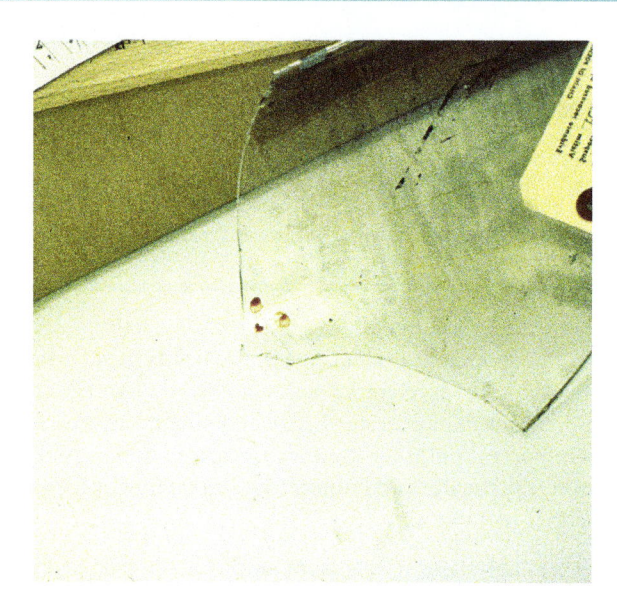

Figure 12.20 Arsonist could have cut himself in the dark while breaking into the scene and left blood evidence.

EXPLOSIVES AND EXPLOSION INCIDENTS

Characteristics of explosives and explosions

One thinks of explosives-related crimes as acts of terrorism or major destructive events that make the nightly news on television (Figure 12.21). There are many additional incidents involving explosives or explosions that are more localized, such as a pipe bomb placed under a car, injury of an individual transporting an explosive device or extortion attempts using a crudely constructed explosive device. Both the Treasury Department's Bureau of Alcohol Tobacco and Firearms (BATF) and the Justice Department's Federal Bureau of Investigation (FBI) have considerable resources that can be dispatched to explosion scenes or in response to possible acts of terrorism. They also assist state and local law enforcement agencies on request. As a result, both the BATF and the FBI have specialized equipment for processing explosion scenes and analyzing a wide variety of explosion-related evidence. However, the less spectacular incidents are usually the responsibility of state and local law enforcement agencies and forensic laboratories. Large local jurisdictions and most states have bomb squads to handle such incidents and ensure that any suspicious object is examined and safely disposed of, if necessary. These units are trained to carefully process explosion scenes and collect any potential evidence for analysis by the appropriate forensic laboratory.

As with arson and fires, some understanding of explosives and of the explosion process is necessary to fully appreciate the forensic aspects of explosive incidents, their investigation, and how the resulting evidence is handled in the laboratory. An explosion is a very rapid chemical reaction that produces heat and gaseous products. A great deal of heat and gaseous product can be produced by a relatively small quantity of explosive material. The one sentence definition really does not give one much insight into what is happening in an explosion. The single most important characteristic of explosions is the production of a great deal of heat. It requires a small amount of energy to get the explosion reaction started, but, once started, it produces heat in large quantities and in a much shorter period of time than even rapid combustion. Although some explosions are really a special type of combustion, they are much more rapid and produce enormous amounts of heat very quickly.

Almost equally important to the process is molecular fragmentation. Larger molecules (with many atoms bonded together), which are usually solids and are typically not very volatile, are violently broken apart, in a tiny fraction of a second, into many smaller molecules that tend to vaporize (become gaseous molecules). Nitrogen, carbon dioxide, water vapor, and nitrogen oxides are typical of the gaseous products produced in an explosion. In an explosion, a small volume of solid material is instantaneously transformed into an enormous amount of gaseous material that must obey the laws of physics and, therefore, occupy thousands or even millions of times the volume that was occupied by the original explosive.

Since this happens in a fraction of a second during an explosion, those molecules are bumping against each other very rapidly trying to obey the gas laws. In addition, heating a gas causes the molecules to move more rapidly and,

Figure 12.21 Oklahoma city bombing. (From FBI, Washington, DC.)

therefore, want to expand further (see Science Sidebar under Relative Vapor Density). Thus, an explosion produces a double effect: the increase in the number of molecules that by itself causes expansion and the heating effect that causes additional expansion. The solid, that occupies a small volume, is converted almost instantaneously into an enormous number of gas molecules that need to occupy a much greater volume, and at a very high temperature,

The rapidly expanding gases from the explosion are retarded by the air surrounding the explosive mass. The air will be compressed by the explosive gases pushing on it rapidly as it tries to expand, resulting in a shock wave. The shock wave is the result of gases from the explosive and surrounding air molecules, crowded together in the rush to expand, and it takes on the character of a physical force that pushes things out of its way violently. This shock wave (Figure 12.22) is what causes much of the damage associated with an explosion. It is not generally the heat, but the shock wave, that causes an explosion's violent destruction. The expanding gases are transformed into a tremendous force that can break windows, knock down trees, and fragment almost anything in its path. A shock wave can have awesome power that can be harnessed to do useful work, such as mine raw materials, prepare building sites, and flatten unwanted structures, or it can cause enormous destruction.

SIDEBAR

It has become fairly common to see slow motion films of an entire housing complex being leveled in a matter of seconds using carefully placed and triggered explosive charges. A few years ago, a major league baseball stadium was leveled in that way to make way for a new one. The actual moment of destruction was carried live on national television and was a highlight on the nightly news for those who had not seen the actual destruction. It is the shock waves emanating from the explosive charges that shatter the reinforced concrete and cause the stadium to collapse from its own weight. The key, of course, is proper placement and timing of the explosive charges. An enormous amount of energy is released in a matter of a few seconds doing the work that would require many hours, perhaps days, of labor with cranes and wrecking balls.

Figure 12.22 The visualization of the initial shockwave associated with an explosion; the shockwave can be seen as the expanding grey ring on the ground surrounding the origination point.

Containment is a major factor in the nature of explosives and explosive incidents.

In the section titled The three major types of explosions the so-called "low explosives" are explosives that burn rapidly but do not detonate, as we shall discuss below, there must be containment to have an explosive event. "High explosives," on the other hand, will detonate even without containment. With low explosives, such as smokeless powder, fireworks, or certain chemical mixtures, one can cause detonation and increase the force produced by trying to contain the products of the rapid reaction. A low explosive might be enclosed in a piece of pipe (Figure 12.23); for example, and the material is ignited with a fuse, blasting cap, or other source of ignition. As the explosive material fragments are heated by the energy released by the reaction, the gases produced violently try to expand. Pressure builds, and the pipe is burst, releasing the gases, and propelling broken pipe fragments in all directions. The released gases also form

Figure 12.23 An exploded pipe bomb and associated pieces recovered.

a shock wave that can damage anything in its immediate path. Generally, the stronger the container, the stronger the force necessary to rupture it. Thus, when the stronger container does rupture, the potential damage is increased.

The three major classes of explosives
Low explosives

Low explosives are materials that burn rapidly and will explode only when contained. Some sort of ignition is needed, such as a fuse or spark, to cause them to start decomposing. This decomposition reaction is a very rapid reaction that results in the explosive molecules being converted into much smaller gaseous molecules. The most common low explosives are smokeless gunpowder and black powder. Black powder is a mixture of charcoal, sulfur, and sodium nitrate or potassium nitrate. Also in this group are most fireworks. The technical name for fireworks is pyrotechnics, and they are made from chemical mixtures of a number of different highly energetic inorganic salts, usually nitrates and perchlorates.

SIDEBAR
Pyrotechnics are most familiar as entertainment on the 4th of July or at sporting events. They produce loud bangs and often brightly colored displays in the sky. Those materials are manufactured by combination of a variety of low explosives and very complex containment schemes. They are similar to the old anarchist's homemade bombs made out of perchlorate and sugar or other such combinations. Unfortunately, one can find on the Internet the formulas for many explosive mixtures. The making of bombs or even fireworks is not for amateurs. Even professional chemists and engineers occasionally make a mistake and blow themselves up. There is a family of pyrotechnic experts named Grucci, who have a large compound on Long Island, where they design and assemble fireworks. They are world renowned and one of the world's most prestigious designers of fireworks and fireworks displays. This family has been in this business for several generations. A number of years ago there was a huge explosion out on Long Island that blew up almost their entire plant, killing several members of the family. Even with their almost encyclopedic knowledge of the chemicals involved in pyrotechnics, the danger inherent in these materials was sadly reinforced for the surviving members of the family.

Fireworks are a large group of low explosives. At one time the only fireworks commonly available were the so-called Chinese firecrackers and cherry bombs (small round devices, about the size of a cherry, painted red). They had a wick and, when ignited, produced a loud bang. In recent years, the most popular "fire crackers" have been M80's and super M80's, and even more powerful cousins. M80s are cylindrical in shape, about three quarter of an inch in diameter with a wick in the center. They produce a perceptible shock wave and are much louder than the old Chinese firecrackers. When one cuts one open, they contain only a rather small amount of pyrotechnic mix but many, many layers of tightly wound heavy paper for containment. The reason for their higher power lies more in the containment than in the explosive mixture.

Primary high explosives

The second class of explosives is primary high explosives. These are the most treacherous explosive materials, and are used as primers or detonators. They are not usually the explosive material that does the major work or damage, but are used to cause other explosive material to explode. Primary high explosives are sensitive to many possible disturbances. An electrical spark, intense heat, mechanical force and many other disturbances, will cause these materials to explode (Figure 12.24). The most well-known primary high explosive is nitroglycerine. It has become widely used because a chemist named Alfred Nobel discovered a simple way to convert it from a primary high explosive to a secondary high explosive, and allow it to do useful work. Nitroglycerine is an oily liquid that explodes violently for a variety of reasons, with potentially deadly effects. It was used as a primary high explosive at one time, but its unpredictability made it too dangerous to handle safely.

The most common primary high explosives are mercury fulminate and lead stybnate, which are used in firearms cartridges as the shock sensitive priming material (Chapter 8), and in blasting caps. In a firearm, when the firing pin strikes the "primer" in the end of the cartridge or shell, it causes a tiny amount of the shock-sensitive primary high explosive inside to explode. This produces a flash of intense heat and a little flame or spark, which then ignites the smokeless powder in the cartridge and "fires" the weapon. The gunpowder burns very rapidly producing rapidly expanding gases that are contained by the combination of the bullet and the cartridge case.

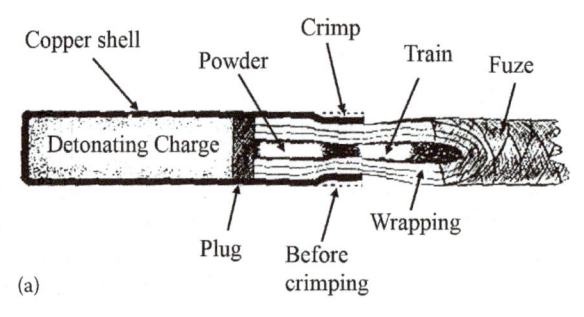

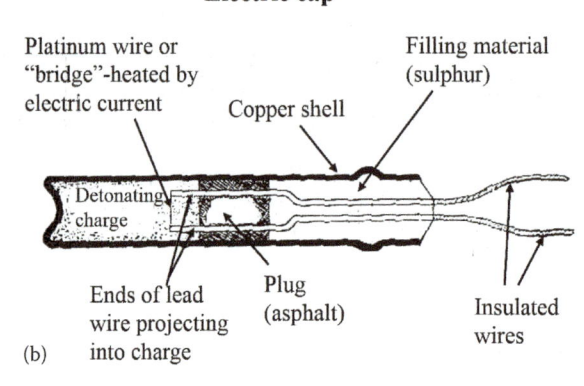

Figure 12.24 (a and b) Blasting caps containing primary high explosive used to trigger a larger amount of secondary high explosive.

Thus, the cartridge is essentially a small explosive device (see section "The explosive train or device", page 317). When containment is breached, the gasses propel the bullet from the gun at very high speed.

Secondary high explosives

Secondary high explosives can be fairly safely handled, but with the proper inducement will explode with great force. The primary distinction between these and low explosives is that high explosives do not have to be contained to explode. Once sufficient initiation is provided, they will explode violently. Containment may be used to intensify or direct the explosion, but is not necessary for it to occur. Depending on the circumstances and the explosive, the "inducement" (often referred to as "initiation") can take many forms. The most common are an electrical spark, a small flame (from a fuse), intense heat, or a sharp blow. Some common military examples are TNT (trinitrotoluene), the active ingredient in military dynamite, PETN, the material used in detcord, and RDX and HMX, which are explosive components of common military explosives usually referred to as "plastic explosives" (Figure 12.25).

Detcord is a thin cord that can be wrapped around an object and when detonated explodes with sufficient force to shatter or cut whatever it enveloped, such as door handles, fence posts, or bridge supports. RDX and HMX are military explosives that are the primary components of the so-called plastic explosives, or "plastique." They are secondary high explosives that have been mixed with other materials to make them into a putty-like material that can be easily formed into any shape. The explosive material can then be molded around anything one wants to destroy. With the placement of a detonator in it, and a timer or remote trigger to set off the blast, and the combination can destroy bridges, buildings, or other military targets.

Secondary high explosives are also important in civilian applications, such as major construction projects, mining, and road building. The most common secondary high explosives used commercially are different types of dynamite (Figure 12.26), based on nitroglycerin and its derivatives, ammonium nitrate, and water gel explosives based on monomethylamine nitrate. With knowledge and care, these materials can be safely handled, and their potential energy harnessed to do useful work. This was not always the situation. It was Alfred Nobel who made the major

Arson and explosives 315

Figure 12.25 A widely used type of military explosive often referred to as plastic explosive; a female soldier inserting blasting caps into blocks of C-4 explosive. (From U.S. Department of Defense.)

Figure 12.26 Bundled sticks of dynamite with a fuse.

breakthrough that allowed the safe use of high explosives to become practical. As indicated above, nitroglycerine is a very sensitive explosive that can explode with little warning and with devastating effect. Nobel worked for years to develop a way to stabilize nitroglycerine, so it could be handled safely, and its power harnessed. He found that nitroglycerine, when absorbed on a high surface area solid, such as diatomaceous earth or later even sawdust, became a material that could be safely handled, but when initiated with a blasting cap or fuse still carried the enormous explosive force of the nitroglycerine. The stabilized nitroglycerine, wrapped in waxy paper and with a fuse inserted, became dynamite. Dynamite could be safely moved, handled, stored for considerable periods and even dropped. It made modern mining, quarrying, and major road and other construction projects possible. Nobel had taken a primary high explosive and converted it to a secondary high explosive. It became such a useful material that the Nobel Company became enormously successful. It is still the largest manufacturer of explosives in the world. The success of his company made Alfred Nobel enormously wealthy.

smokeless powder or any of the military explosives. The main charge may also include a booster. A booster is a material that is not itself an explosive, but which when combined with an explosive will significantly increase the amount of energy released by the explosion. For example, dynamite frequently has sodium nitrate added to it to intensify the explosion. Similarly, in the ANFO material mentioned earlier, the fuel oil is not an explosive but intensifies the ammonium nitrate explosion. These are the three primary components of an explosive train.

In some cases the fourth component of the explosive train is **containment** (confinement). Particularly when the main charge is a low explosive, there must be containment for an explosion to occur. Even when a secondary high explosive is used as the main charge, the force released can be intensified or directed by using properly designed containment. Containment is most often provided by a container that holds the components of the explosive train. Containment is most common in homemade devices, and also in certain military explosives where maximizing the damage, often in a particular direction, is desired. Containment can be provided in many ways. The container can be a piece of iron pipe, like in a "pipe bomb" as mentioned earlier, or it can be provided by tightly wound cardboard as in the small firework called the M80. Containment can be provided by any device that will retain the gases generated by the explosion until the pressure builds to a high level sufficient to rupture the container, allowing the sudden release of the gases. When the containment is breached, there is a very rapid release of the confined gases, producing an intense shock wave and thereby a much more potent explosion than if there were no containment.

The role of the scene investigator

When a device has exploded and caused damage or injury, the scene must be evaluated so that the maximum amount of useful information can be gathered. The investigator is faced with a complex mixture of materials recovered at the scene, most of which are from the scene and not the device. The first step is to try to sort through all this material to recover portions of the device, and any explosives residues that have survived. Careful macroscopic and microscopic examination of debris can yield a great deal of information.

The first step in this process is sifting. Much of this process is usually done by bomb investigators at the scene or in an area provided near to the scene. For example, in a major explosion at an airport, the investigators were given a large section of a hanger for several months to sort through the debris. All the collected debris is first examined by eye and larger pieces that do not appear to have been at the center of the explosion are removed by hand. Next the screening process is begun. The debris is first passed through a screen with fairly large holes (Figure 12.28), and material that does not pass through is examined and put aside if it looks relevant. The material that passed through the first screen is then sifted again, with a smaller-mesh screen. The process is continued with smaller and smaller size sifting devices. At each stage, one looks for pieces that might be parts of the device, or that could have been used to contain or transport it. Items are collected that show evidence of proximity to the seat of the explosion; such as

Figure 12.28   Processing and sifting debris from an explosion looking for pieces of useful evidence.

heavily deformed pieces of metal or contorted debris items. This is a tedious and time-consuming process, but it is quite surprising how much potential evidence material survives even a devastating explosion. Once the debris has been culled to find the most promising pieces, they are sent to the laboratory for more detailed examination, such as chemical identification of the explosive residues, and examination of the recovered parts of the device.

LABORATORY ANALYSIS OF EXPLOSIVES AND EXPLOSIVE RESIDUES

Examination of an unexploded device

Unfortunately most explosion-derived evidence that is recovered after an explosion is highly contaminated and, as we shall see in the rest of the chapter, analysis is a complex process. However, because bombs intended to explode sometimes fail to do so or are seized before they can be exploded, an unexploded device may be recovered. Forensic laboratories receive unexploded devices for three common reasons. First, a bomber may be intercepted before the device is placed or detonated. Second, there may be a defect in the way the device was constructed, and it fails to explode and third, the detonation may not be properly executed, and there is no explosion. For example, the wick may fall out of a Molotov cocktail when it is thrown and before it ignites, or a component of the electrical circuit could have broken during transport, so the circuit is not complete when the detonating switch is closed.

With such a device, there is the immediate problem of rendering it safe. This can be a significant problem, particularly if an amateur or poorly trained individual made it. These devices can be extremely dangerous, and often the safest option is to remove them to a remote area, and detonate them. This practice results in the loss of considerable information about how the device was constructed, but the lives of bomb experts or the public is not jeopardized.

SIDEBAR

A number of years ago, the New York City Police Department received information that some people were going to bomb a particular dance club. On the evening in question, the police staked out the location. A car pulled up in front of the location in the early morning hours, and two individuals jumped out of the car carrying a large piece of pipe. They were promptly arrested before they could set it off. The police picked up the pipe and, assuming it was a simple pipe bomb, sent it to the laboratory for analysis. The laboratory examiner very carefully began to disassemble the device, on the assumption it was filled with smokeless powder or a homemade pyrotechnic mixture (low explosive). These materials must be treated with respect, but if handled with care, are not particularly dangerous. This was a piece of steel pipe three inches in diameter and over a foot long. As the examiner slowly removed the cap from the end of the pipe, he looked inside the device, quietly put the cap down and immediately left the room to call the bomb squad. It turned out that the pipe was filled, not with smokeless powder as assumed, but with plastic explosive. The amount of plastic explosive was certainly enough to blow the top off the building where the lab was located. The individuals involved had apparently stolen the material, probably from a military armory, and did not realize the enormous explosive force in that amount of material. Instead of breaking some windows and intimidating the club owner, they might have destroyed the whole building and killed anyone on the block with the force of the explosion.

When an unexploded device is received, it should be described in considerable detail. This type of information can be of great importance, particularly when there have been a series of possibly related explosions. Successfully making explosive devices and living to tell about it is not a trivial exercise. As a result, most bomb makers develop a scheme that seems to work for them and continue to use that scheme. In many cases, explosive devices carry almost a "signature" that can be compared, and it will usually indicate if different devices have been made by the same bomb maker or perhaps one of his "students." For this reason, a careful description of each of the components used, and how they are connected to make the explosive train, is so important. Sometimes, if a good understanding of the method of a bomb maker has been developed from unexploded devices, his work may still be recognizable even when present only as debris from an explosion that has occurred (see Figure 12.29a and b).

In addition to the way the device is constructed, the actual nature of the explosive used is an important characteristic of a bomb maker or a group of bomb makers. Such people generally have access to a particular type of explosive, or will favor a particular combination of ingredients in a homemade explosive. These features help not only to point out similar construction, but also can sometimes provide investigative leads. Identification of the explosive in an unexploded device is much simpler than trying to determine what it was from traces left behind after the explosion. If the device uses a commercial or stolen military explosive, there will often be markings on the wrapper that can be traced to when it was made and often where it was distributed after manufacture. Even in a pyrotechnic mixture, the

Figure 12.29 (a and b) Circuit board and wiring components recovered from the Boston Marathon bombing indicated the methodology used in assembling it. (From FBI, Washington, DC.)

exact nature of the chemicals used can often give useful information about where they were obtained. Perhaps the most common low explosive used in bomb making is smokeless powder. It is available from a number of commercial sources to sportsmen who reload their own cartridges or shot shells. Although different brands of smokeless powder are chemically similar, each manufacturer has preferred shapes that they make for certain applications, or to distinguish themselves from their competitors. Careful examination of smokeless powder under a stereomicroscope can often be as useful as chemical analysis in distinguishing different types and different manufacturers. Further, each manufacturer uses certain additives that may allow identification of the manufacturer.

Examination of the exploded device and associated debris

Microscopic examination of recovered residues

Microscopic examination and picking are important steps in selecting materials of interest for further examination from recovered materials. If a pipe bomb was involved, one will often find a portion of the pipe cap with a threaded portion of the pipe still inside. Small specks of the explosive may have been trapped in the threads, thereby protected from the heat of the explosion an therefore, can be recovered. The same might be true of homemade pyrotechnics or other low explosives. Sometimes, even small traces of dynamite may survive the explosion. As noted earlier (see section "Use of washings from residues to search for the nature of the explosive", page 321), these unexploded materials are much more easily identified than trace explosive residues recovered from washings. Even if no unexploded dynamite is recovered; pieces of the wrapper might be present. The wrapper is often blown apart into small pieces, but unless consumed by fire, they may survive. Such a piece of wrapper may have a portion of a serial number or other marking on it, enabling investigators to trace the origin and distribution of the material. Tiny pieces of wire (Figure 12.29a and b), parts of batteries, timing devices, and a myriad of other items can survive the explosion, and provide information about the nature of the device, and sometimes important investigative leads, or even evidence that can be used in a subsequent prosecution.

SIDEBAR

In one case, an explosive device was left in the busy lobby of a large bank, and when it detonated, it caused considerable damage to the bank and injured a number of people. The bomb squad was faced with over a half a ton of debris. They meticulously sorted through all this debris, and quite a number of similar pieces of leather were recovered. From these rather assorted fragments of leather, it was possible to reconstruct the kind of leather case (Figure 12.30) in which that bomb had been transported. A similar bag was purchased at a nearby store and used to quiz witnesses as to whether they had seen anyone enter the bank with such a bag. This helped investigators get a description of the individual who had placed the bomb, and later assisted in developing evidence against that individual.

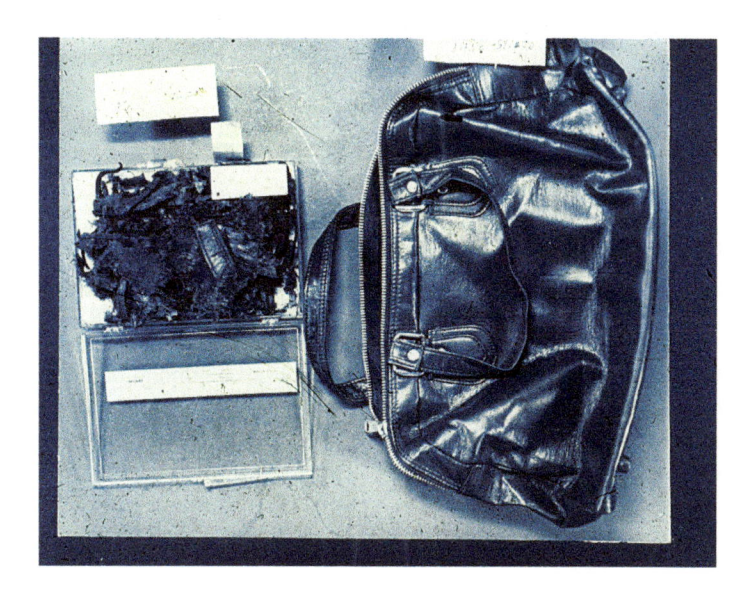

Figure 12.30 Leather fragments collect at bombing scene that could be matched to sample leather bag purchased near scene.

Use of washings from residues to search for the nature of the explosive

Where the information from microscopic examination and picking is insufficient to determine the nature of the explosive or leaves doubt about its nature, the next step would normally be washings. Unfortunately, particularly with high explosives, lack of visible residues is frequently the case. Here, pieces of debris, selected by appearance or screening tests, are washed with a small volume of an organic solvent, such as acetone, to collect any traces of unburned or partially burned residue from the explosive charge.

These samples do not have to have been part of the device itself, but could be material on which the device was resting, or could have been in the immediate path of the hot gases produced. The solvent wash solution is then concentrated and subjected to a variety of screening tests (Table 12.3) to see if any residual explosive might be present. Any washings that show positive screening reactions are then examined more closely using more sensitive and specific instrumental examinations. A second set of washings may be taken, especially where it is suspected that an inorganic explosive, such as ammonium nitrate or a pyrotechnic mixture may have been used, since they are usually not recovered efficiently with the acetone washings. The second washing is usually done with water. It is not easy to concentrate a water wash without destroying the residues, and the washing also tends to be contaminated by salts and other materials from the explosion scene. In cold climates, road salt from ice melting is ubiquitous, near the coast there is salt from ocean spray, and calcium sulfate from plaster will be present in large quantities at a scene in virtually any indoor explosion. Even in the face of the difficulties caused by such contaminants, many laboratories can sometimes successfully characterize the type of explosive used from such washings.

Table 12.3 Common chemical spot tests for explosives

Substances to be tested for	Cupric Tetrapyridine	Diphenylamine	Griess Reagent	J Acid	Alcoholic KOH
Chlorate	NR	Blue to blue/black	NR	Orange/brown	NR
Perchlorate	Purple, ppt	NR	NR	NR	NR
Nitrate	NR	Blue to blue/black	Pink to Red	Orange/brown	NR
Nitrate	Green	Blue/black	Red to yellow	Orange/brown	NR
Nitrocellulose	NR	Blue/black	Pink	Orange/brown	NR
Nitrocellulose	NR	Blue to blue/black	Pink to red	Orange/brown	NR
TNT	NR	NR	NR	NR	Red/violet
RDX	NR	NR	Pink to red	Orange/brown	NR
PETN	NR	Blue	Pink to red	Orange/brown	NR
Tetryl	NR	Blue	Pink to red	Yellow to orange/brown	Red

Source: Parker, R.G. et al., *J. Forensic Sci.*, 20, 133–140, 1975.
NR = no reaction; Ppt = precipitate

With washings, it is usually necessary to try to separate the explosive residue from other contaminants from the scene that were carried along through the washing step. A number of different types of chromatography can be used to obtain a sample sufficiently free of interfering materials to enable an unambiguous identification by instrumental methods. Preliminary screening tests (that are usually color tests) can be helpful in deciding what materials should be examined further, and what fraction from the chromatography may contain the sought-after explosive residues. The Greiss test and the diphenylamine test are useful in looking for nitrate and nitrite residues. If these are not present the item tested, with a few exceptions, does not have identifiable explosive residues, since even organic explosives produce nitrate when they explode.

Thin-layer chromatography is the simplest and most rapid separation technique frequently used with explosives and explosive residues. One obtains characteristic spots on the developed thin layer plates that can be tentatively identified as nitroglycerine, TNT, or the most common organic explosives. Although the TLC results do not provide a rigorous identification, they provide a strong indication, and the identity of the specimen can then be confirmed with other techniques. Other chromatographic techniques can be used with difficult samples and can provide an even stronger indication of the identity of the material. Not every laboratory will use every technique but, gas chromatography, high performance liquid chromatography (HPLC), ion chromatography and capillary electrophoresis are all used to some extent in dealing with the complex mixtures produced by washings from recovered materials from an explosion. There are even special detectors that are used with HPLC that are particularly suited for chromatography of explosive residues. They are designed to detect primarily the types of materials characteristic of explosives and not most extraneous contaminants.

Isolation of even miniscule amounts of the unchanged explosive will usually allow rapid identification using modern instrumentation. Organic explosives such, as TNT or most military explosives, can be identified using infrared spectroscopy or gas chromatography/mass spectrometry (GC/MS). These are instruments available in virtually every forensic laboratory. There are other more specialized instruments that may be available in laboratories that are specifically equipped for explosive residue analysis. Residues from inorganic explosives are examined using infrared spectroscopy, X-ray diffraction or fluorescence, or several other specialized techniques. GC/MS is particularly useful for reasonably volatile organic residues. Unfortunately, many explosive materials are not sufficiently stable to survive a trip through the GC/MS. If one can obtain crystalline material by picking it out or after purification of material from washings, the X-ray diffraction can provide a chemical identification. Many other instrumental techniques can be used depending on the nature of the explosive and the quality and quantity of the material available.

Examination of the device or debris for other physical evidence

Explosive devices are a combination of many objects and materials assembled in a particular way. Careful examination of the type of wire, how the timing devices have been designed to make the timing circuit, or the nature of the pipe or other containment used, and how it was cut or modified, can provide valuable investigative information. Toolmarks, fingerprints, how wire was stripped of its insulation and dozens of other examinations make up the rest of the examination, can produce results often as useful as the analysis and identification of the explosive residues. Even when fairly common everyday items are used to construct the device, a sales person may remember an individual who came in and bought 36 cheap watches. That might turn out to be an important investigative lead, particularly if watches or parts of watches are recovered and are serialized or have other identification of the manufacturer.

SIDEBAR

A number of years ago, there was a small terrorist cell that was trying to get their message into the press and pressure the political system using incendiary devices. The simple devices were designed to start a fire after a time delay of usually an hour. They used a hard cigarette pack, filled with low explosive (insufficient containment to explode), an inexpensive watch, a battery, and a small flash bulb. The watch had a hole drilled through the crystal and a contact inserted so that when the watch hand came around and touched the contact it closed the circuit, thus firing the flash bulb, which then caused the low explosive to burst into flame. (Some may not remember flash bulbs, they plugged into a receptacle on a camera to provide light, much as the built in electronic flash does in more modern cameras.) These devices were usually placed in a pocket in clothing in department stores just before closing time. The delay ensured that they would ignite the clothing and would start a small fire after the store closed. The small fire would set off the department store sprinkler system, which would actually cause more economic damage than the small fires. The idea was to get some news coverage and place economic pressure on the community. Although the devices were easy to make and conceal, they were unreliable when wires broke, or contacts separated and, as a result, sometimes the device did not ignite. Several of these devices were received unexploded at the crime laboratory. It was the careful examination of the physical evidence gathered at the scenes and comparison with that found at the device factory that made prosecution of the device makers for the entire series of incidents possible.

The discovery of a workshop where bombs are being or have been constructed can provide a wealth of information that might allow one to associate related bombings and aid in prosecution of the workers in the bomb factory by tying them to specific explosive incidents. The usual way of discovering a workshop is the result of something going wrong resulting in an explosion and/or fire. In one case, a workshop where pipe bombs were being made was discovered in this way. A pipe vise is usually used for cutting and treading, or even handling pipe in other ways. This is a special type of vise that is designed with jaws to grip the round pipe and hold it tightly. When these vises squeeze down on that pipe, the jaws make indentations in the pipe. These marks are a type of toolmarks and pipe vises often develop a characteristic set of marks. As a result, the marks on a pipe can sometimes be used so say that a pipe was gripped in that particular vise. When the workshop was discovered with its pipe vise, it was possible to associate pipe debris from several explosive scenes with that particular pipe vise and thereby with that workshop and the group that ran it.

Whenever someone performs an anti-social act fairy successfully, many others, who did not think of it or could not have done it, claim credit for it. The false claimants who take credit or confess usually know only what has been reported in the news. It very important to have detailed knowledge that is not given to the press, to help sort out the false claimants. The careful examination of physical evidence allows investigators to sort out bogus claims of responsibility from those of the individual or group responsible.

Key terms

- accelerant
- burn pattern
- carbon absorption strip
- combustion
- comparison specimen
- containment (confinement)
- detonation

exothermic
explosive
explosive residues
explosive train
fire point
flaming combustion
flammable range
flash point
glowing combustion
heated headspace
ignitable liquid
igniter
low explosive
point of origin
primary high explosive
pyrolysis
secondary high explosive
shock wave
sniffer

Review questions

1. Discuss the importance of pyrolysis in the combustion of solid fuels.
2. What role can the careful processing of a fire scene play in the investigation of the incident?
3. What aids are available to the fire investigator for selecting evidence to be sent to the laboratory?
4. How is evidence collected from a fire scene processed in a forensic laboratory?
5. What are the major objectives of a fire scene investigation?
6. Discuss what is happening when an explosive device detonates and why it can cause so much damage.
7. What are the similarities and differences in the processing of a fire scene and an explosive incident scene?
8. What are the differences in the laboratory's approach to examining materials from an incident where the device failed to explode and one where the device exploded?
9. Briefly describe how a pistol or rifle cartridge can be looked at as a simple explosive device.
10. Give some examples of low explosives and some examples of high explosives.

Fill in/multiple choice

1. A low explosive becomes explosive and most dangerous only when its decomposition is _____.
2. For a liquid fuel the lowest temperature at which sufficient volatilization occurs to produce an ignitable vapor at the surface is called the
   a. boiling point
   b. flash point
   c. explosion point
   d. relative vapor density
3. Most solid fuels are able to support flaming combustion only because _____ converts them to volatile materials.

4. A fuel will only achieve sufficient reaction rate with oxygen from the air to produce flaming combustion when it is in the _____ state.
5. The most sensitive and reliable instrument for detecting and characterizing flammable residues in debris from a suspicious fire is the _____.

References and further readings

Beveridge, A. D. "Development in the Detection and Identification of Explosive Residues." *Forensic Science Review* 4 (1992): 17–49.

Beveridge, A. Ed. *Forensic Investigation of Explosions*. London, UK: Taylor & Francis Group.

Bianchi, F. et al. "Cavitand-Based Solid-Phase Microextraction Coating for the Selective Detection of Nitroaromatic Explosivesi Air and Soil." *Analytical Chemistry* 86 (2014): 10646–10652.

Bertsch, W., and Q. Ren. "Contemporary Sample Preparation Methods for the Detection of Ignitable Liquids in Suspected Arson Cases." *Forensic Science Review I* 1 (1999): 14 1.

DeHaan, J. D. *Kirk @ Fire Investigation*. 5th ed. Englewood Cliffs, NJ: Prentice Hall, 2002, 638 pp.

"Forensic Science Committee Position on Comparison Samples." *Fire and Arson Investigator* 41, no.2 (1990): 50–51.

Fultz, M. L., and J. D. DeHaan. "Gas Chromatography in Arson and Explosives Analysis." In *Gas Chromatography in Forensic Science*, ed. I. Tebbett, vol. 109. Chichester, UK: Ellis Horwood Ltd., Chapter 5, 1992; Prentice Hall, 1999.

Lentini, J. J. *Scientific Protocols for Fire Investigation*. Boca Raton, FL: CRC Press, 2005.

National Fire Protection Association. *NFPA 921: Guide for Fire and Explosion Investigations*. Item 92104. Quincy, MA: NFPA, 2004.

Newman, R., M. W Gilbert, and K. Lothridge. *GC-MS Guide to Ignitable Liquids*. Boca Raton, FL: CRC Press, 1999.

Stauffer, E., and J. Lentini. "ASTM Standards for Fire Debris Analysis: A Review." *Forensic Science International* 132 (2002): 63–67.

Chapter 13

Drugs and drug analysis

Lead Case: Bufotenine Poisonings

Four men died in New York City after they ate a purported aphrodisiac sold at New York shops during a period of a little over 2 years. The substance was supposed to be rubbed on the genitals but was not labeled as such. When the victims instead ate the substance, they began to vomit and had erratic heartbeats. In addition, at least two other deaths in major cities were eventually attributed to this material. The story of what happened was eventually pieced together by the U.S. Centers for Disease Control based on information developed by several law enforcement and health agencies, and, as a result, the material was banned from import. Very likely other deaths, where the connection to this unusual material was not made, could be attributed to this substance.

During this period, a dark-brown resinous substance, which was sold in grocery stores and tobacco shops under such names as Stone, Lodestone, and Black Stone, was also being submitted to the New York City Police Department forensic laboratory as suspected hashish. In most cases, the packaging lacked labels listing ingredients or directions for use. The initial recorded deaths occurred in New York City, but the material was also available elsewhere around the country.

These odd looking little, dark-brown resinous cubes were submitted as suspected hallucinogenic substances or thought to be hashish. Initial tests indicated they might contain low concentrations (small amounts) of a hallucinogen called psilocin, most commonly encountered in so-called "magic mushrooms." These mushrooms of the genus *Psilocybe* have been used by native peoples in religious ceremonies for centuries and by drug abusers, particularly, in the 1960s and even today as hallucinogens. More careful analysis disclosed that this component was actually not psilocin, but a material called bufotenine. Bufotenine is chemically very similar to psilocin and is what chemists call an isomer of psilocin. That means that it has the same chemical composition, but the individual atoms in the molecule are arranged slightly differently.

Although this material is a strong hallucinogen, it is not typically very toxic, and it was, therefore, unlikely be have caused the deaths of the individuals who used the material. Nevertheless, its identification was an important clue in unraveling the puzzle of this odd material and the unexplained deaths. That bufotenine is of natural origin, and not man-made, like LSD and phencyclidine, was important in determining what these odd-looking cubes were.

Bufotenine is most commonly found in three different types of natural materials. Its most common source is in the seeds of plants of the *Anadenanthera* genus, and it is also found in certain mushrooms and in skin toxins exuded by some toads. It was important in tracing this material and understanding its properties to discover from which of these sources the dark-brown resinous cubes arose. A visit to the New York Botanical Gardens, to consult with and have the material examined by their Curator of Mycology (study of mushrooms), quickly eliminated mushrooms as the source. Enough was known about the complex chemical constitution of the plant extracts and toad toxins to distinguish material from these two sources based on careful chemical analysis. This analysis determined that this resinous material originated from the toad toxin rather than the plant seeds. This was consistent with the fact that

Chinese herbal medicine used this material in treatment of low sexual potency, whereas the mushroom and plants derived materials were more commonly used as snuffs or extracts to induce hallucinations. One of the traditional folk-uses of these cubes involved topically applying it to the penis to prolong sexual prowess. Unfortunately, several purchasers thought that the material was to be taken orally, and this resulted in vomiting, convulsions, and, as indicated previously, several deaths.

In addition to the bufotenine, these cubes were found to have significant amounts of materials called bufadienolides. These are the most toxic component of the black resinous material. Bufadienolides are very potent cardio-active steroids that cause heart arrhythmias. This means that they cause the heart to beat irregularly, which can prove fatal. These very toxic compounds protect the toads from predators. This property is consistent with the observation that several of the victims had irregular heartbeat and convulsions before they died.

The "1994 Dietary Supplement Health and Education Act" (DSHEA) creates a presumption that dietary supplements are safe, thereby removing them from FDA regulation. This places the burden on the FDA to show that they are dangerous. By proving that these odd-looking brown cubes contained known drug materials that were seriously threatening to human health, they could be brought out from the protections of DSHEA and under the regulatory control of the FDA as drugs. The FDA then banned the cubes from importation. Although there is a toad that is found in the southwestern U.S. that produces similar skin toxins, which have been abused as hallucinogens, the dark-brown cubes known to have been available in New York and several large U.S. cities appear to have been of Chinese origin. For all practical purposes, this stopped their flow into the U.S. as Chinese herbal remedies.

"Legal 'Love Drug' Eyed in Deaths of Three Men"; *The New York Time,* January 18, 1995.

"Cops Declare War on Head-Shop 'Love Drug'"; *The New York Times,* February 20, 1995.

"U.S. Says 4 Died in New York Eating and Alleged Aphrodisiac"; *The New York Times,* November 25, 1995.

"Deaths Associated with a Purported Aphrodisiac – New York City February 1993 – May 1995"; *Morbidity and Mortality Weekly Report,* Vol.44/No.46, p. 863.

Bufotenine – A Hallucinogen in Ancient Snuff Powders of South America and a Drug of Abuse on the Streets of New York City; Chamakura, R.P., *Forensic Science Review,* 6, 1–18.

> **LEARNING OBJECTIVES**
> - Why a substance is called a drug
> - The nature of drug dependency and its two major forms
> - The impact of drug abuse on society and how society reacts
> - Each of the major classes of abused drugs, with examples
> - The rationale behind the controlled substances laws
> - Processing of suspected control substance samples through the crime lab
> - The major analytical steps from initial physical description to unambiguous identification
> - The important distinction between qualitative and quantitative analysis
> - The analysis of body fluid and tissue samples for drugs and poisons (forensic toxicology)
> - The critical role of alcohol and drugs in impaired driving cases

NATURE OF DRUGS AND DRUG ABUSE

Introduction to drugs and drug abuse

Drugs and drug analysis form a particularly important forensic topic since a significant portion of all the scientists, who work in forensic laboratories, are employed in drug related analyses. The drug analysis section is the largest section in the majority of forensic laboratories in the U.S. Interestingly, the Canadians and the British have traditionally given drug analysis responsibility to health department laboratories, perhaps because neither the Canadians nor the British have the same level of controlled substance (drug) abuse that we have in the U.S.

Working definition of a drug

What is a reasonable working definition of a drug? Because of the importance of drugs in our society it would seem trivial to come up with a good definition of a drug, but it is not. A good starting point is a very broad statement—Any substance that produces physiological or psychological effects on the body. A broad definition is necessary because the term drug means different things to different people. Someone who is an expert in physiology might like the above definition. It covers all the possibilities that one might encounter. However, for law enforcement purposes, it would be much too broad to be useful. Under that definition, since almost everything one takes can be thought of as producing **some** physiological or psychological effect on the body, too many substances are included. The real question that interests drug abusers is how strong an effect and how soon after it is taken the effect is felt. Medical science is currently trying to deal with the problems of substances that produce effects only many years after exposure; that is, showing an extended induction or incubation period. In the context of drug abuse and the regulation of controlled substances, such substances are not a concern. For example, is a substance like cholesterol a drug? Much medical research indicates that it does have some physiological effect on the body. Certainly some people, after having above-normal concentrations of it in their systems for an extended period of time, develop heart conditions.

One of the few things in the world that virtually no one considers a drug is water. However, there is a condition where people drink enormous quantities of water, gallons and gallons every day. Although water is generally good for you, these people destroy their electrolyte balance and become quite sick and can die if not treated.

These examples suggest that the definition should be modified by adding three conditions to the definition. One, is that to be a drug, certainly of abuse, the substance must have its effect within a relatively short period of time after ingestion, within minutes or hours at most. Two, is that the dosage is critical because almost everything will have an effect on the body if the dose is large enough. Conversely, with a few exceptions, few substances have much effect if the dose is very low. Thus, a drug is a substance that produces a physiological or psychological effect that is significant, occurs within a reasonable time after dosing, and results from an easily ingested dose. With these added provisions, the overly broad definition becomes much more appropriate.

Nature of drug dependence

Drug dependence is the primary reason that certain drugs are a major law enforcement problem in the U.S. It is also the reason drug chemistry (controlled substance identification) is such a large part of forensic science laboratory efforts.

Law enforcement spends an enormous amount of its resources dealing with the many ramifications of illegal drugs, including their distribution and the many secondary effects they have on society. Drug abuse and drug dependence are important concepts. Although specialists might quibble, the terms dependence and addiction can be used almost interchangeably for law enforcement purposes. If people did not become addicted to, or dependent upon, certain drugs, society would not be so deeply involved in drug enforcement. The fact that some people who take drugs are no longer productive members of society drives the need to try to control availability of such substances. It is well known that the most abused drug in the Western world is alcohol. People have been abusing alcohol since before recorded history. Alcohol and drug abuse have enormous effects on our society in many ways. Their families are often affected. Considerable productivity in the workplace is lost every year. Substance abuse is frequently involved in other crimes, such as theft, murder, manslaughter, and sexual assault. It is also involved in a very high percentage of serious highway accidents that result in injury and death. The concept of dependence is often subdivided onto the categories physiological dependence and behavioral (psychological) dependence.

Physiological dependence

Physiological dependence occurs when one takes a substance, usually in increasing dosages because the body seems to require increasing dosages to get the **same** effect. When one stops taking the drug, many unpleasant symptoms occur. Individuals may become sick to their stomach, have terrible headaches, experience shakes, sweat profusely, or even go into convulsions. Thus, if a drug on which a person is physiologically dependent is withheld, the body strongly reacts. Probably, many adults are, at least mildly, physiologically dependent upon (addicted to) caffeine. Many of us take most of our liquids in the form of coffee, tea, or carbonated beverages. Almost all have caffeine in them, as do many other things we ingest. Most individuals, who carefully avoided caffeine for a day, would almost certainly have a headache before the end of that day. A common physiological effect of caffeine withdrawal is headache. Physiological dependence is fairly easy to understand at a practical level since the withdrawal symptoms are quite tangible. Dependency on heroin or morphine is the prototype with which everyone is familiar. The movies and literature are rife with heroin withdrawal: the sweats, nausea, cramps, and a variety of other unpleasant physical symptoms. Alcoholics become dependent on having a sizeable dosage of alcohol every single day. If they stop or are in the process of trying to break the dependence, they can go into violent fits called delirium tremens (DT's).

Behavioral dependence

Behavioral dependence can be just as strong an effect as physiological dependence. The tremendous increase in the abuse of cocaine in the last 20 years of the twentieth century has changed the way experts think about dependence. Abusing cocaine for a period of time and then stopping does not usually cause physical illness. Often, a person does develop what is usually called an "uncontrollable craving" for the drug. That craving is more than a wish to continue taking the drug; it is a desperate need to continue. People who are dependent on cocaine will do awful things to themselves and others to provide themselves with cocaine. Behavioral dependencies are every bit as difficult to break as physiological dependencies. This brief description is a bit of an over simplification, since physiological dependency certainly has a behavioral component and behavioral dependence has a physical component as well. Moreover, the detailed mechanisms of substance dependence are complex, and remain somewhat unclear.

Drugs and society—controlled substances

Drugs and medications generally have become extremely important in treating and preventing disease and illness. Many drugs are available only by prescription—a physician must order them for the patient. Prescription drugs are generally designed for specific conditions and with specific directions about how they are to be used and proscribed in dosages appropriate to the desired action. In addition, many drugs are also available "over-the-counter;" that is, no prescription is required.

Access to drugs is regulated by the Federal Controlled Substances Act and local state acts in all 50 states. Under these laws, drugs are placed onto various "schedules" (lists) depending on the extent to which they are to be regulated. Some drugs are highly addictive and have no recognized or approved medical use in the U.S. They are the most highly regulated, and mere possession of them is a violation of the law. The Drug Enforcement Administration of the U.S. Department of Justice administers the Controlled Substances Act. In this chapter, we will discuss drugs that are commonly subject to abuse. A few of these have no recognized therapeutic use. However, many of them do. Forensic science laboratories get involved in this area because a drug or substance has been, or can be, abused. Substance abuse affects the abuser directly, but also affects his or her family, and can very well affect the greater society. Some abused substances are not ordinarily thought of as "drugs." As mentioned previously, alcohol is the best example, and, although society and the law do not treat it as a drug, it certainly fits the definition of a drug as provided earlier.

Society through its laws tries to control the abuse of drugs. There is generally a relationship between the degree of regulation of a substance and the amount of harm its abuse is perceived to cause in society. Drug dependence is a serious problem, and not tolerated in most segments of our society. The extent of society's acceptance of an addictive drug depends to some extent on the extent to which the abuser is functional in spite of the drug abuse. For example, drug-abusing physician may go along for years so long as the problem does not obviously affect his/her care and treatment of patients, or is otherwise detected. The doctor likely has the funds to buy the drugs, and does not have to resort to committing crimes to get them. On the other hand, a drug abuser who has to commit robberies to finance a drug habit is having an obvious negative effect on others and, if apprehended, is more likely to be prosecuted.

There is a rather high tolerance for those dependent on alcohol. Alcohol has been a legal drug in the U.S. (except for one short period) throughout its history. However, the fact is alcohol abuse costs society an enormous amount in various ways—traffic accidents, broken families, crime, lost productivity, and much more. By comparison, society has almost no tolerance for people who use or become dependent on cocaine. It is illegal to possess the drug, and selling it carries even greater penalties. Society's attitude toward a particular drug can change over time. An interesting case is tobacco. Tobacco is a "drug" under our definition, but is not treated as a drug under the law. Its growth and sale has been a major positive factor in the U.S. economy since colonial days. It was not only tolerated, but revered even though it users develop a strong dependence. In the past few decades, society's attitude towards tobacco has undergone a major change, primarily because the long-term health effects of chronic smoking have finally made a clear impression on the public consciousness. Although it is still legal, attitudes toward its use have changed significantly.

While smoking was widely advertised and encouraged (at least it wasn't discouraged) a few decades ago, it is now broadly discouraged in every venue of popular culture.

MAJOR CLASSES OF ABUSED DRUGS

The important drugs of abuse can be divided into six basic categories, the first four based on their physiological action.

Analgesic drugs (pain killers or narcotics)

The analgesic drugs (with recognized medical use) are taken primarily to dull pain Analgesic is a better name, but law enforcement tends to refer to drugs such as heroin and others related to morphine as "narcotics." Heroin has no recognized medical use in the U.S. Narcotic drugs, more properly known as analgesics, are used medically as painkillers, and abused as euphoriants (provide a feeling of extreme well-being).

They allow one to better tolerate pain and are beneficial, particularly for those experiencing severe pain from a variety of conditions or the last stages of many illnesses. These are called narcotics or opiates for historical reasons. Morphine is one of the oldest and most effective painkillers. Morphine is the primary active drug in opium. Opium is the dried sap from the opium poppy plant, which produces a small red flower. The seeds from the plant are used on rolls, bagels and in Danish pastry. If one scores the pod (Figure 13.1) with a sharp instrument just before the poppy blooms, gooey material oozes out. This material can be scraped from the pod, dried, and processed to make opium. Opium can be smoked directly, as it has been for centuries in the Orient, or it can be chemically processed to isolate pure morphine and several other drugs.

About 30–35% of the dry weight of opium is actually morphine, and it is the largest single drug component in opium. There are many other chemical compounds in opium and many of them have drug activity. All these compounds, and many others that can be made from them, are referred to as *opiates*, because of their origin in opium. Morphine has been abused for centuries because it has the ability to make users fall into a drugged sleep with accompanying feelings of extreme well-being and blissfulness. This property makes the opiates both physically and behaviorally addictive drugs. The desirable feelings, coupled with the classic need for larger and larger doses to obtain the desired effect, make dependence rapid and strong. Further, the severe physiological withdrawal symptoms also reinforce the dependency.

Morphine's strong dependence-producing side effect has caused scientists to search for other drugs that are as effective as painkillers, but not as addictive. Interestingly, one of the many attempts to find a non-addicting form of morphine produced a fine painkiller by chemically adding acetate groups (combining form of acetic acid) to morphine, the drug produced was called heroin. Heroin, one of the most abused opiates, (Figure 13.2) is not approved for medical use in the U.S., was actually developed by a pharmaceutical company in its search for a non-addicting form of morphine. Heroin is at least as addictive as morphine, perhaps more so.

Figure 13.1 A poppy pod being slit to allow the thick sap to flow out of the pod to be collected and harvested later, when it has solidified, for opium production. (Courtesy of Shutterstock, New York.)

Figure 13.4 Cocaine base is usually taken by inhaling the vapors using a pipe and a heating device. (Courtesy of Shutterstock, New York.)

Figure 13.5 Cocaine salt is not volatile and is inhaled through the nose using a hollow tube or straw.

The popularity of the Internet has spawned novel mechanisms for distribution of drugs of abuse. Characteristic of this trend is the rise of "bath salts." Materials that are sold as "bath salts" (Figure 13.6) on the Internet but are known by sellers and buyers alike as stimulant drugs for abuse. They pose a significant problem for law enforcement because the sellers are difficult to locate, and buyers pass on seller's contact information to other potential buyers. The other major problem is that the sellers are supplied by producers who seem to be able to produce a large number of chemical variation of materials that are similar in their physiological effect. That means they generally stay legal until specifically legislated against. Thus, the buyers feel that they are safe from arrest. The street names for these materials change so rapidly that listing them would be fruitless.

Hallucinogen

This broad group of substances change one's mental state. They primarily affect perceptions of oneself and of environmental stimuli. Most of them have no recognized medical use.

Hallucinogens have been used for centuries to effect one's perceptions. One of the oldest and the most widely used is marijuana. It has been used in the Middle East for thousands of years. No one knows exactly when it was discovered that the leaves of the hemp plant, the stem of which was used for making rope, would also made one feel a little bit woozy and a little strange when smoked or ingested with foods. Many people think these sensations are pleasant. People

Figure 13.6 So-called "Bath Salts" or designer cathinones are a stimulant in which the principle active ingredient MDPV (methylenedioxypryrovalerone) is a highly potent synthetic cathinone derivative. They can often by obtained by drug abusers purchasing through the Internet. (Courtesy of Drug Enforcement Agency, Springfield, VA.)

have smoked, chewed, "brewed" marijuana in teas, and taken it in a variety of other ways. The leaves of the hemp plant (the marijuana plant—*Cannabis sativa* [Figure 13.7]) have a coating of a resinous material. It is this resinous material that contains the physiologically active ingredients, which are called cannabinoids. The most well- known, largest single component and most physiologically active of the cannabinoids is tetrahydrocannabinol, which is called THC for short. Much of the marijuana purchased on the street today is a totally different product than it was 20 or more years ago. The breeding stock has now been improved to produce more resin and, thus, more THC. Premium grades of marijuana now have THC content in the 7–10% range, two or three times more active ingredient than seen in material a number of years ago. Because marijuana is more potent now, it has become a more dangerous drug.

Another form of marijuana is called hashish; it was traditionally a more potent form of marijuana. Originally, it was made by going down the rows of growing marijuana plants, when they flowered, with a large piece of leather strapped on one's arms and hitting the tops of the plants. The sticky stuff and the flowers, particularly, would stick to the leather. When the leather was well covered, the material sticking to it was scraped off. These scrapings were

Figure 13.7 Mature marijuana plant.

then collected and compressed to make the hashish. This material was considered a more potent form because it had more resin and less leafy material. Normal marijuana has a lot of plant material and a little bit of resin. As noted above, plant breeding practices have changed the nature of much marijuana and, as a result, hashish made from "unimproved" marijuana may have a lower THC content than premium marijuana.

A third form of marijuana is called hash oil. It is made by taking a large quantity of marijuana plant material, placing it into a large vessel, and "cooking" it with alcohol or some other solvent. Hash oil processors often use gasoline or kerosene. In this way, the resin, which is very soluble in solvents like gasoline and kerosene, is dissolved off the plant material. The solid material is removed, and the solvent is then evaporated down (concentrated) until it is a thick oily material, which is almost pure resin and has very little vegetable matter in it. This process produces a concentrate that can be more easily smuggled and can also be used to increase the potency of marijuana plant material. Smuggling small bottles of a highly concentrated hallucinogen is much easier than large bales of marijuana plant material. Concentrated hash oil can be reconstituted when mixed with tobacco or other vegetable material, and that makes even a small volume of it valuable.

LSD is probably the most potent and the best known of the true hallucinogens. The great guru of LSD in this country, Timothy Leary, had been a professor at Harvard who advocated the "mind-expanding" nature of the LSD experience in the 1960s. LSD is short for lysergic acid diethylamide, its chemical name. It is an extremely potent hallucinogen. Very minute doses of it cause long and often frightening hallucinations. Obviously, some find the experience satisfying since it has been abused since it was first synthesized.

Unfortunately, many people have done serious harm to themselves under the influence of LSD. It is said to cause visual hallucinations, brilliant colors, and the perception that one is very wise or in touch with a supreme being. The normal LSD dosage is 30–50 micrograms (millionths of a gram), which is a 1000 times less than that required for most other drugs to produce a physiological effect. Because a popular dosage form of LSD is impregnated on brightly colored stamps (Figure 13.8), one of its greatest dangers is that children sometimes are attracted to the dosage forms and cannot tolerate the dosage taken by abusers. Another danger of LSD is that some people will continue to have hallucinogenic episodes even long after they stop taking the drug. Recurring and/or delayed hallucinations can occur even after a single use of this drug.

Phencyclidine, usually known as PCP or "angel dust," is another hallucinogen that was extremely popular for a while. It is a drug that is fairly simple to make, so the vast majority of it is made in clandestine laboratories. The term "laboratory" is actually not very accurate since the clandestine "laboratory" is often someone's bathroom, garage, or barn. It is fairly easy to find out how to make it from inexpensive starting chemicals. Unfortunately, safety is often not a concern in these operations. Even if the people involved in the activities know better, they ignore safe practices. Many

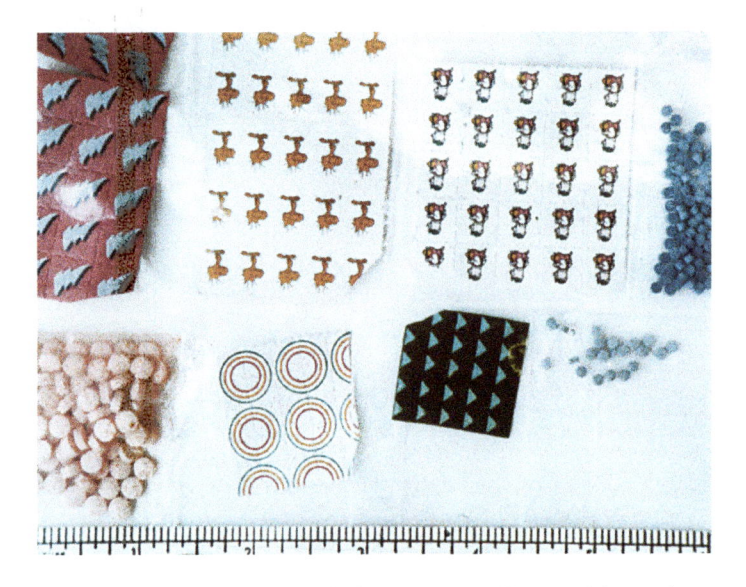

Figure 13.8 LSD dosage forms some of the many forms of LSD encountered in crime laboratories.

times, clandestine laboratories are discovered because the operators blow themselves up or cause a fire. PCP has decreased in popularity, because the hallucinations are frequently not very pleasant and because it tends to make people aggressive. There was a period when PCP was the number one drug of abuse, particularly on the West coast. It was even more popular than cocaine or heroin, in some areas.

It should be mentioned here that clandestine drug laboratories pose potentially extreme chemical and explosion hazards for investigators. The Drug Enforcement Administration (DEA) and many forensic laboratories have "clan lab" teams trained to safely disassemble these makeshift laboratories in a safe manner and properly dispose of any hazardous chemicals.

There are a number of naturally occurring hallucinogens that have been used by native people for centuries, usually as part of religious rituals. Peyote is the bud of a particular kind of cactus. It contains as the main active ingredient a hallucinogen called mescaline. In fact, it is legal to use peyote as part of the ritual of the Native American Church. It is a controlled substance under any other circumstances. One of the disadvantages of peyote is that it is not pleasant to take. It is said to be extremely bitter and unpleasant tasting. User's report that it makes people feel like they are having an out of body experience and are closer to their God. That is a common thread among plant hallucinogens abuse: such plants, or parts of them, causing people to have feelings of being very wise or closer to their supreme being.

So called "magic mushrooms" were also popular hallucinogens in the 1960s and are still being abused. These are mushrooms of the genus *Psilocybe*. The two major active components are psilocin and psilocybin. It is possible to buy spores through the mail that can be placed in humus or some similar growth medium, placed in a dark damp place, and psilocybe mushrooms may emerge. The spores themselves are not illegal because they don't contain any psilocin or psilocybin, which are the controlled materials in the mature mushroom.

There are a number of other synthetic hallucinogens that were also popular in the 1960's and early 1970s. They had abbreviated names like MDA, DOM, STP, and DMT. Interestingly, one of the most heavily abused drugs in the current "club drug" scene, that will be discussed later, is a close relative of MDA known as MDMA, or "ecstasy."

A particularly serious problem with hallucinogens has emerged, along with "Bath Salts" as largely an Internet driven problem. The material sold as so-called synthetic marijuana can be readily obtained through the Internet. Synthetic marijuana was sold originally under the name "KR2 or Spice" but both drugs and names have proliferated. There are literally dozens of different materials that have no connection to marijuana, being widely abused. Some are quite dangerous and subject to overdose and serious health effects. The creativity of the producers, and naivety of the abusers, give law enforcement an extremely complex problem (see "Bath Salts" previously).

Depressants, hypnotics, and tranquilizers

This group of substances tends to dull our senses. In most ways, their effects are the opposite of stimulants. Many are used to reduce anxiety or to help induce sleep. These three different types of drugs are treated together since their abuse is similar and many have actions that are somewhat similar. Depressants, hypnotics and tranquilizers have different medical uses, but abusers tend to use them almost interchangeably, often in combination with alcohol.

Alcohol alone, and in combination with other depressants, is certainly the most common and widely used depressant drug. It is the most abused drug in the western world. Almost everyone knows something about the effects of alcohol. Abuse of depressants usually results in a mental and physical state similar to alcohol-induced intoxication. Barbiturates are compounds that were widely used as sleeping pills. For many years, barbiturates were the sleeping products of choice. They are still prescribed for several different functions and used in hospitals. There are many slight variations on the basic barbiturate structure that allow tailoring the drug to a desired effect. There are fast, slow, and intermediate acting ones. Phenobarbital, given in small dosages, is also used to control seizures in epileptics and others subject to seizures. It is considered a slow acting barbiturate, because it takes significantly longer than the rapid acting barbiturates to induce sleep. Barbiturates are not currently abused extensively in the U.S., but were a major problem in the past and contributed to the death of several famous individuals including Marilyn Monroe. They are highly physiologically addictive and can produce extremely unpleasant withdrawal symptoms and even death when usage is abruptly stopped.

Tranquilizers are drugs designed to relieve anxiety. Valium, which was one of the first and most popular tranquilizers, was the most prescribed drug in the U.S. for a number of years. It may have been over-prescribed because its use was

Several states have now taken the next step and legalized use of small amounts of Marijuana. This conflict will undoubtedly eventually be resolved by Congress or by the U.S. Supreme Court.

The internet has spawned a new category "synthetic marijuana" materials chemically unrelated to THC that are sold as having similar effects. The materials are a considerable legal problem since there are many chemical variants that usually have to be individually added to controlled substances laws.

ANALYSIS OF CONTROLLED SUBSTANCES IN THE FORENSIC LABORATORY

As with some types of evidence already discussed, such as gunshot residue (Chapter 8) and blood and physiological fluids (Chapter 10), the laboratory identification of controlled substances uses screening tests followed by more reliable confirmation tests. The screening tests are designed to be quick, easy to perform, and to allow for screening many specimens in a short time. Confirmatory tests are typically more complicated and time-consuming, but more specific.

Screening tests

An important adjunct to drug enforcement is the utilization of drug screening tests. These are "presumptive" tests. Positive results indicate possible, but not certain, presence of drugs. Most of these tests are simple and based on color changes and are easy to use outside a laboratory setting and are not effected by common diluents in street level drug samples. Nowadays, screening tests are usually found in the form of little packets (Figure 13.13) called drug test kits.

The packet has a sealed glass ampoule of a chemical test reagent. One places a small amount of the suspected drug in a tube or in the outer package of the kit, breaks the ampoule thereby adding the reagent, mixes and looks for a certain color. There are several different ways the kits are designed. Some having several steps, but the result is a quick and relatively simple set of operations that give a fairly reliable indication of the probable presence of the target controlled substance or substances. These screening kits were designed for use by law enforcement officers who, after a short training program, can perform the tests quite successfully. Their importance comes from the requirement for probable cause when making a drug possession arrest. An officer can see someone who appears to be dealing drugs and wants to make an arrest. In many jurisdictions, more than just the officer's experience that the incident looks like a drug deal is required. Using a test kit, to show that it is likely that there is a controlled drug in the material seized, adds another reason to believe it is a drug deal. These field tests or screening tests are very similar to the initial screening tests used in the forensic laboratories. However, they are not a substitute for laboratory identification! Prosecution

Figure 13.13 A test kit of the type used by officers to test suspected controlled substances on the street. This kit is for the commonly abused tranquilizer Diazepam.

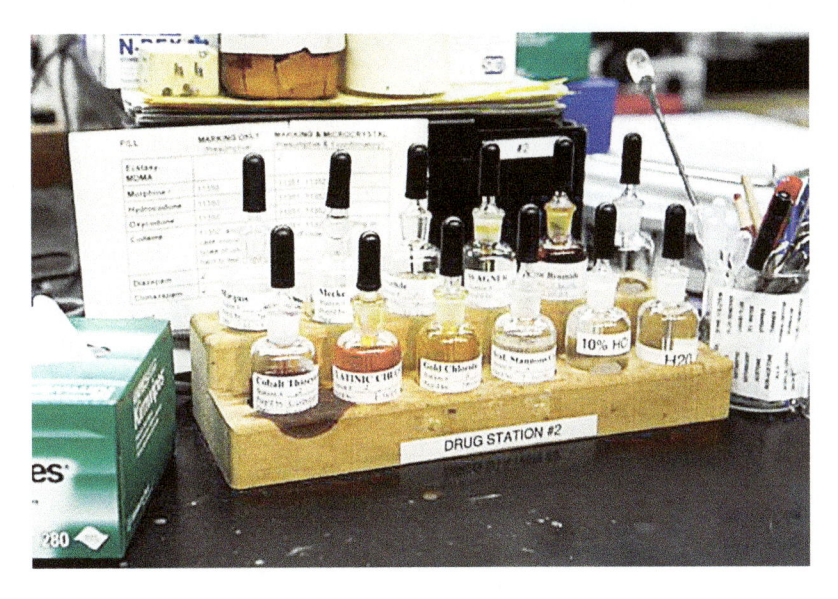

Figure 13.14 A collection of reagents used for screening suspected drug samples.

for possession of a controlled substance requires that the material be submitted to a forensic drug laboratory, and an examiner qualified to testify as an expert verifies the identity of the material.

There are commercially available screening kits for the probable presence of cocaine, marijuana (*Cannabis sativa*), the opiate drugs, amphetamines, and for many hallucinogenic drugs. When seized materials come into a forensic laboratory, the first step is usually testing with a screening tests. These tests are similar, but not identical, to those used in the field by police. The laboratory chemist screens the alleged drug substances using color tests (Figure 13.14) to decide which specimens require confirmation with more definitive methods. One of the problems with commercial screening test packets is dependence on the officer to add the proper amount of sample. Forensic chemists with more experience and understanding of the process can better judge the amount to use and judge the significance of the color. Therefore, the repetition of the screening test in the laboratory is not at all redundant.

Isolation and separation

Each laboratory may follow a slightly different set of detailed steps in controlled substance identification, even though the general analytical scheme used by forensic laboratories is very similar. The steps might include viewing a small portion of the exhibit under a microscope to more closely examine what is present. This microscopic examination gives the examiner some idea of whether there are several different materials present. A purification step, to separate the controlled substance from the other materials present usually, follows the microscopic examination. Liquid/liquid extraction is one of the simplest separation methods and, with proper choice of solvents, the drug component or components can frequently be separated from most diluents. Gas chromatography, high performance liquid chromatography, and several other chromatographic techniques are also commonly used (more on these to follow).

One of the most useful things about a screening test is that it does not require an isolation or purification step before use. Screening tests generally are effective on the kind of mixtures usually encountered in "street" drug samples.

The simplest chemical separation technique is called extraction. For example, one has a mixture of two or more drugs in a body fluid sample. It is necessary to largely remove one of those drugs into a different solution to ease its identification. The mixed drug solution is shaken with an immiscible solvent; that is, a solvent that will form a separate layer when mixed with the first solution (Figure 13.15).

A common example of the principle is oil and vinegar salad dressing, where the oil and vinegar layers can be shaken together to form what appears to be one phase, but upon standing will always revert back to two separate layers.

(a) (b) (c)

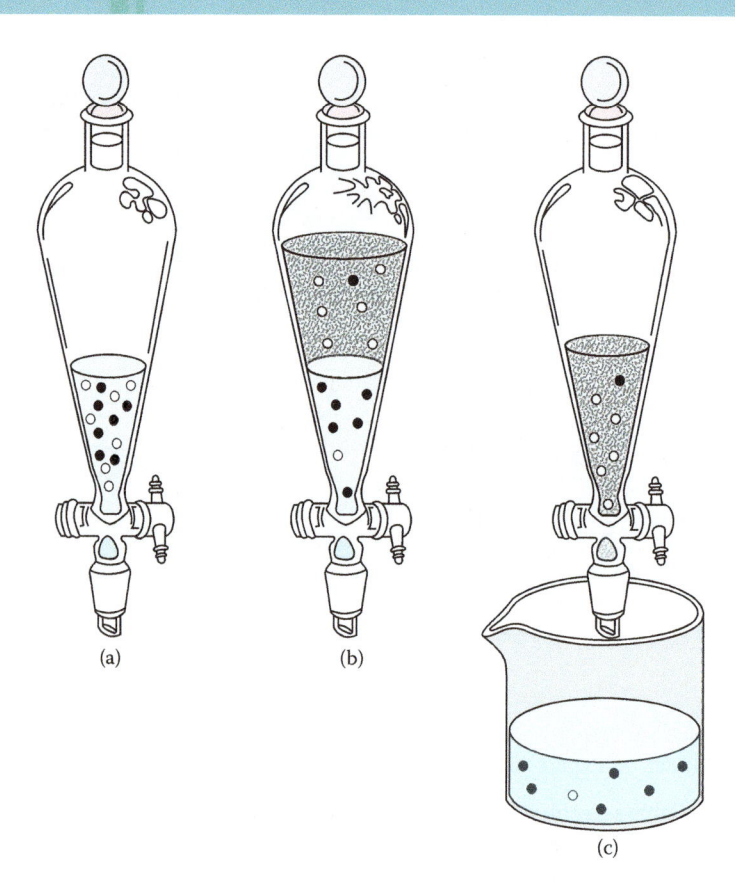

Figure 13.15 This illustrates the process called liquid/liquid extraction where substances can be separated by shaking with two immiscible liquids where one of the compounds is preferentially moved to one of the liquids. In that way when the liquids are separated the moved compound can be isolated.

The immiscible solvent is selected such that much more of one of the two drugs will move into the solvent when the solvent and the drug solution are shaken together. Thus, when the mixture separates, the second solvent will have more of the desired drug and little of the other drug or drugs. The layers are then separated, a fresh portion of the immiscible solvent added, and the process repeated. The separated solvent layers can be combined and, thereby, much of the desired drug is separated from the other drug or drugs. As a result, it can be more effectively analyzed.

Microcrystal tests

A type of confirmatory test that works well on "street" drug samples directly (no separation or purification necessary) is the microcrystal test. They are useful for rapid yet reliable identification of many drugs, and were widely used for many years before some of the more sophisticated spectroscopic techniques became available. The microcrystal test is performed simply by taking a small amount of the street drug sample, placing it on a microscope slide, and adding a drop of a chemical reagent that is known to give crystals of a particular shape (Figure 13.16) with the suspected drug component. The shape (morphology) of the crystals formed is characteristic for that particular drug when combined with the chemical reagent used to form them. Chemists can be trained to recognize these crystals as a way of identifying many drugs.

Crystal tests have fallen out of favor as spectrometric confirmatory techniques have become available, though they are still used in many laboratories as rapid confirmatory tests. One of the problems in forensic laboratory drug identification is the way controlled substance laws are written and interpreted in the courts. If a forensic laboratory receives a seizure of 50 identical envelopes containing white powder, one may not be allowed to just take two or three envelopes randomly and analyze them. In some jurisdictions, the courts will require analysis of a sample from each

Figure 13.16 Cocaine crystals form when a material containing cocaine is mixed with a dilute solution of Gold Chloride on a microscope slide and viewed at about 100 X in a compound microscope. The characteristic shape of these crystals is highly indicative of the presence of cocaine in the material.

of the 50 envelopes. A big advantage of crystal tests, and the reason they are still used in many crime laboratories, is that they are rapid and, in conjunction with other tests, specific. Tiny samples from each of the 50 envelopes thought to contain the drug can be placed on a glass plate. A drop of the appropriate crystal forming reagent is added to each sample and the drop examined using a microscope to see if the characteristic crystals form. Thus, multiple specimens can be processed rapidly and simultaneously.

Chromatography (separations)

All forms of *chromatography* are actually separation techniques. See the science sidebar in Chapter 12 on chromatography.

Thin-layer, or gas, chromatography can be used to separate the components of a suspected drug specimen that contains a mixture of compounds. Originally devised to work on paper, the technique was later extended to thin layers of silica gel or cellulose, as well as to columns lined with high boiling liquids designed to aid in particular chemical separations. All chromatography is based on the principle that the chemical components of a mixture will partition themselves between the phases of a two-phase system (just like in the liquid-liquid solvent extraction method described earlier). If one of these phases is moving while the other is stationary, physical separations can be achieved. In thin-layer chromatography the mobile phase is the solvent that moves up the plate by capillary action and in gas chromatography the mobile phase is the carrier gas flowing through the column.

Chromatographic techniques can also provide a method of tentative identification. If one measures the time that a drug takes to emerges from the instrument after injection, the so-called "retention time" is a characteristic of the drug. In thin-layer chromatography the distance that a drug travels on a thin-layer chromatography plate is characteristic of that drug. For example, a known heroin sample is injected into a gas chromatograph under carefully controlled conditions. The heroin always takes 8.3 minutes to emerge under these conditions. Now, if one injects a suspected heroin sample, and there is a significant amount of that sample that emerges at 8.3 minutes, it provides a strong suspicion that the specimen contains heroin, though it does not prove it. In addition, if there are two additional materials emerging at 7.5 and 7.7 minutes, and street heroin virtually always shows these same two peaks, there are now three different peaks observed that indicate heroin. Thus, although chromatography is not an identification technique as such (it is a separation technique), it *can* be used in connection with other information to assist an examiner in identifying controlled substances.

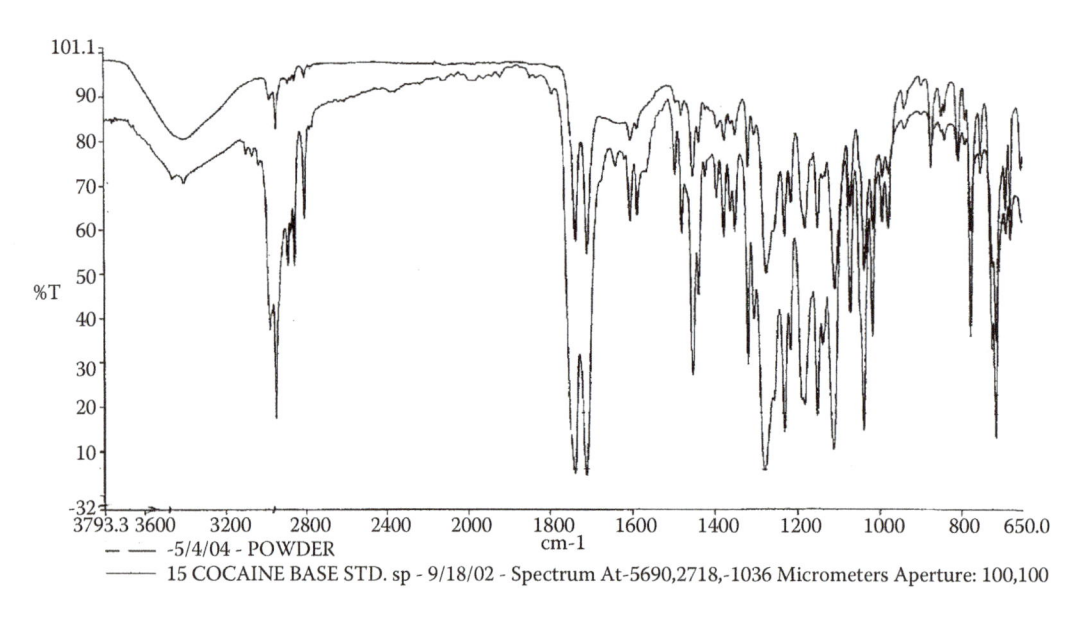

Figure 13.17 Infrared spectral traces of standard cocaine base and evidence seized in a crack bust.

Spectroscopy/spectrometry

Either mass spectroscopy or infrared spectroscopy provides a complex pattern (spectrum) (Figure 13.17) that is characteristic of a particular drug and is different from that obtained from other drugs. If one has a mixture of several drugs or a drug and several diluents, the spectrum produced from the mixture would not be identifiable because there would be many peaks. One could not say which peak belongs to which component of the mixture. Therefore, spectroscopic identifications work best when the sample examined is a relatively pure material. Infrared or mass spectrometry (Figure 13.18) are excellent ways to identify drugs, particularly when combined with a chromatographic technique to separate and isolate the individual pure components.

Although chromatographic techniques can, as noted, give indications of the identity of controlled substances, the identifications must generally be confirmed by mass spectroscopy or by infrared spectrophotometery. For some time, instruments have been available that combine gas chromatography with mass spectroscopy or with infrared spectrometry. Such instruments—and the combination of techniques they perform—are typically called GC-MS or GC-IR. These may also be called "hyphenated techniques."

The so-called "hyphenated techniques" use a single instrument that combines chromatographic separation with a highly specific spectral identification technique. Gas chromatography, for example, can be used in combination with either mass spectrometry or infrared spectroscopy to allow separation of the drug mixture into its pure components. The drugs are then passed one at a time, to the Mass spectrometer or infrared spectrometer for identification. High-performance liquid chromatographs can also be combined with mass spectrometers (LC-MS). The separation instrument provides a characteristic retention time from the chromatographic portion of the instrument, along with an unambiguous spectrum that allows for identification by the other part. Known standard spectra of almost all drugs likely to be encountered, are available to compare with those obtained from case specimens. This combination of a chromatographic retention time and a spectrum provides a significantly stronger identification than either technique alone.

Qualitative versus quantitative analysis

The examination of a street drug specimen, and identification of a particular controlled substance in it, enables an examiner to say that this material contains a particular controlled substance. The analyst is **not** saying that the specimen contains nothing else, nor is s/he saying *how much* of that sample is made up of that controlled substance. That is a **qualitative** analysis. Qualitative analysis means determining whether something is there

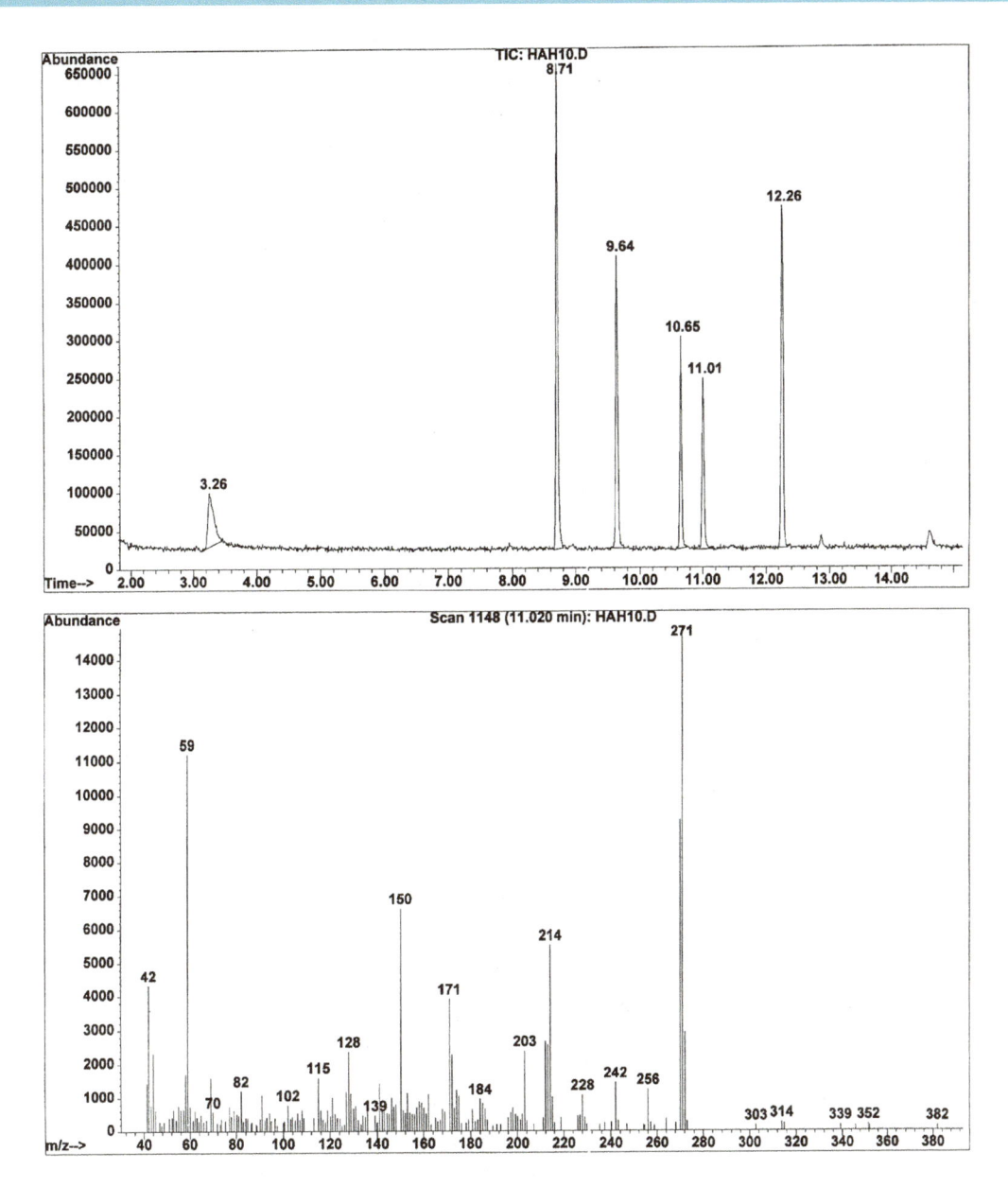

Figure 13.18 A mixture of six common over the counter drugs is separated using gas chromatography and identified using mass spectroscopy. Top half is the total ion chromatogram of the six-drug mixture, and the bottom half is the mass spectrum of Dextromethorphan, the peak at about 11 minutes.

or not there. To charge someone with possession of a controlled substances, under most drug laws, a qualitative identification is all that is necessary. The drug law makes it illegal to possess *any* quantity of the material. The legal term is aggregate weight-based law. If a brick of compressed powder weighing a pound has been seized, and that powder contains cocaine, the defendant will be charged with possession of a pound of material containing cocaine.

The concept of quantitative analysis becomes important when a drug law is written in such a way that the severity of the charge is dependent on the actual weight of the controlled substance possessed. In such cases, a **quantitative** analysis is required. Depending on the law, it may be necessary to establish not only that the seized one-pound brick contains cocaine, methamphetamine, or heroin; for example, but also that such-and-such percent of the total weight of the brick is the controlled drug. ***Qualitative*** **analysis says the substance is there and what it is, and *quantitative* analysis says how much of it is in the specimen**. In most cases, forensic labs do not have to perform a quantitative analysis, since most drugs are subject to aggregate weight laws. Because it takes about 10 times as much effort to do a quantitative analysis as it does to do a qualitative analysis, quantitative analysis is normally done—only when legally required, for intelligence purposes, or for toxicology as described below.

FORENSIC TOXICOLOGY
General description

A toxicologist analyzing body fluid or tissue samples has a significantly different task than a crime laboratory chemist analyzing a street drug sample. As we will see; however, most of the analytical techniques used are the same. Although scientists have been engaged in toxicology for many years, it was not until the 1960s that modern analytical instrumentation made it possible to routinely identify most drugs in most body fluid samples. The availability of gas chromatography and the gas chromatograph/mass spectrometry advanced toxicology enormously and more recently liquid chromatography/mass spectrometry (LC/MS) become very useful in many cases.

For the drug chemist, the sample is submitted in a form of an envelope of white powder, a collection of tablets or capsules, crude looking cigarettes, vegetable matter, or a large seizure that may fill a suitcase or a large carton (Figure 13.19). One can usually see it, touch it, weight it, and easily take a small sample to analyze. Toxicologists have a much more difficult problem. They usually receive blood, urine, or body tissues. The sample may be only 5–10 mL (normally a tube of blood taken from one at a doctor's office contains about 5 mL of blood) and is only a small fraction of the total amount of drug in the donor. There are usually only trace, milligram quantities of the drug distributed throughout all the body fluids and organ systems. For a solid, pure drug a milligram is about the amount that would easily fit on the end of a flat toothpick. Thus, the toxicology specimen provides only an extremely small amount of the drug material for analysis. Not only is the drug in very low concentration, there are many other biological materials contained in the sample. Therefore, the toxicologist must be able to analyze for small amounts of drugs but also be able to separate them from the complex biological matrix. In contrast to the drug chemist, toxicologists generally need quantitative information as well. They must know how much of a drug is in a person's body fluids to determine

Figure 13.19 Four oz of cannabis recovered by law enforcement agents.

if that amount is consistent with a therapeutic (an amount prescribed to be taken by a doctor) dose, an amount consistent with an abuse dosage taken by a drug abuser, or an amount sufficient to cause death whether accidental or intentional (e.g., suicide).

The analysis of body fluids for drugs and poisons is further complicated by the nature of the matrix (body fluid or tissue). The analysis is complicated by three major factors: (1) low concentrations of drugs and poisons, (2) enormous variety of possible drugs and poisons, and (3) the body's metabolism of drugs and poisons. As mentioned previously, any drug or poison taken into the body is quickly distributed throughout the body. Thus, any sample taken for analysis will have only a very small fraction of the amount of drug taken. Since the amount taken is usually quite small, this means that any analytical technique used must be very sensitive to be able to detect such miniscule amounts. To get a feeling for the variety of drugs available in the U.S., look at the many shelves full of drugs that your local pharmacy has in stock the next time you go. This is only a fraction of the full range of pharmaceutical materials available and does not include most illegal substances and poisons that can also be found in body fluids. This alone makes the analysis of body fluids, and corresponding identification of materials of interest, a staggering undertaking.

The third important factor is metabolism. Metabolism is the body's mechanism for getting rid of foreign things (almost anything that the body does not manufacture itself) that come into the body. The simplest mechanism used to metabolize drugs is for the body to chemically change the substance to a more water-soluble form which is readily transported to the bladder for excretion in the urine. As a result, for a time, different chemicals derived from drug or poison ingested are circulating through the system. These materials make identifying a drug or poison ingested even more difficult. For some drugs, the metabolism process is several steps (Figure 13.20) and begins almost immediately after ingestion. A sample taken from someone who has taken cocaine or heroin, even when analyzed a rather short time after they took the drug, will not show the unchanged drug but rather its first metabolite.

An important function of the forensic toxicologist, perhaps the most important, is to aid others in understanding the possible effects of the substances found in an individual's body. For example, was it likely that the drugs or poisons found caused impairment, loss of normal reason, unconsciousness, or even death? This information can have many legal ramifications.

It is convenient to divide toxicology into two major areas. Samples from living individuals (antemortem) and samples collected after death, usually from an autopsy (post-mortem).

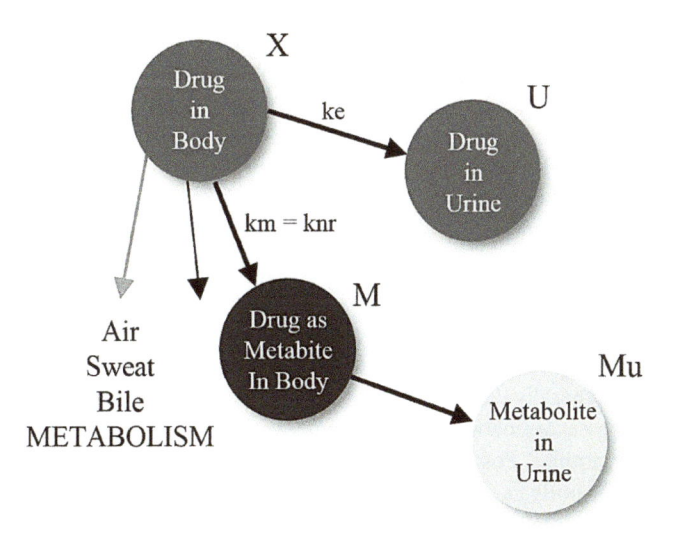

Figure 13.20 Diagram of the pathways common in drug metabolism.

Forensic toxicology on samples from the living

Many forensic toxicology laboratories analyze specimens from living persons, post-mortem samples, or both. Any analysis of drugs or toxins from the body for regulatory or law enforcement purposes can be considered forensic toxicology. A number of laws and rules are in force having to do with the workplace use of drugs. In some occupations, such as truck drivers, law enforcement officers, airplane pilots, and others, drugs of abuse are absolutely forbidden. These rules are enforced by drug testing, sometimes random testing, of the persons involved. Those specimens (usually urine) are sent to forensic toxicology laboratories for analysis. These labs must generally meet high testing and quality-control standards set by the National Institute for Drug Abuse (NIDA), which regulates this type of testing.

Certainly, the most publicized activity involving forensic toxicology on the living is the enormous effort to enforce sanctions against use of performance-enhancing substances by athletes, human, and animal. We all know about baseball players, bicycle racers, and dozens of other athletes who are routinely tested. Horses and dogs that race are also athletes who are tested for performance-enhancing drugs. Many significant advances in analytical toxicology have come from the specialized laboratories that perform these tests.

We have briefly mentioned that incapacitating drugs are sometimes used to facilitate sexual assaults. When an individual reports that they have or may have been sexually assaulted, standard law enforcement procedure is to transport the victim to a local hospital for an examination that almost always includes taking samples for a sexual assault evidence kit. Among the samples needed for that kit is virtually always a blood and a urine sample for forensic toxicology examination. These samples can be quite important when the victim is not able to clearly describe what has happened to them.

Pre- and post-employment screening for drugs of abuse is well established and a growing practice. The critical nature of some jobs, and the high cost of lost time and quality of work product, are of concern to a great many employers. Most employers would like to weed out candidates who are drug abusers before hiring them rather that discharging them after training and other investments have been made. The most sensitive jobs may require employees to take unannounced random drug tests after they are hired. In addition, employees who require security clearance or are in positions where they have access to company trade secrets, can be required to provide samples for testing, since drug dependency is considered a danger for blackmail by those trying to gain such information.

In many cases, a condition of probation or parole is that the individual not use controlled substances. This often entails providing body fluid samples, on a regular basis or on demand, to the agency overseeing the probation or parole. Positive tests are considered grounds for revoking probation or parole. Family courts often require drug testing in case effecting child custody and other family decisions. Of course, the examination of samples from persons involved in a wide variety of serious accidents or suspected of operating a motor vehicle while impaired by alcohol or drugs, is important to the investigation of such incidents.

Post-mortem toxicology

As we have noted in Chapter 1, a medical examiner is responsible for determining the cause and manner of sudden, suspicious, or unattended deaths. In fulfilling that responsibility, the medical examiner must consider the possibility that poisons, or drugs caused or contributed to the death. Post-mortem toxicology is performed primarily to assist the medical examiner in that duty. Larger medical examiner's offices have toxicology laboratories associated with them. In jurisdictions that have coroners, forensic science laboratories often provide toxicology services. A medical examiner can often do an autopsy in a several hours, but analysis of the toxicology samples taken will usually take much longer. A medical examiner or coroner generally does not make a final determination as to cause of death, except in the most obvious cases, until the toxicology examinations are complete. As we noted above, toxicology is made complex by the nature of body fluid and tissue samples. Post-mortem toxicology is further complicated because when the individual dies all normal processes stop, the body can be invaded by other organisms, and decomposition begins. Further, if the death was caused by drowning, fire, serious disease, or many destructive causes, the integrity of the body is breached, and many foreign things can enter.

Since post-mortem toxicology also has the disadvantage of little to no information about possible presence of drugs or poisons, in some cases, the toxicologist must be concerned with a wider variety of possible causative agents. Although homicidal poisoning is not currently a very common cause of death, it must still be considered. In addition, environmental toxic agents must be considered, assuming no other obvious causes are identified. This expands possible causative agent significantly.

Classes of poisons

The presence of a poison or poisons in someone's body fluids or tissues can be critical to many law enforcement and public health investigations and the dosage taken is often critical. One can divide poisons into three basic groups: (1) inorganic poisons; (2) organic poisons; and (3) biological toxins.

> "All substances are poisons;
> There is none which is not a poison;
> The right DOSE differentiates a poison, and a remedy."
>
> —Paracelsus

Some examples of inorganic poisons are metals like arsenic, beryllium, or cadmium, and other compounds like cyanide or hydrogen sulfide. Some examples of organic poisons are the classic plant poisons: the alkaloids, such as strychnine, coniine, curare, as well as digitalis, belladonna, and many others. The biological toxins can have many sources like venoms from snakes or spiders; or biological toxin that comes from microorganisms, such as botulism from foods; bacteria that cause red tides; and many others. These toxins are chemicals produced as a by-product of the growth of the bacteria and can be extremely toxic to humans.

SIDEBAR

Interestingly, the pufferfish (Figure 13.21) produces a material called tetrodotoxin, one of the most toxic materials known to man. That fish is considered a great delicacy in Japan but can only be safely eaten if a little gland that contains the tetrodotoxin is completely removed. Chefs must be licensed in Japan as qualified to perform this task. Nonetheless, a few people die every year from the toxin. A very interesting fact about tetrodotoxin, which has a very complex chemical structure, is that there is a newt that lives in the western U.S. and Canada, far away from Japan, that makes exactly the same toxin.

Figure 13.21 The meat of the pufferfish is a Japanese delicacy that can contain a gland with a very toxic substance.

It is ironic that although many people think rather simplistically that "chemicals" are unhealthy, and that all "natural" things are wholesome, the most highly poisonous things on earth are "natural" biological toxins.

ALCOHOL AND DRUGS AND DRIVING

Driving while impaired by alcohol

In many ways alcohol is one of the easiest substances for toxicologists to find in body fluids. It is the only drug frequently analyzed that is present in such large quantities in body fluids. One has to take a large dose (amount) to obtain the desired effect—grams rather than the milligram doses characteristic of most other drugs.

If one drinks several beers, each is about 4% alcohol, which provides roughly 15 g of ethyl alcohol or several shots of liquor (1 oz of 80-proof liquor has about 12 g of alcohol) close to 30 g of alcohol is being ingested. That is roughly 5000 times as much material as the normal dose, or even an abuse dose, of most other drugs. This high volume consumed per dose causes the alcohol to be present in easily detectable quantities—for a considerable time after the ingestion.

Alcohol is metabolized and excreted from the body fairly rapidly. However, not nearly as rapidly as it can be taken into the body when one is actively drinking. As a result, the drinker's level of impairment will increase for a period after they stop drinking, as the alcohol in the stomach is absorbed and will not begin to decline until absorption is complete. The length of time required after cessation of drinking for the level of alcohol to return to below the legal limit can be from an hour or two to more than 10 hours for someone who has been drinking heavily.

Alcohol's effects on people has been well studied and documented, and there is a good correlation between the amount of alcohol in a person's system and the level of impairment of judgment and motor skills. For that reason, state legislatures defined a level of alcohol content at or above which a person would be considered impaired for purposes of operating a motor vehicle. For many years in most states in the U.S., a driver had to have at least 0.10% (that is 0.10% percent, weight to volume, or 100 mg per 100 mL blood) alcohol in his or her blood to be considered intoxicated. In recent years, under pressure from the federal government, all 50 states have lowered the legal limit of blood alcohol concentration to 0.08%.

Another reason alcohol is easier to analyze than some other substances is that it is volatile. That means it passes readily into vapor phase from liquid phase, especially if warmed. That property allows the alcohol to be fairly easily separated from almost everything else in a body fluid. Other drugs, and almost all other materials found in body fluid samples, are not volatile. Thus, the separation problems mentioned earlier—that make analysis of body fluid samples for most drugs so challenging—are not as challenging for alcohol analysis. As a result, analysis of blood, or other body fluid samples for alcohol in impaired driving cases, can be done in laboratories that are not equipped to do other, more complex toxicology analyses.

As mentioned previously, post-mortem toxicology is the analysis of drugs and poisons in body fluids to assist in determining the cause (and manner) of death. Determining the presence of small amounts of drugs or poisons may be critical to this determination. Although a fatal dose of alcohol is rare, cases where alcohol has contributed to a death are quite common. Further, alcohol potentiates (has a strong effect upon) the actions of other drugs and can make a non-lethal dose of a drug become lethal. Alcohol is a factor in an extremely high percentage of fatal automobile accidents. In automobile accidents where an individual is not killed, but alcohol involvement is suspected, a specimen will often be sent to a local forensic laboratory. There are also cases in which a clinical laboratory in a hospital does an alcohol analysis on an accident victim's body fluid.

The analyses necessary for impaired driving enforcement are quite important because in many large forensic laboratories, particularly those serving the State Highway Patrol, they make up the largest single class of evidence submitted to that laboratory. Caseloads can reach into the hundreds per week. Impaired driving enforcement cases are unusual in that they reach the trial stage more frequently than many other types of cases. Even though they are usually violations of a state law, where the penalty is likely to be a fine or perhaps loss of driving privileges for a time, they are often vigorously litigated. An important reason for this is that many people consider the ability to drive to be closely tied to their ability to get to work and make a living. There are attorneys who make a specialty of defending driving-under-the-influence-of-alcohol cases. Therefore, the laboratory analysis in blood alcohol cases may be put under particularly strong scrutiny.

There are many methods to perform alcohol determinations. In the field, the Breathalyzer, Intoxalyzer, or another field instrument is used. This type of test is administered by a police officer trained to do so, or now some

Figure 13.22 A breath alcohol breath field test being administered.

handheld units (Figure 13.22) can be used by an individual to see if his/her level is below the legal limit. The suspect is asked to blow into the Breathalyzer instrument that allows most of the individual's breath pass through it, while capturing a known volume of that breath that comes from deep in the lungs. That known volume is then bubbled through a chemical solution capable of converting it to a compound that changes color that the instrument can subsequently detect. Other instruments use other physical methods to measure the amount of alcohol. The techniques are simple, yet quite reliable. There is an established correlation between breath alcohol level and blood alcohol level. As a result, the instrument can convert breath readings to legally acceptable blood alcohol levels (which is how the prohibited amounts are defined by law).

In the laboratory, it is generally blood that is subjected to alcohol determination. Such analyses are usually done using gas chromatography (Figure 13.23). Gas chromatography techniques can quite accurately determine the quantity of alcohol present in a blood specimen. A number of checks and balances are built into the analysis to ensure that the concentrations are accurate. The obvious issue with blood alcohol testing is that it requires taking blood from a suspect. While this does happen regularly, it is worth noting that taking blood from someone for alcohol testing entails complex legal search and seizure rules.

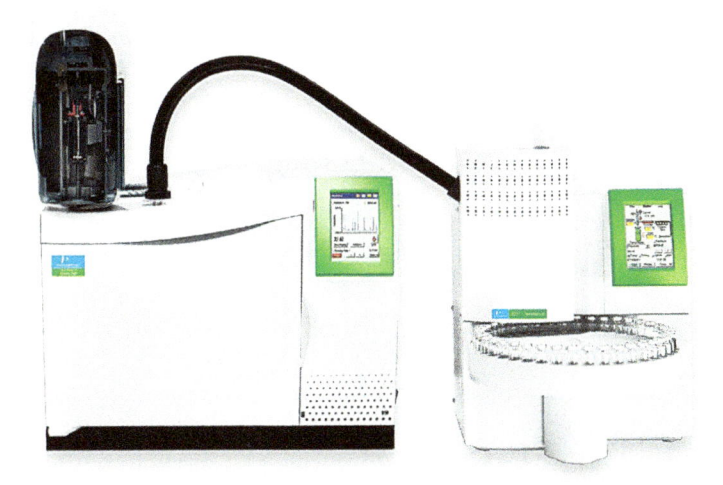

Figure 13.23 A gas chromatography with a heated headspace sampler is the laboratory instrument commonly used for determination of alcohol concentration in blood. (Courtesy of PerkinElmer, Waltham, MA.)

Figure 13.24 A sign warning drivers of a spot check of drivers' ability to assure that they are not impaired to drive by alcohol or drugs.

The situation with field test instruments is not as uniform. There are quite a number of different breath test instruments that use a variety of different analytical techniques, including some portable gas chromatographs. The National Highway Safety Board tests and approves instruments for breath testing. When the instruments are properly maintained and calibrated, and used by properly trained individuals, they produce reliable results. Because of the difficulties of administering such testing programs, they are more often challenged in court. There is a third class of instrument, which is used for screening only. In fact, the public can buy low-cost breath test instruments to test themselves after drinking to see if their alcohol level is within legal limits before driving.

Other drugs and driving

It should be mentioned that all states have laws against driving while impaired by drugs other than alcohol, as well. In most states, these laws are used when a driver is observed to be driving erratically and preliminary tests indicate no alcohol or a low alcohol level. The observation of poor driving practice is used as the primary gauge of the level of impairment, rather than blood levels of the drug. Although a charge of driving while impaired by drugs is primarily supported by the behavioral observations (Figure 13.24), a laboratory analysis showing a "significant level" of a drug, usually a controlled substance, is almost always required for such a charge to be successfully prosecuted.

Driving while impaired by drugs cases are more complex because prohibited levels for each drug have not been developed and placed into state law, as with alcohol. The problem here is the way the body metabolizes and excretes different drugs. Some drugs or their metabolites may be present in tissue or fluids for a long time after ingestion, and a long time after the physiological effects are no longer present. Thus, unlike with alcohol, there is no necessary correlation between body fluid levels and levels of impairment that would affect the operation of a motor vehicle. This situation makes it almost impossible to define impairment levels by law, since levels of different drugs at or above which a person would be considered impaired are highly variable. Thus, there would be a lack of uniformity across the states as to what might constitute a "significant level" of a particular drug.

Key terms

anabolic steroids
analgesic
behavioral dependence
chromatography
club drugs

controlled substance
diluents
driving while impaired
drug
drug facilitated sexual assault
gas chromatography
hallucinogen
illicit
impaired by alcohol
infrared spectroscopy
licit
manner of death
mass spectroscopy
metabolism
narcotic
physiological dependence
psychological dependence
purification
rave
spectroscopy
stimulant
under the influence

Review questions

1. Give a general definition of what makes a chemical a drug.
2. List and give a few examples from each of the major categories of abused drugs.
3. Outline the scheme used by most forensic laboratories for the examination of a sample suspected of containing a controlled substance.
4. What are qualitative and quantitative analyses and when is each important in the analysis of controlled substances?
5. Why is the examination of body fluid samples by a forensic toxicologist more complex than the identification of street drugs in a forensic laboratory?
6. Why is the determination of alcohol and drugs in samples from an impaired driving case so important to the case?
7. What are the general criteria used to decide on while federal schedule a drug is to be place?
8. Discuss the role of separation techniques in accurate identification of controlled substances.
9. Discuss the major forms of drug dependence.
10. Why has the development of the so-called "Hyphenated Techniques" made drug identification much more reliable?

Fill in/multiple choice

1. The two most commonly abused stimulant drugs are cocaine and _____.
2. The development of _____ dependence on a drug is shown by withdrawal symptoms, such as convulsions when the user stops taking the drug.
3. Forensic Toxicology is a specialty that deals primarily with the identification of drugs and poisons in _____ samples.
4. Name three drugs that are commonly abused as part of the "club" drug scene.
5. The most common sequence of steps in the identification of suspected controlled substances evidence is:
   a. AA/NAA, Microcrystal test, color test, quantitative analysis
   b. Color test, crystal test, GC/MS
   c. FT/IR, separation, GC/MS, color test
   d. Color test, FT/IR, separation, GC

References and further readings

Baker, P. B., and G. E Phillips. "The Forensic Analysis of Drugs of Abuse." *The Analyst* 8, 777, (1983).

Bradley, D. "Tracking Cocaine to Its Roots." *Today @ Chemist at Work,* May 2002.

Cole, M. D. *The Analysis of Controlled Substances.* New York: John Wiley and Sons, 2002.

Cole, M. D., and B. Caddy. *The Analysis of Drugs of Abuse: An Instruction Manual.* London, UK: Ellis Harwood Ltd., 1996.

"Drugs and Chemicals of Concern: Oxycodone." www.deadiversion.usdoj.gov/drugs-concern/summary.html.

Drug Enforcement Administration. www.dea.gov.

"Drug Fighter's Turn to Rising Tide of Prescription Abuse." *The New York Times,* March 18, 2004.

Gomm, R. J., J. Humphreys, and N. A. Armstrong. "Physical Methods for the Comparison of Illicitly Produced Tablets." *Journal of the Forensic Science Society* 283 (1976).

Hazarika, P., S. M. Jickells, K. Wolff, and D. A. Russel. Multiplexed Detection of Metabolites of Narcotic Drugs from a Single Latent... Fingermark; *Analytical Chemistry* 82 (2010): 9150–9154.

LeBeau, M. A., and A. Mozayani. *Drug Facilitated Sexual Assault: A Forensic Handbook.* Academic Press, New York, 2001.

Lebel, P., K. C. Waldron and A. Furtos. Rapid Determination of 24 Synthetic and natural Cannabinoids for LC-MS-MS screening In Natural Products and Drug Inspection Applications; *Current Trends in Mass Spectroscopy*; March 2015; 8.

Moffat, A. C. "Drugs of Abuse." *Science & Justice* 40 (2000): 89–92.

National Institutes of Drug Abuse (NIDA) Web site, www.nida.nih.gov.

SWGDRUG Methods & Reports Subcommittee: "Minimum Recommended Analytical Scheme for Forensic Drug Identification," users.erols.com/scitechz/twgm&r.html

"Synthetic Cannabis in Wikipedia.com, the free encyclopedia."

Chapter 14

Materials evidence

Lead Case: Illinois versus Cecil Sutherland

At 9 a.m. on July 2, 1987, an oil field worker discovered the nude body of a 10-year-old girl approximately 100 feet from an oil lease access road in rural Illinois. Her body was lying on its stomach covered with dirt. There were shoeprints on her back and several hairs were found stuck in her rectal area. In addition, a large open wound on the right side of her neck exposed her spinal cord area, and a pool of blood next to the head indicated that the murderer had killed her where she lay.

The victim's clothes—her shirt, shorts, underpants, shoes, and socks—were found strewn along the oil lease road. Within 17 feet from the body, automobile tire impressions were found. Near the tire impressions, a shoeprint impression similar in design to the shoeprint on the body was found. Plaster casts of the tire and shoeprint impressions were made.

An autopsy was performed and indicated that a 14.5 cm wound that ran from the middle of the throat to behind the right earlobe cut through the neck muscles, severed the carotid artery and jugular vein, and cut into the cartilage between the neck and vertebrae. The victim's right eye was hemorrhaged, and there was a small abrasion near her left eyebrow. Her ear was torn off the skin at the base of the ear, and both her lips were lacerated from being compressed against the underlying teeth. There was also evidence of tearing of the rectal mucosa. From the foregoing, it was deduced that the victim was strangled to unconsciousness or death, anally penetrated, her throat slit, and stepped on. The time of death was estimated at between 9:30 and 11 p.m. on July 1, 1987. See Table 14.1 for an overview of the evidence gathered.

The plaster casts of the tire print impressions made at the scene of the crime were examined, and it was reported that the tire impressions left at the scene were consistent in all class characteristics with only two models of tires manufactured in North America; the Cooper "Falls Persuader" and the Cooper "Dean Polaris."

Several months later, the police at Glacier National Park in Montana called the county Sheriff's office in Illinois regarding Cecil Sutherland's abandoned car, a 1977 Plymouth Fury. At the time of the murder, Sutherland had been living in Illinois, quite close to the murder scene. It was determined that the car in question had a Cooper "Falls Persuader" tire on the right, front wheel. Inked impressions of the right front wheel of Sutherland's car were taken.

The plaster casts of the tire impression at the scene were compared with the inked impression of the tire from Sutherland's car. It was concluded that they corresponded with Sutherland's tire and could have been made by that tire. The examiner could not positively exclude all other tires due to the lack of comparative individual characteristics, such as nicks, cuts, or gouges.

Table 14.1 A summary of the major pieces of materials evidence gathered in the case against Cecil Southerland

Evidence in the amy schultz murder case—July 1987

Physical evidence items	Correspond to …
Tire print—At scene	Cooper tire "Falls Persuader"—Defendant's Car
Boot print—Victim's back	Timberland boot—Owned by defendant
2 Pubic hairs—Victim's rectal area	Cecil Southerland's pubic hair
32 Dog hairs—Victim's clothing	Black lab—Owned by defendant
Numerous dog hairs in defendant's car	Black lab—Owned by defendant
28 Gold fibers on victim's clothing	Carpeting in defendant's car
12 Cotton fibers—Front seat of Defendant's car	Victim's shirt
4 Polyester fibers—Front seat of Defendant's car	Victim's shirt
3 Polyester fibers—Front seat of Defendant's car	Victim's shorts

The forensic hair examiner microscopically compared the two pubic hairs recovered from the victim's rectal area with Sutherland's pubic hair, as well as with pubic hairs from members of the victim's family and those from 24 other possible suspects in the case. He concluded that the pubic hairs found on the victim did not originate from her family or the 24 suspects, but "could have originated" from Sutherland.

The hair examiner also examined the approximately 34 dog hairs found on the victim's clothing and concluded that the hairs were consistent with and could have originated from Sutherland's black Labrador retriever. He further found that the dog hairs on the victim's clothes were dissimilar to her family's three dogs, her grandparents' dog, and three neighbors' dogs. Numerous dog hairs found in Sutherland's car were found to be consistent with the hairs from the black Labrador.

The victim's clothing was examined for the presence of foreign fibers. A total of 29 gold fibers were found on the victim's socks, shoes, underwear, shorts, and shirt. It was concluded that all but one of the gold fibers found "could have originated" from the defendant's auto carpet, but the examiner could not state that the fibers originated from Sutherland's auto carpet to the exclusion of all other auto carpets. It was determined that the one remaining gold fiber found could have originated from the upholstery of Sutherland's car.

Twelve cotton and four polyester fibers found on the front passenger side floor of the defendant's car were compared with cotton and polyester fibers from the victim's shirt. These fibers from the car displayed the same size, shape, and color as the fibers from the shirt, and it was concluded that they could have originated from the shirt. Furthermore, three polyester fibers found on the front passenger seat and floor of the defendant's car were compared with the fibers from the victim's shorts and found to be consistent in diameter, color, shape, and optical properties. They could, therefore, have originated from the shorts.

A forensic expert for the defense agreed with the State's expert's conclusions on all the comparison evidence, except as to the comparison of the cotton fibers found in defendant's car to the victim's shirt. He did not agree that the cotton fibers were consistent, as he noticed some differences in size and color.

The victim had last been seen alive at approximately 9:10 in the evening of July 1, 1987, while walking alone in the town of Kell, Illinois. The defendant's sister-in-law testified that on the evening of July 1, 1987, the defendant was visiting his brother and her at their home. She testified that on the night of the murder, the defendant left her home at approximately 8–8:30. It was established that the distance from Kell to the crime site was 12.1 miles and took 14 minutes to drive.

The jury found the defendant guilty of aggravated kidnapping, aggravated criminal sexual assault, and three counts of murder. The jury found the defendant eligible for the death penalty under the felony murder aggravating factor, and subsequently returned a verdict of death. The verdict was upheld on appeal by the Illinois Supreme Court, who found that some overstatements by the prosecution were not sufficient to affect the verdict in the face of the overwhelming physical evidence.

The saga did not end here, however. Because of a number of serious cases of prosecutorial misconduct and/or extremely poor quality defense representation were found to have resulted in a number of serious miscarriages of justice, the Governor suspended the death penalty in the State. The next governor established a large fund to provide adequate defense for those charged with capital crimes and reinstated the death penalty.

A new appeal by the defendant in this case again reached the Illinois Supreme Court and this time the case was sent back for retrial. The court cited the failures of the defense at the trial to present possible mitigating evidence that might have reduced the weight of the footprint and tire track evidence and more importantly the defense's failure to bring out the fact that the victim's grandfather, at whose house she was before her disappearance, was a convicted sex offender.

The case was retried, almost 18 years after the crime, and although the weakening of the footprint and tire track evidence and the grandfather's record were presented, the second trial resulted in a second guilty verdict. In the intervening years, DNA and particularly mitochondrial DNA on hair shaft became possible. Reanalysis of the dog hairs and, more importantly, a hair found in the victim's rectum using DNA technology supported the original hair examiner's conclusions. The dog hairs from the victim's clothing were found to be consistent with the dog hair from the defendant's truck and inconsistent with that of dogs owned by the victim's family. The hair from the victim's rectum was consistent with the defendant and inconsistent with the grandfather as a possible source. The conviction was upheld through all the appeals. The defendant was convicted largely on the basis of the large number of items of materials evidence that when taken together convinced two juries of the defendant's guilt.

LEARNING OBJECTIVES
- Materials evidence is used primarily for indicating possible connections
- Explain the nature of and difference between transfer and trace evidence
- Indicate the most common sources of materials evidence
- Outline the major categories of materials evidence and provide examples
- Discuss the process of examination of materials evidence
- Explain the major techniques for collecting materials evidence
- Discuss in more detail the five most important types of materials evidence
- Discuss the range of fibers encountered as evidence
- Explain the structure and growth of human and animal hair
- Discuss the proper collection of hair control standards
- Outline the laboratory examination and comparison of fiber and hair evidence
- Discuss the nature of paint and the importance of architectural and automotive paint evidence
- Explain the collection of proper paint control standards
- Outline the laboratory analysis and comparison of forensic paint evidence
- Explain the nature of glass and its manufacture
- Explain the proper collection of glass evidence
- Outline the laboratory analysis and comparison of forensic glass evidence
- Discuss the composition of soil
- Outline the common forensic occurrences of soil evidence
- Explain the proper collection of soil evidence
- Outline the laboratory analysis and comparison of forensic soil evidence

limitation of infrared spectroscopy was that one had to have a fairly pure chemical and a reasonable amount of that material to successfully identify it. In addition, the sample preparation required was time consuming. With the micro FTIR, one has a special microscope connected to an infrared spectrometer, and all the time consuming sampling and need for larger quantities of sample largely disappears. One places a tiny amount of a sample on a special microscope slide (transparent to infrared light), finds the item or crystal one wants to identify in the microscope field, and using a diaphragm, like on a camera, close down on just the particular object of interest and one can obtain a spectrum in a matter of a minute or two. This means that one can take a paint chip that one can barely see, or a fiber that is equally small, and obtain an infrared spectrum in a matter of minutes.

IR spectrometry (and its more sophisticated variant, FT/IR) have been called "chemical fingerprinting." This terminology means that a pure compound could be identified through its infrared spectrum. There are large libraries of spectra available for this purpose. Many of the materials that come into forensic labs for analysis are not; however, pure chemical compounds. Take a chip of paint, or fingernail polish, or a particle of cosmetic powder as examples. All of them are mixtures, not pure compounds. But an examiner can still obtain an IR spectrum. That spectrum will be more complex than one of a pure compound, because it represents a sort of "mixed" spectrum containing features of all the compounds present in proportion to their relative abundances. So what good is it? The IR spectrum from a paint chip by itself may indeed not be very useful from a chemical identification standpoint. But FT/IR is an excellent *comparison* tool. As we will discuss below, most materials evidence examinations consist of comparisons between a "questioned" (evidentiary) item and a "known control." If the IR spectrum of a questioned paint chip very closely matches that of a suspected paint source, then the examiner can form a real conclusion—that the potential source *could be* the actual source. Similarly, if they did not match, an examiner could conclude that the evidence did not come from the suspected source.

Another powerful instrument occasionally useful for trace evidence analysis is the scanning electron microscope (Figure 14.5) (usually abbreviated SEM). This very powerful microscope has the capability of not only visualizing very small samples like a normal light microscope, but with much higher magnification than with a light microscope. But, in addition, when it is equipped with a special analyzer known as an energy-dispersive X-ray analyzer (EDX), the

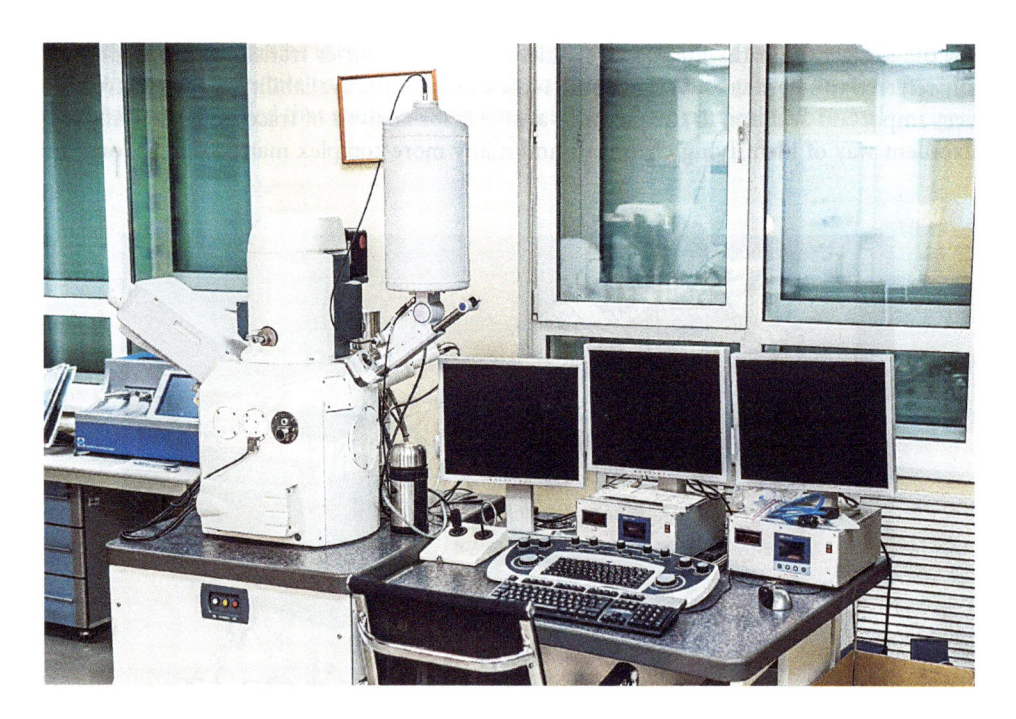

Figure 14.5 A scanning electron microscope is an instrument for high magnification and elemental analysis if tiny particles.

elemental composition of the specimen can be determined. The combination is known as a SEM/EDX instrument. It is a complex and very expensive piece of equipment. It has been quite useful for certain specific analyses, such as gunshot residue identification (Chapter 8). The instrument can also be used for the identification and comparison of other particles and biological materials, such as pollen. Because the SEM with EDX yields information on the elemental make up of a single particle, it is particularly useful for complex samples made up of different particles. Knowing the elemental composition of the particles in a material often help in determining what the material may be. Because the instrument has limited applicability to forensic evidence, many laboratories have not acquired one, because it cannot be justified as cost effective.

Materials evidence comparisons—individualization, inclusion, or exclusion

Before discussing several specific, regularly encountered categories of materials evidence, it is important to reemphasize (Chapter 1) the concepts of individualization, association (comparison), and exclusion in connection with materials evidence.

As we have previously noted, materials evidence has potential value when transferred, because its presence on the target surface or object can indicate contact with the source surface or object. Once in a while, the objective of an examination might be to identify what the material is. But most of the time, the objective is comparison between an evidentiary specimen and a known control. Investigators are nearly always responsible for providing the lab with the known control for comparison. Could this red fiber have come from the victim's red sweater? Could the soil on the suspect's shoes have come from the scene? Could the paint smear on the hit-and-run victim's bicycle have come from the suspect's vehicle? Could the head hair on the victim's coat have come from the suspect? All of these are relevant forensic questions. Trying to answer them requires a comparison—a comparison between the evidence specimen and a known control.

An important point about materials evidence: it generally **cannot** be individualized. This fact means that the questions above cannot be answered definitively, unless the answer is "No." If questioned and known control specimens match in every property and aspect compared, the most the examiner can say is that the "questioned specimen could have come from the suspected source." Another way of saying the same thing is that the "questioned specimen could not be excluded as having come from the suspected source." Criminalists sometimes refer to this process as "partial individualization." Many items in a class of materials can be excluded, but the questioned specimen still cannot be attributed exclusively to the object from which the control was taken. A questioned red nylon fiber with certain microscopic properties can be consistent with the known fiber in all compared characteristics. Here, the source of the questioned fiber must be one with red nylon fibers having those same properties. Other fibers can be excluded as consistent with that source. However the analyst cannot say that the questioned fiber came from this particular known, because there may be many potential sources. If properties do not match, then an examiner can say definitively that the "questioned specimen did not come from the suspected source." For this reason, materials evidence is sometimes said to have primarily **exclusionary** value.

THE MOST COMMON TYPES OF MATERIALS EVIDENCE

Immediately below are discussed examples of the some of the most commonly encountered types of materials evidence. Although these categories are the most common, they are by no means the only types of materials that will be encountered as evidence. At scenes, evidence is not found with a little sign that says it is evidence. Collecting everything is not a viable solution, because the critical trace evidence might be obscured in the enormous amounts of submitted material. As we have noted, what really distinguishes a quality investigator or crime scene specialist is the ability to **recognize** what may turn out to be useful evidence. This skill requires a systematic, scientific approach to the scene and the case, along with a realistic appreciation of the value of various types of evidence. Investigators need to understand what information the lab can provide based on an analysis or comparison of different types of evidence. How significant is the evidence? Is the information that can be provided by the piece of evidence relevant to the case? Investigators must also be familiar with various control specimens (Chapter 3) that are required in order to make laboratory analysis or comparison possible. One thing all have in common is the importance of microscopical examination, particularly using the comparison microscope. The comparison microscope is critical for careful comparison of microscopic or near microscopic evidence objects for both color and morphology. The type of comparison microscope usually used for materials evidence uses transmitted light to look through the objects, rather than the reflected light used for bullet and cartridge case comparisons.

Natural and synthetic fibers

Fibers, both natural and synthetic, are common pieces of evidence. Many natural fibers, such as cotton, wool, and silk, are used in clothing, while others, such as jute, manila, and hemp are used in cordage and ropes. If one finds cordage fibers on a victim's body or their clothing, it might indicate that they were tied up. Those fibers could be very useful in identifying what type of rope was used to immobilize the victim. Synthetic fibers now have become major components of many types of clothing and many other things as well. Draperies, bedding, and carpeting, to name just a few, are now are largely composed of synthetic fibers. Because fibers are so widely used in commerce, and because they are easily broken, and they tend to stick to things, they are very commonly transferred during contact. They are light and will catch on rough surfaces of garments and many other things. A nylon jacket, which is very smooth, will seldom retain fibers (lint), while a flannel shirt, a sweater or a wool jacket, will have fibers tangled in the weave and thereby retained on its surface. There are many situations where such retention of fibers can provide forensically important information. Fibers are usually divided into two major groups, natural fibers and artificial or manufactured fibers. At one time, it would have been only one group, the natural fibers. Development of manufactured fibers really became important with the need to find a substitute for silk for making parachutes during World War II.

Natural fibers

Natural fibers can be subdivided as in the game of 20 questions: Is it animal, vegetable, or mineral? All animal fibers are protein, all vegetable fibers are cellulose, and mineral fibers are not very common and seldom of forensic interest (with the possible exception of asbestos). Yet, there are dozens of different commonly encountered natural fibers.

There are really only two mineral fibers of forensic significance: asbestos and mineral wool fibers. Asbestos is an unusual mineral that has a naturally fibrous structure as it is mined from the earth. It was used for many years as insulation and in many other applications. Because it is essentially rock, it is non-flammable, and this is an important advantage for an insulation material. It was used in fire safes to protect the papers in the safe from even prolonged fire outside the safe. For that reason, it could be encountered when safe robbers were forced to drill or use explosives to blow a safe. They invariably disturbed the layer of insulation between the inner and outer wall of the safe and released asbestos fibers into the surroundings including onto themselves. There are a number of different forms of asbestos, and several tend to produce fibers that are so small and light that they can remain suspended in air for extended periods. Prolonged inhalation of such asbestos fibers can cause asbestosis, which is an eventually fatal disease of the lungs. Because of this danger, asbestos is no longer used in any application where it might be released into the atmosphere. There is still a great deal of asbestos around, but it is slowly being removed, covered up or disposed of in one-way or another. Forensic microscopists and materials analysts who perform work for civil cases or regulatory enforcement matters may still do asbestos cases. Mineral wool is a material made by melting glassy minerals and forcing them through a fine die in close analogy to the regenerated fibers discussed below. Its uses are quite analogous to asbestos and primarily of interest forensically because of its use as safe insulation.

Different varieties of vegetable fibers, because they are primarily cellulose, cannot be easily differentiated by chemical analysis. Because they are natural and formed in plants by biological processes, they have complex three-dimensional shapes (morphology). They can usually be recognized by careful microscopic examination, and the morphology is often more easily observed in the microscope if the fibers are stained with a biological stain that preferentially colors a particular component of the fiber. Therefore, vegetable fibers are usually identified in the forensic laboratory by a combination of visual and microscopical examinations. These days, cotton is the most important natural fiber, and linen is also a popular vegetable fiber. In the laboratory, one encounters a great variety of cotton fibers (Figure 14.6) in many different cases. Unfortunately, undyed white cotton is often not of much use forensically, because so many things, from undergarments to bed sheets, are made out of white cotton. Dyed cottons, although still very common,

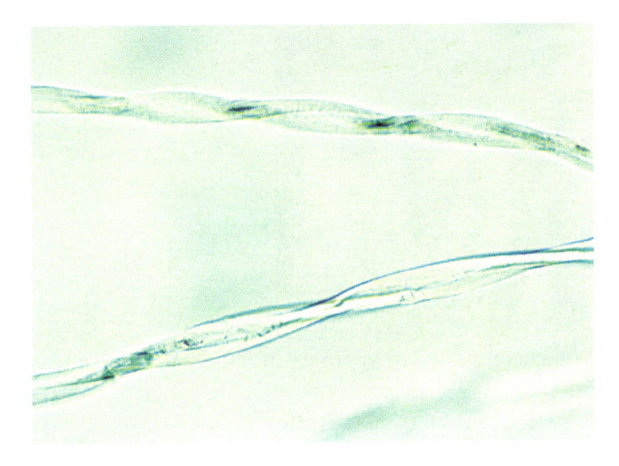

Figure 14.6 A photomicrograph of a cotton fiber. Notice the very characteristic twisted structure used to recognize cotton fibers.

can be more useful forensically because of the many colors and types of dyes used. Manila, hemp, and jute are coarse natural plant fibers that can be used in fabrics, but are much more commonly encountered in packaging twine, rope and door mats, or in the padding used under carpeting.

The most important animal fibers are wool (Figure 14.7) and similar animal hair materials. Wool, of course, comes primarily from sheep, but it also comes from goats, llamas, and a number of other related animals, such as alpacas. All these animals have fine hair that can be shaved off of the animal, then spun into thread, and used to make fabrics. One can usually easily tell whether a fiber is a vegetable fiber or an animal fiber just by looking at it. Animal fibers, which are essentially hairs (see below), have an outer layer of scales that overlap like shingles on a roof to form a waterproof protective layer. This structure is often visible to the unaided eye or under slight magnification. Because they are spun to make thread before being woven into cloth, one has the strength of many strands in each thread and, therefore, the resulting cloth is very durable. Although cotton is currently the most popular fiber for clothing, wool is still very important and a premium fiber for many applications, such as pants and sport jackets, sweaters, blankets, and coats.

Figure 14.7 A photomicrograph of a blue wool fiber. Notice the very characteristic scaly appearance of wool used to recognize wool as animal fiber.

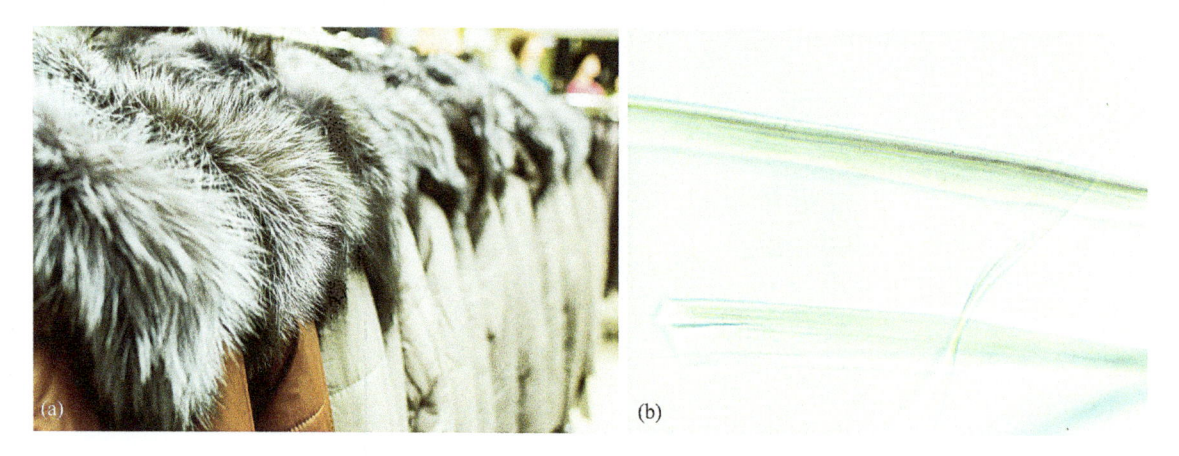

Figure 14.8 (a) A photo of fur coats; fur evidence arises in a surprising number of cases. (Courtesy of Shutterstock, New York); (b) A photomicrograph of a silk fiber. Notice no scale pattern but striations and other structure in the fiber characteristic of natural fibers.

Many animal hair products are not spun and woven into fabrics, but used as the animal did—as fur coats (Figure 14.8a). Fur is the term used when the animal hair is not removed from the pelt and is still embedded in the animal's skin. Coats made from seal, mink, beaver, and a variety of other animal pelts have been popular for centuries. Natural fur coats are both warm and very durable and, oftentimes, trim can be fur even if the whole coat is not. While currently it may not be deemed "politically correct" to own or wear fur coats, there are many such garments still in use by people. Animal hair is used in other products as well. Rabbit hair is used as the soft inner lining in gloves and is woven with other natural and synthetic fibers into mixed fabrics. As a result, fur is encountered in forensic cases more often than one would expect.

CASE ILLUSTRATION (SIDEBAR)

Interestingly, pink-dyed rabbit hair in a mixture with acrylic fibers played an important part in a murder case a number of years ago. When last seen, the victim was wearing a very fluffy, pink sweater with that unusual combination of fibers. Finding traces of those fibers in a suspect's house cast serious doubt on his denials of having any contact with the victim after she dropped off her car for some service. He was eventually arrested, and fiber evidence played a prominent role in his trial and subsequent conviction for murder.

The animal fiber that looks least like an animal fiber is one of the oldest and most widely used in high quality fabrics. Silk (Figure 14.8) is definitely an animal fiber, but it's not an animal hair—rather, it is the unraveled cocoon of a silkworm caterpillar. Silkworm caterpillars, during the part of their lifecycle (biologists call it the pupae stage) where they are transforming themselves into a butterfly, spin cocoons. Silk is obtained by unraveling those cocoons. The cocoons are gathered and boiled before being unraveled into silk thread. The silk thread is not formed like a hair, but rather is exuded from a little organ called a spinneret in the abdomen of the silk worm. Silk is more like a spider web than hair, and does not have a scale pattern like the other animal fibers. Silk has been made into excellent fabrics for thousands of years in the Orient. The use of silk had been moderated by its high price in the Western World. However, with the increased entry of China into world markets, the availability of silk has increased, and its price has dropped. Silk has many excellent properties and is highly prized for clothing, undergarments, and a variety of other applications.

Manufactured fibers

An important fact about synthetic, manufactured, and some natural fibers is that, chemically, they are all *polymers*. Polymers are compounds made of repeating units. The repeating units may be the same, or there may be several different chemical types. See the sidebar on polymers nearby.

SIDEBAR

Polymers are both nature's and man's construction materials. They are chemically built up of large numbers of repeating units called monomers. A polymer may be made up of up to hundreds of thousands of monomer units, as with the common plastic called polyethylene. Polymers may also be more complex, such as the natural polymers we call proteins, which are also made of repeating units. Here the units (monomers) are chemically similar, but not necessarily identical repeating units, called amino acids.

Cellulose, the construction material of the vegetable world is made up of repeating units of simple sugar molecules. Forensic science's most talked about polymer is DNA, which is made of repeating sugar units also, but ones that have phosphate ester groups and organic bases attached to the sugar backbone. It is the sequence of the four different organic bases found in this polymer that make up the genetic code.

These repeating units can be visualized as links in a chain. Different polymers are made up of different types of links. Further, the chains can be of different length and can be lined up parallel or can be mixed together randomly or tangled. This means that the same polymer (type of link) can have different physical properties depending on chain length or arrangement. A polymer material made up of many polymer strands (chains) can sometimes be modified after its initial formation by chemically causing different polymer strands (chains) to have points of attachment. This process is called cross-linking and is very important in adding strength or rigidity to polymeric materials. When a paint film "cures" on the painted surface, it becomes stronger because of cross-linking of chains caused by reaction with oxygen in the air.

Man imitated nature in developing synthetic polymer materials using a very wide variety of building blocks (links). We call most of these materials plastics, man-made polymers include synthetic fibers, synthetic rubber, and a great many other simple or very complex polymers. The very common polymer polyethylene terephthalate, usually referred to as polyester, is used in many forms from clothing fibers to soda bottles to engineering polymers used to make many other useful objects.

Synthetic and manufactured fibers, which became generally available primarily after World War II, have come to dominate the fiber business. One can divide man-made fibers into two basic categories. One small, but historically important category, referred to as *regenerated* fibers, and the now much more important category of *synthetic* fibers. Regenerated fibers were the first manufactured fibers. Rayon was developed as a man-made version of vegetable fibers like cotton or manila. Rayon, like the vegetable fibers, is cellulose. The key step in the manufacture of regenerated fibers was devising a way to dissolve a source of cellulose, like sawdust or cotton waste. This is not an easy thing to do, since cellulose is not soluble in many things that do not destroy it. Several different rather ingenious processes were developed to accomplish the task. Once the cellulose is in solution, it can be forced through a die, which is usually a piece of steel with many tiny holes (Figure 14.9), into a different chemical solution that immediately causes the cellulose to deposit from the solution as a solid "fiber." In this way, many long, thin fibers are formed, and they can be recovered and dried to form the final manufactured fiber. This process is used for rayon and acetate fiber production. Fibers are produced having the diameter of the holes in the die. Rayon and acetate fibers are called regenerated

Figure 14.9 The different dies used to make fibers by forcing a liquid through the fine holes in a die and solidifying it as it emerges into fine strands. (Courtesy of Shutterstock.)

fibers, because the starting material is a natural cellulose source. The regenerated fibers were based on cellulose but processed in a way to produce a fiber different from any natural fiber. A variation on the solution process for producing man-made fibers is a melt process. A starting material for a fiber that can be melted more easily than dissolved, is heated above its melting point, and then the liquid material forced through fine holes in a die, and rapidly cooled it as it emerges from the die, causing it to instantly solidify into the desired fibers.

Although rayon found a solid market in tire cord and clothing, the real impetus behind the development of synthetic fibers was WW II, and the control of silk production at that time by the Japanese. The importance of aviation to the war effort, and the fact that early parachutes were made from silk, made it critical that some substitute for silk be found. Silk was the only material available that was strong enough and light enough to make a canopy large enough to support the weight of a pilot and compact enough to be carried in a pouch on the fliers back when they bailed out of an airplane. When Japan came into WW II and quickly took control of most of the Far East, silk became unavailable in the West, prompting an urgent search for an alternative material.

A chemist named Caruthers, who worked at the large chemical company, DuPont, led a team that developed a polymer they called Nylon. Nylon proved to be a material with many important properties quite similar to silk. One can think of nylon as imitation silk. It is chemically similar, and very light in weight for its strength. As a result, the U.S. was able to manufacture parachutes for our pilots using this new synthetic fiber.

When synthetic polymers are extruded through dies to form filaments they have a round cross section, relatively uniform diameter and a smooth surface—in contrast to most natural fibers, which have complex morphology (shape) and uneven diameter. This is one of the major differences between natural fibers and manufactured fibers. Again, as mentioned earlier, silk was the one exception to this rule-of-thumb, appearing more like a synthetic fiber than a natural one.

Modern technology has allowed synthetic fibers to be manufactured with more complex cross sections (Figure 14.10), because, when making the extrusion dies, instead of simply drilling round holes one can now use a laser to create very fine holes of any desired shape. As a result, there are fibers manufactured that have a square, pentagonal, or even clover leaf shaped cross sections but consistent diameter. They still do not have the complex morphology and variable diameter that one sees in the natural fibers, which helps forensic scientists to differentiate natural from manufactured fibers. Over the years since WWII, a great variety of other synthetic fibers have become available.

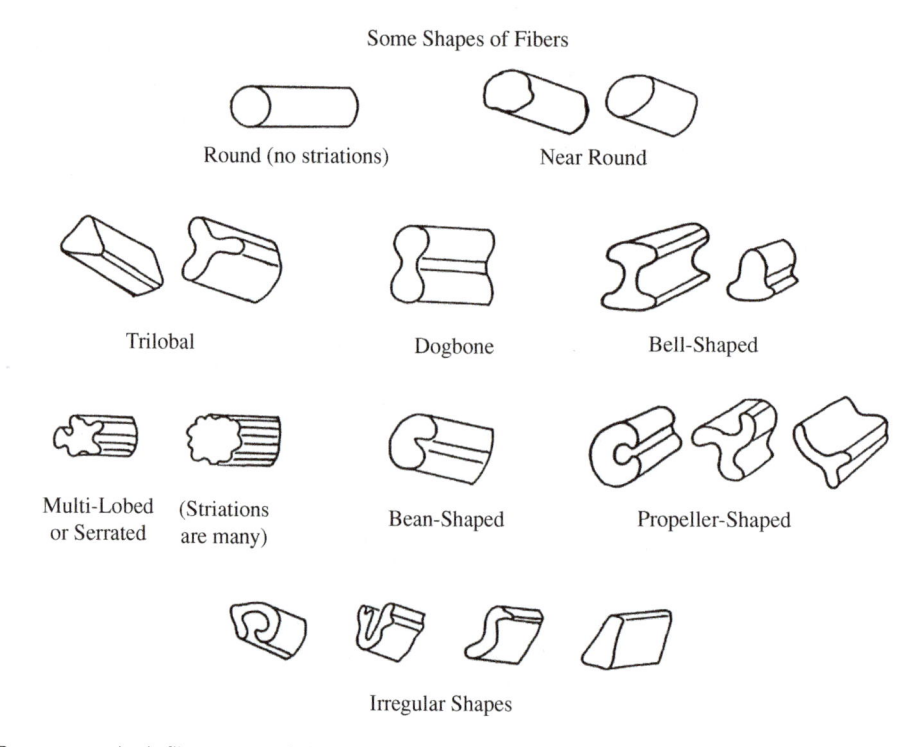

Some Shapes of Fibers

Round (no striations)

Near Round

Trilobal

Dogbone

Bell-Shaped

Multi-Lobed or Serrated

(Striations are many)

Bean-Shaped

Propeller-Shaped

Irregular Shapes

Figure 14.10 Because synthetic fibers are made by being forced through fine holes in dies, they can be formed with any cross-section shaped hole that a laser can make in the die.

Because these different fibers are made from various synthetic polymers, they can be differentiated using chemical and physical tests—in stark contrast to the natural fibers, all of which are formed from either cellulose (vegetable) or protein (animal) and can best be differentiated using their differing morphology.

Each of the different types of synthetic polymers used in fiber manufacture has been optimized for particular applications. For example, polyester has become the major synthetic fiber in the world economy. It has a great variety of desirable properties and it is used in an enormous variety of applications. It is relatively inexpensive to make, quite strong; has a fairly high melting point, and is easily processed. At one time, men's suits were made from pure polyester (and it was once fashionable to make fun of men who wore such suits), a mixture of wool and polyester makes a popular material that is much more desirable. Nylon replaced rayon in tire cord, because it is much stronger, and, of course, became the standard for women's stockings (replacing silk).

Polyolefins (polyethylene and polypropylene) are the low price leader of the commercial fibers, perhaps best known as the fiber of Astroturf, or more mundanely, as that of indoor/outdoor carpeting. Polyethylene fibers are inexpensive not only because of low raw material cost, but because polyethylene is used for so many other non-fiber applications. It is well-suited to applications where its resistance to bacterial action and water are advantageous, and its low melting point is not a problem. Acrylics are important fibers that are used in clothing, particularly sweaters, blankets, and scarves. In fact, one is hard pressed to find a blanket that is not made from acrylic fibers these days. They are bulky and light and fluffy, which gives them their insulating power and thereby their warmth. There are many other types of synthetic polymers that find use as fibers. One particularly interesting one is spandex, a stretchy fiber, which is used for many types of form fitting clothing, such as bicycle shorts, gym outfits, and to give a little stretchiness to some jeans.

Another rather important fiber, particularly to law enforcement is polyamide, which is a rather expensive fiber, but is an extremely strong one. It is used to make bulletproof vests for law enforcement officers. It is better known by its brand name, Kevlar, and is widely used for applications where strength is absolutely critical. If one is an avid sailboat racer, and wants to have the "racer's edge," one purchases Kevlar sails.

Another specialty synthetic fiber is fiberglass. It is made by forcing molten glass through dies. It was used for curtains and some kinds of fabrics, but is not as common now in fabric type applications. It is not used in clothing because of the obvious scratchiness problem, but is used particularly in draperies. It is strong, chemically resistant, and it rejects dirt fairly well. One of the largest applications for glass fiber is as insulation. The attics of most homes are insulated from the main floors with glass fiber and it is often used in the walls as well. In addition, enormous amounts of glass fiber are manufactured to be incorporated into fiberglass objects. The glass fibers are woven into cloth and used to make a whole range of very durable composite materials generally called fiberglass. These items are made by combining a fiberglass cloth with a liquid polymer resin that sets to a hard, strong plastic material. Frequently many layers of cloth and resin are used to build up the final fiberglass structure. It is found in furniture, car bodies, boats, and a myriad of other products. When these materials wear down or are broken, glass fibers can be released and occasionally become evidence.

Laboratory examination of fibers

Successful laboratory examination of materials evidence, including fibers, depends upon the collection of proper control samples. Questioned (or evidentiary) fibers must be compared with known controls, and those controls must normally be obtained by investigators and submitted to the laboratory. Most examiners agree that an adequate reference sample must take into consideration the various types of fibers at the fiber source, including different colors. Also, where fibers have been exposed to sunlight for extended periods, the color may fade or otherwise change. This possibility needs to be considered in collection of standards. Because modern laboratory fiber analysis techniques require only miniscule amounts of fiber, only a small amount of control sample need usually be taken.

The initial laboratory examination of fibers, because of their generally small size, is usually microscopical examination, regardless of whether they are natural or man-made fibers, but particularly for natural fibers, which are usually identified almost exclusively from their morphology. Natural fibers tend to have complex and distinctive internal and external structure. Each different type of natural fiber has its own characteristic appearance. This applies to all of the vegetable fibers, such as cotton, linen, ramie, manila, sisal, and most of the animal derived fibers, such as wool, rabbit, mohair, and cashmere. The prime exception is silk, which has little structure and looks more like a manufactured fiber than an animal derived fiber. Because all these fibers are cellular in origin they are built up from many millions of units and thereby have complex structures. Most of the animal fibers are actually hairs, and have scale patterns and medullas in addition to the cortex (see section "Hair-both human and animal"). In fact, in animal hair the medulla is a particularly important feature, because it usually makes up a much larger portion of the hair than with human hair and often shows interesting and complex structure. In addition, plant and animal fibers often have natural color and even show things like banding patterns, highly variable diameters and quite variable cross sections. Because of their complex morphology, natural fibers may not absorb dyes evenly and this too can be a useful property to observe.

Manufactured fibers provide much less information when viewed in the microscope. The main reason is that most man-made fibers are formed by being forced through a die. As a result, they have a uniform diameter and cross section. The diameter is determined by the size of the hole in the die and the cross section by the shape of the hole in the die. In addition to this diameter and cross section, the key property of manufactured fibers is the color, because the enormous variety of colors that are now produced. Unlike hairs, the color in fibers is, as a rule, fairly uniformly distributed throughout the fiber. In fact, when a manufactured fiber shows a non-uniform color, it is an important property for comparison because it is not common. This consistent color distribution makes the comparison of two fibers on a comparison microscope much easier. Thus, a fiber from a crime scene and a control fiber, from the carpeting in a suspect's vehicle can be reliably compared in terms of color and diameter. Because Western society is very consumer oriented, and people are concerned about having their possessions look just right, the fiber industry has created an enormous variety of colors. There are hundreds of different dyes, dye combinations, and pigments that are used to produce those colors. This adds a major dimension to the comparison of manufactured fibers, which would otherwise be largely restricted to chemical analysis to determine type of polymer, as we shall discuss shortly.

Another important property of manufactured fibers is their shininess, more accurately called luster. The first man-made fibers, the regenerated fibers rayon and acetate, tended to be rather shiny and fabrics made from them took on a consumer perception of being cheap, even though shininess of silk always had a luxury connotation. Therefore, manufacturers of synthetic fibers sought to control this shininess. It was discovered that addition of the basic white pigment titanium dioxide to the fiber added depth to the color and reduced the shininess. As a result, many man-made fibers now have varying amounts of titanium dioxide added. This is referred to as a delusterant, and when delustered fibers are viewed microscopically (Figure 14.11), they appear to have many dark specks distributed throughout the fiber. The delusterant appears dark when viewed with transmitted light and white when viewed in reflected light. The appearance is somewhat similar to the pigment granules observed in hair but the delusterant appears black rather than brown or red. These specks of delusterant provide another characteristic in comparing manufactured fibers. The amount of delusterant, and the way it is distributed within the fiber, is a useful characteristic in comparing some manufactured fiber evidence. The delusterant may be evenly distributed, clumpy or perhaps more pronounced around the outside portion of the fiber.

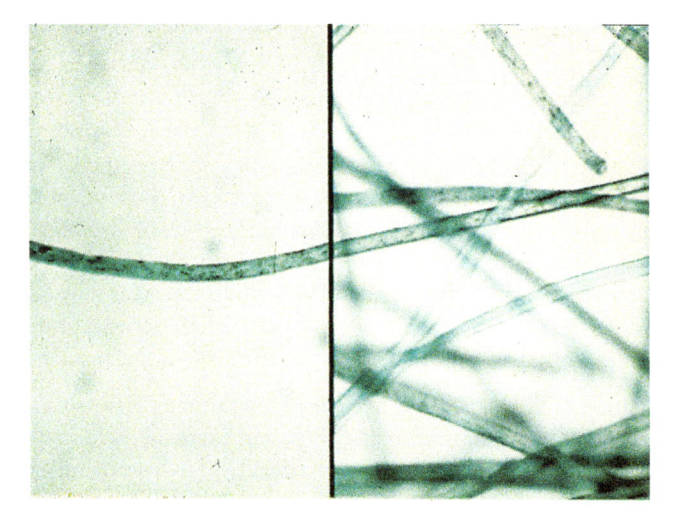

Figure 14.11 Delustered fiber: a synthetic fiber with tiny specks of titanium dioxide (delusterant) added to reduce the shininess of the fiber.

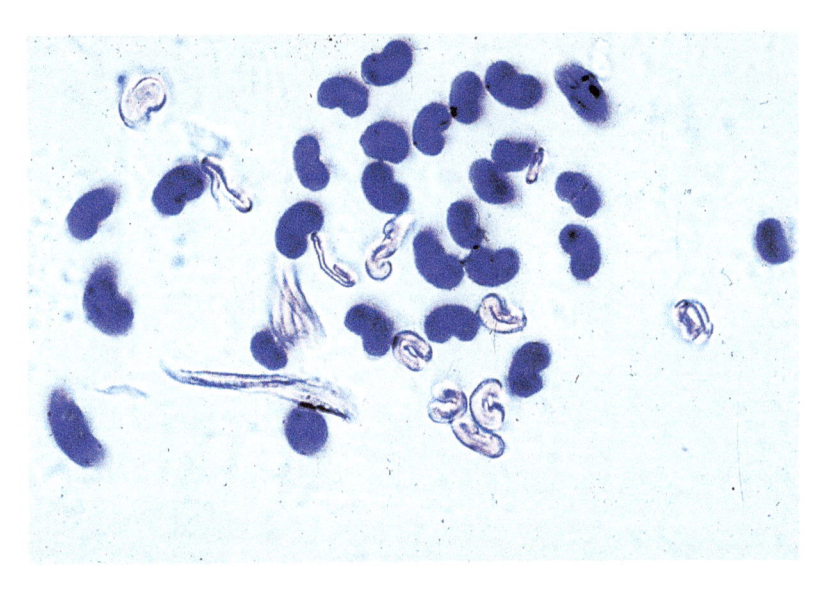

Figure 14.12 Two different types of fibers, one natural and one synthetic, shown in cross section. The synthetic fibers have a consistent bean shaped cross-section whereas the natural fibers have varied cross-sections.

Size and cross section are key fiber characteristics. The diameter of a fiber is easily measured under the microscope. Observing fibers next to each other in the field of a comparison microscope also allows one to compare the relative diameter of the fibers. If, as in the first man-made fibers, the cross section were always round there would not be much information in the cross section. Modern technology allows one to manufacture dies with holes of virtually any shape. As mentioned above, one now sees fibers with cross sections that are oval, square, pentagonal, shaped like a three-leaf clover (Figure 14.12), and many other variations. The various shapes give the fibers different physical properties and, as a result, are often favored for particular applications. As a result, the cross section of a fiber has become a useful characteristic for examination and comparison of manufactured fibers.

Several optical properties can also be used in the comparison of manufactured fibers. Fiber examiners have several ways of measuring the refractive index of a fiber. The refractive index is defined as the ratio of the speed at which light passes through the fiber versus the speed at which light passes through a vacuum. The speed of light through a fiber depends on which direction it is passing through the fiber and, as a result, most fibers have two different refractive indices. There is the refractive index along the length of the fiber and the refractive index across the diameter of the fiber. The difference between these two refractive indexes is called the dispersion of the fiber. Although this sounds rather complicated, both these properties can be fairly easily measured for most fibers and are important characteristics for comparison of fibers.

Finally, chemical properties can be useful in the identification or comparison of fibers by allowing one to identify the type of polymer in a manufactured fiber. As we mentioned earlier, natural fibers are either cellulose if of vegetable origin or protein if of animal origin. Development of manufactured fibers has utilized the skills of countless polymer chemists so that there are many different polymer types found in manufactured fibers. This gives forensic scientists an additional property to examine when comparing fibers. Using either infrared spectroscopy (Figure 14.13) or pyrolysis gas chromatography it is usually possible to determine the chemical composition of even the tiniest fibers. These techniques are used to determine if a fiber is nylon, rayon, acrylic, olefin or any one of several dozen other types of fibers. Again our consumer driven society has caused development of a myriad of different fibers, each optimized for its application. For example polyolefin type fiber melt at a lower temperature than most other fibers so they would not be used in applications where heat is a problem, whereas nylon melts at a quite high temperature but is more expensive. Similarly different fibers types have different chemical reactivity. Where chemically resistant properties are desired Tyvex or Teflon would be used. Forensically, this means is that useful information can often be obtained by looking at the type of fiber collected at a scene or from comparing a fiber from a scene to a fiber from a suspected source. Both the identification of the fiber type and the direct comparison of evidence and control fibers can be quite important. The techniques available in the modern forensic laboratory to make such examinations are quite powerful and can often provide either important investigative or highly probative information, or both.

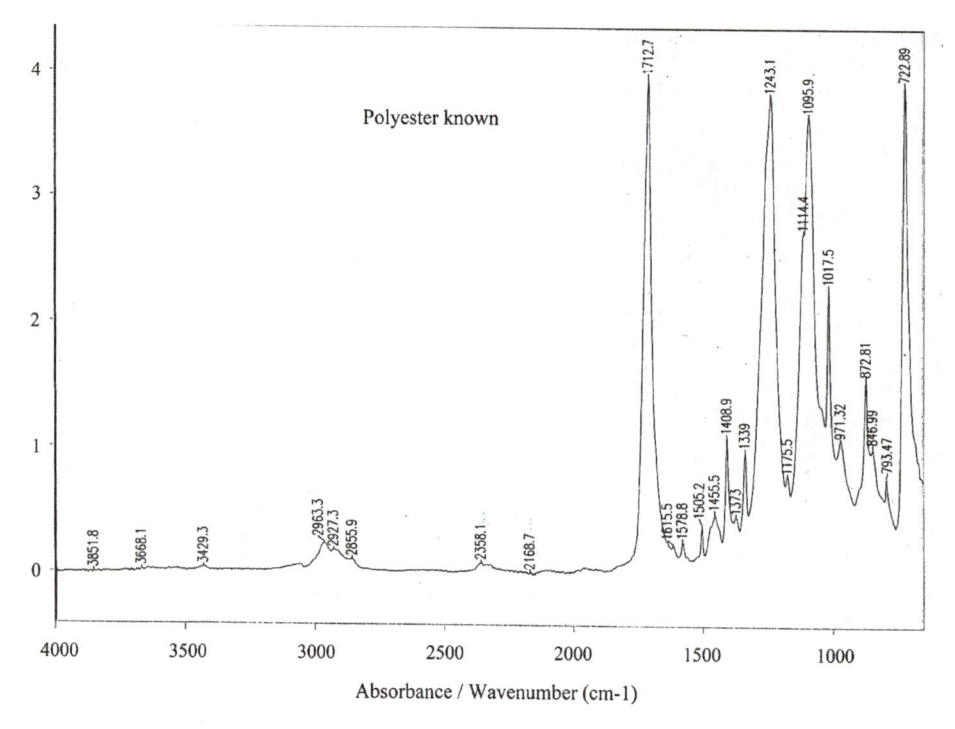

Figure 14.13 Infrared spectrum of a polyester fiber.

Hair—both human and animal

Hair is also an important type of fibrous evidence. The basic physical structure of human and animal hair is very similar. Hairs are shed from both humans and other animals as part of the natural growth cycle. People who are in and out of houses where people have pets often walk away with samples of shed animal hair on their clothing and shoes. These shed hairs are everywhere, and even if not readily visible, microscopical examination can quickly disclose the presence of these pet hairs. In fact it can be difficult not to walk away from a pet owner's home without removing a sample.

Hair structure

Human and animal hair is formed from keratin, which is a protein left when the cellular material produced in the hair follicle dies. Although the cells have died, the hair retains structure and considerable strength. Hair, although obviously biological, is considered in this "materials evidence" category because it is actually a special category of fiber, and the methods used for examining hair are similar to those used for fibers.

Hair has a structure consisting of three major components: cuticle, cortex, and medulla (Figure 14.14). The cuticle, the outer layer, is formed of a very thin sheath of overlapping cells called scales. These scales are usually compared to the shingles on a roof. They overlap, and are strong and waterproof. As a result, hair is difficult to destroy. A hurriedly buried body, perhaps a murder victim, may still have a full head of hair, even after all of the flesh has disappeared and the body is skeletonized. It is largely these cuticular scales (Figure 14.15) that make hair so durable. They keep out the water and bacteria, which fairly quickly decompose the rest of the soft tissue. There are some insects that can actually take little bites out of the hair, and thereby disrupt the cuticle, but it is otherwise largely impervious to other types of decomposition. As mentioned above, it is this scaly cuticle that allows one to easily distinguish hair and fur fibers from vegetable and synthetic fibers.

The major portion of the hair is called the cortex. It is formed from long strands of keratin. There are some other materials present, such as the little specks of pigment that give the hair its color. The third major structure is the

Materials evidence   377

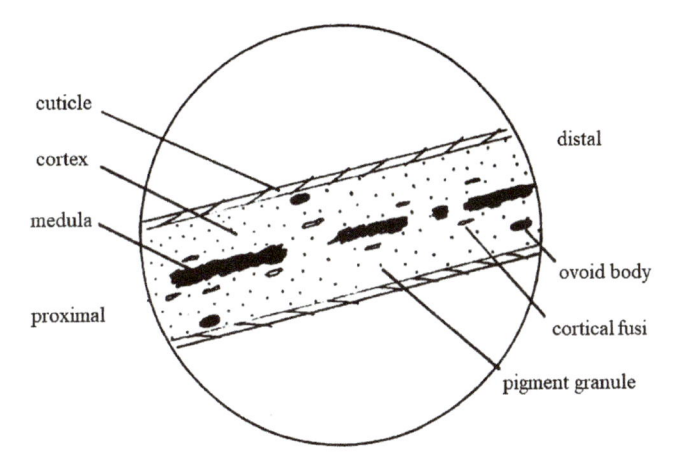

Figure 14.14 A cross-section of a hair—hair has three major areas: the cuticle made up of scales, which surrounds the hair; the cortex, which is the main portion of the hair; and the medulla, which is a central air channel that is in the middle of the hair.

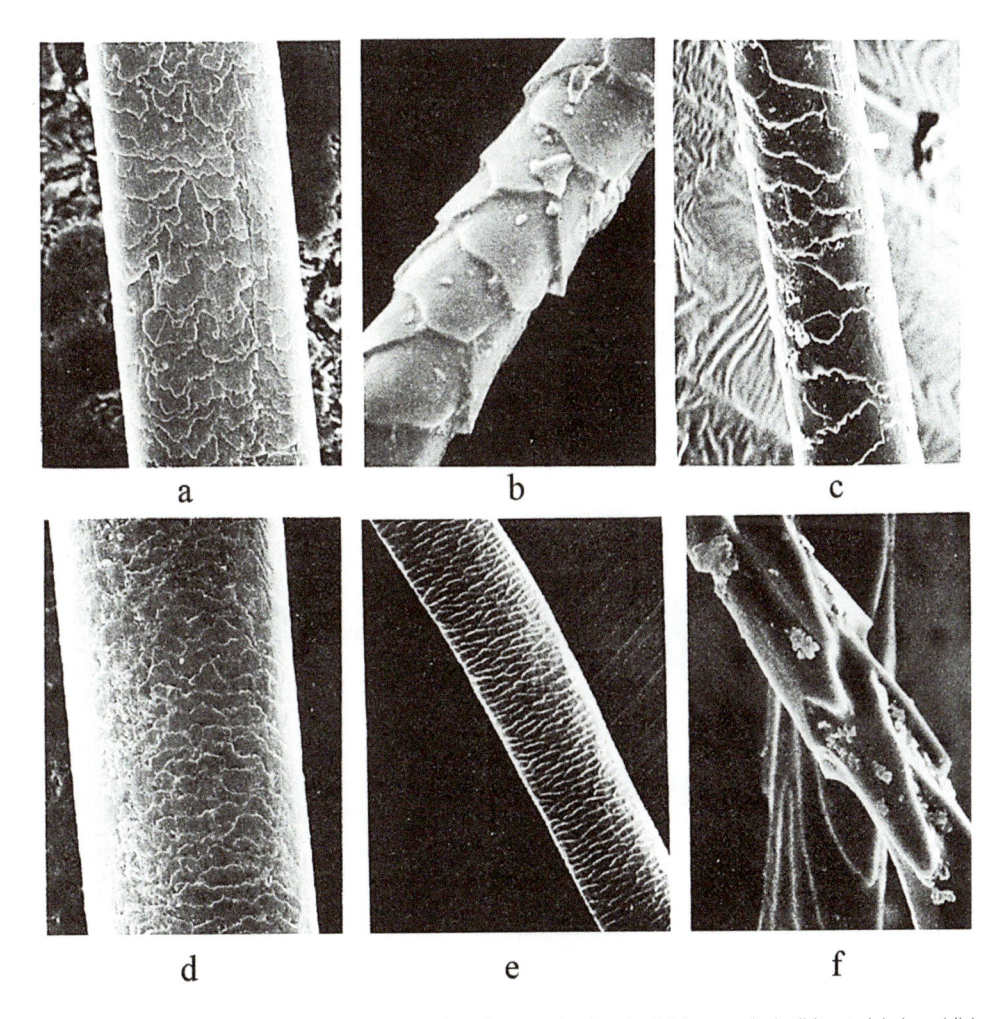

Figure 14.15 The scale patterns on a human hair and hairs of several animals. (a) human hair, (b) cat, (c) dog, (d) horse, (e) deer, (f) rabbit.

central canal, that, when the hair was growing and alive before it emerged from the skin, acted like a canal that transported liquids and nutrients to the growing cells. This area in the center of the hair is called the medulla and is usually an air filled channel that goes up the middle of the hair. If one looks at a hair through a microscope with transmitted light, the medulla usually appears as dark line going up the center of the hair.

The color we see on hair is derived from a pigment called melanin. Hair can be from light blonde to almost black, and this gradation of color is caused by varying amounts and distribution of this pigment. Those whose hair appears in the reddish shades have a different form of melanin, and the amount and distribution determine the shade. Those whose hair has become white have little or no melanin in their white hairs, which appear largely transparent in transmitted light.

Growth phases

Hair does not grow continuously, but rather passes through three different growth phases. The longest is called the anagen phase, which is the active growth phase. When hair is in the anagen phase new cells are being produced in the hair follicle in the skin and are pushing the older cells out, which lengthens the hair. The second phase is the telogen, where the hair follicle stops making new cells and hair growth stops. During the telogen phase, which is relatively short, the little bulb of live cells in the follicle begins to shrink and dry up, destroying the hair's attachment to the scalp, and allowing it to fall out. The third phase is the catagen phase, which is a resting period for the hair follicle. After a resting period, which is quite variable depending on the part of the body and the individual, the hair follicle again enters the anagen (growth) phase and a new hair starts to grow. The forensic significance of the cycle is that, at any given moment, a certain percentage of one's hairs are ready to fall out (telogen phase). Everyone sheds many hairs every day. Further, because telogen hairs are about to fall out, even slight pressure will remove them. As a result, a struggle or other strenuous activity is likely to cause telogen hairs to be shed. These hairs may become important evidence, particularly if transferred between a victim and an attacker.

An understanding of the growth phases, and what the root end looks like during the three phases, allows one to determine, in many cases, whether a hair was forcibly removed or just fell out as part of the normal cycle (Figure 14.16). If someone is hit in the head with a hammer, or hairs are pulled out during a violent altercation, the hairs have a bulb of live tissue clinging to the root end. Thus, a hair examiner can often look at an evidence hair and determine whether it was forcibly removed or whether it fell out as part of the normal hair life cycle. This life cycle is also the underlying reason that there are often many loose hairs in an inhabited dwelling that an intruder may carry away from a scene on his/her clothing, shoes or body.

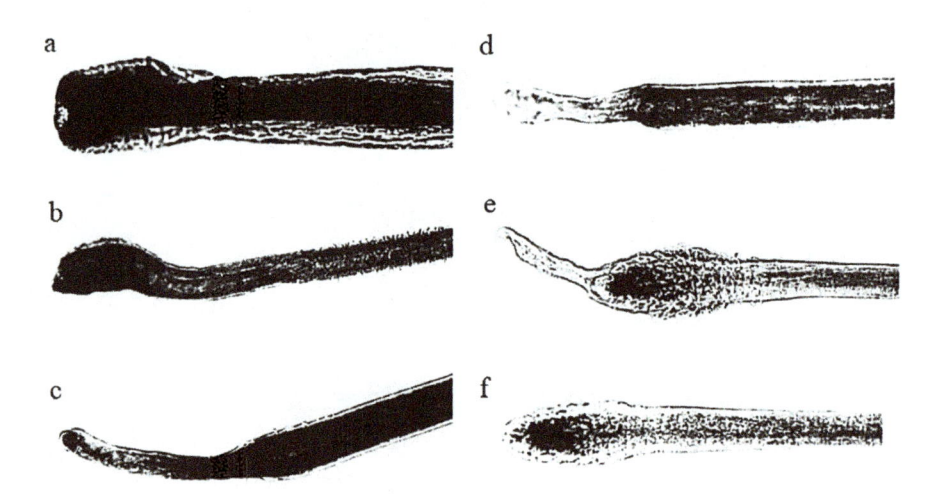

Figure 14.16 The process by which a hair progresses through its growth stage with a large live root to the final released stage where only a rounded club remains of the root tissue.

Racial characteristics of human hair

Traditionally, humans were divided into three "major" races (Caucasoid, Negroid, Mongoloid) that have significantly different looking hair, especially head hair. The vast amount of genetic information available through the extensive research on DNA has shown this classification scheme to be significantly over simplified. As a first approximation visual properties show three major groupings of properties based on historic family origin. Thus those primarily of European origin, African origin, or far eastern origin show significant different in the appearance of their hair. Obviously since the world has become much smaller and there is more inter-marriage between peoples from different continents, these difference are slowly disappearing. Because these differences are even more apparent by microscopical examination, it is often possible for a hair examiner to "guestimate" the origin of an individual who was the source of a particular hair (Figure 14.17). The visual characteristics of the hair, as detailed in Table 14.2, may not be consistent with the hair donor's actual ancestry. A hair examiner might still offer a "guestimate" of the race of the hair donor to investigators, but most hair examiners would not claim to be able to determine the race of a donor from his or her hairs. However, even a "guestimate" could be investigatively useful.

Collection of hair standards

As we have noted, successful laboratory examination of materials evidence, including hair, depends upon the collection of proper control samples. Questioned (or evidentiary) hair must be compared with known controls, and those controls must normally be obtained by investigators and submitted to the laboratory. The proper collection of a control sample of hair from known persons is critical to the success of hair comparison. Several factors should be kept in mind when collecting known control hair. Hair examiners often ask that known control hairs be plucked (pulled), not cut. Examiners want to be able to examine the entire length of the hair, including root. The detailed structure of the root end of the hair can vary significantly in different individuals. Further, the length of a hair can only be determined if the full hair from root to tip is present. It must be said; however, that there is not universal agreement on whether known control hairs must be plucked, or whether they can be cut close to the scalp. In addition, there is not uniform agreement on how many hairs are required to constitute a usable known control specimen. Examiners agree that an "adequate sample" is necessary; that is, hairs from different parts of the scalp, in sufficient number, usually at least 20) to show the full range of variation between different hairs in an individual head. Another factor is that

Figure 14.17 Although there is a great deal of genetic mixing between the three races of humans, each have some significant differences in the appearance of their hair. The three hairs pictures are from individuals with primarily Caucasian, Negroid, or Mongoloid hair characteristics.

Table 14.2 Hair characteristics by race

Race	Diameter	Cross section	Pigmentation	Cuticle	Undulation
Mongoloid	90–120 um	Round	Dense auburn	Thick	Never
Caucasoid	70–100 um	Oval	Evenly distributed	Medium	Uncommon
Negroid	60–90 um	Flat	Dense and clumped	—	Prevalent

known control hairs must be taken from the same part of the body as the questioned hairs. Most of the time, hairs in question will be from the head. But pubic or other body hairs can be evidence in some cases. It is normal to look for transferred pubic hairs and to collect known control public hairs from sexual assault complainants and suspects when identified.

Laboratory examination of hair evidence

Hair examinations can be broadly divided into two categories today. The traditional method, microscopical comparison of hair morphology, is now widely regarded by experts as valuable primarily as an exclusionary tool. Just as we noted above about materials evidence in general, hair cannot be individualized by microscopic morphological comparison. Even hair samples that are microscopically consistent can be said only to potentially be from the same source. If a questioned hair is sufficiently different from the known standards, an examiner can say it did not come from the source of the standards. That's what is meant by saying hair comparison has primarily "exclusionary value." In the past few years, cases have surfaced where examiners testified to greater degrees of hair individualization than most experts think the science can support.

The other method of hair comparison is DNA analysis (see in Chapter 11). Most forensic DNA analysis involves nuclear DNA, as noted in the earlier chapter. With hair, nuclear DNA is found only in the root sheath cells, and the root sheath is present only if the hair was pulled out of the scalp or skin during the anogen or early telogen growth phase. Shed hairs have no root sheath cells, and, therefore, no nuclear DNA. However, there is mitochondrial (mt) DNA in the hair shaft. As explained in Chapter 11, mtDNA does not individualize like nuclear DNA does. The majority of evidentiary hairs are telogen (shed) and, thus, not amenable to nuclear DNA analysis. If present in sufficient quantity, they may be amenable to mtDNA typing. Anagen hairs, again if present in adequate quantity, and if properly recognized, collected and preserved, may be suitable for nuclear DNA profiling.

Most examinations begin with a microscopical inspection. Hairs are usually examined using a transmitted light (compound) microscope. Although most hairs do not look transparent, they will allow sufficient light to pass through them if brightly illuminated. Placed on a glass slide on the stage of a microscope, and with appropriate illumination, one can actually see into the structure of the hair. Although the hair shaft is translucent, sometimes very heavily pigmented hairs, particularly very dark (Asian) far east-type hair, may not let much light pass through and are difficult to examine microscopically. This problem may be overcome with a very bright microscope light source. Interestingly, hairs that have no pigment granules appear nearly transparent with transmitted light under the microscope, but white to our eyes under normal (reflected light) illumination. It is the pigment granules that give hair its color. Hairs can range from blond to black in color by having differing amounts of the same pigment. Only red hair has a different pigment. Usually, even in very dark hairs, the root and tips ends are thinner and allow more light to pass, thus they can be observed more easily.

Because most evidentiary hairs are telogen, microscopical examination for morphology is still an important part of many laboratories approach to hair analysis. It is just not practical at present to think about doing mtDNA typing on every submitted hair without prescreening to eliminate hairs from an obviously different source.

Hair examiners usually examine up to 30 or 40 different characteristics when doing a full microscopical hair comparison. Color again provides a useful point of comparison, but there are a great many other important characteristics, such as pigment distribution, the appearance of the medulla, a variety of structures in the cortex, and tip characteristics (Figure 14.18). The area where the hair is emerging from the skin, the root to shaft interface, has a number of important characteristics commonly included in hairs comparisons. Examiners can generally tell whether a hair was pulled or cut, and whether certain cosmetic treatments have been applied to it, such as bleaching or dyeing.

It should again be noted that hair cannot be individualized by microscopic morphological comparison, no matter how many features the examiner compares. If the evidentiary hair is consistent in its observed properties to a known control specimen, it could have come from that person. However, the combination of both microscopic consistency and a consistent mtDNA profile greatly elevates the evidentiary value of hair comparison.

Microscopic Examination of Animal Hair—Animal hair examination in forensic laboratories is usually primarily to differentiate human and animal hair, and to determine what type of animal an evidence hair could have come from. There are some good reference collections of hair from many different types of animals available. Some laboratories have developed their own, even more extensive collections. Although more detailed comparison between evidence

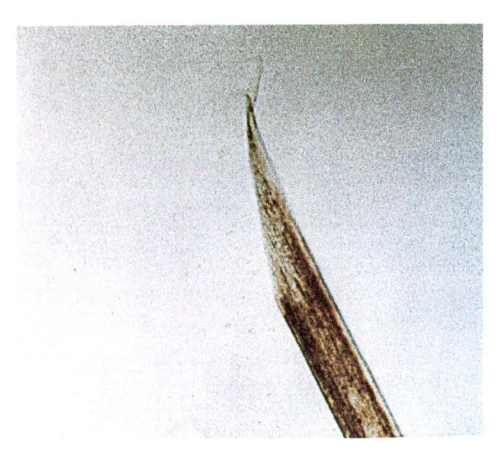

Figure 14.18 The tip of a hair that was trimmed with a razor.

and control hairs from a particular animal can be done, such comparisons are usually restricted to color comparison and consistency of major features. It is usually possible to exclude an animal as a possible source, but strong conclusions that a particular evidence hair came from a particular animal based on microscopic characteristics are unlikely. There has been some work done attempting to use DNA technology to make more positive association, but such work is highly specialized and not yet fully validated.

Automotive and architectural paint and other coatings

Most forensic laboratories would list paint as one of the three or four most commonly encountered types of materials evidence, probably right after the top two, hairs and fibers. The types of paint used to color and protect buildings are different from those used on automobiles and other vehicles. Automotive paint is often encountered not only as tiny paint chips on the clothing of a hit-and-run victim, but also as the result of an automobile having hit a building, rural mailbox, wall, tree or other stationary object, including another vehicle (Figure 14.19). Architectural paint is encountered in breaking and entering cases, on burglar's tools, on shoes, and may even be picked up from recently painted surfaces. Specialty paints from tools, boats, truck beds, and a variety of other sources are also encountered. Paint has the dual function of protection and beautification. In the U.S., it seems that virtually every surface, both vertical and horizontal, is painted. This is not necessarily a bad thing from a forensic point of view, since paint can provide useful transfer evidence.

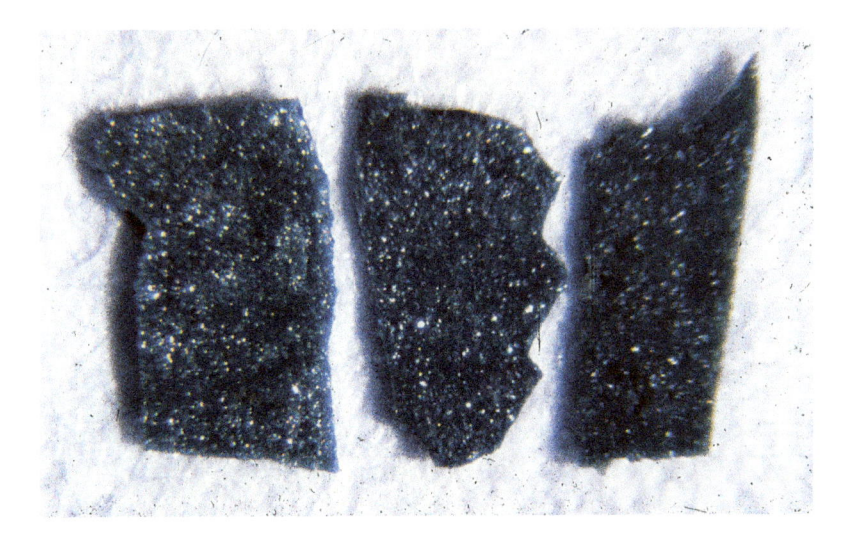

Figure 14.19 Three paint chips collected from three different sources in a multiple murder case.

On the basis of composition and application, paint can be divided into two major categories: architectural paints, used to paint commercial buildings, bridges, houses, and a variety of other structures; and automotive paints, used for cars, trucks, and other vehicles. There are dozens of other types of paint as well, but forensically architectural and automotive paints are the most important because they are the most commonly encountered. It is useful to discuss them separately because they are quite different in composition and easily distinguished.

Architectural paint

The variety of paints used for both interior and exterior painting of buildings is staggering. Walking through the paint displays in a large home improvement store will quickly convince you of this proposition. Most paints have four basic components: *binder*, the polymeric resin to hold it to the surface; *coloring agent* or pigment; *fillers and additives*; and a *carrier solvent* to allow it to be applied easily. The combinations of ingredients used in these four basic components are large, and they make different paints distinguishable and, therefore, make analysis of evidentiary paint worth the effort.

Architectural paint can become evidence; for example, in the form of paint chips transferred to a pry bar during a break-in, chips on the floor being picked up by an intruder, and from smears of still damp paint from a fence. With architectural paint on the outside surfaces of houses, commercial buildings (Figure 14.20), bridges, and many other applications, weathering can loosen paint and cause it flake or bubble. Such loose paint must be removed to prepare the surface for re-painting. This puts paint chips into the nearby environment.

If there are several layers of paint, the layer structure (Figure 14.21) may add weight to the value of a comparison over and above the color and type of paint. Layer information is obtained by a cutting across section of a paint chip and observing the cross section in a stereomicroscope. Particularly in an older home, or building that has been repainted many times, the layer structure can become nearly unique. Those who restore historic buildings often look at paint chip cross sections to discover what original or early paint colors were used.

Automotive paint

Automotive paints are useful not only because there are many different compositional varieties, but also because of the variety of colors. There are an astonishing number of colors of automotive paint found on cars driving around the U.S. A number of years ago, the National Bureau of Standards (now National Institute of Standards and Technology) began collecting samples of all the automotive paints used on American made automobiles. Each year they added to the collection all the new paints coming into use that year. They collected samples of each paint and painted small metal squares. Each square had a code on the back that identified the make and model of vehicles on which it had been used. Most laboratories had these reference sets, and arranged them by color and, as the collection grew, it

Figure 14.20 Applying paint to the exterior of an old warehouse building.

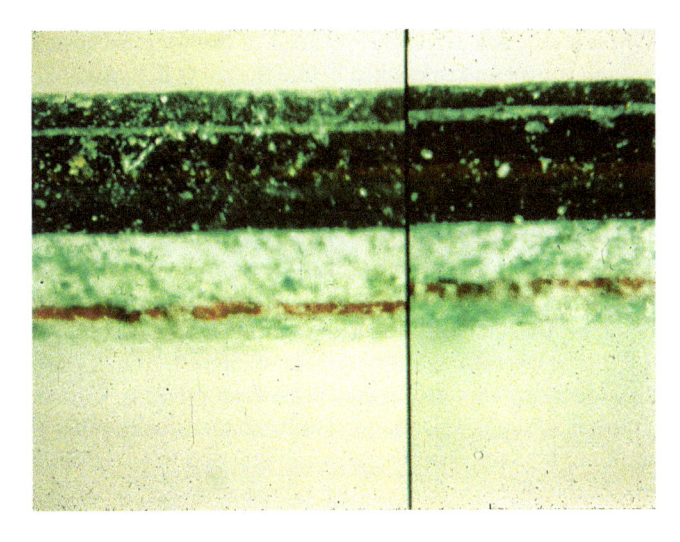

Figure 14.21 A multilayer paint chip of the type encountered in architectural (buildings) paint cases.

became clear how many shades there were of each color. Towards the end of the nearly 20-year existence of the collection, there were more than a dozen pages (20 paint squares per page) of just shades of white. This was a handy little volume when paint chips from a hit-and-run case were examined, because the lab could often give an investigator a list of the makes and models of cars on which that particular shade had been used. This information was often an aid in their investigation. Unfortunately, the collection was too expensive to maintain and was discontinued a number of years ago, although several large investigative agencies continue to collect reference paint information.

One important fact about the painting of automobiles and usually trucks that makes paint chips from them particularly useful, is that the paint is always multilayered (Figure 14.22). Generally, there is a prime coat to protect the metal from rusting, then on top of the prime coat, at least one color coat. Often, the color coat has metallic flecks added to it or a separate metallic coat is added. Then usually another layer, which is clear and designed to make the sheen on the car appear deeper. That means that when one receives a chip of automotive paint from a case, at least three different kinds of paint and usually three different colors on a single paint chip can be identified. The net result is that there is a great deal of often quite useful information in that tiny chip of automobile paint. Of course, if an automobile has been damaged and touched up or repainted, usually over the original paint, it has two or three more layers of paint. It is not uncommon to find a paint chip from a hit-and-run case that has five or more distinct layers of paint. In fact, even a

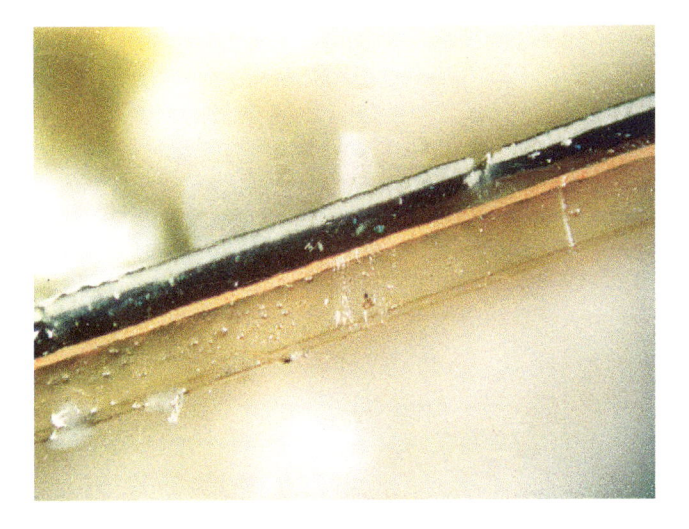

Figure 14.22 A multilayer paint chip of the type encountered in automotive paint cases.

new car often has more than three layers, at least on certain areas of the body, because a surprisingly high percentage all new cars have been damaged in transit and retouched before the car is delivered to its "first" owner.

Collection of known control paint samples

As we have discussed, most paint examinations are, like other materials examinations, comparisons. Having an adequate and representative control sample with which to compare the evidence material is critical. With both architectural and automotive paints, known control samples should be taken in the immediate area where the evidence material is thought to have arisen. The sample should be taken close to that point, but not directly from the suspected area. The reason for this procedure is that on occasion, a physical fit match of a paint chip and the area from which it came is possible. Nothing should be done in collecting a known control that might destroy that possibility. Further, control samples should include all the layers of paint in the area of interest. Even if the evidence sample appears to be a smear involving only the top layer, there may be traces of other layer present. Another important reason for taking the control sample close to the area of damage is the possibility that retouching has changed the number of layers in that immediate area. In addition, prolonged exposure to sunlight and weathering can affect the color. Because of the sensitivity of modern analytical techniques, the control sample does not have to be large. Quality of the control sample (location and completeness) is more critical than quantity.

Laboratory analysis of paint evidence

The first step in examining paint evidence is usually a simple visual examination, frequently using a stereomicroscope to see surface details and color. One should observe inclusions and scratches as well as the sheen and other visual characteristics. Next, the color and the cross section for the layer structure will be examined under appropriate illumination. A razor blade can be used to make a clean cut through a paint chip if necessary, and then the cut piece turned onto its side to look at the layer sequence and any inclusions in the paint. In addition, one can occasionally physically fit a paint chip into the area from which came. As indicated in Chapter 5, a direct physical match may produce an individualization. Since examinations based on other characteristics and features produce only associations (inclusions) or exclusions, the prospect of a physical match is highly desirable when possible.

The single most important examination is the careful comparison of the color between the evidence sample and the known control sample (Figure 14.23). This comparison is done using a comparison microscope that allows the evidence and control paint samples to be viewed simultaneously, adjacent to each other in the microscope field of view, using carefully controlled lighting. The human eye is able to perceive even subtle differences in the shade of colors. There are certainly cases where the chemical analysis is needed to differentiate paints that have virtually identical colors, but in the vast majority of cases if the well-trained human eye indicates that two chips appear to be an identical color, further testing seldom contradicts that conclusion. There are also microscopes equipped with visible spectrophotometers, similar to the microscopes equipped with FT/IR spectrometers discussed earlier. Microspectrometers

Figure 14.23 An evidence automotive paint chip from a truck bed compared for color and texture similarity with a known sample of paint from a large collection of automotive paint samples.

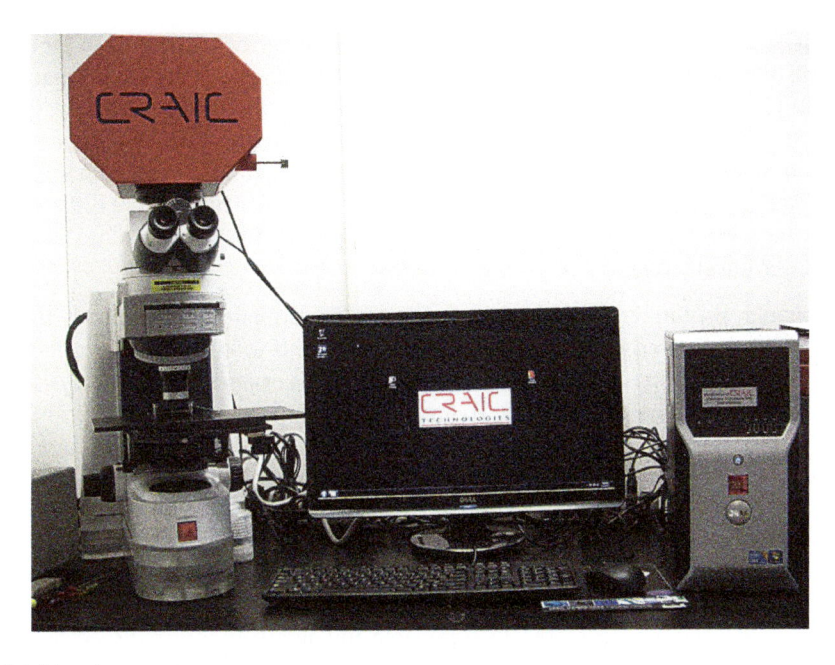

Figure 14.24 A UV/visible microspectrometer.

(Figure 14.24) can produce a visible light spectrum of the specimen. Because color is a function of the absorption, reflection, and transmission of different wavelengths of visible light, the visible spectrum trace provides an analytical record of the "color." These instruments can be used to compare known and questioned specimens to see if the visible spectra are truly identical. They can confirm the visual examination, or sometimes "see" differences too subtle for the human eye. Many forensic laboratories have invested in such instrumentation to add further weight to their visual comparisons.

Paint evidence usually arrives in the forensic laboratory in one of three forms, and it most often arrives riding on another object. It may be in the form of tiny fragments or chips of paint resulting from a forceful shattering of a paint film. This is the common result of a hit-and-run case where paint fragments are found on the victim's clothes or at the scene of the impact. It may arrive as a paint spatter or transfer from liquid paint or from drying paint. Finally, and most difficult to work with, is as a smear (Figure 14.25). Smears result from a painted surface moving across another object.

Figure 14.25 Scraping an evidence paint smear from an auto bumper to obtain a sample.

A common and rather sad example is a smear on the bicycle of a youngster who has been hit a glancing blow by an automobile. There will often be paint from the automobile smeared on the area of the bicycle that was hit, and there can also be paint from the bicycle smeared on the portion of the automobile that hit it. Paint smears are very challenging evidence to examine because of the very small amount of paint transferred and the mixing of the top layers if the two painted objects. It is often not possible to remove the smear, so the examiner must work with the material in place. That makes the instrumental identification of the paint type and even true color much more difficult. As indicated above, the existence of multiple layers can add considerable weight to comparisons when they are found. Smears, of course, seldom can provide layer information.

The basic type of polymer, the binder portion of the paint, can be estimated fairly quickly using solubility or spot color tests. These may be considered preliminary classification tests. If more specific identification of the binder is needed, or if additional comparison is to be done, chemical analysis is necessary. Such analysis is complicated by the fact that many paint samples are multiple layered. To make the information useful, chemical data must be obtained for each of the layers. This process can be attempted by "peeling off" individual layers using a scalpel or razor blade with a steady hand. With a multiplayer chip one can sometimes gather data on the top layer and then scrap that layer away and then look at the next layer, and so on. The instrument of choice for identification of the film-forming polymer is the infrared spectrometer. These instruments have seen some important improvements in recent years. The newest ones can look at a very thin surface layer and obtain high-quality data using an ATR (Attenuated Total Reflectance) or Drifts (Diffuse reflectance) accessory. There are also microscope sampling devices for infrared spectrometers that allow one to locate a microscopically small area and obtain an infrared spectrum. With such a device, a chip can be turned on its side, the microscope focused onto each layer, one at a time, and a spectrum collected to identify the polymer in that layer. In that way, multi-layer paint chips provide the chemical composition of each layer, to supplement the microscopic observation of layer color from a cross section. In some cases, these techniques can allow determination of the chemical nature of the organic or inorganic pigments used to color the paint as well as the binder type (Figure 14.26).

There are other instrumental analysis techniques that can highlight subtle differences in the chemical composition of the polymer that the infrared spectrometer cannot see. Further, one can using other techniques to look at the elemental composition of the inorganic pigments in the paint, even though the pigment makes up only a small percentage of the paint layer.

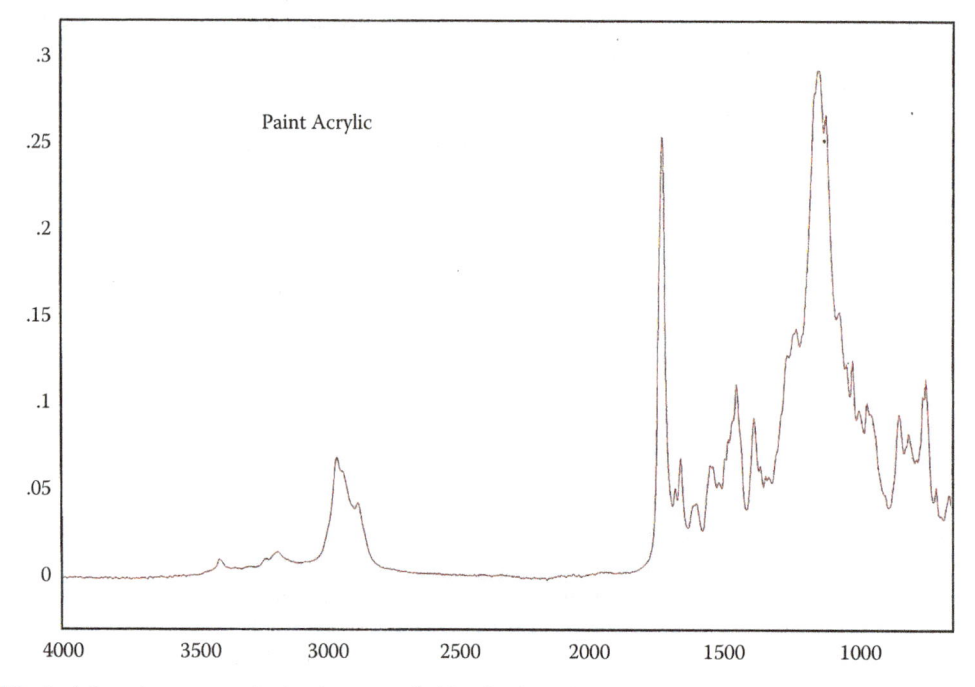

Figure 14.26 An infrared spectrum obtained on a small chip of paint using a diamond ATR based infrared spectrometer showing it to be an acrylic based paint.

Soil and dust

Soil evidence can be recovered from shoes or from muddy footprints left behind on doors, casts of impressions taken at a scene, and many additional sources. Automobiles too can provide an excellent source for soil evidence. When an individual drives off the road, perhaps to dump a body out in the woods, the automobile will begin to pick up soil or debris characteristic of that particular area as soon as it leaves the pavement. Perhaps someone drives down to a lake to throw a body into the lake. The vehicle may leave tire tracks at the edge of the lake, and probably take good bit of that soil away on the tires or in the wheel wells. If one looks inside the wheel wells on any automobile that has been driven off paved roads even for a short distance, one usually finds soil that has been thrown from the tire onto the inside of the wheel well. If a sample of that soil were dislodged, and examined for its layers of soil, one might find a "history" of the different soils through which the vehicle has driven. There is a lot of potential for soil as useful forensic evidence, but because it is difficult to examine and requires considerable special expertise, many laboratories do not extensively examine soil evidence.

A subcategory of soil is dust, and there are experts who specialize in examining dust. Dust is very much a part of our environment. It can often tell quite a lot about where any item that has collected that dust has been. Dust is nothing more than fine particles that are deposited on objects when they settle. Anything left undisturbed either in or out of doors will soon have a coating of dust on it. An item left exposed to the air for a while will collect a history of what has been falling from the air for the period it was exposed. There may be a considerable amount of information in that dust for one who knows how to interpret it, so a dust sample may prove to be useful evidence. Every area has dust characteristic of that particular area. It may be due to pollen from the local plants, fly ash from industrial furnaces, or just airborne fine soil particles. The variety of things settling out of the air, particularly in highly populated areas, is great and ever changing as well. Therefore, it may be useful in determine where something was lying, or if it has recently been moved, as well as many other things.

Soil can be encountered under a wide variety of circumstances that can make it useful evidence in a criminal or civil case. In fact, soil evidence would be much more widely used as evidence if more forensic laboratories were proficient in analyzing it. Analysis and comparison of soil evidence can be a complex and time-consuming problem. As an example of the utility of soil evidence, someone might leave behind footwear impressions in a flowerbed or some other area at or near a crime scene. This is clearly a contact process, where, as Locard postulated, there is likely to be an exchange of material between the shoe or boot and the ground. That is a formal way of saying, "You step in the mud, your shoes get muddy." Muddy shoes or tires provide potential soil evidence. Cases often arrive at the forensic laboratory where the question is: Could a particular shoe have made a footwear impression found at the crime scene? Examination of the pattern of the impression may disclose that it is not sufficiently clear to allow it to be compared to the suspected shoe. There is, however, sufficient soil on the submitted shoe for a comparison to be made with a known control sample from the scene. Unfortunately, it can happen that no known control soil evidence was collected for comparison, because the investigator submitting the evidence knew about the potential value of shoe print pattern evidence but did not realize the potential value of the soil on that shoe. Similarly, soil on the tires or in the wheel well of an automobile may be compared to that found at a location where a body was dumped.

Cases have arisen where someone is missing but the investigators cannot find the body. There is a suspect, and a soil sample is obtained from the suspect's car. The investigators ask for assistance from a local agricultural station. The soil expert examines the soil and, based on their extensive knowledge of local soil types, may suggest an area to search. Even if it is a fairly large possible area, it may focus the search and significantly aid the investigator's ability to locate the body.

Soil evidence provides a challenge for forensic laboratories, because the composition of soil is so highly variable. Soil is generally composed of material from the physical, chemical, and biological weathering of the uppermost layer of the exposed rock, plus decomposed biological material and miscellaneous material that falls from the sky including, in many places, man-made materials. It is certainly a complicated material. Forensically, this complexity can be useful. It means that soil from different places is likely to be different. Anyone who has dug in a backyard, a flowerbed, and in adjacent woods knows that the soil usually looks quite different in the three places, even if they are not far from each other. These differences have potential forensic utility, but may not be easy to exploit because of the difficulties of soil analysis. There are, however, some relatively simple examinations of soil evidence that can yield useful information.

Collection of soil evidence

The collection and preservation of soil evidence is not trivial, especially collection of appropriate known control samples. If the soil is on a moveable object, it should be left undisturbed and the intact object submitted for analysis. As we noted above, submission of intact items bearing materials evidence is the best method of collecting materials evidence. If the soil evidence is adhering to a footwear impression cast, as mentioned earlier, the cast should be submitted with clinging soil because removing it may result in the loss of useful information. Further, there may be layers of soil and removing the soil would destroy the layer information. Wrap the object carefully or place it in a paper bag. If the soil dries and some falls off, it will still be present in the wrapping material. If the soil evidence is not on an object, than the soil should be placed in a plastic or glass vial, or other similar container and protected as above. Finally, and essential to any subsequent comparison, one must collect control samples from the suspected origin and areas around it. If one has an area where the crime may have been committed, one should sample that area and also take soil samples from four or five surrounding areas, about 20–50 feet away, depending on how uniform the surrounding area appears. A simple way would be to take samples about 20 feet from the suspected point of origin at the compass points, north, south, east, and west. This practice makes sense if there is the potential for soil variability that is not obvious. If the evidence soil is consistent with the soil at the suspected location, the variability of soil in the immediate area can be a useful piece of information. If the control samples taken from areas around that suspect location prove to be inconsistent, then that makes the observation of consistency between evidence and control sample more useful.

Laboratory examination of soil

Because soil can be so enormously variable and the experience of most forensic laboratory examiners with soil analysis is quite limited, there is little standardization of procedure for soil analysis. The first step is almost always simple physical observation and examination of the soil. Prior to any examination, the sample must be properly prepared. It should be dried, gently broken up, and any stones or large pieces of extraneous material removed and put aside. Those larger pieces may provide some useful information, but they would interfere with the preparation of the soil. The purpose of the initial preparation is achieving homogeneity of the specimen insofar as possible. One technique for obtaining a homogenous sample called "coning and quartering." The origin of this technique is not clear, but it has been used in both analytical and pharmaceutical chemistry applications. The purpose is to homogenize a particulate specimen, so that any portion removed for analysis will be nearly identical to any other portion.

SIDEBAR

Coning and quartering is the process where a representative sub-sample of a fairly large sample that consists of a mixture is desired. After the material has been generally mixed, the sample is piled into a cone shape with a flattened top, and the cone divided into quarters. Two opposite quarters are put aside, and the remaining quarters are mixed together to form a second cone. The process is repeated until the desired sample size is reached. The process is based on techniques originally developed in the pharmaceutical industry to insure the homogeneity of drug solid dosage forms prior to packaging.

In many cases, simple visual inspection of the color (Figure 14.28), texture, and general appearance of the soil is all that is required. If one looks at the evidence sample and the known control sample(s) and they are a different color or texture, they did not come from the same location. One may think of soil as simply brown, but there are actually hundreds of different shades of brown from nearly yellow to nearly black. Careful observation of the color under proper lighting conditions frequently can unambiguously distinguish properly prepared soil samples in a few minutes. Proper preparation is important, particularly with respect to moisture content, because the color and general appearance of soil are very dependent on how wet it is. Soil may have to be dried by placing it in a warm oven over night or until its color has stabilized.

Another simple examination that can very often include or exclude a soil sample from further consideration in a comparison, is simple microscopic examination for miscellaneous inclusions. As noted above, one might see items that are not natural components of soil, such as coal, paint, rubber, or one of hundreds of items indicative of nearby human activity. Another quick examination is determining the texture of the soil. It may be very uniform, very fine, have lots of clumps or appear to have quite a high content of organic residues. Even if the color is consistent between two samples if the texture is significantly different they are very unlikely to be from a common source.

Figure 14.28 Soil from five different locations showing the wide variation in color often found in soil samples.

If the above examinations on evidence and control samples indicate they are similar, further analysis is required to attempt to distinguish them. One of the common things that can be done by many forensic laboratories is particle size distribution. A weighed amount of dried, homogenized soil is placed into the top sieve of a stack of sieves with different size holes, ranging from course to fine. The soil that is placed into the top sieve is then allowed to distribute itself through the series of sieves as the whole stack is shaken. By using a standard set of sieves, one can separate a soil into different fractions based on the size of the particles. The soil distributes itself with the largest particles staying on top, the slightly smaller particles fall to the second sieve and so on with the finest (clay fraction) passing through all the sieves into a collector. If the sieves have been pre-weighed, the analyst then weighs each one again after the shaking step, and calculates the percentage of the soil in each fraction. In a random selection of different soils, a very different distribution of the particle sizes is observed. Thus, if two samples show very similar particle size distributions, it is an indication of possible common origin.

Another technique for examining soil samples is called density gradient (Figure 14.29). One creates a density gradient in a glass tube by mixing liquids of differing density and allowing the mixture to equilibrate over hours to even days. Once equilibrated, the bottom of the column will correspond to the densest liquid and the top to the less dense one. Density will vary roughly linearly along the length of the column. When a small soil sample is introduced into the tube and allowed to distribute itself, the heavy particles tend to settle toward the bottom of the tube and the lighter particles fall only to the level that matches their density. A few may even float on the top. What is usually observed is little patches of material along the length of the tube and the distribution of these patches of material is used to compare soil samples. There are a number of additional instrumental approaches including measuring the acidity or alkalinity of the soil, the elemental composition and finally comparing the different minerals present in the soil.

The combination of minerals in the soil is also quite distinctive if its origin. There are a number of ways to determine the major minerals in a soil sample and this is an area where research is continuing. Most soil samples have at least 10 different minerals in detectible concentrations, which provides a great deal of differentiating power between different soil samples. One can obtain useful identification of those minerals on individual small grains using a polarized light microscope, infrared microspectroscopy, Raman microspectroscopy, and X-ray diffraction. These analyses can be quite time consuming, but where the soil is a key piece of evidence in a case it is worth the effort.

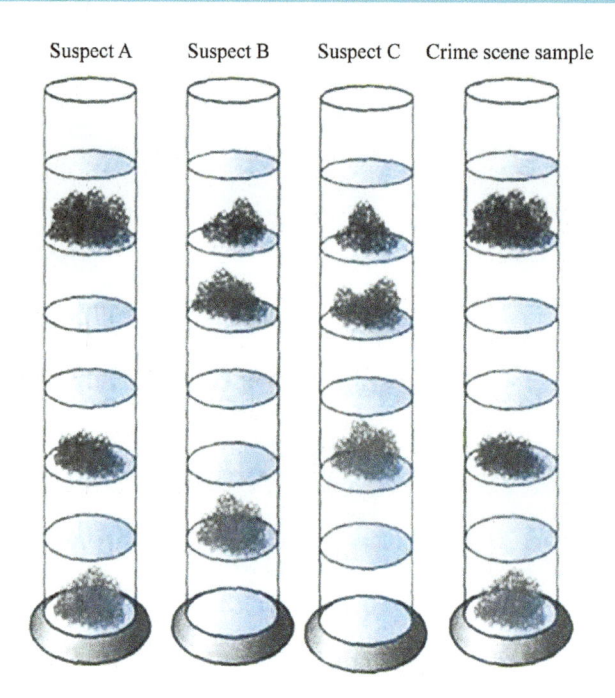

Figure 14.29 Density gradient tubes are designed to separate soil samples based on the different density of components in the soils. Soils from different locales will produce a unique pattern. They can also be used in comparing other samples including glass.

Forensic glass evidence

Glass is another familiar product that can be an important type of materials evidence. A scientist's definition of glass is that it is a melted mix oxide that does not crystallize on cooling. The most common types of glass have three primary chemical components: silicon dioxide, sodium carbonate, and calcium oxide. In common language that means it is largely sand, with a little lime (calcium oxide) and some washing soda (sodium carbonate). If one heats sand until it melts and then allows it to cool down, it will become solid again, look opaque and appear as combined sand grains. If, on the other hand, the correct amount of soda ash and lime are added to the melted sand, it will now not crystallize when it cools, and the resulting material will be solid and have a glassy appearance. This process was discovered, probably by accident, in ancient times. When one looks out a window, it is not obvious that the material was once sand. To a chemist, glass is not a crystalline solid but a "super-cooled liquid." That means it has a structure more like a liquid than it does like a crystalline solid even though it is quite solid. Since most liquids are fairly transparent, it is this structure that allows light to pass directly through it rather than be diverted by crystal boundaries.

Types of glass

Although we now think of glass primarily as a material for making windows and drinking glasses, many useful and decorative things were made of glass before people discovered how to make relatively flat sheets to use for windows. One important early use of glass was to make a wide variety of bottles and containers. There are many different kinds of glass that have been developed for different applications. The most common type is what is called soda lime glass. It is a very inexpensive form of glass made mostly of sand, with a little soda (sodium carbonate), and some lime (calcium oxide). All three starting materials are readily available and inexpensive. Almost all window glass is soda lime glass as is most bottle glass. Specialty bottles, such as perfume bottle, or crystal pieces, which will be mentioned shortly, are made from quite different glass formulations.

A second important type of glass is low modulus glass. It is formulated so that when one either heats it or cools it, it does not shrink or expand significantly. One problem with soda lime glass is that if you heat it or cool it too rapidly, it wants to either expand or shrink, which causes internal strain that may cause it to shatter. If one has a drinking glass, for example, pouring boiling water in it too quickly may cause it to shatter. Folk wisdom said that one should always put a spoon in the glass before you poured the boiling water in it. The metal spoon helped to dissipate the heat. Low modulus glass was developed to solve that problem—it allows glass containers to be placed directly on a

Figure 14.30 Glass that is manufactured with a special formulation to allow it to withstand rapid changes of temperature is used in labware and coffee flasks that can be place directly on intense heat.

heating device like a stove or hot plate. Silex and Mr. Coffee carafes (Figure 14.30) would not be possible without low modulus glass. It was found that adding just the right amount of borax, another fairly inexpensive mineral, to the glass formulation greatly reduces the material's tendency to change size upon heating or cooling, and allows this whole range of new uses.

The beautiful material we call "crystal" is a special form of glass, which is much clearer and appears more brilliant than normal glass. It is made from starting materials that are specially purified and has small amounts of a number of additional materials, such as lead, added. Each manufacturer of crystal has its own formulation and manufacturing process. In most cases, these are carefully guarded trade secrets. Famous manufacturers, such as Steuben in the U.S., Waterford in Ireland, and Lalique in France, command high prices for the beauty and artistry of their crystal objects. Optical glasses are also highly specialized glasses that have been developed to have very specific optical properties. They are used to manufacture lenses for eyeglasses, cameras, telescopes, microscopes, and many other specialized optical devices. The chemical composition of optical glasses is often quite different than that of most other glasses.

When one looks through window glass it appears quite clear. However, if one looks at window glass carefully there is a bit of a tint visible. Most window glass has a bit of a red, blue or a green tint to it, caused by the impurities that were present in the sand that was used to manufacture it. For letting light into our houses and buildings, absolute clarity is not critical. For optical applications, the ability to pass as much light as possible, and at all visible wavelengths equally (to be colorless), is much more important.

Glass as evidence

One reason glass can be important as evidence is because it is used for so many different purposes. Secondly, in addition to the many varieties of glass and the large amount produced each year, when glass is not treated with proper respect, it breaks or shatters producing numerous pieces of potential evidence. When there is a violent confrontation of some sort, if there is glass around, it is very likely to end up in pieces. Because there are many different types of glass and even subtle differences can be detected, it has forensic value in determining whether two pieces of glass could have the same or a different source. Automobile accidents contribute more than their share of glass evidence. Windshield glass is a special type of glass—safety glass, designed to resist breakage, but when it breaks it is designed to disintegrate into small pieces to minimize injury. The windshield and side window glass are different types of glass, as are the headlight glass, the mirror glass, and the glass in light bulbs. For example, materials evidence from a hit-and-run case may include broken headlight glass, windshield glass, side window glass, mirror glass, and even optical glass (from someone's eyeglasses).

Collection of glass evidence

As noted earlier, materials evidence seldom can be individualized using microscopic, chemical or instrumental techniques. This is as true of glass as of other materials. However, because glass fractures randomly, physical fit matches are sometimes possible between glass fragments. Physical fit matches can often lead to individualizations—such a fit proves that the pieces were originally part of the same structure. This fact is an important consideration in discussing glass fragment collection. Chipped edges or further breakage will lower the probability of making a jigsaw match. In a hit-and-run case if the side view mirror is broken, pieces collected at the scene may be jigsaw matched to any mirror fragments remaining in the frame on the damaged car. Jigsaw matches of such evidence are a real possibility. Further, glass is a wonderful substratum for fingerprints. Fingerprints can often be developed readily on glass surfaces (Chapter 6). A fingerprint on a piece of glass can identify (individualize) the person who left the print. Both physical fit matches and the possible presence of fingerprints, must be considered first when collecting glass evidence. Among other things, these considerations mean that larger pieces should be carefully collected and cushioned against further breakage during transport. Investigators should always wear latex or other protective gloves when handling this evidence (and all evidence for that matter).

A second important concern in collection of glass evidence is prevention of loss from clothing or other surfaces. Tiny glass chips that are on clothing or other objects can easily be lost if the objects are not properly handled and protected. When glass is broken, often little pieces of it are propelled backwards and land on the clothing of the one doing the breaking. Under a microscope one can find those tiny pieces, pick them out, and compare them to control samples from a scene. These tiny pieces can easily be lost, unless the object on which they sit is handled gently and properly packaged. Any evidence with the potential of containing glass fragments should be carefully wrapped in paper and put in a bag.

Laboratory examination of glass

The examination of glass evidence is usually a three-step process. An initial observation and measurement of the glass is the simplest but often the most critical step. It involves looking at properties, such as color, thickness, surface characteristics, and the shape and size of the pieces. Although one thinks of glass as being clear, placing it on a white surface under proper illumination usually reveals subtle shades of color. Only fine crystal or optical glasses are likely to be almost colorless. Multiple pieces of glass from an incident can be quickly sorted just using color. Similarly, thickness is an important property of flat glass. One can measure the thickness if any of the pieces are large enough. If two pieces of glass were broken out of the same larger piece of flat glass they clearly will have the same thickness. Surface characteristics, such as pebbling, surface grinding, or imperfections caused by being exposed to the elements for 10 or 20 years can be helpful in examining glass evidence With window glass, determining "inside" and "outside" may be important, and is usually possible because of, in many cases, exposure of the two sides to different environments. Once the initial observations are completed, the physical properties of the glass may be examined. An important optical property of glass, the fact that light passes through it, is the basis for much of its utility. One of the important physical properties of glass is its *refractive index*. That is normally abbreviated RI and is defined as the speed of light as it passes through the glass divided by the speed of light in a vacuum. One can measure this ratio for any transparent material, and it is a useful property for distinguishing different kinds of glass. It is affected by both the chemical composition and the density of the glass. There are several simple ways to measure refractive index, even on very small glass chips. It is often the first optical property measured because it is a good way to determine that two glasses could not have had the same origin since their refractive indices are significantly different. If, on the other hand, their refractive indices are very close, then they could have come from the same source.

A second important physical property is density. One usually measures the density of a glass chip using liquids of known density. Placing a little chip of glass into a liquid and observing whether it floats, sinks or is just suspended (is floating in the liquid and not on top of the liquid) tells one if the class is the same, more or less dense than the liquid. When an object (like a glass fragment) is placed in a liquid of greater density than itself, it floats; when placed in a liquid of less density than itself, it sinks; and when in a liquid of the same density, it is suspended. One can adjust the density of the liquid by mixing a heavy liquid and a light liquid in different proportions until the glass is suspended, which means the liquid mixture and glass have the same density. The density of the liquid mixture is then determined, and the density of the glass chip is known. Since again different kinds of glass have different densities, this becomes another characteristic to help determine if two glass samples could have had the same origin. Many glasses have the property of giving off visible light when illuminated with ultraviolet light, called *fluorescence*. When illuminated, the glass may give off a colored glow (fluoresce) when observed in the dark. Fluorescence is useful physical property of glass that can be used in comparing glass samples. Flat glass is made by floating on molten tin. This leaves tin traces on the side in contact with the tin and makes that side fluorescent. Sometimes when trying to jigsaw match pieces of flat glass, knowing which side is up can greatly assist in putting the pieces together correctly.

There was a crime laboratory proficiency test a number of years ago where the laboratories were sent three small pieces of glass. The question was: Could any of the three have a common origin? All three of the pieces of glass matched each other very closely in refractive index, density, and even chemical composition. One, however, fluoresced strongly and the others did not. Some laboratories indicated possible common origin because they failed to check for fluorescence. A lab could have immediately eliminated common origin of those pieces, and saved many hours of additional analysis, by spending the relatively short time required to observe the pieces with ultraviolet illumination and look for fluorescence. This anecdote illustrates another important point as well, one that we have tried to emphasize above. When questioned and known control specimens of materials evidence are compared, and found to be substantially the same *in the properties compared*, they still *may not have* had a common origin. Other ways of saying this same thing are: They are consistent in properties compared with having had a common origin; or the possibility of common origin cannot be excluded. Note how important the words *in the properties compared* are in the proficiency testing story. Labs that did not compare fluorescence, and said that the specimens could not be excluded (could have had a common origin) were not wrong. Their conclusion was correct based on the properties they compared. With any "match" result in materials evidence, it is always true that comparison of another property could result in exclusion. This is one reason materials evidence comparisons are difficult to evaluate—there is no basis for knowing whether one more measurement might produce an elimination or further strengthen the association.

Some of the major elements present in glass and their relative concentrations are usually the result of materials added to provide desired properties. Therefore, it is often possible to determine the type of glass an evidentiary sample falls into, if a database of major elemental composition of many different types of glass is available. The major elements in glass are usually silicon with some calcium, some sodium, and a little iron or other elements added to tailor properties as discussed above. If one looks very carefully there are trace amounts of many other elements. This is because glass is made from sand and sand being formed from naturally occurring minerals, will have traces of other elements. More importantly, where the sand comes from will determine its trace element composition. Since sand is cheap to mine, but expensive to ship because it is heavy, companies who manufacture glass tend to use sand from nearby sources. Therefore, glass manufactured in different places will likely have different trace element composition, as shown in Table 14.3. Even glass manufactured in the same place but of different types will have different trace element compositions due to different proportions of sand and other materials used in the glass making process. Further, as sand is mined its trace element composition may change depending on where in the sand deposit it was found. One can determine elemental composition, measuring as many as 15 different elements, and be able to use that information to determine if two glass samples have a similar or different distribution of trace elements. Determining the elemental composition profile of glass specimens is time consuming and requires specialized instrumentation. Specimens that do not have similar elemental compositions can be distinguished even where other physical properties are consistent.

Miscellaneous types of materials evidence

Because of the huge variety of materials possible it is impossible to describe them all in any detail. We can group them in categories for brief discussion.

Biological evidence other than human and animal derived material discussed earlier can be quite important. One example of a biological material is pollen. Information that may be available from examination of pollen is usually dependent on whether an object was left outside at a particular time of year or how long it was left out. Pollen falls out of the air at particular times of the year depending on the life cycle of the plants that released it. Those who have hay fever are acutely aware of the life cycle of golden rod or whatever plant pollen to which they are sensitive. Because pollen is biological and has very characteristic shapes, it can be recognized as pollen and associated with a particular plant source. Two other examples are plant residues (usually collected from soil samples) and feathers. Feathers can be rather interesting evidence in some circumstances, because in cold climates many people wear down coats. Down coats are supposed to be stuffed with duck down (feathers). "Supposed to" is the operative phrase, because unless one buys down coats from a highly reputable source, the stuffing is frequently not actually duck down. The stuffing is either totally or partially chicken feathers. Since there are many more chickens than ducks sold at the grocery store, chicken feathers are much less expensive. When someone wearing a down coat is stabbed,

Table 14.3 Use of minor and trace elemental composition of glass fragments to connect them to a particular broken automobile headlight

	Headlight fragments					
	HL H6024		HL 4656		HL 5006	
Element	Mean ppm	St. Dev.	Mean ppm	St. Dev.	Mean ppm	St. Dev.
Na	3500	90	34900	540	3590	80
Mg	41.78	0.37	70.6	0.96	62.8	0.55
Al	1200	0.04	1000	0.01	1040	0.01
Ca	153	29.23	221	14.59	200	34.29
Ti	71	5.19	46.3	0.97	44.5	0.71
Cr	1.26	0.29	2.19	0.36	2.13	0.26
Mn	2.67	0.06	1.29	0.1	1.2	0.03
Fe	96	1.85	234	4.15	237	4.23
Ni	0.43	0.06	0.32	0.05	0.27	0.08
Zn	1.44	0.13	1.01	0.1	0.89	0.16
Rb	0.38	0.01	0.38	0.01	0.4	0.01
Sr	4.08	0.11	5.16	0.1	3.95	0.1
Y	9.42	1.27	0.92	0.03	0.86	0.03
Zr	5099	711	119	4.62	97	5.95
Mo	3.28	0.11	0.69	0.08	0.53	0.06
Ba	4.42	0.09	1.86	0.05	1.82	0.04
Ce	3.09	0.15	3.79	0.06	3.57	0.06
Hf	113	15.85	2.96	0.13	2.23	0.15
Pb	0.41	0.03	0.42	0.01	0.36	0.02
Th	1.74	0.22	0.29	0.01	0.25	0.01

there will be feathers flying everywhere. Under such circumstances identification of the type of feathers transferred to a suspect's clothing could turn out to be a very useful piece of evidence.

Additional categories of biological evidence are wood and paper. Paper is made from highly processed wood and both are composed largely of cellulose fibers. Most other natural fibers are made of cellulose as well, but can easily be differentiated, because nature makes things that have interesting and identifiable shapes. In virtually every household, no matter how well kept, it is likely that careful examination will disclose a few little balls of fluff underneath a bed or hiding in an inaccessible corner. There are several names for these little dust balls, such as "dust bunnies" or "woollies," depending on what part of the country one comes from. They are made of intertwined fibers and hairs. Interestingly, one of the main components in these bundles is wood fiber. There are no trees in most homes so why are there so many wood fibers? The simple reason is that wood fibers are used to make paper and in every home there are many different types of paper; for example, facial quality tissues, toilet tissues, and paper towels.

A wide variety of building materials can, under some circumstances, provide useful evidence. Should a criminal gain entry to a building or room by breaking through a wall, one will likely find plaster dust all over that individual's clothes and shoes. If one commits a homicide near a highway, for instance, and moves the body, examination for trace evidence may disclose tiny pieces of rubber. As mentioned earlier when they are thrown up in the air from tire wear they eventually settle near the road. As a result, fine tire rubber particles can be fairly useful pieces of evidence to indicate that an object was at one time on the ground near a road or highway. There are many types of foam-based materials used as building materials, in furniture and as insulation. Foam rubber is familiar to everyone, but much more common now is polyurethane foam. It is the material that is most often used in pillows and cushions, especially the seats of automobiles and furniture pillows. There have been many cases where tiny specks of polyurethane foam have proven to be useful as transfer evidence. Nearly everyone has seen an old couch or a very used automobile seat that "leaks" little bits of fine yellowish powder. When one examines this "powder" with a microscope, there are tiny bubbles visible that indicate that it is a foam, probably polyurethane foam. When a chair seat begins to leak this fine powder through seams in the fabric, anybody who sits in that chair will leave with some of that powdered foam on their clothing.

CASE ILLUSTRATION

There was a multiple murder case on Long Island several years ago where the bodies were dumped in dumpsters. After considerable investigative effort, the detectives developed a strong suspect. Interestingly, examination of the bodies and clothing of the victims showed that one type of materials evidence, found on several of the victims, was a particular kind of yellow powder that when observed microscopically was clearly foamed material. The investigators went to a suspect's apartment to interview him. He and his brother had a successful business and were making a good living, but he was living in a filthy, cluttered apartment. The furnishings were old and decrepit. There was one particularly decrepit chair, which was clearly his favorite chair. That chair was leaking fine dust from several places, and there was powdered foam all over the floor near that chair. This observation alone reinforced the idea for the investigators that they probably had the correct individual. They obtained a search warrant for the apartment, and subsequent analysis showed that this was exactly the same type of yellow-powdered foam found on the bodies, and it was useful in linking the victims to this particular apartment.

Because metal is used so widely in building and manufacturing there are many forensically important metallic traces encountered, from lead residue left behind by a bullet, to rust, to magnesium residue from a flare used to start a fire. When someone forces open a door with a pry bar, the Locard transfer principle says that some of that pry bar metal is left on the door frame, and some of the door frame material (paint or wood) is transferred to the pry bar; thus an exchange of traces. If one finds residue of the metal on a tool, it may be possible to determine the composition of that metal. Steel is mostly iron, but also has chromium in it and easily detectable amounts of four or five other elements. Besides steel most other metals have small amounts of other metallic elements added during manufacture to give them desirable properties. It is possible, for example, to compare the elemental composition of a piece of material recovered from a tool that gouged it out of a cash box to a control sample taken from that cash box, and have a pretty good indication of whether or not the cash box **could** have been the source of the tiny trace of metal found on a particular tool.

Although cosmetics and beauty products are not a frequent type of evidence receive by forensic laboratories, there are cases where a lipstick stain left behind on a glass; for example, may prove to be a useful piece of evidence. One can both analyze the composition and match the color of a lipstick stain to a known control sample from a suspect or victim. The variation in color alone is enormous. A visit to a large department store can quickly convince one that the variety of lipstick shades available exceeds even that of automobile colors. Similarly nail polish can be chipped off at a violent scene and again the variety of colors is enormous. Keep in mind that color and even composition of these materials are class characteristics, and matching class characteristics does not mean individualization, but can indicate or eliminate possible common origin. There was analysis of nail polish in the Crafts case, presented in Chapter 1.

STRENGTHENING MATERIALS EVIDENCE

As we have mentioned earlier because most materials evidence can only be associated and not individualized its primary purpose is for comparison. This means that when presenting this kind of evidence the most one can normally say is that two items could have come from the same source. This is not a very strong statement and many attorneys would like a much stronger statement but that of course is not possible with most types of materials evidence.

However, the strength of materials evidence in court can be greatly enhanced under certain circumstances. There are a number of things that can strengthen the testimonial value of materials evidence and make it more useful for the jury in reaching their decision. If the particular type of evidence is something relatively uncommon or rare it will perhaps carry more weight if found to be consistent. The circumstances under which the evidence was found or where it was found may also be relevant. If material evidence is found in places that one would normally not find such evidence or if there are two-way transfers, the evidence's weight may be enhanced. Thus, if one finds materials evidence from a suspect on the victim and from the victim on the suspect this is more significant.

Perhaps the circumstance that can most strengthen the effect of materials evidence is finding many different kinds of materials evidence that are consistent in the same case. Thus, if we have a case where we find hairs, several different kinds of fibers, paint chips, footprints, pollen grains, and a large number of different kinds of materials evidence, that all draw the inference that they all came from the same individual or location, this makes that kind of evidence significantly stronger.

There are a number of well-known cases that can illustrate this particular situation. The famous Wayne Williams case from Atlanta Georgia where a number of bodies were recovered from the local river and when they were examined for trace evidence a whole series of different fibers were found. There were dog hairs as well as textile fibers, rug fibers from the bedspread, carpet fibers from both the suspects bedroom, and from the suspect's automobiles and all of these were shown to be consistent with locations associated with the suspect. There were multiple victims and many of the same traces were found on each. In that case so many different items of trace evidence consistent with the suspect strongly implicated the suspect and carried a much more weight, than any one of the items could possibly have provided alone.

In a second such case is the lead case in this Chapter (see above). It involved the murder of a 10-year-old girl. The body was found in the rather isolated area. In that case there were fibers from the victim's clothing in the suspect's car, fibers from the car on the body, there were dog hairs found on the victim's body that matched a dog that was known to have ridden in that car often, there were tire tracks that match the tires on that car, there was a footprint on the back of the victim that was made by a shoe similar to one that the suspect was known to own. Perhaps the most telling piece of materials evidence was a pubic hair found in the rectal area of the victim. This case occurred before DNA was being done on hairs. The defense felt that hair examination was not reliable. This case was appealed several times for a variety of different reasons and perhaps the last time was after mitochondrial DNA analysis had become available. This is the kind of a case where multiple different types of materials evidence were found to be consistent with the defendant and thereby strongly implicates the defendant. Such cases provide a powerful recommendation for doing materials evidence, because under the right circumstances it can provide very strong evidence.

Key terms

- anagen phase
- architectural paints
- attenuated total reflectance (ATR)
- automotive paints
- catagen phase
- cortex
- cuticle
- delusterant
- diffuse reflectance infrared Fourier transform spectroscopy (DRIFTS)
- exclusionary value
- float glass process
- keratin
- layer structure
- materials evidence
- mechanical dislocation
- medulla
- melanin
- natural fibers
- optical properties
- paint smears

particle size distribution
pyrolysis gas chromatography
refractive index (RI)
regenerated fibers
soda lime glass
synthetic fibers
tape lift
telogen phase
trace evidence
transfer evidence
visible spectrometers

Review questions

1. What are the differences and similarities between transfer and trace materials evidence?
2. List and give examples of the major techniques used by investigators and in forensic laboratories to collect materials evidence.
3. Name some things that could be considered materials evidence.
4. Why do fibers have such importance in the materials evidence class?
5. What are the major types of fibers that can become evidence.
6. Why does an analyst need large control samples for successful hair comparison and how should these control samples be collected?
7. What types of incidents are likely to produce forensically important paint evidence?
8. Briefly discuss the nature of glass and why it is so often important forensic evidence.
9. What are the major components of soil?
10. Discuss the proper collection of soil evidence and the importance of proper control samples.

Fill in/multiple choice

1. The Locard exchange principle concerns evidence resulting from physical _____.
2. The laboratory device most commonly used in the *search for* trace and transfer evidence is the _____.
3. The preferred method for submitting an object that may have trace evidence adhered to it is:
   a. submission of the object without prior removal
   b. submission of the object and the removed trace material
   c. in a clean metal paint can
   d. in a clean plastic bag
4. Soil samples can be _____ by careful determination and comparison of the particle size distribution of the questioned and known samples.
5. Hair forcibly removed from the body
   a. is known as telogen hair
   b. has cellular material adhering to its root, and may be analyzed for nuclear DNA
   c. is much more common as evidence than hair that just falls out of the skin
   d. is analyzed primarily with an instrument known as the FTIR

References and further readings

"Application of Infrared and Raman Spectroscopy in Paint Trace Examination." *Journal of Forensic Sciences* 58 (2013): 1359–1363. doi:10.1111/1556-4029.12183

Caddy, B., ed. *Forensic Examination of Glass and Paint.* London, UK: Taylor & Francis Group, 2001.

Convicted Killer Seeks Execution; St. Louis Post-Dispatch, June 15, 2004, Five Star Late Lift Edition.

D'Andrea, F., R. Francoise, and R. Coquoz. "Preliminary Experiments on the Transfer of Animal Hair during Simulated Criminal Behavior." *Journal of Forensic Sciences* 43 (1998): 1257–1258.

Deedrick, D. W. "Hairs, Fibers, Crime, and Evidence." *Forensic Science Communications* 2 (2000). Available at: www.fbi.gov/hq/lab/fsc/current/descript.htm.

Deedrick, D., and S. Koch. "Microscopy of Hair Part 1: A Practical Guide and Manual for Human Hairs." *Forensic Science Communications* 6, no. 1 (2004).

Demmelmeyer, H. "Forensic Investigation of Soil and Vegetable Materials." *Forensic Science Review* 7 (1995): 120.

Dettman, J. R. et al. "Forensic Discrimination of Copper Wire using Trace Element Concentrations." *Analytical Chemistry* 86 (2014): 8176–8182.

The Forensic Examination of Hair. London, UK: Taylor & Francis Group, 1999.

Houck, M. M., ed. *Mute Witness: Trace Evidence Analysis.* New York: Academic Press, 2001.

Houck, M. M., and B. Budowle. "Correlation of Microscopic and Mitochondrial DNA Hair Comparison." *Journal of Forensic Sciences* 47 (2002): 964–967.

"The Ingredients of Paint and Their Impact on Paint Properties." Available at: www.paintquality.com/under specifiers

Junger, E. P. "Assessing the Unique Characteristics of Close-Proximity Soil Samples: Just How Useful is Soil Evidence." *Journal of Forensic Sciences* 41 (1996): 27–34.

Kurouski, D., and R. P. Van Duyne. "In Situ Detection and Identification of Hair Dyes Using Surface-Enhanced Raman Spectroscopy (SERS)." *Analytical Chemistry* 87 (2015): 2001–2006.

"Microscopy of Hair Part 2: A Practical Guide and Manual for Animal Hairs." *Forensic Science Communications* 6, no. 3 (2004).

Scientific Working Group on Materials Analysis. "Forensic Fiber Examination Guidelines." *Forensic Science Communications* April 1999, Archived at: https://archives.fbi.gov/archives/about-us/lab/forensic-science-communications/fsc

Scientific Working Group on Materials Analysis. "Forensic Paint Analysis and Comparison Guidelines." *Forensic Science Communications* (1999).

Keto, R. A. "Analysis and Comparison of Bullet Leads by Inductively-Coupled Plasma Mass Spectroscopy." *Journal of Forensic Sciences* 44 (1999): 1020–1026.

Lesney, M. "Eyeing the Glass Past." *Today's Chemist at Work* (February 2004) 55–56.

Ogle, R. R. Jr., and M. J. Fox. *Atlas of Human Hair: Microscopic Characteristics.* Boca Raton, FL: CRC Press, 1998.

Palenik, S., and Palenik, C. "Microscopy and Microchemistry of Physical Evidence." In *Forensic Science Handbook*, 2nd ed. Ed. R. Saferstein, Chap. 4. pp. 175–230, Upper Saddle River, NJ: Prentice Hall, 2002.

The People of the State of Illinois, Appellee v. Cecil Sutherland, Appellant; 155 Ill.2d. 1; 610N.E.2d 1; December 4, 1992.

The People of the State of Illinois, Appellee v. Cecil Sutherland, Appellant; 194 Ill.2d. 289; 742 N.E.2d. 306; November 16, 2000.

Robertson, J. ed. *Forensic Examination of Fibers.* 2nd ed. London, UK: Taylor & Francis Group, 1999.

"Weighing Bullet Lead Evidence," National Research Council Monograph, 2004.

Appendix A
J. Methods of forensic science—The scientific tools of the trade

As we have frequently noted, forensic science is science in service of the law. All sciences make use of various methods and techniques, many that involve sophisticated instruments. Throughout the book, various methods and instruments will be discussed in the context of particular evidence types and their analysis and comparison. Here, we introduce and briefly discuss a few basic scientific terms, methods, and instruments that are extensively used in forensic science laboratories. Many of them are methods that come from analytical chemistry.

Forensic science and criminalistics tend to borrow techniques and methods from the basic sciences, like biology, chemistry, physics, geology, and so on, and adapt them to resolving forensic questions. Basic sciences are not, for the most part, directed toward forensic questions. Forensic scientists are frequently asked if two objects could have had a common source or were once part of a single object. The forensic scientist normally then adapts chemical, biological, or physical science techniques to try to answer the question. This Appendix outlines some very basic scientific concepts that are is helpful to be familiar with in understanding forensic methods.

1. PHYSICAL PROPERTIES, STATES OF MATTER, AND THE METRIC SYSTEM

Physical properties of matter are properties that are intrinsic. They do not depend on any actual or potential chemical reaction. They are also generally independent of quantity. Some examples are mass, density, temperature, melting point, and boiling point.

Mass may be thought of, roughly, as the quantity of a substance. It is often confused with weight. And, although they are not identical, scientists do weigh things to determine their masses. Mass is usually expressed in grams in the metric system, and in pounds or ounces in the Avoirdupois (English) system.

Density is the mass of a material per unit volume. Usually expressed in units of grams per cubic centimeter. Denser liquids or gases will sink if placed in less dense ones. A cork floats on water because it is less dense. A lead weight sinks in water because it is denser.

Temperature is a measure of how cold, warm, or hot something is. The two main temperature scales in use are the Fahrenheit, used throughout the U.S., and the Celsius scale, used almost everywhere else. A conventional freezer operates at about −2°C or 28°F. Water freezes at 0°C or 32°F. Water boils at 100°C or 212°F. Normal human body temperature is about 37°C or 98°F. The melting and boiling point temperatures of a substance are characteristic physical properties and may sometimes help in identifying the substance. The "°" is a universal symbol for degrees of temperature in both systems.

Scientists use the metric system for measurement as does everyone in Europe and much of the rest of the world. But in the U.S., the English (or engineering) measurement system has persisted.

There are three main states of matter: solid, liquid, and gaseous. The state of a substance is temperature dependent. If we heat water, it will boil and turn to steam. If we cool it, it will freeze and become ice. Other substances behave the same way. There are some other states of matter too, such as colloids and plasmas, but they won't come up in this book.

2. LIGHT, THE EM SPECTRUM, COLOR

Another physical property is color, and it results from the interaction of the material with the light that is hitting it and how the light it reflects interacts with the light sensors in our eyes. Light—the light that we see called the visible spectrum—is part of a much larger spectrum of different wavelengths called the electromagnetic spectrum (Figure A.1).

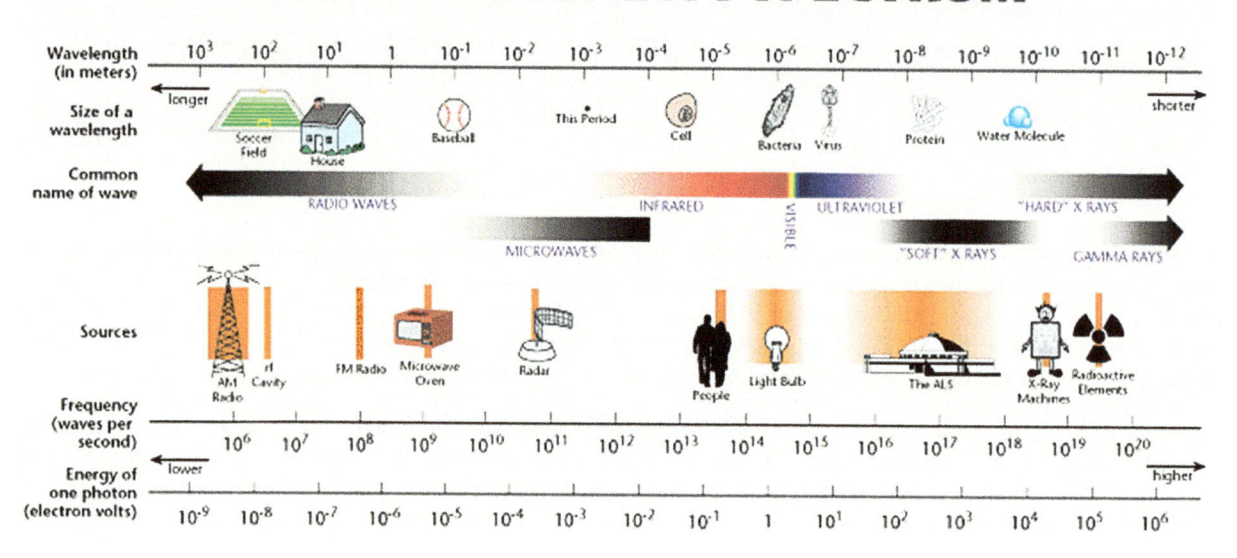

Figure A.1 An illustration of the full electromagnetic spectrum with the common application ranges indicated. (From Lawrence Berkeley National Laboratory.)

The different forms of electromagnetic energy along that energy spectrum are distinguished by their wavelength. At the very short wavelength end (high energy) are cosmic rays, gamma rays, and X-rays. Toward the center are infrared light, visible light, and ultraviolet light. And at the long end are radio, microwave, and television broadcast waves (low energy).

All color and many instrumental methods used in forensic science rely on measurements of light's interactions with substances and materials. When electromagnetic radiation strikes a substance, three things can happen: it can be reflected; it can be "absorbed"; or it can be transmitted. Reflected means essentially bounced. There is no effect on the substance. A mirror is an efficient reflector of visible light. Transmitted means the radiation passes through the substance unchanged. Window glass transmits most visible light. Absorbed means that the molecular structure of the substance absorbs some energy from the radiation and undergoes changes or heating as a result. The differential absorption of electromagnetic radiation is the basis of all spectrometric and some spectroscopic methods.

Sometimes, when a substance absorbs electromagnetic radiation, the molecules can go into an excited state briefly, then relax. As they relax, light of a longer wavelength (lower energy) is emitted. This phenomenon is called fluorescence, and forensic scientists use it frequently. Usually, ultraviolet light (which is of shorter wavelength than visible) is impinged onto a specimen. If the specimen is fluorescent, bluish, or other colors of visible light will be given off.

In the visible light portion of the spectrum—the part we see with our eyes—light consist of a mixture of many wavelengths, which we see as different colors. Mixed together, it looks white. A mnemonic for remembering the colors in a rainbow (visible spectrum) is ROY G. BIV—Red, Orange, Yellow, Green, Blue, Indigo, and Violet is the order of the colors, from long to short wavelength visible light. Something looks green to our eye because the green wavelength light is reflected by that object. For partially or fully transparent objects one sees their "color", when back-lighted, as the wavelengths that are passed through the object.

3. SPECTROMETRIC AND SPECTROSCOPIC METHODS

Spectrometric methods of analysis are all based on measuring the wavelengths of light that are absorbed, transmitted or reflected by a specimen. All the different wavelengths of the electromagnetic spectrum have been exploited to design a wide variety of analytical instruments of great utility to scientists. We will discuss scanning electron microscopy and energy-dispersive X-ray analysis in connection with the identification of gunshot residue, for example. The most common analytical instruments rely on the infrared (infrared spectrometer), visible (visible spectrophotometer), and UV (ultraviolet spectrophotometer). For purposes of this book, you can think of IR (or its more sophisticated relative FTIR) as a method of chemical "fingerprinting" of pure substances. IR and FTIR are used in the analysis of drugs, toxins, and many materials comparisons. Visible spectrophotometry is, if you will, an analytical method for determining color. And UV spectrometry is the same as visible, except it uses the UV wavelength range, which our vision does not see as color.

Spectroscopy and some spectroscopic methods are somewhat different. They do not rely on differential interaction with electromagnetic radiation. The most important forms of spectroscopy are atomic absorption (AA), which can be used to identify which elements are present in a substance, and mass spectroscopy. Mass spectroscopy is probably the single most important analytical tool in a forensic chemist's arsenal.

In a mass spectrometer, the molecules in a specimen are first ionized by bombardment with electrons and they usually break into smaller fragments. The ions formed are specific to a given compound. Magnetic fields in the instrument influence the movement of the ions so that they migrate through the field and strike the detector in a way that allows identification of their molecular weight. The results appear as a series of peaks on a chart. A mass spectrum of a compound is usually unique, and many compounds can be unequivocally identified from their mass spectra. It is sometimes said that mass spectrometry, when coupled with a gas chromatograph (see just below), is the "gold standard" for drug and toxicological compound identification.

Atomic absorption (AA) spectrometry and several X-ray techniques are used to identify the chemical elements present in a specimen. The most common use for AA in forensic laboratories has been to screen specimens for elements present in gunshot residue (Chapter 8).

4. CHROMATOGRAPHY AND CHROMATOGRAPHIC METHODS

Chromatography is a name given to certain chemical separation methods. The word itself comes from a Greek word meaning "color," because the procedure was originally used to separate colored pigments from plants.

There are several different kinds of chromatography. The most widely used in forensic science are: thin-layer, gas (once called "gas-liquid"), and high-performance liquid, and capillary electrophoresis (Figure A.2).

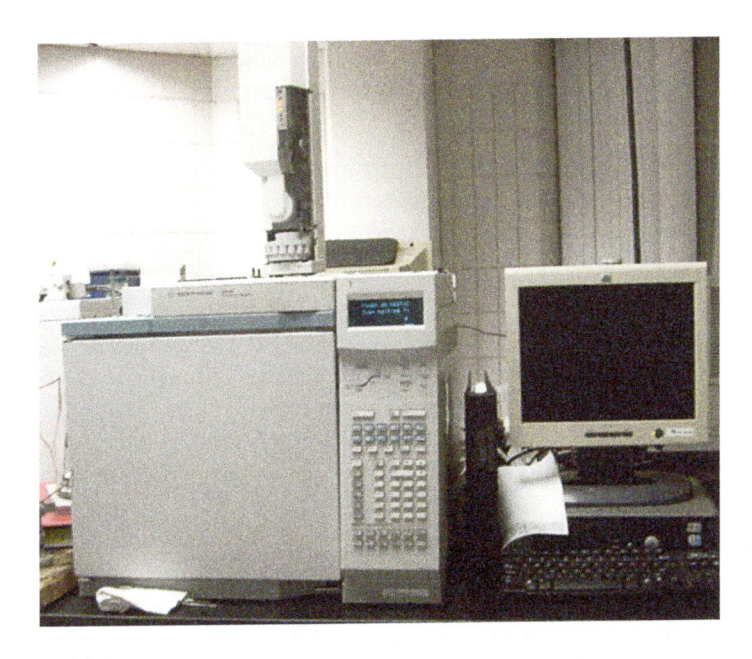

Figure A.2 Gas chromatographic instrument.

They are typically abbreviated TLC, GC, and HPLC. Chromatography brings about separation of substances based on their differing affinity for one or the other of the chromatographic "phases." In thin-layer chromatography (TLC), for example, there is a stationary solid phase and a mobile liquid phase. A plate, coated with a solid layer of adsorbent, is placed vertically into a tank so that it just contacts a small amount of liquid (the mobile phase) in the bottom of the tank. By capillary action, the liquid "climbs up" the solid phase. Substances that have been spotted near the bottom and absorbed by the solid phase ahead of time may have greater affinity for the liquid phase as it moves, and be carried along with it. Or, they may have more affinity for the solid phase, and move more slowly. Because different substances have differing affinities for the phases, they are separated.

In gas chromatography (GC), the mobile phase is a gas (often helium). The "liquid" phase is a high molecular weight liquid absorbent that coats the inside of a long very narrow bore column. The substances one wants to separate are heated to high temperatures causing them to vaporize. The vapors then pass onto the column and are carried along by the gas. The more affinity the substances have for the liquid phase, the slower they travel. Over the years, chemists have designed many different liquid phases and discovered many different instrument conditions for separating a variety of different compounds.

The gas chromatograph allows complex mixtures, such as concentrated fire residue sample to be separated into its many components. The sample is injected into the instrument, vaporized, and swept through a very fine glass or more commonly pure silica column by the inert carrier gas. The carrier gas is usually either helium or hydrogen. The flowing gas moves the sample through the column, which has a thin coating of a high boiling liquid on the walls. The thin layer of the liquid on the walls causes the components of the sample to be partitioned between the liquid (stationary phase) and the flowing gas (mobile phase). This partitioning is a dynamic process in which the different chemical compounds dissolve in the liquid coating and then are re-vaporized many thousands of times during their passage through the column. The different components are moved along through the column while in the gas phase but are stationary when dissolved in the liquid phase. Since each different chemical compound has a slightly different tendency to move from the gas to the liquid phase, they take different amounts of time to pass through the column. The more time a compound spends in the liquid layer the less time it is moving and the longer it will take to make the trip through the column. Conversely, the more time a compound spends in the gas phase the sooner it will emerge. The net result is that even very similar chemical compounds with only a slight difference in their partitioning between the mobile and stationary phase will take slightly different amounts of time to pass through the column. Any slight differences in their boiling points or their attraction to the stationary phase will cause them to emerge at different times. At the end of the column, there is a detector that tells the operator when something other than the carrier gas is emerging from the column. Accordingly, the length of time each component of a mixture takes to traverse the column can be measured accurately. In addition, the sizes of the peaks produced by the detector are proportional to the quantity of the component represented by that peak in the mixture.

One of the most significant developments in analytical instrumentation was figuring out how to interface chromatographs (for their separation ability) with mass spectrometers (for their chemical identification ability). These instruments (and the associated techniques) are sometimes called "hyphenated." Gas chromatograph-mass spectrometry (GC-MS) is very commonly used for drug and toxin identification, and generally for the identification of organic compounds (Figure A.3).

Sometimes, a compound is difficult to volatilize (get into the vapor state). A compound must be volatilizable for GC. In those cases, liquid chromatography can be interfaced with a mass spectrometer (LC-MS). In GC-MS and LC-MS, the MS is acting as the "detector," if you will, for the chromatographs.

Figure A.3 Gas chromatograph/Mass spectrometer Instrument is a very useful combination of separation and identification. (From Library of Congress.)

5. MICROSCOPY AND MICROSCOPICAL METHODS

Because much physical evidence is microscopic or near microscopic, forensic scientists make use of microscopes in many ways, including for finding, separating and examining evidence. Forensic scientists primarily use four different types of microscopes—each for a different type of activity. The four discussed below are the stereomicroscope, compound microscope, comparison microscope, and the scanning electron microscope.

A point of terminology: We use the term "microscopic" to mean small—small enough to require a microscope for viewing clearly. We use the term "microscopical" to mean "of, or having to do with, microscopes." Thus, a forensic scientist uses microscopical methods to examine microscopic dust particles.

The stereomicroscope (Figure A.4) actually has separate optical paths, one for each eye of the viewer. In that way it gives the viewer a three-dimensional view of objects in the field of view that allows the examiner, with a steady hand, to touch, grasp, and move objects while viewing them. These are also called dissecting microscopes, because biologists use them to carefully observe objects while they are being taken apart; dissected. These microscopes are designed for relatively low magnification and a long working distance. Working distance is the distance between the objective lens of the microscope and the object being viewed. This is critical so that a probe, forceps, or scalpel can be manipulated without the lens being in the way. The magnification range of a stereomicroscope can be from life size to about 60 times life size. It is the product of the magnification of the eyepiece and the objective. The objective lens is the one closer to the object. More expensive stereomicroscopes have a zoom objective, which allows continuous magnification change, usually between 0.7 and 3.5 times (some have an even larger zoom range). With a 10-times eyepiece lens, this produces a magnification range of 7–35 times, which is quite adequate for most forensic applications. The most important forensic application of the stereomicroscope is searching for evidence; for example, clothing for foreign hairs or fibers, and initial examination of potential evidence to gain information about what it might be.

Figure A.4 Scientist using a stereomicroscope outfitted with a digital imaging pick-up and fiberoptic illumination.

The compound microscope (Figure A.5) is also known as the biological microscope. It is the most widely used microscope, combining an eyepiece lens with one or more different objective lenses to provide a wide range of magnification. The objective lens is usually on a rotating turret and must be quite close to the sample to provide proper focus. As with the stereomicroscope, the total magnification is the product of the magnification of the eyepiece and the objective lens. For example, a 10-power eyepiece is often combined with a 4-power objective to provide a total magnification of 40 times. One can usually rotate the objective turret to a 10-power objective and obtain a magnification of 100 and so on. The compound microscope is usually used in the range of magnification of about 40 times to about 1000 times. Unlike the stereomicroscope, it has a single optical path, even when provided with binocular viewing, and produces a flat field with a very small depth of focus. Its optical design provides a very short working distance, particularly at high magnification.

Most compound microscopes used forensically are designed for transmitted light applications. The object to be viewed is usually placed on a transparent glass microscope slide and the light is passed through the sample. This process, of course, works best with very thin items that will allow enough light to pass for observation. Hairs, fibers, small paint chips, and most small crystals, are suitable for transmitted light examination. As we will discuss below, items like bullets, cartridge cases, and toolmarks can be best examined using reflected light and usually are examined at lower magnification than that customary with a compound microscope.

The comparison microscope (Figure A.6) is a particularly forensic microscope. It is designed to optimize the examiners ability to make careful comparisons on a microscopic scale. It is actually just two compound microscopes joined together by an optical bridge. Using some simple controls on the optical bridge, the examiner can look into the optical bridge and view either what is on the left or right microscope or can split the field of view to allow viewing both simultaneously. Thus, one the can place a control hair from the victim on the right microscope and bring a particular area of that hair into focus. On the left microscope a corresponding portion of an evidence hair from

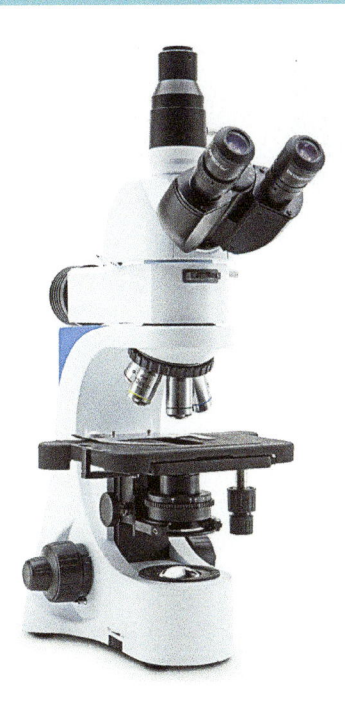

Figure A.5 Compound microscope is used for examining small evidence at high magnification using transmitted light.

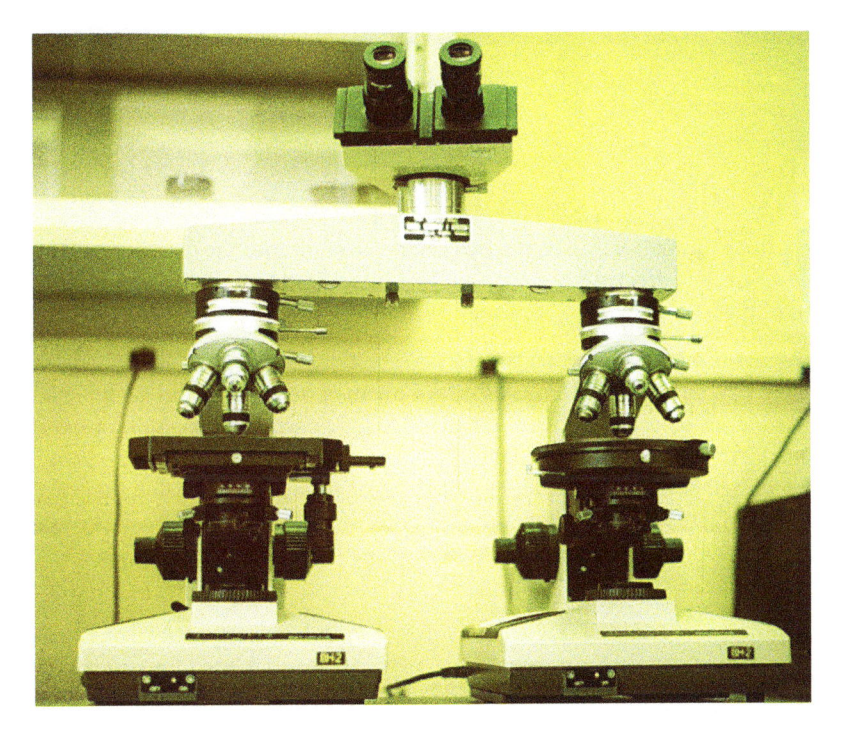

Figure A.6 Two compound microscopes connected by an optical bridge for direct comparison of microscopic or near microscopic evidence.

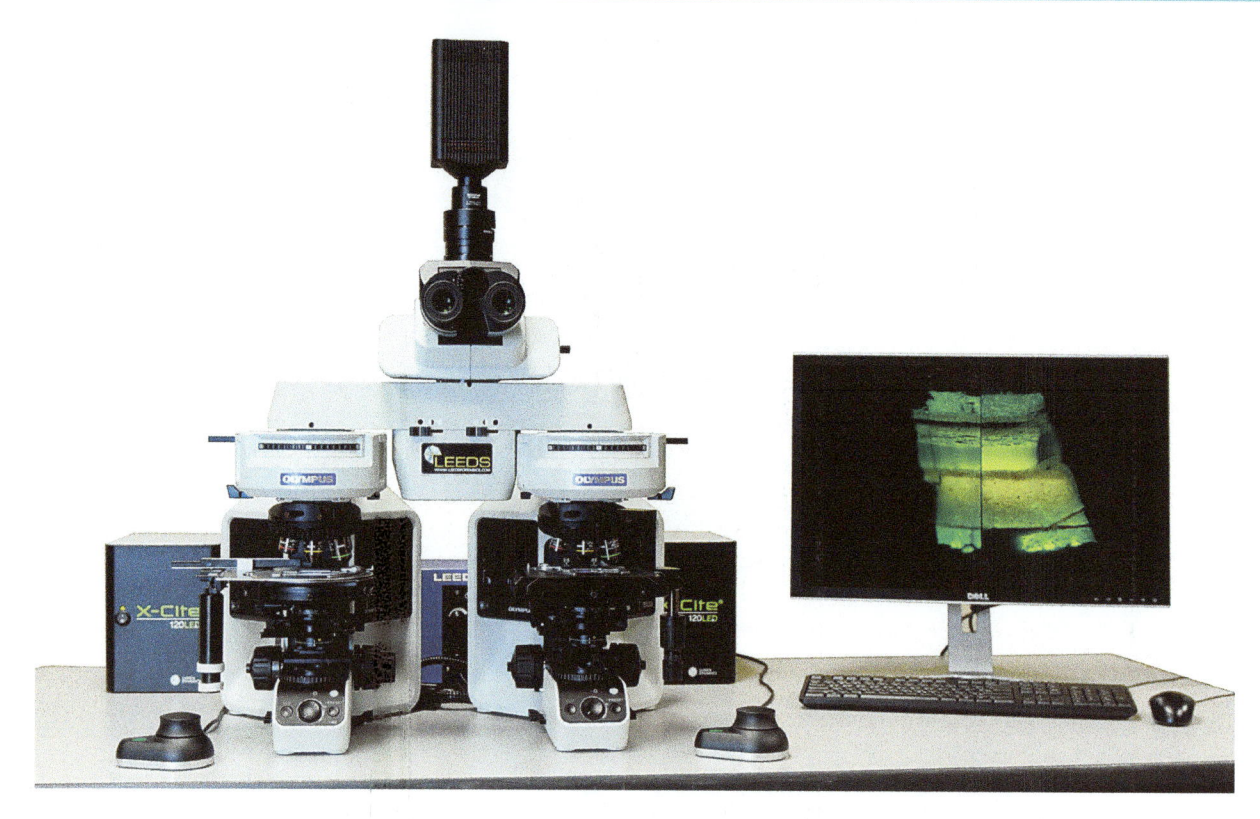

Figure A.7 Comparison microscope designed for both reflected light and transmitted light comparison (Used with permission courtesy Leeds – www.leedsmicro.com.)

the suspect's clothing can be located, and then the two can be brought into the field of view together, each on its half of the split field of view, to allow precise comparison. The transmitted light comparison microscope is particularly useful for small items of materials evidence (Chapter 14).

For bullets or cartridge cases (Chapter 8) or other opaque objects, a different type of comparison microscope (Figure A.7) is used. It works the same way, by linking two microscopes with an optical bridge, but is used with microscopes that can use both light reflected from the object being observed as well as light passing through the object (transmitted) methods. For the observation of bullets, they are mounded with sticky wax to a holder that can be rotated in all three dimensions and are usually lighted with fiber optic light cables to allow highlighting or shadowing the particular areas of interest. Reflected light comparison microscopes usually also have other stages to hold other types of samples. Again, areas of interest are selected by observing the objects on each microscope independently and then closely compared using the split field.

The scanning electron microscope (SEM) (Figure A.8) uses a focused beam of electrons, instead of visible light to image objects. It is a much more complex and expensive instrument and is used for special applications. It has magnification capabilities much greater than an optical microscope, being capable of magnification from only a few times to several hundred thousand times. Its other advantage is its ability to perform elemental analysis on the objects it is magnifying. The disadvantage is that it requires that the objects being observed are in a high vacuum chamber. Therefore, there is no working distance since the objects must be in a sealed chamber. One important forensic application is visualization of gunshot residue particles collected from the hands of one suspected of firing a weapon (Chapter 8). Because the SEM can allow both visualization of the rather characteristic appearance of such particles, and the ability to determine that they contain the characteristic elements of barium, antimony, or lead, the reliability of their identification is raised to a much higher level. The SEM has application to many other types of trace evidence as well.

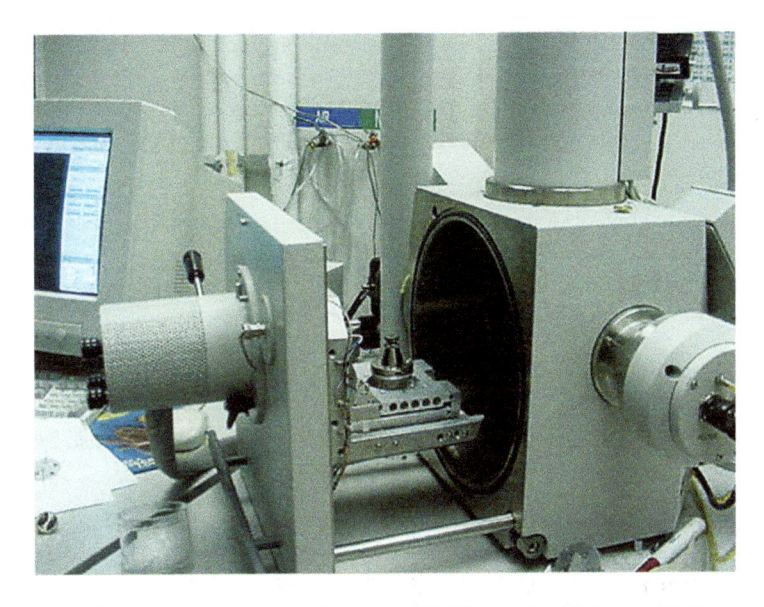

Figure A.8 The open chamber of a scanning electron microscope (SEM) used for high magnification and micro-elemental analysis (Image credit: User Olaboy-. Used per Creative Commons Attribution-Share Alike 2.5 Geneirc https://creativecommons.org/licenses/by-sa/2.5/legalcode.)

6. BIOLOGICAL METHODS

There are a number of methods in biological sciences that are employed in the examination of biological evidence. Probably the most prominent and worthy of mention are enzymatic assays, histochemical staining, and DNA manipulation. Enzymes are nature's catalysts—catalysts speed up a chemical reaction without entering into it. All the enzymes are proteins, and we'll discuss more about this in Chapter 10. Some methods involve trying to find an enzyme in a specimen, while others involve using enzymes to look for some other component of it.

Histochemical staining (Figure A.9) is regularly employed when cells are to be examined under a compound microscope. The stains are organic compounds, which can differentially color various components of cells and make it easier to see and study them microscopically. One stain commonly used in the forensic lab, for instance, is called

Manila 40x Hematoxylin

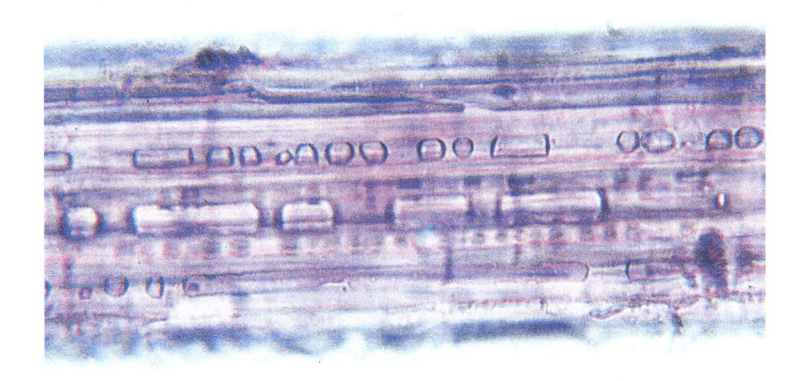

Figure A.9 A slide with a natural fiber stained with a biological stain to help visualize internal structure of the fiber.

"Christmas tree." It makes sperm cells have red heads, green midpieces, and blue tails. This staining makes it easier for examiners to locate sperm cells in smears of vaginal swabs from sexual assault complainants. This is discussed further in Chapter 10, as well. Another potential type of biological evidence fibers come from plants (see Chapter 14). Although botanical traces will typically be examined in a manner similar to other trace/materials, histological staining could play a role in that analysis.

In Chapter 11, we will describe how DNA can be isolated and manipulated so that it can be typed for forensic purposes. The very active research and applications of molecular biology have spawned a whole range of new instrumentation designed specifically for molecular biology and, specifically, DNA manipulation and analysis. Several of these instruments are described and pictured in Chapter 11.

7. COMPUTERS AND COMPUTER PROGRAMS

As computer technology has overwhelmed society in the past several decades, it has also become an essential feature of scientific instruments and laboratory practice.

Every major analytical instrument is now attached to a computer, and the instrument makers typically design proprietary software to go with it. The instrument is controlled by the software, which also captures the instrument's output signals and translates them into useable formats, like gas chromatograms or mass spectra. Since chromatograms and spectra can be represented fairly easily by a series of numbers, large libraries have been built. And, since computers are fast at comparing numbers, these libraries can be searched quickly and efficiently to see if the spectrum of the specimen being run is in the library (Figure A.10). This type of search helps analytical chemists enormously in their work and speeds up identification of compounds.

There are also computer programs designed to manage laboratory operations, inventories, and so forth. And, so-called LIMS (laboratory information management systems) are big, integrated systems for managing a laboratory's entire operation, from evidence accession, through analysis, evidence storage, evidence return, and report writing.

As described in Chapter 9 many highly, specialized programs have been developed for application to forensic Digital Evidence cases. Quite a number of these are described in Chapter 9.

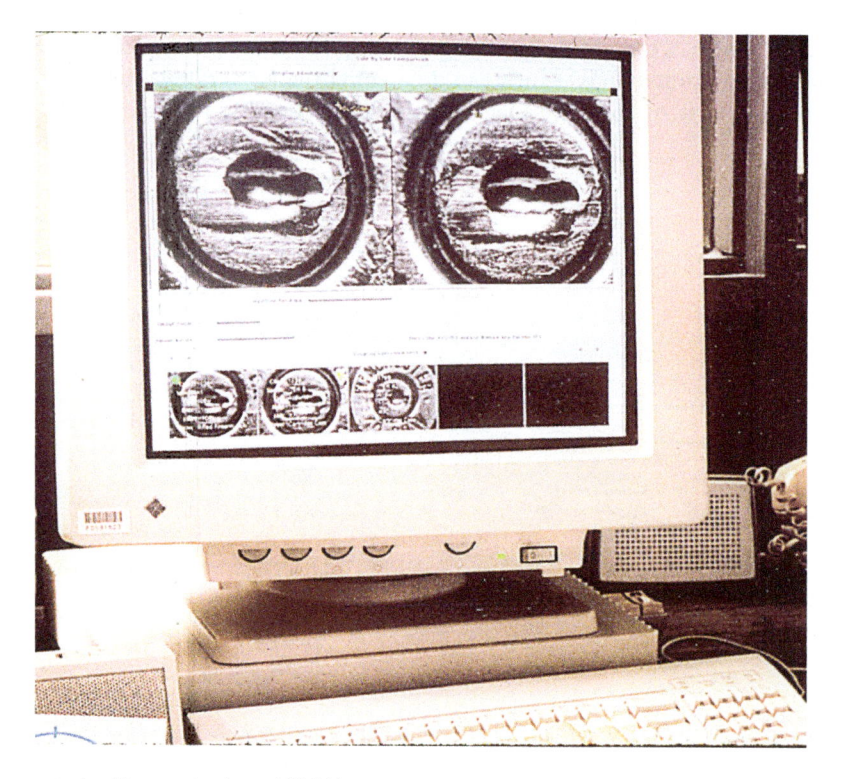

Figure A.10 The computerized image database NIBIN is used to search bullet stria and cartridge breach face images.

Index

Note: Page numbers in italic and bold refer to figures and tables respectively.

AA (atomic absorption) 194, 403
AAFS (American Academy of Forensic Sciences) 12, 25
ABC (American Board of Criminalists) 12
ABO blood groups 223, 249
absorbed, EM spectrum 402
accreditation 12
ACE-V (analysis, comparison, evaluation, verification) 134, *135*
acid etching technique 197–8, *198*
"acid phosphatase" test 238
acrylics fibers 373
adenosine triphosphate (ATP) 261
admissibility *vs.* weight 50
Adobe Photoshop 66
AFDIL (Armed Forces DNA Identification Laboratory) 263
AFIS *see* Automated Fingerprint Identification System (AFIS)
AFTE (Association of Firearms and Toolmark Examiners) 198
alcohol and drugs and driving 352–4
alcoholic beverage *294*
aliasing, analog-to-digital conversion 207
alibi known controls 230
alterations and erasures, detection: chemical separation of inks 160; color of ink 159; lotteries 158, *159*
alternate light sources 128, *129*, 130, 132
American Academy of Forensic Sciences (AAFS) 12, 25
American Academy's ethics code 25
American Board of Criminalists (ABC) 12
American Society of Crime Laboratory Directors (ASCLD) 12, *12*
amino acids 259, 371
ammonium nitrate mixed with fuel oil (ANFO) 316
anagen phase 378
analgesics 331
analog signal 206–7
analysis, comparison, evaluation, verification (ACE-V) 134, *135*
ANFO (ammonium nitrate mixed with fuel oil) 316
angel dust 331
animal fibers 369
anonymous DNA 264
antemortem X-rays 15
anthropometry 137
anticoagulant 226
antigens and antibodies 234
"anti-human" serum 235
antisera 235
architectural paint 381, *382*, *383*
arc swing patterns 89
Armed Forces DNA Identification Laboratory (AFDIL) 263
arson dog 300, *300*

arson fires 297
arterial spurt pattern 89
asbestos fibers 368
ASCLD (American Society of Crime Laboratory Directors) 12, *12*
Ashbaugh, D. R. 134
Association of Firearms and Toolmark Examiners (AFTE) 198
association process 21
atomic absorption (AA) 194, 403
ATP (adenosine triphosphate) 261
Attenuated Total Reflectance (ATR) 386
audio recording 70
authenticated collected writing 154
authentication, biometrics 137–9, *138*
Automated Fingerprint Identification System (AFIS) 5, 8, 41, *41*, 136; applications 123; fingerprint match *125*; forensic file 124
automobile windshield glass 91–2, *93*
automotive/architectural paint *381*, 381–4; collection 384; laboratory analysis 384, 384–6, *385*, *386*; multilayer paint chip *383*
azoospermic 239, 282

ballistics 95
BATF (Bureau of Alcohol Tobacco and Firearms) 310
bath salts/designer cathinones *335*
bench-top DNA purification/liquid handling systems *264*
benzodiazepines 338
Bertillon, A. 10
bertillonage 137
biological evidence 38; blood/buccal swab 226–7, *227*; examination 231–2, *232*; from scenes 227–9; stain from window frame *39*
biological methods *409*, 409–10
biological microscope 406
biological stains 228
biometrics 119; anthropometry 137; authentication 137–8; identification 137–8; identifying human remains 139–40; individualization 139; scanner 204
bitemarks 15
black powder 313
blank controls 76, 230, 233
blood/buccal swab 226–7, *227*
blood cells 235, 249
blood evidence: ABO blood groups 223; forensic identification 232–4; nonsecretors 223; secretors 223; sexual assault 223–4; species determination 234–5, *235*–*6*
blood nature 225–6, *226*
blood separation *226*

411

blood spatter patterns 33, *33*, 61, 88; arterial spurt 89; calculation of angle 87–8; cast-off 89; characteristics 87; contact deposit 88; high-velocity spatter 85, *86*; low-velocity spatter 85, *86*; medium-velocity spatter 85, *86*; running 89, *90*; secondary 90; shape 87; straight line reconstruction from *88*; wipe and swipe 89, *89*

Bloodstain Pattern Analysis (BPA) 85; blood spatter patterns 88–90; factors affecting and interpretation 90–1; identification 9, *25*, 58, 81–2; interpretation 85; velocity, energy, and force 85–8, *86*, *87*

bloody fingerprints 130
BPA *see* Bloodstain Pattern Analysis (BPA)
breach face *187*, 188, *189*, *191*
BTK Killer 201–2
bufadienolides 328
buffer solution 272
bufotenine 327
bullet lands 177
bullet recovery water tank 187
bullets, deformed evidence 192
Bureau of Alcohol Tobacco and Firearms (BATF) 310
burn pattern 298, *299*

CAD (computer-aided drawing) tools 71–2
caliber 177
California v. Gerald Mason 117–18
cannabinoids 335
Cannabis sativa (marijuana) 7, *7*, 34–5, *335*
capillary electrophoresis (CE) 270, *271*, 272
carbon strip 304, *304*
carrier gas 404
cartridges 174–5, *175*
case study: CODIS matches 276; DNA exclusion 279–80; DNA inclusion 279; Narborough rape murders 267; sexual assault, parentage testing 281; soldier identification, Vietnam war 262–3; Tommy Lee Andrews 268
cast-off patterns 89, *89*
catagen phase 378
catalyst 238
cause/manner of death 13
CDMA (Code Division Multiple Access) **214**
CE (capillary electrophoresis) 270, *271*, 272
cell phones 205
Central Processing Unit (CPU) 205, *205*
certification programs 12
CFTT (Computer Forensics Tool Testing) 210
chain of custody 207–8
charcoal 292
charred documents 161, *161*
chemical fingerprinting 366
chemical method 126
chemical microscopy 364, *364*
chemiluminescent 233
Christmas tree 410
chromatography/chromatographic methods 403, *403*, 405
clarification/contrast improvement techniques 112–13
class characteristics 22, 105
classification (identification) process 21, 38; *see also* identification process
clothing and article/object patterns 94

clusters 208
coal, flaming combustion 293
cocaine base/salt *334*
Code Division Multiple Access (CDMA) **214**
CODIS *see* Combined DNA Indexing System (CODIS)
"cold case" unit 43
collected writings 154
collecting/preserving physical evidence 73; chain of custody 76; collection method 73–4; controls 75–6; laboratory submission 76; numbering/evidence description methods 74; packaging 74–5
collection method 73–4
collection methods, material evidence 362–4; at laboratory 362–3; without sampling 362
collection, packaging, and preservation 44, 44–5
Combined DNA Indexing System (CODIS) 5, 8, 41, 275, 277
combustion 291
comparison microscope *187*, 406; bullet comparison 188; cartridge case *189*; firing pin 188, *189*; positive identification 189; semiautomatic firearms 190; shot shell cases 190
comparison samples 309
computer-aided drawing (CAD) tools 71–2
computer forensics 17, 204, 210
Computer Forensics Tool Testing (CFTT) 210
computer hardware components *205*, 205–6
computers/computer programs 410, *410*
computer software programs 72
computer technology 5, 15
cone-shaped pattern 91, *91*
confirmatory tests 233–4, 239
coning/quartering 390
contact deposit patterns 88
containment/confinement 318
contamination 230–1
contracting/expanding spiral 62, *62*
controlled experiment 18
controlled substance laws: chromatography 345; forensic laboratory 342; isolation and separation 343–4; microcrystal tests 344–5; qualitative *vs.* quantitative analysis 346–8; screening tests 342–3; spectroscopy/spectrometry 346, *346*
controls in forensic testing: alibi known 76; blank 76; known 75–6; substratum 76
control specimens 45
conventional genetic markers 249
copiers 205
copy machines 158
core CODIS STR loci 273
coroner system 14
corpus delicti 7, 21, 43, 242
corroboration 24, 32, 42
CPU (Central Processing Unit) 205, *205*
crack cocaine 333
Crafts, H. 1
credit card skimmers 205
crime scene(s): actions and security 58–60; block 47–8; indoor/outdoor 58; mnemonic guide to 59; reconstruction 20; responder task **59**; searches 60, 62, *62*; survey and evidence recognition 61; types 57–8
crime scene pattern analysis 83–4, **84**

crime scene photography 65; forensic aspects 67–70, *68*, *69*; technical aspects 66–7
crime scene processing and analysis: actions and scene security 58–60; collecting/preserving physical evidence 60, 73–6; documentation 60, 62–73; processing *vs.* analysis 57; and reconstructions 61, 76–8; and scientific method **61**; searches 60, 62, *62*; *State of Connecticut vs. Duntz* 55–6; steps in 60–1; survey and evidence recognition 60–1; types of scenes 57–8
crime scene recording technology: CAD 71–2; data analysis computer software 72; drone technology 72, *72*; 3D imaging and scene mapping 70–1; total station mapping technology 71
criminal identification method 10
criminalistics 10, 16
criminal justice system and process: adjudicative agency role 48; evidence pathway through system 46, *46*; investigative agency role 46–8; physical/scientific evidence, importance *49*, 49–50
criminal profiling 97
cross-linking 371
cross-projection sketch 65
crystal methamphetamine *333*
crystal tests 234
cursive writing *153*, 155
cutting method 228
cyanoacrylate fuming 128
cyber crime *see* internet crime

DA (district attorney) 31
data authentication 203
data integrity 203
date rape drug 248, *248*, 339
dating documents 163–5, *165*
Daubert case 51–2
Daubert criteria 51; and National Academy of Sciences 107–8
Daubert rules 51
debris: evidence samples and packing 301–2; patterns, sniffers, and arson dogs 300, 300–1; samples for analysis 302–3
defacing 196
degradation 206
delirium tremens (DT's) 330
delusterant 374
density 401; gradient 391, *392*
deposits, dispersions, and residues 34–5
depth of field 67
detcord, military explosives 314
DFSA (drug-facilitated sexual assault) 248–9
differential extraction 265
digital camera 204
digital evidence 17; analysis *see* electronic data; BTK Killer 215–16; collection 206–8; computer forensics 204; definition 203; devices/attachments 204–6, *205*; DME 203; features 203; internet crime 210–15, *211–13*, **214**; processing 206
digital (multimedia) evidence collection: chain of custody 207–8; forensically sterile conditions 208; hashing 207; originals *vs.* copies 206–7
digital fingerprint 207

digital forensic science 17, *17*
digital image 66
digital multimedia evidence (DME) 203
digital photography 66
digital-related devices/attachments 204–6, *205*
digital watches 204
direct confirmatory tests 234
direct physical match 104
dispersion 375
dissecting microscopes 405
district attorney (DA) 31
DME (digital multimedia evidence) 203
DNA 258; alleles *vs.* profiles **23**, 41; finding in body 261–3; nature and functions 258–60; polymerases 259; profile 274; *vs.* conventional genetic testing 251
"DNA fingerprints and fingerprinting" 119
DNases 263
DNA typing: biological evidence, collection and preservation 263; databasing and CODIS system 274–6; development and methods 263–4; forensic applications 277–82; isolation 264–5; PCR 268–70; power 272–4; RFLP 265–8; strengths, limitations, promise, hype 284–5; STRS 270–2; technology 282–4
documentation 60, 62; audio recording 70; crime scene recording technology 70–2; duty to preserve 72–3; marking for identification 43; notes 62–3; photography 65–70; sketches *63–4*, 63–5; video recording 70
document evidence types 145–7, *146*
dongle device 204
down coats 395
drive slack 208
DriveSpy 216–17
drone technology 72, *72*
drug-facilitated sexual assault (DFSA) 248–9
druggist fold package 74, *75*
drugs and drug analysis 328; behavioral dependence 330; classes 331–2; club drugs 338–9; controlled substances 330–1, 340–2; depressants, hypnotics, and tranquilizers 337–8; drug dependence 329; hallucinogens 334–7; nature 328; performance enhancement 339–40; physiological dependence 329–30; state legislatures 352; stimulants 333–4; working definition 329
drug test kits 342
"drunk driving" case 7
dumpsite 60
duty to preserve 72–3
dye 151, 160, 165
dye stains 128
dynamite 315

ecstasy, MDMA drug 337, *338*
EDX (energy-dispersive X-ray) analyzer 195, 366
electromagnetic spectrum (EM) 160–1, 401–2, *402*
electronic data: deleted/un-deleted process 209; file slack 208–9, *209*; forensic tools 209–10; unallocated file space 208
electrophoresis 272
Electrostatic Detection Apparatus (ESDA) 162, *163*
elution (dissolving) method 229

EM (electromagnetic spectrum) 160–1, 401–2, *402*
e-mail forensics 210–13
e-mail header 211, *212, 213*
e-mail messages examination 211
e-mail tracing 213
Encase Forensic 216
energy-dispersive X-ray (EDX) analyzer 195, 366
entomology 14
enzymes 238
ESDA (Electrostatic Detection Apparatus) 162, *163*
ethyl alcohol/ethanol 293
European justice systems 10
evidence recognition 21, 57, 60–1
exclusionary value 367, 380
exclusions/inconclusive/insufficient detail 107
exempt preparations 341
exothermic reactions 291
expert witness 45
explosive device *317*
explosives and explosion incidents: characteristics 310–13; classes 313–16; device examination 322–3; exploded device and associated debris 320–1, *321*; laboratory analysis 319–23; low explosives 313; microscopic examination 320, *321*; primary high explosives 313–14; secondary high explosives 314–16; train/device 317–18; unexploded device 319–20; use of washings 321–2
extortion note 147
extraction process 343
eyewitness testimony 7

facsimile machine 205
familial searching technique 276
Federal Bureau of Alcohol, Tobacco, and Firearms (ATF) 196
Federal Bureau of Investigation (FBI) 12, 310
Federal Rules of Evidence 50
fiberglass 373
file slack 209, *209*
fingerprint classification 123
fingerprint evidence: comparison and identification 134–6, *135*; initial search and detection 126; physical types 126; visualization *see* latent print visualization
fingerprint identification 134–6; profession 136
fingerprint patterns *122*
fingerprints 5, 8, 38; AFIS *41*, 123–5; arch pattern *122*; classification 123; composition 120, *121*; at crime scene 41; diagnosis and identification 10, *10, 11*; elimination 60; features 120; hand gun with *39*; history and development 122–3; level I/II/III, 135; loop pattern *122*; with pores *120*; ridge characteristics *122*; sebum composition **121**; systematic approach 132–3, *133–4*; three-dimensional *120*; two-dimensional *119*; from weapons 40; whorl pattern *122*
Finger Prints (book) 122
fire and arson: characteristics fuel 294–6; combustion nature 291–3; fuels used 293–4; pyrolysis of solid fuels 296; sidebar 293

firearms 313; cartridge 174–5, *175*; comparison microscope *see* comparison microscope; components 174; definition 174; physics 178–9; professional 198; rifling *see* rifling; types 179–84, *181–4*
firearms evidence: bullets 184; cartridge 184; collection 185; comparison microscope *187*, 187–90; databases 190–1, *191*; examination and comparison 185–6; reconstruction 193–5, *195*; safety issues 185; serial number restoration 196–8, *197–8*; test for functionality *186*, 186–7
firearms types: semiautomatic 180–1; shotguns 182–3; single-shot handguns 179
fire burn patterns 96, *97*
"fired evidence" 185
fire triangle/tetrahedron/pentagon *291*
firing pin 174–5, 188, *189*, 190
firing pin impression 108, 188, *189*, 190
"firing train" 175
first responder 59–60, 62
flammable range 294, **295**
flash point tester 294
fluorescence 394, 402
footprints 36, 38; three-dimensional pattern *67*
forceps use 362–3, *363*
forensic anthropologists 15
forensic anthropology 15, *15*
forensic biology 225, 240
forensic databases, suspects from 41–2
forensic engineers 16
forensic entomologists 14, *14*
forensic evidence analysis elements 20–1; classification 21; evidence recognition 21; individualization 22–4; reconstruction 24–5
forensic file 124, 275
forensic glass evidence 392; collection 393–4; glass as evidence 393; laboratory examination 394–5, **396**; types 392–3, *393*
forensic identification, blood: confirmatory tests 233–4; identification 232; individualization 232; preliminary tests 233–4; reconstruction 232
forensic identification, body fluids *see* physiological fluid evidence
forensic medicine *see* forensic pathology
forensic odontologists 14–15
forensic odontology 114
forensic pathology 13–14
forensic psychiatrists and psychologists 16
Forensic Replicator 217
forensic science: applications 9–10, *10*; color 401–2; to criminal justice system 7–8; databases 5; definition 5–6, *6*; elements 20–5; ethical concerns and professional practice 25–7; faces 13–17; human biological and medical sciences 13–16; identify persons 8; identify substances/materials 8; investigation 4; laboratories 6–7; laboratories and professional organizations 10–13; light 401; metric system 401; natural sciences 16; physical properties 401; role in justice system 4–9; and scientific method 17–20; technology and engineering 16–17; television programming 57; value to society 6–8

forensic scientist's role, sexual assault 241–2
forensic sculptors 15
forensic sculpture 15
forensic serology 226
Forensic Toolkit (FTK)/FTK imager 217
forensic toxicologists 15–16
forensic toxicology 15; laboratories analyze 350; overview 348–9; poisons 351–2; post-mortem 350–1
forgery 146–7, 164
fountain pen 150
"44 Caliber Killer" (Son of Sam case) 169–70
friction ridge skin 120, 139–40
"Frye rule" 50
Frye v. United States case 50
f-stop/*f*-number 67
fur coats 370, *370*

Galton features 122
gaming consoles 204
gas chromatograph-mass spectrometry (GC-MS) 404; GC-IR 346
gas chromatography (GC) 305, 305–6, 404; ASTM standards, ignitable liquids **308**; highly evaporated gasoline *307*; individual peak, identification 307–9; instrument *403*; kerosene and fuel oil standards *307*; pattern recognition 306–7; physical evidence 309–10, *310*
General Rifling Characteristic (GRC) 178, 191
genetic code 259
genetic markers 225, 249–50, 257
genetic polymorphism 257
genetics 257
genome 258
genomic DNA 261
glass fracture patterns 91; force applied 91–2; gunshots fired through 92–3
glass fracture 3R rule 92
global positioning satellite (GPS) 65
Global System for Mobiles (GSM) **214**
glowing combustion 292
Goddard, C. 11
Google mail e-mail header 211, *211*, *212*, *213*
GPS (global positioning satellite) 65
GRC (General Rifling Characteristic) 178, 191
grooves: and helical lands *176*; lands 177–8, *179*; rifling 176
Gross, H. 10
GSM (Global System for Mobiles) **214**
gunshot residue (GSR) patterns 94–5, *95*, 193–5

hackle marks 92, *92*
hair (human/animal) 376; growth phases 378, *378*; hair standards, collection 379–80; laboratory examination 380–1, *381*; racial characteristics 379, **379**, *379*; structure 376–8, *377*
hallucinogens 334–7
Handbook for Coroners, Police Officials, Military Policemen (book) 10
handgun 175; breech face 189; categories 180–1; restoration *197*; semiautomatic operation 180
handprinting examination 155

handwriting comparison: and handprinting examination 155; individual's handwriting 151–2, *152*; legal status 156–7; mechanical and pictorial characteristics *153*, 153–4; signatures 155–6, *156*; standards 154–5, **155**
Hard Drive (HD) 206
Hardware Write-Blockers 210
hashing, digital files 207
hashish *see Cannabis sativa* (marijuana)
hash oil 336
head stamps 190
heated headspace 303
Helix, Unix-based tool 218
Hematrace® card *236*
heroin 331
high performance liquid chromatography (HPLC) 322
high-velocity blood spatter patterns 85, *86*
hit-and-run cases 40
Hit-and-Run Death of a State Police Lieutenant 101, *102*, *103*
"Hitler Diaries" 148
HLA-DQA1 typing system 224, 269–70
Holman, F. 31–2
HPLC (high performance liquid chromatography) 322
human biological and medical sciences 13–16
human dentition analyses 14
hypervariable region (HV) 261
hyphenated techniques 346, 404
hypothesis testing 18–20

IAFIS (Integrated AFIS) 124–5
IAI (International Association for Identification) 12–13, 136
identification process 21, 38, 105, 232; comparison, instrumental methods for 365–7, *366*; testing 43
ignitable liquid 300
Illinois v. Gary Dotson 223–4
Illinois vs. Cecil Sutherland 357–9
ILooKIX, forensic software packages 216
immunochromatographic test strips *239*
immunological test 234–5, 240
impressions 108; collection/preservation *109*, 109–11, *110*, *111*, *112*; footwear, tire, and 112; imprint/striation comparison 104; plastic prints 108; *vs.* striation mark 108–11
imprints/indentations 36
inconclusive 107
indentation 108
indented mark *171*
indented writing 162–3, *162–3*
indirect confirmatory tests 234
indirect match 104, *104*
individual characteristics 105
individualization 22–4, *23*, **24**, 232; evidence 250; reconstruction patterns and 83
individual's handwriting 151–2, *152*
informants 31
infrared spectroscopy 365
inheritance 257; rule 256
initiator/igniter 317
ink compositions 151, 165
ink eradicator 159–60
"Innocence Project" 279

Integrated AFIS (IAFIS) 124–5
International Association for Identification (IAI) 12–13, 136
Internet browsing history 210
Internet cache 210
Internet cookies 210
Internet crime: computer forensics 210; e-mail forensics 210–13, *211–13*; mobile device/smartphone forensics 213–15, **214**
inverted cone pattern 298, *299*
investigating suspicious fires: arsonists' motives 296–7; burn patterns 298; debris 300–5; evidence samples 301–4; examination of specimens 304–5, *305*; ignitable liquid residues 299–300; possible accidental cause 298–9; role of investigator 318–19
"investigation" block 48
iodine fuming 131
IP address 211, *213*
iPhone forensics 215
iPod/MP3 player 204
isoenzymes 249
isolation/extraction DNA 264
IXImager, forensic software packages 216

Jeffreys, A. J. 266
jigsaw direct physical match *106*
jigsaw matches 394
junk DNA 264

Kennedy document, case 164
keratin 376
kerosene, liquid fuel 293
known controls 230
known (K) specimens 45

laboratory analysis 45; *see also* forensic science; and evidence comparisons 76–7
laboratory information management systems (LIMS) 410
land impressions 177, *188*
laser illumination 128, *129*, 130, 132
latent fingerprint cases 5
latent fingerprints on non-absorbent surfaces: cyanoacrylate fuming 128; illumination and combination methods 128, 130; physical methods *127*, 127–8
latent print 126; on absorbent surfaces 130–2
latent print visualization 126; absorbent surfaces 130–2; non-absorbent surfaces *127*, 127–30, *129*; systematic approach 132–3, *133–4*
layer information 382
legal status, writing comparison 156–7
LeGrand, D. 31
The LeGrand Case 31–2
LIMS (laboratory information management systems) 410
Lindbergh Kidnapping case 143–4
linkages/exclusions, victim 40
liquid phase 404
Locard, E. 10–11, *11*, 33
Locard Exchange Principle 33–4
Locard's Exchange Theory 204
Locard Transfer Principle 34, 360, 397
"love drug" 338
low copy number 285

low-velocity blood spatter patterns 85, *86*
luster 374

magazine 180–2, 185, 190
magic mushrooms 327
magnetic brush technique 128
main charge, explosive device/train 317
Maresware software 216
marijuana (*Cannabis sativa*) 7, *7*, 34–5, *335*
mass 401
mass spectrometry/spectrometer/spectroscopy (MS) 322, 346, 403
materials evidence: automotive/architectural paint 381–6; clothing and vehicles 361–2; collection methods 362–4; deposited without contact 361; forensic glass evidence 392–5; hair (human/animal) 376–81; laboratory examination 364–7; miscellaneous types 395–7; overview 357–60, **358**; soil/dust 388–91; transfer 360, *361*; types 367–75; *vs.* individualization/inclusion/exclusion 367
Mathison, Y. 81
MDMA/ecstasy drug 337–8, *338*
mechanical agitation technique 363
mechanical and pictorial characteristics, handwriting *153*, 153–4
mechanical dislocation 363
mechanical printing *157*, 157–8
medical examiner system 14
medico-legal institutes 9
medium-velocity blood spatter patterns 85, *86*
melanin 378
melting/denaturation 259
memory card 204
Mendel, G. 256, *257*
Mendelian principles 256
Merrill Dow Pharmaceuticals v. Daubert et ux. 51
mescaline 337
message body 211
Message Digest 5 (MD5) 207
message header 211
messenger ribonucleic acid (mRNA) 259
micro-FT/IR 365, *365*
microgram bulletin 338
microliter syringes *303*
microsatellites 270
microscopy/microscopical methods 405–8, *406*, *407*, *408*, *409*
mineral wool fibers 368
minisatellites 270
minutiae, fingerprint 122
mitochondrial DNA (mtDNA) 261, *261*
mitochondria/mitochondrial genome 261
mitosis 259
mitotyping 262
MO *see modus operandi* (MO)
mobile device/smartphone forensics 213–15, **214**
mobile forensics tools 214–15
modems 204
modus operandi (MO) 16; leads and suspects 40–1; and profiling 97
monomers 371
mRNA (messenger ribonucleic acid) 259
mtDNA (mitochondrial DNA) 261, *261*
muzzle-to-target distance 193

NAA (neutron activation analysis) 194
Narborough rape murders 267
narcotics/opiates 331
National Center for Forensic Sciences 306
National Institute of Justice (NIJ) 203
National Institute of Standards and Technology (NIST) 210
National Integrated Ballistic Identification Network (NIBIN) 8, 41, 189–90; database 190, *191*
natural law 18
natural sciences 16
natural/synthetic fibers 368; laboratory examination 373–5, *374, 375, 376*; manufactured fibers 370–3, *371, 372*; natural fibers 368–70, *369, 370*
negative association 21
network cables and connectors 205
network components 204
neutron activation analysis (NAA) 194
NIBIN *see* National Integrated Ballistic Identification Network (NIBIN)
NIJ (National Institute of Justice) 203
ninhydrin 131
NIST (National Institute of Standards and Technology) 210
nitroglycerine 315
non-handwriting examinations: copy machines 158; typewriter and printer *157*, 157–8
normal variation 152–4
nuclear DNA 261
nucleotides 258
nylon, polymer 372
Nyquist's Theorem 207

oblique lighting *162*
Oklahoma city bombing *311*
opiates 331
optical microscopy 364; chemical microscopy 364, *364*; polarized light microscopy 365
optical properties 375
Ouchterlony method 235, *236*

p30 protein 239
PA (prostatic antigen) 239
pagers 205
paint smears 386
paraffin test 194
particle size distributions 391
pathology 13, *13*
patterned injuries 113
pattern evidence, type 83
pattern identification 83
pattern interpretation 83
pattern recognition 83
pattern reconstruction 83
PCP/angel dust 336–7
PCR-STR protocols 230
peroxidases 233
Personal Digital Assistants (PDA) 204
photographic techniques 113
photo log 63
photoshop and image pro 113
physical anthropology 15
physical/chemical techniques 113
physical developer 132

physical evidence in legal system: change at crime scene 33, *34*; classification and uses 38–43; damage/tears/cuts/breaks/jigsaw matches 35–6; deposits, dispersions, and residues 34–5; exchange of material 33–4; imprints/indentations 36; *The LeGrand Case* 31–2; mechanism for 33–7; scientific evidence, importance 49–50; striations marking 36–7, *37*
physical evidence process 43; collecting/preserving 73–6; collection, packaging, and preservation *44*, 44–5; documentation/marking for identification 43; laboratory analysis 45; recognition 43; reporting and testimony 45
physical fit/direct fit match/matches 103, *103*, 106, *106*, *107*
physical pattern evidence, examination: classification 103–5; principles 105–8
physical types, fingerprint 126
physiological dependence 329–30
physiological fluid evidence: saliva 240; semen 237–9; sexual assault cases *see* sexual assault cases; urine 240; vaginal "secretions" 240
physiological/psychological effect 329
pigment 151, 159, *161*, 165
plasma proteins 226, 249
plastic/impression print 126
platen 158, 162–3
polar coordinate system 71
polarized light microscopy 365
pollen material 34–5, *35*, 395
polyester fiber 371; infrared spectrum *376*
polyethylene 371
polygraph 50
polymarker 270
polymerase chain reaction (PCR) 224, 268, *269*
polymers 370
polyolefins 373
population genetics 250
positive identification 105, 189
post facto 7
powder dusting 127, *129*
powder pattern 193
pre-DNA systems 249–50
preliminary tests 233–4
primary/first-order matching 103
primer 174–5, 183, 188–9, 194, 259, 268
primer/detonator 317
printers 205
"probable cause" concept 48
processing *vs.* analysis 57
professional organizations: and codes of ethics 25–6; and ethical concerns 26; ethical responsibilities as scientists 26; for forensic scientists 26–7; and functions 25, *25*; laboratories and 10–13
profilers 16
profiling 77–8; MO and 97
projectile (bullet) 175, 177, 184
projectile trajectory patterns 95–6, *96*
prostate specific antigen (PSA) 239
prostatic antigen (PA) 239
psilocin 327
pyrolysis gas chromatography 375
pyrolysis process 296

questioned document examination 145; alterations and erasures, detection 158-60, *159*; charred documents 160-1, *161*; dating documents 163-5, *165*; handwriting comparison 151-7, *152, 153*, **155**, *156*; indented writing 162-3, *162-3*; non-handwriting examinations *157*, 157-8; recognition, collection, and preservation 147-8; types 145-7, *146*; writing process 148-51, *149-51*
questioned (Q) specimens 45

Random Access Memory (RAM) 206; slack 208-9, *209*
rayon 371
Read-Only Memory (ROM) 206
recognition 43; collection, and preservation 147-8; process 105
reconstruction 24-5, 232; *see also* reconstruction patterns; conducting 77; crime scene analysis and 76-8; human remains 140; on pattern evidence 81; *vs.* reenactment/profiling 77-8
reconstruction patterns: are crime scene patterns 83-4, **84**; documentation importance 84-5; and individualization 83
recovered firearm 193
reenactment 24, 77-8
refractive index (RI) 375, 394
regenerated fibers 371
relevance/reliability, requirement 50
reporting and testimony 45
requested writings 154-5
residue prints 108
resolution 66
restriction endonucleases 265
restriction fragment length polymorphism (RFLP) 224, 265, *266*
retention time 345
revolver *179*, 180, *181*, 182, 184
RI (refractive index) 375, 394
ribose 258
ridge characteristics *122*
rifling: broach *176*, 176-7; bullet lands 177; caliber 177; GRC 178; land impressions 177; lands 177; projectile 176
ROM (Read-Only Memory) 206
rough sketches 63-4
routers, hubs, and switches 205
RSID® (Rapid Stain Identification) test card 240
running patterns 89, *90*

Safe Streets Act of 1968 12
safety, firearms 184-6
safety plate glass *see* automobile windshield glass
saliva, identification 240
sampling/digitization 207
SANEs (Sexual Assault Nurse Examiners) 240-1
SART (Sexual Assault Response Team) 240-1
scales, overlapping cells 376
scanners 205
scanning electron microscope (SEM) 195, 408; EDX instrument 367; gunshot residue particle with *195*
scientific method 17, *18*, 108; applications 19; careful observation 18; controlled experiment 18; crime scene analysis and **61**; forensic science and 17; formulations 17; hypothesis testing 18; to investigation/forensic science 20; logical suppositions to observations 18; refine hypothesis 18
scientific/technical evidence: admissibility *vs.* weight 50; Daubert case 51-2; relevance/reliability, requirement 50
Scientific Working Group on Imaging Technology (SWGIT) 113
scraping method 74, 228-9
secondary physical matches 106
secondary/second-order matching 104
Secure Hash Algorithm version 1 (SHA-1) 207
security log 63
SEM *see* scanning electron microscope (SEM)
semen identification: "acid phosphatase" test 238; confirmatory test 239; enzymes 238; p30 239; seminal plasma 237; spermatozoa 237, *237*; UV light 238
semiautomatic firearms 180-1
semiautomatic pistol 181
semiautomatic weapon *182*
seminal plasma 237
serial number restoration: after acid etching *198*; ATF 196; cotton-tipped swab (Q-tip) 197; "Crafts" case 197; handgun *197*; shotgun tracing 196
serological analysis 55
serum 226
servers 205
sexual assault cases: DFSA *248*, 248-9; evidence collection 243, **244-6**, 247; forensic scientist's role 241-2, *242*; initial investigation 241; medical examination 242-3; SANE 240-1; SART 240-1; types and investigation 247-8
sexual assault evidence collection kits 243, **244-6**, 247
Sexual Assault Kit *242*
Sexual Assault Nurse Examiners (SANEs) 240-1
Sexual Assault Response Team (SART) 240-1
SHA-1 (Secure Hash Algorithm version 1) 207
shaking method 74
shape, pattern, marks/form: comparisons 113; determination 104-5
shockwave, explosion 312
short tandem repeats (STRs) 270
shotguns 182-3; ammunition (shells) 183, *184*, 190; for hunting birds/skeet shooting *183*; slug 183
signatures 155-6, *156*
silicone rubber casting material *172*, 172-3
single nucleotide polymorphism (SNP) 282
skeletal remains 140
sketches *63-4*, 63-5
small particle reagent (SPR) 128
smart card 204
smartphone 213-14
smokeless powder 175, 178
smooth sketches 64
sniffers 300
snow print wax 111
SNP (single nucleotide polymorphism) 282
soda lime glass 392
Software Write-Blockers 210
soil/dust 387; collection 390; investigative value, example 388-9; laboratory examination 390-1, *391*, *392*; major components 388
soldier identification, Vietnam War 262-3

solid fuels 296
solvent wash 304
Southern blotting 265
species determination: antisera 235; ELISA method *236*; Hematrace® card *236*; immunological methods 234; Ouchterlony method *236*
species of origin 234–5
spectral comparator 160–1
spectrofluorometers 131
spectrometric/spectroscopic methods 402–3
spermatozoa 237, *237*
sperm cells 237, 239; from nonsperm cells, extraction process *266*
spinneret 370
standards, handwriting: authenticated collected writing 154; requested writings 154–5
State of Connecticut vs. Duntz 55–6
State of Hawaii vs. Mathison 81–2
states of matter 401
State v. Richard Crafts 1–4, **4**
statutory rape 247
stereomicroscope 405
sticky side powder 132
stimulants 333–4
striated mark 171
striations 102, 108
striations marking 36–7, *37*
STRs (short tandem repeats) 270
substances/materials, identification/analysis 42–3
substratum controls 76, 230
super-cooled liquid 392
Super Glue (cyanoacrylate esters) 128, *129*
surface coating 150, 160
swabbing methods 228, *228*
synthetic fibers 368

tablet PC 204
tandemly repeated sequences 264
tangential fracture lines 92–3, *93*
tape lifting method 73, 109, 127, 363–4
TATP material 316
Technical Working Group for the Examination of Digital Evidence (TWGEDE) 206
Teichmann's and Takayama's tests 234
telogen phase 378
temperature 401
test controls 230
test firing *186*, 186–7
"Test Floppy for WPD review" 215
tetrahydrocannabinol (THC) 335
thermal power plant extinguishers *296*
The Sleuth Kit (TSK) and Autopsy 218
thin-layer chromatography (TLC) *161*, 404
threatening letter 147
three-dimensional (3D) imaging and scene mapping 70–1, *71*
tire and skid mark patterns 93–4
TLC (thin-layer chromatography) *161*, 404
toolmarking/toolmarks 56, 108; class and individual characteristics 171–2, *172*; collection 172–3; comparison 173, *173*; examination 173; residue 171; screw driver and indented impression, blade *171*; types 171
touch DNA 278
toxicology *see* forensic toxicologists
trace evidence 360
track and trail patterns 93
trajectory 176, 193
transcription 260
transfer evidence 360
translation process 260
transmitted, EM spectrum 402
"Trial" block 49
triangulation method 65
trier of fact 7, 50
trigger pull 185, 193
tube absorption 304, *304*
twist, direction *177*, 177–8, 185
two-dimensional (2D) scene diagramming 70
typewriter *vs.* printer, mechanical impressions *146*, 149, 157, *157*

ultra-violet (UV) illumination 126, 238; visible microspectrometer *385*
unallocated file space 209
uncontrollable craving 330
urine examination 240
U.S. Supreme Court 51, *51*
UV illumination *see* ultra-violet (UV) illumination

vacuuming method 74, 362
vaginal secretions 240
Variable Number of Tandem Repeat (VNTR) 264
victim/suspect: identification *39*, 39–40; linkages/exclusions 40
video recording 70
visible/patent print 126
visible spectrophotometry 403
visible spectrum 401
VNTR (Variable Number of Tandem Repeat) 264
Vollmer, A. 11

watermarks 164, *165*
weapon, tool/object marks 113
Web Cache 210
Web Cookies 210
windshield glass 393
wipe and swipe patterns 89, *89*
Wood Chipper Case 1–4, **4**
working hypothesis 61
wound ballistics 95–6
wound, injury, and damage patterns 98
write-blockers 209–10
"writing circuits" 156
writing instrument 148–51, *149*, 159
writing mechanics 148, 154
writing process: ballpoint pens 150–1, *151*; clay tablets *149*; documents and forms *149*; matted paper fibers *150*

X-ways Forensics (or WinHex) 218

Y chromosome, human inheritance *283*